BIOENERGETICS

Molecular Biology, Biochemistry, and Pathology

BIOENERGETICS

Molecular Biology, Biochemistry, and Pathology

Edited by

Chong H. Kim
Rensselaer Polytechnic Institute
Troy, New York

and

Takayuki Ozawa
University of Nagoya
Nagoya, Japan

PLENUM PRESS • NEW YORK AND LONDON

Library of Congress Cataloging-in-Publication Data

Bioenergetics : molecular biology, biochemistry, and pathology /
edited by Chong H. Kim and Takayuki Ozawa.
    p. cm.
  Papers presented at the International Symposium on Bioenergetics:
Molecular Biology, Biochemistry, and Pathology, held in Seoul,
Korea, Agu. 18-21, 1989, sponsored by International Union of
Biochemistry and Ewha Woman's University.
  The symposium was held in honor of Prof. Kunio Yagi to commemorate
his 70th birthday.
  Includes bibliographical references.
  Includes indexes.
  ISBN-13: 978-1-4684-5837-4   e-ISBN-13: 978-1-4684-5835-0
  DOI: 10.1007/ 978-1-4684-5835-0
  1. Energy metabolism--Congresses. 2. Mitochondria--Congresses.
3. Mitochondrial pathology--Congresses. I. Kim, Chong H., 1943-
. II. Ozawa, Takayuki. III. Yagi, Kunio. IV. International
Symposium on Bioenergetics: Molecular Biology, Biochemistry, and
Pathology (1989 : Seoul, Korea) V. International Union of
Biochemistry. VI. Ihwa Yŏja Taehakkyo.
  [DNLM: 1. Energy Metabolism--congresses. 2. Mitochondria-
-metabolism--congresses. 3. Mitochondria--pathology--congresses.
QH 603.M5B615 1989]
QP176.B46 1990
574.19'121--dc20
DNLM/DLC
for Library of Congress   90-7851
                                              CIP

Proceedings of an International Symposium on Bioenergetics:
Molecular Biology, Biochemistry, and Pathology,
held August 18-21, 1989, in Seoul, Republic of Korea

© 1990 Plenum Press, New York
Softcover reprint of the hardcover 1st edition 1990
A Division of Plenum Publishing Corporation
233 Spring Street, New York, N.Y. 10013

Printed in the United States of America

FOREWORD

The emergence of the Biochemical Sciences is underlined by the FAOB symposium in Seoul and highlighted by this Satellite meeting on the "New Bioenergetics." Classical mitochondrial electron transfer and energy coupling is now complemented by the emerging molecular biology of the respiratory chain which is studied hand in hand with the recognition of mitochondrial disease as a major and emerging study in the basic and clinical medical sciences. Thus, this symposium has achieved an important balance of the fundamental and applied aspects of bioenergetics in the modern setting of molecular biology and mitochondrial disease. At the same time, the symposium takes note not only of the emerging excellence of Biochemical Studies in the Orient and indeed in Korea itself, but also retrospectively enjoys the history of electron transport and energy conservation as represented by the triumvirate of Yagi, King and Slater. Many thanks are due Drs. Kim and Ozawa for their elegant organization of this meeting and its juxtaposition to the FAOB Congress.

Britton Chance
April 2, 1990

PREFACE

This book contains the contributed papers presented at the "International Symposium on Bioenergetics: Molecular Biology, Biochemistry and Pathology", held in Seoul, Korea, August 18-21, 1989, sponsored by International Union of Biochemistry (as IUB Symposium No.191) and Ewha Womans University, Seoul, Korea.

The symposium was held in honor of Professor Kunio Yagi to commemorate his 70th birthday. Professor Yagi has not only made many scientific accomplishments in biochemistry, but also enormous contributions to the international biochemical communities such as International Union of Biochemistry and the Federation of Asian and Oceanian Biochemists. His unusual pleasant personality and ability to get many scientists together, have helped many biochemical societies of different countries to grow.

In the last several years, advances in bioenergetics, in particular molecular genetics, have facilitated detailed study of the structure-functional mechanism of mitochondrial electron transfer and energy tranducing system, at the molecular level, and has been directed into the study of the molecular basis of mitochondrial disease, which is no longer a new field now. This is the very reason why the broad subtopics are chosen to be discussed in this symposium.

The purpose of this symposium was to bring together the scientists from those various disciplines of bioenergetics field to discuss recent developments in their fields so that they can exchange new ideas and interact each other to promote collaborations. It was our intention to provide an opportunity to discuss a wide range of bioenergetics topics that has previously been all too rare in Asian biochemical communities. Hopefully, a meeting like this can be continued regularly, in Seoul, Korea or in Nagoya, Japan.

The symposium was structured in a similar way to the previous "Symposium on Membrane Biochemistry and Bioenergetics", held in Rensselaerville, NY, August, 1986 (Kim et al., Plenum Press, 1987). Four lecture/discussion sessions and one poster/discussion session covered a variety of subjects from the molecular evolution of mitochondrial DNA to bioenergetic consequences of genetic disease. Drs. Y. Anraku, A. Azzi, M. H. Han, C. H. Kim, Q.-S. Lin, A. W. Linnane, T. Ozawa, S. Papa, P.S. Song, and C.-L. Tsou served as the session chairmen. All the participants were energetically involved in discussions of the subjects presented, giving us a lesson to plan a day more for the next meeting. All invited participants contributed a chapter to this volume except for a few, who could not make it with regret, due to various reasons.

This symposium was made possible with financial support from sponsors. We are grateful to sponsors, the Symposium Committee of the International Union of Biochemistry, Ewha Womans University, Korea Science Foundation, MIWON Co. Ltd, Seoul, Korea and Institute of Applied Biochemistry, Gifu, Japan. We express our special thanks to Mr. Chae Bang Kim, President of MIWON Co. Ltd. for his generous support to this meeting without hesitation. We also thank to Dr. Young Bog Chae, Director of Korea Research Institute of Chemical Technology for his support.

We wish to express our warm thanks to Professor Hea Chung Yun, Ewha Womans University, for her constant help throughout the organization of the symposium. We give thanks to Ms. Kum Hee Cho, Jung-Sun Kim, Hae-Jin Kim, Hye-Gyu Lee, Yong-Nam Roh, Sang-Hee Shin and Sung-Hee Shin, from Ewha Womans University for their efficient assistance during the meeting. We also thank Ms. Akiko Hashimoto in Nagoya, Ms. Donna Hilbert and Stacey Hood in RPI for their assistance throughout the organization of the symposium. Our special thanks extended to Mr. Michael Seaman for his assistance in editing and Ms. Laura Waelder for her word processing service for this volume.

Finally, we are thankful for all the enthusiastic participants who made this symposium successful and hope that we can get together in the next meeting with further interesting results and ideas in the field of bioenergetics.

Chong H. Kim
Takayuki Ozawa
March 1990

CONTENTS

Part I. 2. Energy Coupling and Ion Transport

Part I. 3. Other Related Topics

Part II. Molecular Biology

Part III. Mitochondrial Pathology

Appendix

Opening Session

INTRODUCTORY REMARKS

Chong H. Kim

It is my pleasure to open the "International Symposium on Bioenergetics: Molecular Biology, Biochemistry and Pathology" in honor of Professor Kunio Yagi.

I welcome all of you to Seoul, Korea! I think that most of you - probably all of you who is not resident here - come to Seoul for the first time. Before I introduce Professor E.C. Slater, President of International Union of Biochemistry, for opening lecture, I would like to say a few words for having this symposium in Seoul.

It is a very exciting moment for me to have all of you here at the Symposium on Bioenergetics. Because there is a reason. It has been fifteen years since I left my homeland, Seoul, Korea, to pursue advanced studies in Biochemistry in the United States. At that time, it was very difficult to do any advanced biochemical experiments in Korea. I had already done a few rather simple biochemical experiments by that time. However, I had to run from one institute to another in order to accomplish even those simple experiments. Unfortunately, I could not find a good teacher to lead my enthusiasm towards biochemical research. Therefore, I decided to study biochemistry in the States, where I had already visited for a month in 1973, on the way of my return from Copenhagen, Denmark, where I spent a few months for the advanced courses in Clinical biochemistry.

Fifteen years later now, Korean Biochemical Society has impressively grown to the level, at which they can freely exchange and discuss their scientific progress with leading scientists from all over the world. Some of you must have seen such an example at the 5th FAOB (Federation of Asian and Oceanian Biochemists) Congress, which was held last week just before this symposium. I am very proud to see that such an excellent international scientific gathering has been held in Seoul.

Our decision to hold a Symposium on Bioenergetics in Seoul, Korea, came about a year ago in Prague, at the 14th International Congress of Biochemistry. It is my honor to have a "Symposium on Bioenergetics" participated by all you, leading scientists from 13 different countries, in the country I was born and educated, with the support of the institution at which I was educated. I am grateful to those of you, my teachers, friends and colleagues, who have nourished my knowledge in bioenergetics, and helped me to open this successful, I believe, symposium in Seoul. Above all, I thank my co-organizer, Professor Ozawa, who is an excellent partner. I must also mention about the person behind me to make the smooth running of this symposium, Professor Hae-Jung Yun, Dean of College of Pharmacy, Ewha Womans University. Her unaccountable help made my task for this meeting easier.

I hope all of you can have a joyful time during this meeting even though our schedule is too tight and this can be a momentum to bring another Bioenergetics meeting in Seoul near future. I especially invite you to experience the beauty of the city of Seoul at the end of this meeting, and now I would like to introduce Professor Slater for the opening lecture.

OPENING LECTURE

E. C. Slater

Thank you very much, Dr. Kim and Professor Yagi and Dr. Ozawa and to other colleagues. There are, of course, many reasons why I am happy to say a few words at the opening of this symposium. First, and foremost, it is my pleasure to be able to attend this meeting to honor Professor Yagi on the occasion of his 70th birthday. I shall have more to say about you and to you, Kunio, tomorrow evening, and it suffices now to congratulate you and welcome you to the select band of septuagenarians present at this meeting. Now another reason why it is a great pleasure for me to say a few words here is that this is a IUB symposium, it is #192 as a matter of fact and I have attended all of the IUB congresses since the first one in 1949. I, of course, cannot say the same thing about IUB symposia, even during the period in which I am president of IUB. This year alone IUB is sponsoring no less than 28 symposia through its two committees-Symposium Committee which is chaired by Arnost Kotyk who is here at this meeting and our so-called Interest Groups Committee. But, although we have allocated more than one-third of our total budget - our total budget is only about $300,000 to support our symposia, a quick calculation will show you that our support is spread out pretty thinly. Actually, Dr. Kim has been most skillful in utilizing our resources efficiently. Since quite apart from direct subvention from our Symposium committee – three of the speakers are members of the executive committee of IUB who are meeting here in Seoul, earlier in the week, and have had their travel expenses paid by IUB under a different budget heading, namely of the executive committee. Moreover, others have attended the FAOB Congress which ended yesterday, which also received some support from IUB under yet another budget heading, and I am sure that our Treasurer, who is Tony Linnane, will be very pleased at your efficient use whereby money is spent twice.

This is actually the second IUB sponsored symposium on bioenergetics held in the FAOB area. I suppose those who were not at this very wonderful FAOB meeting earlier this week – really extremely well organized meeting with a good program indeed, and some of you who have just flown in may not know what FAOB is – well let me tell you – it is the Federation of Asian and Oceanian Biochemists. Now the first IUB symposium in Asia was held in Beijing, China in 1984, in Bioenergetics. I just want to sketch because Dr. Kim asked me to fill up the program up to 9:00 o'clock.

The Progress that I see or the sort of changes that I see have come into the field in the last five years – during much of the time which I have no longer been active in bioenergetics research. I think, of course, a lot of progress was made particularly by the application of techniques introduced from molecular genetics. Of course, cloning and DNA sequencing had already, five years ago, given us the amino acid sequences of many subunits of the electron transferring and ATP-synthesizing proteins. This has been largely completed now in the intervening period. And the powerful tool of site-directed mutagenesis has been directed to the study of the function, mechanism of action and particularly in the biosynthesis also of the subunits. I think that we can say that the mechanism of electron transfer or the pathway, in any case, of electron transfer within the large electron transferring proteins of the mitochrondiron has been largely worked out. Also, we now know rather more than we did five years ago about which electron transferring reactions are energy conserving, that is, they contribute to the charge separation across the membrane. Although I am not sure this will be discussed in this symposium − I hope it will be −. We now have a deeper insight into the forces which drive electron transfer between different well separated centers in the protein. I do not think we have progressed very far in our understanding of how an electric field across a membrane or protonmotive force, if you will, catalyses the synthesis of ATP or the dissociation of already formed ATP but I may be wrong − maybe you will prove me wrong at this meeting.

Perhaps the most striking and unexpected development − unexpected for me in any case − in recent years has been in the molecular basis of mitochondrial pathology, which is the last topic that will be dealt with in this meeting. For me, my first surprise came from Dr. Kadenbach's demonstration of the tissue-specific expression of cytochrome oxidase genes. In any case, this made it more understandable why certain inborn errors may lead to lesions in specific tissues. I have also been surprised by the change of attitudes to mitochondrial disease, since I left the laboratory. Where it was once thought to be very rare, it now appears that the cases examined are the tip of an iceberg or perhaps we should say for the investigators, the exposed reef of a gold mine. It is not only the genetically inherited disorders of the respiratory chain that are attracting attention, but also those that arise owing to mutations of the mitochondrial DNA after birth. Indeed, Tony Linnane and Takayuki Ozawa have suggested that the failures in function associated with aging − which is something, that is of interest to you and I, Kunio − are due to a progressive loss of respiratory capacity of tissues as a result of random mutation of the mitochondrial DNA. I must say that it is of some comfort to know that the field, in which I have worked during most of my scientific life, might now be providing me with an explanation of why I find it more difficult to follow the work of my younger colleagues nowadays − who, of course, still have beautifully operating mitochondria producing kilograms of ATP per hour, and even the increasing number of mistakes I make in proofreading papers and letters.

In the late 1940s and in the 1950s, bioenergetics belonged to the cutting edge of biochemistry, if you can excuse the jargon. New fields opened up by the discovery of the structure of nucleic acid superseded bioenergetics in the center of interest. In the middle of the 1960s, a new era opened up with the discovery that mitochondria have their own genetic apparatus distinct from that present in the nucleus. However, this did not have much impact outside of mitochondria − others considered it of secondary importance − a sort of evolutionary fossil, retained after bacteria established themselves as symbionts in the animal cells and evolved into mitochondria. For a short period in the 1970s bioenergetics became popular again with the granting agencies when it was suggested that the combination of photochemical reaction centers and hydogenase could be utilized on a commercial scale to provide hydrogen gas to provide fossil fuel. This is not yet possible, but the relative priority given to this study, did I believe, lead to the discovery of the role of nickel in hydrogenase and related enzymes, and that of manganese in the oxygen evolving reaction of photosynthesis a little more rapidly than otherwise might have been the case, because up till then support in the very rapidly productive field of photosynthesis was declining.

The concept that mitochondrial diseases are not only more important than we thought, but perhaps universal since all of us age – even you younger bioenergeticists age –, opens new possibilities and perhaps it will become a major field of medicine.

To those in the field – even if not always to those outside the field-bioenergetics, has always been exciting. To paraphrase Arthur Kornberg in another context, "I have never met a dull electron transferring protein."

The fundamental importance of the electron transport chain and oxidative phosphorylation is, of course, obvious to everyone. But some might have felt that the main features have long been established and have been incorporated into textbooks – just like the Krebs cycle has – and it only remains to dot the "i"s and cross the "t"s. However, the recent developments in the importance of mitochondrial disease, for example, would never have taken place if the filling in of details – the dotting the "i"s and the crossing the "t"s, particularly in the connection of the mechanisms of biogenesis of mitochondria had not been pursued.

I would like to conclude this opening remark by thanking Dr. Kim and Dr. Ozawa for putting together this interesting symposium and inviting many of us. Practically everybody here visits Seoul for the first time. I can assure you that we are extremely impressed by what we have seen here in Seoul and how you organized the extremely successful meeting. I also would like to thank Ewha Womans University for sponsoring this meeting together with IUB and also those co-sponsors.

Thank you!

Special Session

MEMORIES OF MY FLAVOPROTEIN RESEARCH

Kunio Yagi

Institute of Applied Biochemistry, Yagi Memorial Park
Mitake, Gifu 505-01, Japan

Friends! Ladies and Gentlemen!

It was, and still is, a great pleasure and honor to me when I learned that the present symposium was to be held in honor of myself. Before continuing further, I want to express my heartfelt thanks to the members of the organizing committee and to all participants in this symposium.

Generally speaking, I am a very future-oriented person and do not often focus on past events. However, on very rare occasions, such as the present one, I am prompted to do precisely that. So, during the next few minutes I want to share with you some of the historical background of my past research, especially that on flavoproteins, not failing of course to also bring in the element of human relationships.

When I was a middle-school student in Tokyo, sometime around 1932 to 1936, I had aspirations of becoming a novelist or poet, and so I entered the literary program at Tokyo Metropolitan High School. During the three-year term of high school, however, I began to have second thoughts about my choice of future career. At some point I changed my mind and decided that I would become a medical researcher instead, since medicine seemed to me a more exciting endeavor than pen and ink. So, in 1936, I entered medical school. Since Tokyo University as well as Kyoto University did not allow literary students to apply, I decided to apply to the medical school at Nagoya University. Shortly after entrance there, my excitement waned considerably, especially when I attended the lectures on anatomy, a topic that seemed to me very similar to geography. In addition, I realized that I would not be able to pass the biochemistry examination, for my knowledge of chemistry was so poor. Thus, I decided to spend most of my time during the first year of medical school studying chemistry. Fortunately, the professor of biochemistry was very considerate of me and allowed me the use of a bench in his laboratory. Thereafter I conducted primitive experiments in chemistry on a daily basis until one eventful day. On that day the dean of the medical school happened to visit the laboratory and asked me, "What are you doing?" "Chemical experiments," I answered very politely. Then he asked me, "Are you a student?"; and I replied, "Yes." He then continued, "In this medical school, students must attend the lectures. What lecture is being given now?" Promptly I answered, "Hygiene." He next inquired,

Fig. 1. A portrait of Prof. Leonor Michaelis.

"Why aren't you attending this lecture?" I answered with all honesty,
"A lecture on hygiene does not interest me." In response he displayed a
most unpleasant face and then asked me to come to his office later.
When I entered his office at the appointed time, he persuaded me to
attend all of the lectures, but, to my relief and great joy, he also
informed me that he was giving me official permission to conduct experiments in the biochemical laboratory at any time, as long as it was
between evening and morning! And then he added, "Night is not only for
sleeping!"

Thus, having been granted a position of sorts in the biochemical
laboratory, I carried out experiments every night. When I was tired, I
often slept in the lab. Fortunately for me and to my surprise, there in
the laboratory was of all things a very beautiful, large, black-painted
bed decorated with golden designs! So I gratefully put it to good use,
but thought it odd that such a disharmonious thing should be present in
a laboratory! One day I asked a senior staff member of the laboratory
about this most interesting bed. He enlightened me, saying "This was
the bed of Professor Leonor Michaelis. He is the founder and was the
first professor of this laboratory. As you will learn in the biochemistry course, he is the very person who proposed the Michaelis-
Menten theory (1). Also, you must read this book." Then he showed me a
book written by Prof. Michaelis entitled "Die Wasserstoffionen-Konzentration" (2). Thus, this was my first contact with Prof. Michaelis.
Figure 1 shows a portrait of this very distinguished gentleman (3).

Prof. Michaelis is, of course, famous for the theory that bears his
name along with that of his colleague, in addition to his many other
pioneering works, but few people know about his contribution to the
Japanese biochemical community in its early days. In 1922, Prof.
Michaelis came to Japan from Germany at the invitation of Aichi Medical
College, now the University of Nagoya, and was asked to found a department of medical chemistry, which is now known as the Institute of Biochemistry in the Faculty of Medicine of the University of Nagoya.
Figure 2 shows a photograph of Prof. Michaelis with his family and
colleagues in front of the college (4).

Fig. 2. Prof. Leonor Michaelis (fourth from the left) with his family and colleagues in front of the college.

In 1923, one year after his arrival, he began lectures for medical students; and, in addition, he began conducting some highly inspiring and stimulating special courses for young Japanese biochemists in this new department. Figure 3 is a photograph depicting the Professor in the process of demonstrating an experiment to his eager class (5).

Fig. 3. Prof. Leonor Michaelis (fourth from the left), conducting a physico-chemical experiment.

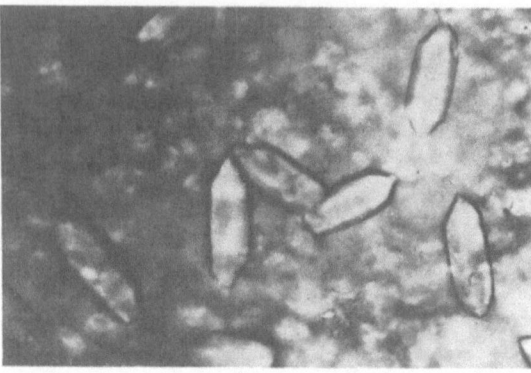

Fig. 4. **Crystals of the enzyme-substrate complex of D-amino acid oxidase.** From ref. 9, courtesy of Elsevier Publishing Company.

When I was studying as a medical student in the Department of Biochemistry, School of Medicine, Nagoya University, during the period 1939-42, some years had passed since Prof. Michaelis had left Nagoya, but many things he had brought from Germany still remained and somehow strongly influenced me. In addition, when I started my research work just after the Second World War, the mechanism of enzyme action was one of the most challenging targets for young biochemists. Feeling somewhat close to Prof. Michaelis and excited by his now famous theory, I decided to study the enzyme-substrate complexes (6).

Prof. Michaelis had predicted a single enzyme-substrate complex, that is, the Michaelis complex, to exist during an enzymatic reaction:

$$E + S \rightleftharpoons E \cdot S \rightleftharpoons E + P$$

However, many people thought later that a series of intermediate complexes could exist in the reaction sequence:

$$E + S \rightleftharpoons E \cdot S \ldots\ldots E \cdot P \rightleftharpoons E + P$$

While contemplating just how to demonstrate such intermediate or intermediates, I thought that an enzyme having characteristic properties such as a color, as is the case for flavin enzymes, would be advantageous. I finally selected D-amino acid oxidase, a dissociable flavin enzyme, as a model enzyme, since I presumed that the flavin chromophore would serve as an indicator for the oxidation/reduction state, as well as for the formation of any complex. As a working hypothesis, I reasoned that if an enzyme were to form an enzyme-substrate complex, which would then result in the enzyme and product, and if the enzymatic reaction were reversible, then a certain amount of the complex should become detectable in the presence of adequate amounts of the substrate and product. Furthermore, I assumed that the complex might become stabilized under reaction conditions other than those giving the maximal reaction and that the complex might thus be crystallizable under some appropriate conditions (7). With these thoughts in mind, I carried out extensive research, and after much hard work of long duration, I succeeded, in collaboration with Dr. Takayuki Ozawa, in the crystallization of a beautiful purple-colored enzyme-substrate complex, which

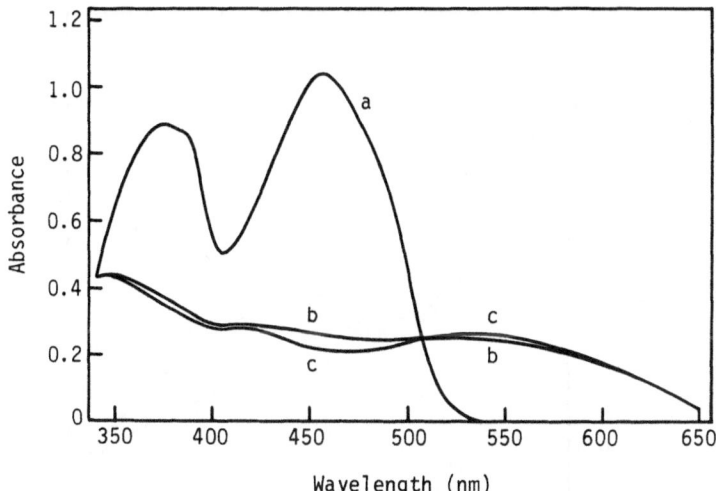

Fig. 5. Identity of the purple intermediate of D-amino acid oxidase. Changes in the absorption spectrum of D-amino acid oxidase after the addition of D-alanine under the anaerobic condition. a, Oxidized enzyme; b, 300 msec after mixing; c, 2 sec after mixing. From ref. 17, courtesy of the Japanese Biochemical Society.

was composed of equimolar amounts of D-amino acid oxidase and the substrate D-alanine. Figure 4 shows the crystals of this complex, which, incidentally, was the very first enzyme-substrate complex to be crystallized. The year was 1962.

We submitted a short paper describing this exciting finding to *Biochimica et Biophysica Acta* (8). Professor E. C. Slater, who is among us tonight, accepted our paper for publication in this journal after he had kindly edited the manuscript in his own handwriting. After the publication of this paper, which was followed by a series of related communications (9-12), I rapidly became recognized by many leading biochemists and was given many opportunities to present our work at various symposia around the world. At one such symposium, "Mechanism of Enzyme Action," at the 5th Congress of the International Union of Biochemistry, held in New York in 1964, my presentation was reported by *Chemical and Engineering News* (13). Their report resulted in even wider recognition of our findings. However, opposition soon arose. When I spoke at the First Symposium on Flavins and Flavoproteins, organized by Prof. Slater and held in Amsterdam in 1965 (14), a serious question about our results was raised by Dr. Vincent Massey (15), who, incidentally, is also present here tonight.

By the stopped-flow technique, he and his colleagues (16) had measured the absorption spectrum of the purple-colored intermediate that appeared during the reaction of D-amino acid oxidase with D-alanine under the anaerobic condition. By their procedure they discovered that the spectrum measured at 150 ms after the anaerobic mixing of the enzyme and substrate differed from the one measured at 1.8 s after the mixing. As the spectrum of our crystals was different from theirs measured at 150 ms, but identical to theirs measured at 1.8 s, they therefore argued that our crystallized species was meaningless in the catalytic sense because a 1.8-s species could not explain the high molecular activity of this enzyme.

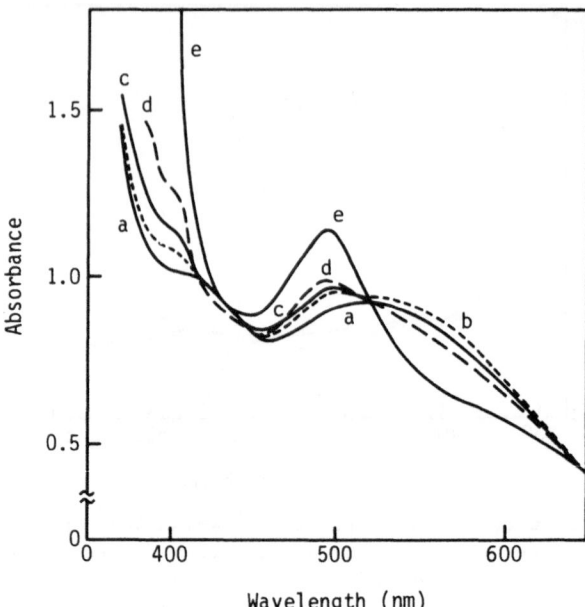

Fig. 6. Changes in the absorption spectrum of the purple intermediate complex during its aging in the dark at 5°. a, Purple intermediate complex; b, after 3 days; c, after 5 days; d, after 10 days; e, after 1 month. From ref. 10, courtesy of Elsevier Publishing Company.

When we looked carefully at their results, we found them to be correct and quite reproducible. Figure 5 shows the data we obtained on this point by rapid-scanning spectroscopy (17). These new findings naturally had a strong impact on me, and so we set out to examine this problem extensively. When we obtained the absorption spectrum of the "rapidly appearing intermediate" and that of the "slowly appearing one" and compared the two, we found that the difference spectrum between them agreed with that between the oxidized enzyme and the "slowly appearing intermediate". We viewed this phenomenon as indicating that the "rapidly appearing intermediate" was not homogeneous but rather was consisted mostly of the purple intermediate and, to a minor extent, of the oxidized enzyme. This indicates that most of the enzyme changes rapidly to the purple entity, whereas a lesser portion does so only slowly. As we were already aware of the monomer/dimer equilibrium of D-amino acid oxidase (18,19), we interpreted these data to indicate that these two forms of the enzyme were responsible for these two entities; that is, the dimeric form reacts more rapidly with the substrate to form the purple intermediate than does the monomeric form of the enzyme. The quantitative analysis supported this assignment (17). Therefore, we concluded that the crystalline purple intermediate is, in fact, the intermediate appearing during the enzyme catalysis.

This different catalytic activity between monomer and dimer of D-amino acid oxidase generated additional interest in the monomer/dimer equilibrium of enzymes in general. Furthermore, this finding provides a warning that the adjustment of parameter values obtained with a low enzyme concentration by those obtained with a high concentration of

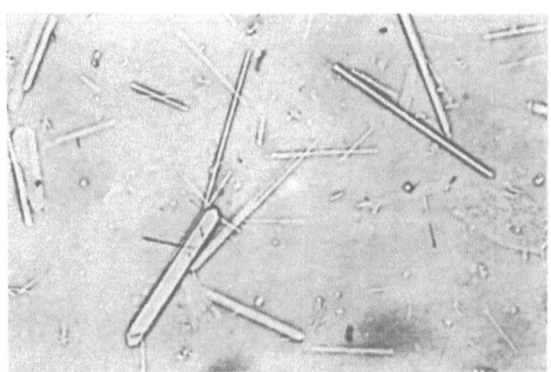

Fig. 7. Crystals of the semiquinoid form of D-amino acid oxidase. From ref. 11, courtesy of Elsevier Publishing Company.

enzyme is crucial, because a low concentration of enzyme contains a larger amount of monomer than does a high concentration of enzyme, as evidenced by D-amino acid oxidase (20,21).

Another question that arose from the crystallization of this enzyme-substrate complex was the nature of the complex itself. Judging from the purple color of the crystals, I thought initially that the coenzyme portion of the entity would be in the semiquinoid form, especially since it was already established that the oxidized form of the enzyme has a yellow color whereas the fully reduced form is colorless. To examine this possibility, we conducted an electron-spin resonance study on the crystal.

Although we did detect a signal (22), quantitative analysis revealed that only a minute portion of the coenzyme, flavin adenine dinucleotide (FAD), was a radical species. Additionally, the amount of radical contained in the crystals varied with time. When we looked into the matter more closely, we found that the crystalline complex initially had no signal but that the signal gradually appeared and increased, with the spectrum of the complex gradually changing from that of a purple to that of a red species (10). Figure 6 depicts the spectral changes. These findings indicate the initial complex to be a diamagnetic species and the gradually transformed one to be a paramagnetic one. By exhaustive examination of the phenomenon we finally concluded that the complex is a strong charge-transfer complex that dissociates gradually into radical cation and radical anion (10,23). The substrate radicals disproportionate to yield non-radical species, whereas the flavin radicals remain unchanged owing to their stabilization by the apoprotein. By starting with the purple complex, we were able to crystallize the semiquinoid enzyme of D-amino acid oxidase (11). Figure 7 shows such crystals, the first ever prepared of a semiquinoid flavin enzyme.

Since the purple intermediate complex was found to be the very one that reacts with oxygen (24), we assumed that the slow dissociation of the complex, that is, the long life time of this strong charge-transfer complex, is responsible for giving the complex sufficient time to react with the oxygen. The reason for this complex being somewhat stabilized

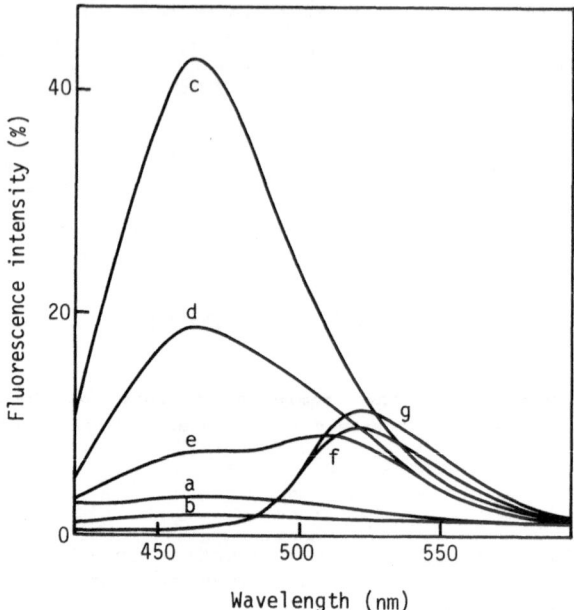

Fig. 8. Changes in the fluorescence spectrum of the apoenzyme-MBAS complex upon the addition of FAD. a, Apoenzyme; b, MBAS (92 μM); c, a + b; d, c + FAD (21.8 μM); e, c + FAD (43 μM); f, c + FAD (108 μM); g, reconstructed holoenzyme. Excitation at 362 nm. From ref. 26, courtesy of the Japanese Biochemical Society.

became the next problem for us to tackle. We predicted that the microenvironment surrounding the charge-transfer interaction might be hydrophobic in nature. To test our prediction, we synthesized several hydrophobic probes that fluoresce when they combine with hydrophobic regions of proteins (25). Among these probes, we found the novel compound monobenzoylated 4,4'-diaminostilbene 2,2'-disulfonate (MBAS) to be suitable, for this dye can substitute for the coenzyme FAD and combine with the apoenzyme of D-amino acid oxidase (26).

As shown in Figure 8, when the dye combines with the apoenzyme, the former fluoresces; and this fluorescence diminishes upon the addition of FAD. The fluorescence spectrum finally becomes identical to that of the D-amino acid oxidase holoenzyme, indicating that the coenzyme resides in a hydrophobic pocket of the apoprotein (26). Additionally, we found that another hydrophobic probe, 1-anilinonaphthalene-8-sulfonate, can combine with the apoenzyme in competition with the substrate D-alanine (26). Therefore, we concluded the microenvironment surrounding the charge-transfer interaction to be hydrophobic. I presented these results just mentioned at the Second International Symposium on Oxidases and Related Redox Systems, which was held in Memphis in 1971 (27). I will never forget the witty comment made by Dr. W. E. Blumberg (28) after my presentation. He said, "In case anybody missed the point of Dr. Yagi's talk in relevance to the discussion earlier today, let me reiterate that one could paraphrase and say that anybody who is talking about flavin water chemistry in this enzyme is all wet!"

From another point of view, such a hydrophobic environment for the coenzyme was also found to be important for the catalytic activity of the enzyme, for our molecular orbital calculations demonstrated the

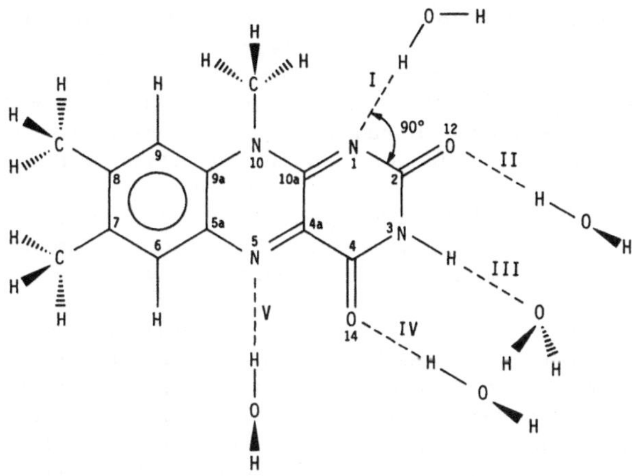

Fig. 9. Geometry of oxidized lumiflavin-water complexes.
I, II, III, IV, and V indicate hydrogen-bonding sites. From ref. 31,
courtesy of the Japanese Biochemical Society.

importance of hydrogen bonding at particular heteroatoms of the iso-
alloxazine nucleus of flavin for its reactivity (29-31). Figure 9 shows
the geometry of various oxidized flavin-water complexes. I-V indicate
sites of hydrogen bonding. In the oxidized flavin, the N(5) atom most
probably accepts electrons (32). The atomic orbital coefficients in the
lowest unoccupied molecular orbital at N(5) of oxidized lumiflavin-
water complexes indicated that hydrogen bonding at all heteroatoms

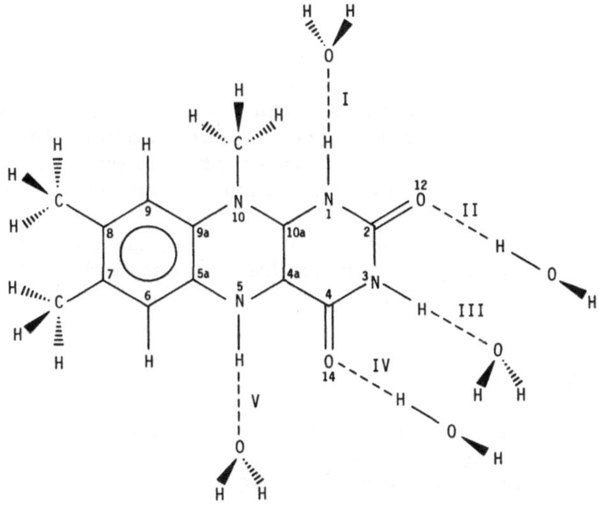

Fig. 10. Geometry of reduced lumiflavin-water complexes.
I, II, III, IV, and V indicate hydrogen-bonding sites. From ref. 30,
courtesy of Elsevier Publishing Company.

except N(3) increases the electron acceptability of N(5). Figure 10 shows the geometry of reduced flavin-water complexes. In the reduced flavin, C(4a) is the atom that donates electrons to a neutral electrophile such as oxygen (33). The atomic orbital coefficients in the highest occupied molecular orbital at C(4a) of reduced lumiflavin-water complexes indicated that hydrogen bonding at N(1) or N(3) increases the electron-donating ability of C(4a). If the microenvironment of the coenzyme were aqueous, all heteroatoms would be expected to form hydrogen bonds with water molecules, and such hydrogen bonding at particular heteroatoms would be impossible. However, hydrogen bonding between selected heteroatoms of flavin and amino acid residues in a hydrophobic region of the apoenzyme would explain the increased reactivity of the coenzyme noted when flavin combines with the apoenzyme. Such hydrogen bonding in a hydrophobic environment answers, at least in part, the question raised by Professor Hugo Theorell, who, incidentally, invited me to collaborate with him in his laboratory back in 1957. In his Nobel Lecture (34) in 1955 he said, "We still do not quite understand how through its linkage to the coenzyme the enzyme-protein 'activates' the latter to a rapid absorption and giving off of hydrogen."

Before concluding my lecture, I would like to mention some of the human aspects related to the work I have just presented to you. Professor Slater, who accepted our paper in his capacity as an editor of *Biochimica et Biophysica Acta*, came to Nagoya in 1973 at my invitation, and attended the symposium that we dedicated to the late Prof. Michaelis on the occasion of the 50th anniversary of the commencement of his lectures in Nagoya. In that meeting, Prof. Slater gave a warm address (35) that afforded considerable pleasure to those colleagues of Prof. Michaelis who were present at that time, though they have all passed away by now. Professor Massey, who had argued initially about the purple crystals, became a very good friend of mine thereafter; in fact, in 1985 he even collaborated with me at my institute in Japan. Doctor Ozawa was a post-graduate student at the time of the crystallization of the enzyme-substrate complex. Today he is a Full Professor of Biochemistry at the University of Nagoya, and is one of the organizers of this symposium.

I view science as a human enterprise that has no end, whereas the scientist himself or herself, during an all too brief period of time, adds his or her contribution and then is no more. During the plying of our trade, we make personal contact with a multitude of like persons: we compete, we argue, we cooperate, and we even become intimate friends. Truly we are a lucky or fortunate segment of society. This present symposium is yet another example of such interactions among us scientists!

Ladies and Gentlemen! Thank you once again for your kind attendance, and I wish each of you much future success and good luck!

ACKNOWLEDGEMENT

The major portion of this paper was reproduced, in paraphrased form, from the following article by the same author that appeared in *Biochim. Biophys. Acta* **1000** (1989) 203-206: Commentary by Kunio Yagi on 'Crystallization of Michaelis complex of D-amino acid oxidase' by K. Yagi and T. Ozawa, *Biochim. Biophys. Acta* **60** (1962) 200-201. The author thanks Elsevier Publishing Company for its permission to reproduce this information.

REFERENCES

1. Michaelis, L. and Menten, M. L. (1913) *Biochem. Z.* **49**, 333-369
2. Michaelis, L. (1922) *Die Wasserstoffionen-Konzentration*, Springer-Verlag, Berlin
3. This portrait of the late Prof. Michaelis was kindly furnished by Prof. Michaelis's daughters, Mrs. Ilse Michaelis Wollman and Mrs. Eva Michaelis Jacoby, upon their coming to Japan in 1973 on the occasion of the symposium dedicated to the late Prof. Michaelis in commemoration of the 50th anniversary of the opening of his lecture in Aichi Medical College
4. This picture was kindly furnished by the late Prof. I. Ogawa of the University of Nagoya, who was a staff member of Prof. Michaelis
5. This picture was also donated by the late Prof. I. Ogawa
6. This story was described in *Reactivity of Flavins* (The proceedings of the symposium dedicated to the late Professor Leonor Michaelis under the auspices of the Japanese Biochemical Society) (Yagi, K. ed.), Univ. Tokyo Press, Tokyo, 1975
7. Yagi, K. (1965) in: *Advances in Enzymology* (Nord, F. F. ed.), Vol. 27, pp 1-36, Interscince, New York
8. Yagi, K. and Ozawa, T. (1962) *Biochim. Biophys. Acta* **60**, 200-201
9. Yagi, K. and Ozawa, T. (1964) *Biochim. Biophys. Acta* **81**, 29-38
10. Yagi, K., Okamura, K., Naoi, M., Sugiura, N. and Kotaki, A. (1967) *Biochim. Biophys. Acta* **146**, 77-90
11. Yagi, K., Sugiura, N., Okamura, K. and Kotaki, A. (1968) *Biochim. Biophys. Acta* **151**, 343-352
12. Yagi, K., Okamura, K., Sugiura, N. and Kotaki, A. (1968) *Biochim. Biophys. Acta* **159**, 1-8
13. *Chemical and Engineering News* 10, 34, 1964
14. Yagi, K. (1966) in: *Flavins and Flavoproteins* (Slater, E. C. ed.), pp 210-222, Elsevier, Amsterdam
15. Massey, V. (1966) in: *Flavins and Flavoprotein* (Slater, E. C. ed.), p 222, Elsevier, Amsterdam
16. Massey, V., Palmer, G., Williams, C. H. Jr., Swoboda, B. E. P. and Sands, R. H. (1966) in: *Flavins and Flavoproteins* (Slater, E. C. ed.), pp 133-158, Elsevier, Amsterdam
17. Yagi, K., Nishikimi, M. and Ohishi, N. (1972) *J. Biochem.* **72**, 1369-1377
18. Yagi, K., Ozawa, T. and Ohishi, N. (1968) *J. Biochem.* **64**, 567-569
19. Yagi, K. and Ohishi, N. (1972) *J. Biochem.* **71**, 993-998
20. Shiga, K. and Shiga, T. (1972) *Biochim. Biophys. Acta* **263**, 294-303
21. Yagi, K., Sugiura, N., Ōhama, H. and Ohishi, N. (1973) *J. Biochem.* **73**, 909-914
22. Yagi, K. and Ozawa, T. (1963) *Biochim. Biophys. Acta* **67**, 685-687
23. Yagi, K., Nishikimi, M., Ohishi, N. and Takai, A. (1971) in: *Flavins and Flavoproteins* (Kamin, H. ed.), pp 239-260, Univ. Park Press, Baltimore
24. Massey, V. and Gibson, Q. H. (1964) *Fed. Proc.* **23**, 18-29
25. Kotaki, A., Naoi, M. and Yagi, K. (1971) *Biochim. Biophys. Acta* **229**, 547-556
26. Naoi, M., Kotaki, A. and Yagi, K. (1973) *J. Biochem.* **74**, 1097-1105
27. Yagi, K. (1973) in: *Oxidases and Related Redox Systems* (King, T. E., Mason, H. S. and Morrison, M. eds.), pp 217-225, Univ. Park Press, Baltimore
28. Blumberg, W. E. (1973) in: *Oxidases and Related Redox Systems* (King, T. E., Mason, H. S. and Morrison, M. eds.), p 225, Univ. Park Press, Baltimore
29. Nishimoto, K., Watanabe, Y. and Yagi, K. (1978) *Biochim. Biophys. Acta* **526**, 34-41

30. Nishimoto, K., Kai, E. and Yagi, K. (1984) *Biochim. Biophys. Acta*
    802, 321-325
31. Nishimoto, K., Fukunaga, H. and Yagi, K. (1986) *J. Biochem.* **100**,
    1647-1653
32. Song, P.-S., Choi, J. D., Fugate, R. D. and Yagi, K. (1976) in:
    Flavins and Flavoproteins (Singer, T. P. ed.), pp 381-390, Elsevier,
    Amsterdam
33. Kawano, K., Ohishi, N., Takai, S. A., Kyogoku, Y. and Yagi, K.
    (1978) *Biochemistry* **17**, 3854-3859
34. Theorell, H. (1964) in: *Nobel Lectures Physiology or Medicine 1942-
    1962*, pp 480-495, Elsevier, Amsterdam
35. Slater, E. C. (1975) in: *Reactivity of Flavins* (Yagi, K. ed.), pp
    VII-VIII, Univ. Tokyo Press, Tokyo

Part I. Biochemistry

1. Mitochondrial Electron Transfer

AN OVERVIEW OF UBIQUINONE PROTEINS

Tsoo E. King

Institute of Structural & Functional Studies
University City Science Center and Department of
Biochemistry & Biophysics, University of Pennsylvania
Philadelphia, PA 19104 USA

SUMMARY

Ubiquinone (coenzyme Q) may serve as a respiratory carrier only when it is bound to a protein. Q may be considered as a prosthetic group, i. e. coenzyme, which can freely dissociate and associate with protein. So far three classes of Q-proteins have been found. QP-S is the immediate electron acceptor of succinate dehydrogenase (SDH). The reconstitution of succinate-ubiquinone reductase needs only SDH and QP-S independent of any cytochrome. QP-S has been isolated to pure form but its high hydrophobic nature prevents the complete determination by chemical means of primary structure.

QP-C acts in the cytochromes b and c_1 region, has been purified and amino acid sequence determined. We do not know whether there is second QP-C to fulfil Q_o and Q_i theory. QP-N exists in the NADH-ubiquinone reductase segment which is free from SDH and cytochrome oxidase and nearly free (<0.06 nmol per mg protein) of cytochromes b and c_1. The protein moiety of Q-protein not only stabilizes the ubisemiquinone but also dictates the site or sites where the QP acts on the respiratory chain.

INTRODUCTION

Professor Richard A. Morton in Liverpool working on vitamin A deficient animals and yeast, got a fraction, called SA (1). Through a lengthy procedure he and co-workers eventually, isolated SA in crystalline form in 1956. In the collaboration led by Dr. O. Isler of Hoffmann-La Roche at Basal with 37 g crystalline substance isolated from pig heart under the help of the results of Professor G.N. Festenstein the structure of ubiquinone was *quickly determined*. The synthesis was made. Ubx of Fig. 1 which represents the general formula and the x signifies the number of carbon atoms in the isoprenoid side chain such as 5,10,15,...50. Now it is as high as 65 and as low as 15 carbon atoms has been found in nature. There are so far no natural occurrence for 5 and 10. Compounds with x=5 to 50 were synthesized by the Swiss group from 1958 to 1960.

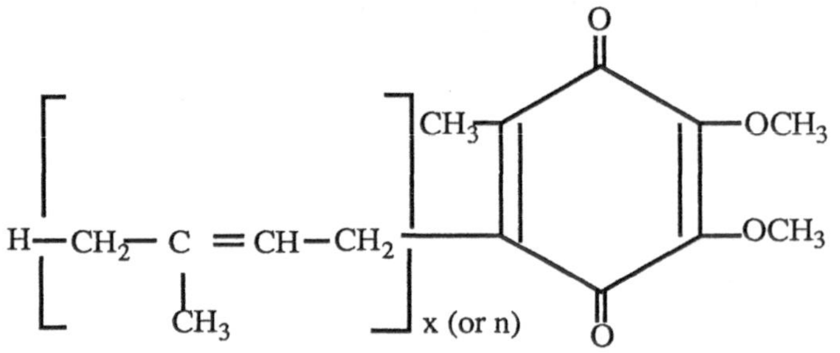

x denotes the number of carbon atoms   (ubiquinone)

n denotes the number of isoprenoid units   (coenzyme Q)

FIG. 1. Ubiquinones or Coenzyme Q

Across the Atlantic, then Professor David E. Green leading a much larger group with scores of post doctoral fellows especially Dr. Fred Crane , Dr. Carl Widmer and Dr. Jos Hatefi in the Enzyme Institute, Madison, Wisconsin for the studies on phospholipid, a substance absorbs light at 275 *nm* was isolated from the unsaponifible fraction of mitochondria. It was called Compound 275. With the collaboration of Dr. K. Folkers' group at Merck Sharp and Dohme Research Laboratories at Rahway, N.J. Compound Q-275 was also crystallized and the structure determined in 1958. The synthesis also made for various homologues. The new names was eventually, called by these investigators as coenzyme Q$_n$ and subscript *n* denotes the number of isoprenoid units such as 0,1,2,...10. Later the International Nomenclature Committee has adopted ubiquinone as the official name and this subscript *n* is used for the number of isoprenoid units instead of the number of carbon atoms, *n* was found from 0 to 10. Recently, *n* as high as 13 and as low as 3 in microorganisms has been found in nature.

An admirable point must be mentioned here that the discovery, structure and synthesis of coenzyme Q-ubiquinone were accelerated by close correspondence across the ocean between Liverpool and Madison during the progress. One must feel somewhat sad that nowadays the preprinted manuscripts are numbered and personal communication of new observations is very rare before appearance in a journal.

FREE UBIQUINONE

Free ubiquinone shows a structure as Fig. 1, R denotes the isoprenoid unit. One electron oxidation-reduction scheme of Q is shown in Fig 2.

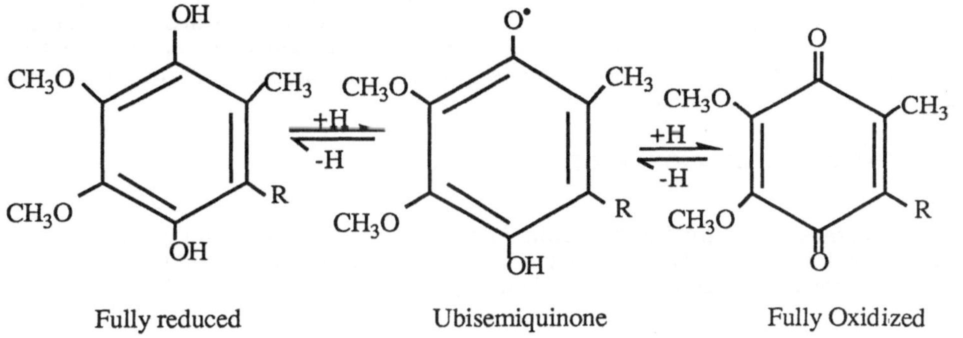

Fully reduced	Ubisemiquinone	Fully Oxidized

(or other electronated or protonated forms)

R=isoprenoid group, in mammals it is usually 10 except in rat liver

FIG. 2. One-Step-Oxidation of Ubiquinone

Q_{10} has absorption at 275 *nm* with an absorption coefficent of 14,000 in ethanol. Upon reduction (for example by borohyride) there is hypochromism and red shift. However, ubisemiquinone can exist only in a very small amount because the dismutation constant of about 10^{10} or higher is much in favor of non-radical form.

First, I heard a talk by Morton and his colleagues at Biochemical Society meeting at the Royal College of Surgery in London on December 13, 1957 (1a). Because of its wide occurrence the term ubiquinone was proposed to replace the provisional name, SA (1), before the name of the Wisconsin-Merck group's coenzyme Q. Actually on October 25, 1957, Morton *et al.* submitted a note with the opening sentence: "The name *ubiquinone* is proposed for a substance which has been studied at Liverpool for some years and has been called '272 mμ substance and later SA'" (2). After the London meeting, the Liverpool group, especially Eric Redfearn, discussed with me a long time because they knew I was working with Professor Keilin. They think ubiquinone might have a function in the respiratory chain as a carrier although only implied in their verbal presentation in the meeting. I did not put too much weight on this subject despite of Eric was very enthusiastic. Few months later at the Biochemical Society meeting in the University of Sheffield in 1958 they formally presented the experimental results that Q acts as a respiratory carrier (3). More private conversations were held. Really they found quite convincing proof of Q as a mediator in the respiratory chain. Now I became more seriously interested in the subject. I asked Professor Keilin why Q, a small molecule can act as a respiratory carrier while all others, such as cytochromes *b*, and *c* are macromolecules with protein attached. The term respiratory carrier was first used by Professor Keilin but he did not answer my question immediately. Not until after our usual early evening conversation in that period of the time he almost ordered me to find out (cf. 4, p. 125). American Federation Proceedings of 1958 eventually arrived in Cambridge and

I found out that Hatefi *et al.* reported similar finding of Q275 (the name coenzyme Q was not yet coined) as a respiratory carrier at the Federation Meeting in Philadelphia the same year (5).

A Ciba Foundation symposium held May 11 to 13, 1960 is entitled "Quinones In Electron Transport" and published by Churchill Press, 1961. This meeting was attended by many prominent bioenergetists including Britton Chance, Hans Krebs, Bill Slater; Alexander Todd, etc in addition to a number of creative organic chemists . More than a dozen of schemes for electron transport and oxidative phosphorylation with free Q in them were presented. Now to read these schemes gives us considerable historical perspective not only for Q itself but also other components.

EARLY WORK ON UBIQUINONE PROTEIN

Anyhow, since Sheffield meeting I continually thought it would be unique, if not impossible, for a small molecule to act as a respiratory carrier. But, I did not have time to touch this problem immediately in the laboratory.

Until 1964, Dr. Eric Redfearn came to America to work with us. We spent inordinate time to ponder this problem: whether Q really is an electron transfer carrier on the mitochondrial respiratory chain. Eric was somewhat hesitant to make a conclusion as he was not exactly convinced Q was on the "main" chain. But, I am adamant that if Q is truly an electron carrier regardless on the side or main chain then it must link with protein. With this premise we worked very hard on this uncharted adventurous water. Not much useful result as you might have expected was gotten.

After Dr. now Professor Shigeki Takemori and I applied sequential resolution of the respiratory chain and got "pure" succinate-c reductase. The reductase can catalyze the oxidation of succinate by cytochrome c without additon of a terminal inhibitor, such as cyanide. The reductase was further resolved to SDH and the b-c_1 complex (6). Either of these components could not catalyze the oxidation of succinate by cytochrome c. Addition of Q is stimulatory. Does this contradict my thinking? Absolutely not, because all of Q and phospholipid (PL) are almost lost during the purification. Anyhow, after their combination or reconstitution the succinate-c reductase is physically reformed and the oxidative capacity is recovered (7). This b-c_1 complex is insoluble in aqueous media but only in suspension. Since 1971 Drs. C.A. Yu and L. Yu joined our laboratory, the project was gradually expanded.

Then we did many experiments as summarized in Table I (8), which is very important in our progress of the subject.

These results show there is a difference between Green's Complex III and our the b-c_1-II complex. Green's Complex III can reconstitute with his Complex II. But, Complex III is different as our b-c_1-II complex despite the fact both can catalyze QH$_2$ oxidation by c because it does not reconstitute to SDH. Green's Complex II is different from SDH although the Complex II also contains SDH plus other component (s) (presumably inpurities) and one mol cytochrome b. The second point is that after our b-c_1 complex is treated with a mercuric reagent at one-half of its total SH content or by control digestion with chymotrypsin, the products gotten are no longer active in reconstitution with SDH to form succinate-c reductase. We pondered these points repeatedly. Then immediately it occurred to me that something is destroyed by the mercuric reagent or chymotrypsin.

Table I. Comparison of the oxidation of succinate by Q and oxidation of QH_2 by c and the reconstitutive activity (with SDH) of Green's Complex II and the b-c_1-II complex before and after certrain treatments (4)

Compound	Succinate $\rightarrow$ Q	$QH_2 \rightarrow c$	Reconstitutive activity (succinate $\rightarrow$ c)
(1). SDH	-	-	-
(2). b-c_1-II complex	-	+	-
(3). (1) + (2)	+	+	+
(4). Complex II of Green	+	-	-
(5). Complex III of Green	-	+	-
(6). (4) + (5)	+	+	+
(7). (1) + (5)	+	+	-
(8). (2) treated by 1/2 P-HMB	-	+	-
(9). (1) + (8)	-	+	-
(10). (2) controlled digestion by chymotrypsin	-	+	-
(11). (1) + (10)	-	+	-

This something may not exist in Green's Complex III but may instead in his Complex II. Moreover, this something looks like most probably a protein. From that point on we started isolation work.

QP-S

The starting material by then chosen for isolation was the b-c_1-II complex. By trial and error a protein was isolated and called QP-S (9,10). Later it has been found that either the QP-S prepared by the original or latter by the "greatly improved method" (11) is a mixture containing another ubiquinone-protein known as QP-C which exists in the b-c_1 region (*vide infra*). Both QP-S and QP-C have the similar molecular weight of about 15,000. By further examining Table I, it is obvious to me that Complex II of Green or the succinate-Q reductase would be a more suitable choice for the starting material. Our succinate-Q reductase (SQR) is even better because it does not contain cytochrome. Sure enough, using SQR, the QP-S prepared differs the amino acid composition, electro phonetic pattern and is quiescent to the antibody of QP-C in contrast of the old preparation even prepared by the greatly improved method (11).

Succinate dehydrogenase cannot catalyze the oxidation of succinate by Q or by dichlorophenolindophenol (DCIP). In the presence of QP-S the reaction can proceed as (Eq.1):

(1)   succinate $\rightarrow$ SDH $\rightarrow$ QP-S $\rightarrow$ Q (or DCIP)

In this reaction no cytochrome is needed. It should be emphasized that Complex II contains about 1 mol cytochrome b_{560} per mol of SDH based on covalently bound FAD. Similar compounds which can mediate electron transfer from SDH to Q have been reported from other laboratories (12-14). But, all their starting materials contains high concentrations of cytochrome b.

Our new QP-S is prepared from succinate-Q reductase which contains very low, almost free from cytochrome in contrast to Complex II. New QP-S (15) shows two bands on SDS-PAGE column with apparent molecular weight of 15,000 and 13,500. Many methods used attempted to separate them but failed. We are inclined to believe the two bands are resulting from different number of detergent molecules bound to the protein as suggested by Professor Charles Tanford of Duke University (personal communications). Literature has such observations for many truly monodisperse proteins (16,17). Anyhow, the amino acid composition of these bands are almost the same within the experimental error by two different methods of determination: viz by cutting each band off the column with gel attached or eluates from cut-out bands so without gel. Chemical sequencing of QP-S is interrupted by its extremely hydrophobic nature. The fragments of the protein spontaneously aggregate and resists for clean purification therefore, not useful for further sequencing.

Reconstitution of QCR with QP-S and SDH gave the result 1 mol QP-S per mol of SDH based on molecular weights of 28,500 and 97,000. The gel pattern of the reconstitution is shown in Fig. 1 of ref. 18. This value is done by two methods of titration i.e. one uses QP-S and the other takes SDH as the limiting which give the same result. Results from numerical methods show that the reconstitution is physical, structural reconstitution.

Stable QP-S radical can be shown only in the presence of SDH, the flavin radical is excluded by poising the E_m of the system at +85 mv to stop the complication of from SDH. The signal shown is shown as Fig. 1 of ref 15. It has g = 2.0026 with a line width of 10-10.5 Gauss and TTFA abolished the signal. The E_m values for QP-S are determined in the presence mediators and corrected by a small contribution of flavin. Two E_m values are determined E_{m}-1 ($QH_2/QH^{\bullet}$) = about -5 mv and E_{m}-2 ($QH^{\bullet}/Q$) = about 125 mv or other properly electronated or protonated forms.

The EPR signals of cytochrome b_{560} could be detected in a very concentrated sample of our SQR, the starting material for new QP-S. It gives signals with g = 3.46 and 3.07. Even addition of 3% of ethanol denatures the cytochrome and g values shift to 2.92. However, the enzymic activity is not affected (18). Also, the cytochrome b_{560} in SQR can be denatured by 1% Zwittergent in the presence of ß-mercaptoethanol followed by dialysis. The Soret absorption band decreased greatly but the activity of QP-S is affected only very slightly. These observations strongly suggest cytochrome b_{560} has no relation whatsoever to the reconstitution of QP-S with SDH to form the succinate-Q reductase.

From Complex II Hatefi and Galante (14) have isolated cytochrome b_{560} which is claimed to be QP-S, i.e. in the presence of SDH, succinate can act as the electron donor and Q as the acceptor. Their low cytochrome content 14 nmol per mg protein has been attributed to the under estimation of the heme. Recently, Yu *et al* has isolated the so-called pure b_{560} but contains only 25 nmol b (14a) per mg protein instead of the calculated value of 35 nmol. This b_{560} claimed has been as QP-S (14a) paradoxically our QP-S (15) is nearly cytochrome free but shows the similar amino acid composition and SDS-gel pattern as those reported by b_{560} of Hatefi and Galante (14) and however completely different from those prepared by the "greatly improved" method of Yu and Yu (11).

QP-C

Dr. Shunji Nagaoka of Giftu Medical School in our lab now at the Japan Space

Center, Tokyo found a distinct EPR signal in the b-c_1-III complex using catalytic amounts of SDH and QP-S and succinate/fumarate as the electron donor.

The signal is not readily saturated even at microwave power as high as 200 mW at room temperature but at 77 K started to saturate from 300 µW. There is no resolved hyperfine structure. Cautious determinations of this QP-C give g=2.0046 ± 0.0003 with line width estimated between the derivative extrema is 8.1 ± 0.5 Gauss at room temperature and 8.4 ± 0.5 Gauss at 77 K. We have called this new QP as QP-C because it acts in the cytochrome b and c_1 region. These signals cannot be due to QP-S or SDH because only catalytic amounts are in the system. The other paramagnetic species in the b-c_1-III complex is the Rieske protein but, which cannot be shown at room temperature.

Under the same conditions, the b-c_1-complex is examined in a Varian model V 4502/4503 spectrometer at 35.0 GHz (Q band) at room temperature, O° or 77 K. A prominent anisotropic EPR spectrum is observed with g values of 2.0064, 2.0054 and 2.0051. It shows a field separation between derivative extrema of 26 ± 1 Gauss at ice (~0°) temperature. The spectrum at ~ 0° is the same as that at room temperature (~ 23°) but disappearance of the signal is slower. The EPR results are in agreement that the ubisemiquinone radical is immobilized and bound to a protein as proposed. The ubisemiquinone signal is very sensitive to antimycin and can be completely abolished by addition of this antibiotic at equimolar concentration of cytochrome c_1.

The radical concentration of ubisemiquinone depends on the ratio of fumarate to succinate as the substrate. The maximal concentration of the Q radical has been found when the ratio of fumarate to succinate is in the system at 100 at pH 8.0. The midpoint potential can be estimated to be about +60 mV by taking Em of succinate/fumarate half cell to be O mV. The redox titration in the presence of proper mediator dyes (19) shows a typical divalent redox system in which an intermediate is an EPR detectable species. The reductive and oxidative titration gives the same result. The midpoint potential of the ubisemiquinone has been shown to be about 70 mV from this figure. These data with more computer simulation by taking several sets of mid potential and fractional spin concentration has been made. Analysis of the data yields E_{m}-1 = 51 mV for ($QH_2/QH^\bullet$) and E_{m}-2 = about 83 mV for (QH°/Q) half cells (or their other electronated or protonated forms).

Next step for QP-C project is to isolate a pure preparation (20). First, we have to find a simple, convenient methods of assay. Considerable time was spent to develop it. Chymotrypsin digestion of the b-c_1-III complex produces a complex with less than a quarter of original ubiquinol-c activity staying and it suitable for the assay purpose. We also have found a method to prepare QP-C using Lubrol PX as the detergent. The preparation produced is not only just SDS-PAGE pure but truly monodisperse except the contamination of the solublizing agent, Lubrol. The last traces of Lubrol is almost impossible to be removed from QP-C preparations to the stage without decreasing the amplitude of EPR signal after reconstitution with the cytochrome b-c_1-IV complex. The concentrations needed for EPR studies are much higher than those in activity assay. So two other methods have been developed without the use of this detergent. Anyhow, all samples isolated from these methods do not contain such usual so-called prosthetic groups as heme, flavin, iron, copper, etc. Another characteristic is free from cystein or SH group and the primary structure of QP-C has been worked out (21). The only absorption peak of the isolated, pure QP-C is 278-9 nm which is decreased upon reduction. The decrease is dependent upon which method is used for preparation. Q_{10} content is different: QP-C (I), QP-C (II) and QP-C (III)[1] contains 1 mol Q_{10}, 0.4 mol and less than 0.1 mol Q_{10}

---

[1] QP-C (1), QP-C (II) and QP-C (III) are prepared from three different methods.

respectively. The phosphalipid content of QP-C (I) is about 100%, i.e. 1 mg PL per mg protein whereas as QP-C (II) and QP-C (III) contains only less than 0.4 mg.

The cytochrome b-c_1-IV complex increases the ubiquinol-c reductase activity gradually by addition of QP-C. To a certain point the activity increase is plateaued. In fact, the point is very sharp. Average of 20 to 30 batches of QP-C and the cytochrome b-c_1-IV complex have been so far tested and give a range of the recovery of 90 to 95% of the activity of the original, i.e. before the chymotrypsin digestion. The reconstitution is very sensitive to antimycin (about 1 mol antibiotic per mol of c_1) just as the activity of the reconstituted.

We also have done some immunochemical determinations for various components, I will not belabor the details of the method and only tell you that the antigen, QP-C, is in a most rigorously monodisperse form, the IgG fraction of the antibody is extensively purified, the pre-serum is always used parallel with other experimental controls. The antibody reacts with any of QP-C preparations, old QP-S (prepared by the "improved" method) (11), the b-c_1-III complex, cytochrome b-s (22) and weakly with the b-c_1-IV complex, but not cytochrome b-L, cytochrome c_1, the Rieske protein, the hinge protein, succinate dehydrogenase, or cytochrome oxidase.

Although different methods produce QP-C with different Q content but the reconstitutive activity is the same. These facts prompt us to incorporate Q to these preparations. Several methods in different media has been used. Sure enough Q is really incorporated to QP-C (II) and QP-C (III). Unexpectedly the QP-C (1) which contains 1.0 mol Q per mol protein also takes up Q. The incorporated Q cannot be separated by gel filtration chromatography but travels with the protein. Moreover, other proteins such as cytochrome c and bovin serum albumin, under these conditions Q is also incorporated. The amount of Q incorporated is dependent on the conditions. Naturally, these artifical Q proteins do not have QP activity whatsoever. To us based on the Q content does not seem to be very reliable.

Lastly, from a genetic view, QP-C sequences from some organisms are shown in Fig. 3. You may notice there is very high homology among them, from more than 45% in yeast to 85% in human tissue based on bovine heart. This fact can hardly be a biological aberration. Q may be considered as a coenzyme and the protein as apoenzyme. That consideration does not imply Q is always bound to protein as perhaps heme a in cytochrome oxidase not mentioning heme c in cytochrome c. Dissociation and association happen always. Some Q-proteins may have higher association with the protein than others. But, generally the assocation of Q proteins is low with a certain exception.

QP-N

While still working on QP-C and QP-S, we already thought about whether any QP occurs in the NADH segment as a counter part of QP-S. The comment made by Professor Keilin rang in my ear clearly. The suitable choice of material to test is complex I of Green. Dr. Hiroshi Suzuki of Nagoya University has done the beautiful work (23). We made a considerable change to prepare and prefer to call it NADH-ubiquinone reductase, or simply NADH-Q reductase. The preparation eventually, purified up to 0.9-1.2 FMN, 3.6-4.4 Q_{10}, < 0.05 total cytochrome b, < 0.01 cytochrome c_1 in the unit of nmol per mg of protein and about 20% phospholipid. But, it is completely free of FAD, succinate

```
              10        20        30        40        50
1. AGRPAVSASSRWLEGIRKWYYNAAGFNKLGLMRDD-TIHENDDVKEAIRRL
              10        20        30        40        50
2. MAGKQAVSASGKWLDGIRKWYYNAAGFNKLGLMRDD-TIYEDEDVKEAIRR
                    10        20        30        40        50
3.         RSKWLDGFRKWYYNAAGFNKLGLMRDD-TMHETEDVKEAIRRLPENL-YNDR
              10        20        30        40        50        60        70        80
4.MPQSFTSIARIGDYILKSPVLSKLCVPVANQFINLAGYKKLGLKEDDLIAEENPIMQTALRRLPEDESYA-RAYRIIRA

              60        70        80        90        100       110
1.   PENL-YDDRVFRIKRALDLSMRQQILPKEQWTKYEEDKSYLEPYLKEVIRERKEREE----WAKK
              60        70        80        90        100       110
2.   LPENL-YNDRMFRIKRALDLNLKHQILPKEQWTKYEEENFYLEPYLKEVIRERKEREE----WAKK
              60        70        80        90        100       105
3.       MFRIKRALDLSMRHQILPKDQWTKYEEDKFYLEPYLKEVIRERKEREE----WAKK
              90        100       10
4.      HOTELTHHLLPRNQWIKAQEDVPYLLPYILEAEAAAKEDELDNIEVSK
```

FIG. 3. Homology of QP-C (1. Bovin 2. Human 3. Rat 4. Yeast)

dehydrogenase and cytochrome oxidase. The activity shows 3.5-4.0 μmol NADH oxidized per min per mg protein at room temperature with Q$_2$ as an acceptor in a same assay system of Hatefi *et al*. These figures compare favorably or higher than their preparations after temperature correction (24).

In the first trial use NADH as the substrate, a distinct organic ubisemiquinone was observed. Repeated experiments show the same EPR behavior. The radical shows g=2.0042 at 9.49 GHz without resolved hyperfine structure. The line width of the signal is 6.8 ± 1 Gauss at room temperature. The signal amplitude reached maximum after additon of NADH and was constant about 5-10 min before slow decay. The ubisemiquinone radical is not detected by adding a mixture in various ratios of fumarate and succinate or even 500 μM of reduced Q$_2$ as these observations conclusively show the ubisemiquinone radical formed by NADH is resulting from another species of QP which we call QP-N. It is completely different from QP-S or QP-C.

The Q band EPR spectra of the radical give *g*-anisotropy at 232 K. Its spectrum also shows a field separation between derivative extrema of 24 Gauss with g=2.0060, 2.0051, and 2.0022. These g values are very close to those of QP-C . This similarity may show the inherent nature of ubiquinone radical. The temperature is chosen because the radical is not as stable as that of QP-C while adjustment of equipment takes sometimes more than 10 minutes. A summary of EPR characteristic is shown in Table II.

Table II. Comparison of the EPR characteristic of mitochondrial
Q-proteins and sensitivity to respiratory inhibitors of
the radicals (14)

Radical	Microwave frequency GHz	g=value	Line width derivative extrema Gauss	Sensitivity to		
				Rote-none	Anti-mycin A	TTFA
QP-N	9.49	2.0042	6.8± 0.1	+, -	-	-
	34.2	2.0060 2.0051 2.0022	24			
QP-C	9.49	2.0046 2.0064	8.1 ± 0.1	-	+	-
	35.0	2.0054 2.0051	26.1			
QP-S	9.49	2.0026	10.2±0.2	-	-	+

The dependence of the EPR signal amplitude on the microwave power was studied at room temperature. The power saturation curve is biphasic. One of the EPR signals starts to show power saturation at a microwave power of about 5 milliwatts and the other start at 20 milliwatts. The biphasic behavior is not affected by protein concentration. Giving to a model system of power saturation the biphasic power saturation curve show that two kinds of populations of the ubisemiquinone radicals are present (25) . Responses of the ubisemiquinone radicals to rotenone, TTFA, antimycin A, and pH were studied at two different levels of microwave powers, 5 and 20 milliwatts, at room temperature. The results are the same.

The amplitude of the signal is varied with the concentration of NADH. The first maximal amplitude is at 40% total Q concentration. Further increase does not change the amplitude until the NADH reaches the total redox equivalent of the reductase. The Hill coefficients of 0.9 and 2.3 . So rotenone biphasically abolishes the EPR. The Hill coefficient is 1.1 showing the lack of cooperativity between these two sites. (see original figures of ref 23)

The signal is affected by pCMS concentration. Increase of pCMS up to 500 uM decreases the signal about 43%. Essentially half of the ubisemiquinone radical still remains even when about 0.9 mM pCMS added. In other words, half of the ubisemiquinone radical is pCMS-insensitive. The inhibition by pCMS does not increase with preincubation of the enzyme with the reagent. It has been claimed that the quinone is reduced by two nonidentical sites in the reductase and that rotenone binds to two different sites in the enzyme and inhibits the reduction. These rotenone-binding sites differ in the chemical composition. One of them contains an -SH group.

The fact that it can be concluded that in NADH reductase, ubisemiquinone is involved in electron transfer to form a stable radical. The observations of rotenone sensitivity, the behavior of microwave power saturation and pCMB inhibition actions of the radical suggest there are probably two populations of radicals. This hypothesis explains or agrees with the claims proposed by other laboratories.

From all the experimental results we suggested that a Q cycle may exist in the NADH-Q reductase but is only in pair in *modus operandi* similar in principle to the protonmotive force Q cycle originally proposed by Mitchell in the b-c_1 region (26). Although it is not reliable to use Q content as the sole criterion as a genuine QP. The ratio of FMN to Q in reductase about 4 seems agreeing with our hypothesis.

Unfortunately, the preparation of NADH dehydrogenase is more involved than that of succinate dehydrogenase. Because of this fact and the more complicate for NADH-Q reductase than the counter part of succinate-Q reductase, we have not developed a simple assay method, functional or structural, for QP-S useful for the systematic isolation.

Q IN MITOCHONDRIAL ELECTRON TRANSFER

Since the function of Q was proposed almost simultaneously with its discovery. The mechanisms or even a reasonable hypothesis was not proposed until many years later. Its link with protein, to my knowledge, was only mentioned by Green in 1959 (27). According to him by then a "Q-lipoprotein" exists with 90% PL, 8% neutral lipid, 0.42% Q_{10} also of c_1, and flavoenzymes therefore, less than 0.5% available protein for binding with other components such as Q in his complex . Maybe resulting from its lipophillic property and small molecular size, Green and co-workers proposed Q and cytochrome c as the mobil components to transfer electron by Q from his Complex I or II to Complex III then by c to Complex IV. This mobile theory seems very plausible. If a small molecule like Q can really swim freely in the lipid millieu then the bilayer model was not known yet. Cytochrome c is soluble in lipid. However, whether those 4 Complexes really exist or only an operational definiton for convenience in conversation is to be rigorously substantiated.

Later, Dr., now Professor Achim Kroger and Professor Martin Klingenberg of Munchen have done extensive work on specially designed instruments, and studied the oxidation and reduction of Q including Q analogues in almost all conditions. They found scores of rate equations for the oxidation-reduction of free Q in the submitochondrial particles. The Q content is usually determined by methanol-petroleum ether extraction. In these extracts, 10 to 20% is inactive form or as they have expressed in equation 2 and 3:

$$(2) \qquad Q_t = Q_a + Q_i$$

$$(3) \qquad Q_a = Q_{ox} + Q_{red}$$

Q_t is the total; Q_i, inactive; Q_a active; Q_{ox}, oxidized; and Q_{red}, reduced forms of ubiquinone. Succinate and NADH are used as substrates in the presence and absence of inhibitors. They have spent many years on the study published half a dozen papers and eventually, crystallized into the two papers in 1973 (32,33) and proposed the pool theory, of which QH_2 is formed from various dehydrogenases and the QH_2 oxidase system. Generally, this pool theory may not be contradictory to our Q-proteins results. Although the proposal of involvement of Q radical was made by others such as Wikstrom and Berden in 1972 (34) they were not aware of dismutation of free ubisemiquinone with a stability constant much in favor of non-radical form with the constant of 10^{10} nor mentioned the importance of protein bound to Q. Dr. Peter Rich (25) has ably summarized the Q-pool theory as:

> "This concept adequately explains the first order behavior of quinone-pool oxidations, sigmoidal inhibition with tight bound inhibitors; inhibition profiles in double inhibitor experiments; extraction-reconstitution [Q is extracted by organic solvents and Q in the free form is added back] studies, and provides a mobile H-atom carriers as needed for a protomotive loop. In this view, the quinone species are not protein bound and essentially freely diffusible in the lipid phase."

Two years later Peter Mitchell went back after deep meditation announced its Q cycle (26)which becomes immediately an intellectual breakthrough if not polemics. The essence of the cycle is through one electron transfer to form Q radical and the system has two centers i.e., Q_o and Q_i. Mainly using inhibitors for the kinetic studies, and thus changed Q cycle have appeared such as by Trumpower (28), by Slater (29) and by Von Jagow (30) and their coworkers. Recently, Mitchell has critically scrutinzed the various possibilities of protonmotive Q cycle and compared the *b* cycle (31). Unfortunately space and time limitation prevents me to discuss more.

It should be emphasized that although in some of these published schemes (cycles) Q or Q radicals is presented, all reactions with Q *can occur only* with protein linked. Trumpower emphasized "no reaction in the Q cycle involve Q without protein attached" (personal communications, August 3, 1989).

CONCLUSION -- UBIQUINONE PROTEINS

It should be emphasized that, in our concept a respiratory Q protein must fulfill the function as a true respiratory carrier and the protein moiety will stabilize the Q radicals and the protein also dictates the specificity of the site or sites of where QP acts on the respiratory chain. This is in agreement of the Q cycle or even straight chain formulation. Is there another QP-C in the Q cycle or is the same with QP-C reported after isolation. I just cannot answer now but it is not QP-S. Anyhow the report of the Q cycle greatly strengthens the Q proteins concept including the stabilization of Q radicals by protein and *vice versa.* The first reluctant acceptance of chemiosmotic theory with electron transport in the loop was that of the placement of Q after cytochrome *b* in the respiratory chain (cf. an excellent critical review by the late Grenville ref 36). The multiple sites of Q removes this obstacle and support the concept of the Q cycle. It is a well-known fact that free ubisemquinone can exist only in a very small amount in mitochondria. Without the formation of Q radical, the promotive force Q cycle can hardly exist, and the original chemiosmotic theory has probably been objected.

Abbreviations used -- *b-c*$_1$ complex or *b-c*$_1$-I complex, the complex catalyzes the oxidation of succinate through the respiratory chain by cytochrome *c*; *b-c*$_1$-II complex is in "soluble" form; *b-c*$_1$-III complex is the *b-c*$_1$-II complex free of QP-S, *b-c*$_1$-IV complex is deficient of QP-C about 20-25% remaining PL, phospholipids. Q$_n$, ubiquinone, the subscript *n* denotes the number of isoprenoid units; Q, ubiquinone, fully oxidized form; QH$_2$ ubiqunione, fully reduced; QP ubiquinone protein; QP-C, QP acts in the *b-c*$_1$ region; QP-S, direct electron-hydrogen acceptor of SDH; QCR ubiquinone-*c* reductase (*b-c*$_1$ complex); pCMS p-chloromercuric sulfonate; SDH succinate dehydrogenase; SDS-PAGE polyarylmide gel electrophoresis in sodium dodecyl sulfate solution ; SQR succinate-ubiquinone reductase; TTFA theonyltrifluoroacetone.

ACKNOWLEDGEMENT

First, I am most grateful to Drs. Kim and Ozawa to award me the privilege of participating in this elegantly organized symposium and to have my first visit of the wonderfully rapidly progressing of the prosperous metropolis. Especially Dr. Kim's unusual hospitality I acknowledge greatly. Usually, I think in this type of meeting the time limit of reports is more strict for earlier papers. If I will be overtime, please forgive me for my good intention.

I want gratefully to acknowledge my present and past collaborators: C. H. Kim, H. Matsubard, S. Kuramitsu, S. Nagaoka, N. Reimer, J. C. Salerno, C. P. Scholes, Y. Shimonishi, H. Suzuki, T. Takao, J. Q. Tai, S. Takemori, A. D. Vinogradov, S. Wakabayashi, T. Y. Wang, Y. H. Wei, W. R. Widser, Y. Xu, K. T. Yasunobu, S. Yoshida, C. A. Yu, L. Yu and Z. P. Zhang. The work is supported by NIH-GM 16767 and NIH-HBL-B37-12576.

REFERENCES

1.   Cain, J. C., and Morton, R.A. (1955) *Biochem. J.*, **60**, 274-283
1a.   Morton, R.A., Wilson, G.M., Lowe, J. S., and Leat, W.M.F. (1958) *Biochem, J.* **68**, 16p
2.   Morton, R.A., Wilson, G.M. Lowe, J.S., and Leat, W.M.F. (1957) *Chem. and Ind.*, 1649
3.   Pumphrey, A.M., Redfearn, E.R. and Morton, R.A. (1958) *J. Biochem.* 70, 1P.
4.   King, T.E. (1980) in Ozawa *et al* .*Eds New Frontier of Biochemistry* p.121-134.
5.   Hatefi, Y., Lester, R. and Ramosarmae, T. (1958) **17**, 607 (Abstract no. 607)
6.   Takemori, S and King, T.E. (1964). *J Biol. Chem*, **239**, 3546-3558
7   King, T.E., and Takemori, S. (1964) *J. Biol. Chem.* **239**, 3559-3569.
8.   King, T.E. (1982) *In Function of Quinones in Energy Conserving Systems* (B.L. Trumpower, ed.) Academic Press, New York, pp. 3-25.
9.   Yu, C.A. Yu, L., and King, T.E. (1977) **78**, 259-265; Yu, C. A., Yu, L., and King, T.E. (1979) *Fed. Proc.* **38**, 638.
10.   Nagaoka, S., Yu, L., and King, T.E. (1981) *Arch. Biochem. Biophys.* 208, 334-343
11.   Yu, C.A. and Yu, L. (1980) *Biochemistry* 19, 3579-3585
12.   Ackrell, B. A. C., Ball, M. B., and Kearney, E. B. (1980) *J. Biol. Chem.* 255, 2761-2769

13. Vingradov, A.D., Gavrikov, V.G. and Gavrikova, E.O. (1980) *Biochim. Biophys. Acta,* **592**, 13-27
14. Galante, Y.M. and Hatefi, Y. (1980) *J.Biol. Chem.,* **255**, 5530-5537
14a Yu, L., Xu, J. X., Haley, P., and Yu, C. A. (1987) *J. Biol. Chem.* **262**, 1137 1143
15. Xu, Y., Salerno, J. C., Wei, Y. H., and King, T. E. (1987) *Biochem. Biophys. Res. Commun.* **144**, 315-322
16. Tanford, C. (1980). *The Hydrophobic Effect 2nd Ed.*, John Wiley, New York, NY
17. Steinhart, J., and Reynolds, J. A. (1960). *Multiple Equilibria in Protein,* Academic Press, New York
18. King, T.E. and Xu, Y. (1988) in *Cytochrome Systems* (S. Papa, *et al* . eds) pp. 503-508, Plenum Press, New York
19. Yu, C. A., Nagaoka, S., Yu, L., and King, T.E. (1980) *Arch. Biochem.* **204**, 59-70
20. Wang, T. Y., and King, T. E. (1982) *Biochem. Biophys. Res. Commun.* **104**, 591-596
21. Wakabayashi, S., Takao, T., Shimonishi, Y., Kuramitsu, S., Matsubara, H., Wang, T. Y., Zhang, Z. P., and King, T. E. (1985) *J. Biol. Chem.* **260**, 337-343
22. Yoshida, S., Zhang, Z. P., and King, T. E. (1982) *Biochem. Intl.* **4**, 1-8
23. Suzuki, H., and King, T. E.(1983). *J. Biol. Chem.* **258**, 352-358
24. Hatefi, Y., and Haavik, A.G. and Griffiths, D.E. (1962) *J. Biol. Chem.* **237**, 1676-1680
25. Salerno, J. C., Lim, J., King, T.E. Blum, H., and Ohnishi, T. (1979) *J. Biol. Chem.* 4828-4835
26. Mitchell, P. (1975) *FEBS Lett.* **56**, 1-6
27. Green, D.E. (1959), *Discussion of Faraday Soc.* **27** 206-216
28. Bowyer, J. R. and Trumpower, B. L. (1981) in *Chemosmotic Proton Circuits in Biological Membranes* (V.P. Shubacher and P.C. Hickel eds) pp. 105-122, Addison-Wesley, Reading, Mass.
29. Zhn, Q. S., Berden, J. A., DeVries, S., and Slater, E. C. (1982) *Biochem. Biophys. Acta* **680**, 67-79
30. von Jagow, G., and Link, T. A. (1986) *Methods Enzymol,* **126**, 253-271
31. Mitchell, P. (1987) in *Advances in Membrane Biochemistry and Bioenergetics,* C.H.Kim *et al* . eds) pp 13-52, Plenum Press, New York, NY
32. Kroger, A., and Klingenberg, M. (1973) *Eur. J. Biochem.* **34**, 358-368
33. Kroger, A., and Klingenberg, M. (1973) *Eur. J. Biochem.* **39**, 313-323
34. Wikstrom, M. and Berden, J. (1972), *Biochim. Biophys. Acta* **283**, 403-420
35. Rich, P. R. (1982) in *Functions of Quinone in Energy Conversion System* (B. Trumpower ed) pp. 73-83, Academic Press, New York
36. Greville, G. D. (1969) in *Current Topics in Bioenergetics* (Sanadi, D.R. ed) pp. 1-156, Academic Press, New York

STUDIES ON THE MECHANISM OF ACTION OF THE MITOCHONDRIAL
ENERGY-LINKED NICOTINAMIDE NUCLEOTIDE TRANSHYDROGENASE

Mutsuo Yamaguchi and Youssef Hatefi

Division of Biochemistry
Department of Molecular and Experimental Medicine
Research Institute of Scripps Clinic
La Jolla, California 92037

INTRODUCTION

This is a brief summary of the recent work of our laboratory on the structure and function of a unique energy-transducing enzyme of mammalian mitochondria, the nicotinamide nucleotide transhydrogenase (TH) (see 1,2 for reviews).

1. GENERAL FEATURES OF TH

The mitochondrial nicotinamide nucleotide transhydrogenase is located in the inner mitochondrial membrane, and catalyzes the transfer of a hydride ion between NAD(H) and NADP(H) in a reaction that is stereospecific for the 4A hydrogen of NADH and the 4B hydrogen of NADPH, and is coupled to proton translocation across the inner membrane with a H^+/H^- stoichiometry (n) close to unity (eq. 1; H_c^+ and H_m^+ protons on the cytosolic and the matrix sides of the inner membrane, respectively) (3,4).

$$NADH + NADP + nH_c^+ \rightleftharpoons NAD + NADPH + nH_m^+ \qquad (1)$$

The amino acid sequences of the bovine mitochondrial (5) and the *Escherichia coli* (6) transhydrogenases and the amino acid sequence of the signal peptide of the bovine enzyme (7) have been determined. The bovine transhydrogenase appears to be a homodimer of monomer $M_r = 110,000$. The *E. coli* enzyme is composed of two subunits, α with $M_r = 54,000$, and β with $M_r = 48,700$ (6). The NAD and the NADP binding domains of the bovine transhydrogenase, as determined by affinity labeling with *p*-fluorosulfonylbenzoyl-5'-adenosine (FSBA), are located in hydrophilic regions of the molecule near the N- and the C-termini, respectively (8). The central one-third of the molecule is hydrophobic, and appears to be composed of about 14 membrane-spanning clusters of about 20 amino acid residues each (5). Inhibitory covalent modification of the dimeric bovine enzyme with [^{3}H]FSBA and [^{14}C]*N,N'*-dicyclohexylcarbodiimide (DCCD) indicated that complete activity inhibition occurred upon incorporation of 1 mol of inhibitor per dimer, thus suggesting half-of-the-sites reactivity (9-11). Furthermore, unlike most energy-transducing enzymes in which DCCD binds to a glutamic or an aspartic acid residue in a hydrophobic, membrane-spanning segment (12), in the bovine transhydrogenase the DCCD-reactive glutamic acid residue is located in a hydrophilic stretch in the NAD-binding domain, 12 residues downstream from the FSBA binding site (11). Shown in Figure 1 are the amino acid sequences of the bovine mitochondrial

```
      1                                                                                                    98
Bovis   CSAPVKPGIPYKQLTVGVPKEIFQNEKRVALSPAGVQALVKQGFNVVVESGAGEASKFSDDHYR-AAGAQI-QGAKEVLASDLVVKVRAPMLNPTLGVHE
              * * *  *** ** ***   **  *    *                  *        *   *        **  *    ** ** *       *
E. coli        MRIGIPRERLTNETRVAATPKTVEQLLKLGFTVALESGAVNWQVLTI--KRLCSGREIVEG-NSVWQSEIILKVNAP-LD-----DE
               α

                                                                                                         197
Bovis   ADLLKTSGTLISFIYPAQNPDLLNKLSKRKTTVLAMDQVPRVTIAQGYDALSSMANIAGYKAVVLAANH-FGRFFTGQITAAGKVPPAKILIVGGGVAGL
         **  ** **** ******  **  *    ** *    **  *    *********** ***  *  * *  ********** ********  *  * *****
E. coli IALLNPGTTLVSFIWPAQNPELMQKLAERNVTVMAMDSVPRISRAQSLDALSSMANIAGYRAIVEAA-HEFGRFFTGQITAAGKVPPAKVMVIGAGVAGL

                              #3         #4                                                                296
Bovis   ASAGAAKSMGAIVRGFDTRAAALEQFKSLGAEPLEVDLK-ESGEGQGGYAKEMSKEFIEAEMKLFALQCKEVDILISTALIPGKKAPILFNKEMIESMKE
         * ***  ****** **** **  **   **  **  ** **  ** **  ** **  *** *** *  ***** ******* ** *  ** *****
E. coli AAIGAANSLGAIVRAFDTRPEVKEQVQSMGAEFLELDFKEEAGSGDG-YAKVMSDAFIKAEMELFAAQAKEVDIIVTTALIPGKPAPKLITREMVDSMKA

                                                             #2                                            392
Bovis   GSVVVDLAAEAGGNFETTKPGELYV-HKGITHIGYTDLPSRMATQASTLYSNNITKLLKAISPDKD-NFYFEVKDDFDFGTM-GH-VIRGTVVMKDGQVI
         *** *****  ***  *  *       * ***  ******   *  ** * **   ***  *** **   ** *    ** *   ***   * *
E. coli GSVIVDLAAQNGGNCEYTVPGEIFTTENGVKVIGYTDLPGRLPTQSSQLYGRNLVNLLKLLCKEKDGNIT--V--DFDDVVIRGVTVIRA------GEIT

                                                                                                         487
Bovis   FPAPTPKNIP-QGAP-VKQKTVA-ELEA-EKAATITPFRKTMTSASVYTA-GLTGILGLGIAAPNLAFSQMVTTFGLAGIVGYHTVWGVTPALHSPLMSV
         ***  *      *     *  *  * *     *  *  *   *       *    * *  **     * *  *  *  *** ** *  *** ****  *
E. coli WPAP-PIQVSAQ--PQAAQK-AAPEVKTEEK-CTCSPWRKYALMA-L--AIILFGWMA-SVA-P-KEFLGHFTVFALACVVGYVVWNVSHALHTPLMSV

                                                                                                         585
Bovis   TNAISGLTAVGGLV-LM-GGHLYPSTTSQGLAALATFISSVNIAGGFLVTQRMLDMFKRPTDPPEYNYLYLLPAGTFVGGYLASLYSGYNIEQIMYLGSG
         ******  **  *          *          *    *   *                                                 **       *
E. coli TNAISGIIVVGALLQIGQGGWVV----S-FLSFIAVLIASINIFCGFTVTQRM                        MSGG-LVTAAYIVAA
                                                                                     β

                                                                                                         684
Bovis   LCCVGF-LAGLSTQGTARLGNALGMIGVAGGLAATLGGLKPCPELLAQMSGAMALGGTIGLTIAKRIQISDLPQLVAAFHSLVGLAAVLTCIAEYIIEYP
         *  ***** ** ***  *          **  ** *  **  **  ****   *    **   ***  *
E. coli ILFI-FSLAGLSKHETSRQGNNFGIAGMAIALIATIFGPDTG-NV-GWILLAMVIGGAIGI-LAKKVEMTEMPELVAILHSFVGLAAVLCGFNSYLHHDA

                                                                                                         784
Bovis   HFATDAAANLTKIVAYLGTYIGGVTFSGSLVAYGKLQGILKSAPLLLPGRHLLNAGLLAGSVGGIIPFMMDPSFTTGITCLGSVSALSAVMGETLTARIG
         *          ** ***** *  *  * *** *  *  **  *  * *      *       *     **                *     *  * *
E. coli GMAP-IIVNIHLTEVFLGIFIGAVTFTGSVVAFGKLCGKISSKPLMLPNRHKMNLAALVVSFLLLIVFVRTDSVGLQVLALLIMTAIALVFGWHLVASIG

                                                                                                         884
Bovis   GADMPVVITVLNSYGSWALCAEGFLLNNNLLTIVGALIGSSGAILSYIMCVAMNRSLANVILGGYGTTSTAGGKPMEISGTHTEINLDNAIDMIREANSI
         *******       *  * *       ***  ***** * *****  ***** ** ** ***  *   *   *
E. coli GADMPVVVSMAELVLRLGGCGCGLYAQQRPVIVTGALVGSSGAILSYIMCKAMNRSFISVIAGGFGTDGSSTGDDQEV-GEHREITAEETAELLKNSHSV

                                                                                                         984
Bovis   IITPGYGLCAAKAQYPIADLVKMLSEQGKKVRFGIHPVAGRMPGQLNVLLAEAGVPYDIVLEMDEINHDFPDTDLVLVIGANDTVNSAAQEDPNSIIAGM
         *******  *  * *  *    *  ***********   ***  ** ** **********   *   *** ********** ** ** *** *
E. coli IITPGYGMAVAQAQYPVAEITEKLRARGINVRFGIHPVAGRLPGHMNVLLAEAKVPYDIVLEMDEINDDFADTDTVLVIGANDTVNPAAQDDPKSPIAGM

        #1                               1043
Bovis   PVLEVWKSKQVIVMKRSLGVGYAAVDNPIFYKPNTAMLLGDAKKTCDALQAKVRESYQK
         *******  *** **   **   *** ** * * *  ** **
E. coli PVLEVWKAQNVIVFKRSMNTGYAGVQNPLFFKENTHMLFGDAKASVDAILKAL
```

Figure 1. Comparison between the amino acid sequences of the bovine and E. coli transhydrogenases. Asterisks indicate sequence identity; dashes are spacers introduced for better alignment. The underlined segments 1,2,3 and 4 show the tryptic peptides that were isolated from the bovine enzyme and used for construction of oligonucleotide probes for isolation of cDNA clones. For other details, see (5).

M A N L L K T V V T G C S C P F L S N L G S C K V   25

↓

L P G K K N F L R T F H T H R I L W C S A P V K P   50

G I P Y K Q L T V G V P K E I F Q N E K R V A L S   75

Figure 2. Amino acid sequence of the signal peptide (first 43 residues) and the N-terminal region of TH. The vertical arrow shows where the signal peptide ends and the mature TH begins. For other details, see (7).

and the *E. coli* transhydrogenases. Appropriately labeled vertical arrows in Figure 6 below mark the modification sites on the bovine enzyme of FSBA and DCCD. Figure 2 shows the sequence of the signal peptide of the bovine transhydrogenase, which is composed of 43 amino acids and contains 6 basic (4 Lys, 2 Arg), 7 hydroxylated (4 Thr, 3 Ser) and no acidic residues.

2. UNIQUE FEATURES OF TH

Several features distinguish the transhydrogenase as a unique energy-transducing enzyme, and make it a valuable system for study of the mechanism of proton translocation.

(i) - Unlike other energy-linked reactions which do not proceed to any measurable extent in the absence of energy, the forward energy-promoted transhydrogenation reaction depicted in eq. (1) does proceed in the absence of energy, albeit slowly, to the equilibrium point (K = 0.79) dictated by the reduction potentials of NADH/NAD and NADPH/NADP ($\Delta E_o'$ = 5 mV) (13,14). This unique feature has allowed accurate measurement of the effect of energy on the kinetic and the thermodynamic parameters of the scalar reaction. Thus, it has been shown that energy increases the rate of the forward reaction (see eq. 1) by about 12-fold and displaces its equilibrium toward product formation, with K' = 500 (2). The increase in V_{max} is accompanied by decreases in apparent K_m for both substrates, resulting in a large V_{max}/K_m increase with an optimum at pH = 7.5 (15).

(ii) - Unlike the scalar reactions catalyzed by the respiratory chain and the ATP synthase complexes, the scalar reaction catalyzed by the transhydrogenase involves no release or uptake of protons. The enzyme has no prosthetic groups, and hydride ion transfer between NAD(H) and NADP(H) is direct. Therefore, it appears that in the reverse reaction, where H$^-$ transfer from NADPH to NAD is coupled to $m \rightarrow c$ (matrix to cytosolic) proton translocation, the protein itself takes up a proton on its matrix side and releases a proton on its cytosolic side.

(iii) - As stated above, ΔE_m between NADPH/NADP and NADH/NAD is only 5 mV. Yet, the reverse reaction shown in eq. (1) results in $m \rightarrow c$ proton translocation, and this conserved energy can be utilized, among other things, to synthesize ATP (16). It is thus clear that the only available energy source to drive the uphill translocation of protons in the reverse transhydrogenation reaction is the difference in the concentrations of substrates (NADPH and NAD) and products (NADP and NADH). The question is how this energy is transduced by TH into a protonmotive force. Described below are the results of experiments addressed to this question.

3. SUBSTRATE/PRODUCT-INDUCED TH CONFORMATION CHANGE

Clues regarding mechanistically relevant enzyme conformation changes were provided by two sets of results. The first followed the discovery in our laboratory that, although TH does not catalyze transhydrogenation from NADH to a reducible NAD analog, it does catalyze an energy-promoted transhydrogenation from NADPH to NADP analogs, such as 3-acetylpyridine adenine dinucleotide phosphate (AcPyADP) and thionicotinamide adenine dinucleotide phosphate (thioNADP) (17). We then asked whether the transhydrogenation reaction shown in eq. (2), which we expected to be catalyzed by TH, was also energy-promoted.

$$NADPH + [^{14}C]NADP \rightleftharpoons NADP + [^{14}C]NADPH \qquad (2)$$

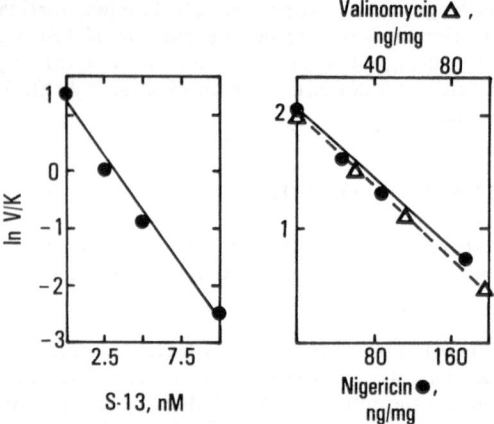

Figure 3. Effect of partial uncoupling on $\ln(V_{max}/K_m)$ for NADH to AcPyADP transhydrogenation. Where indicated, 0.5µg nigericin/mg SMP protein plus variable amounts of valinomycin (top abscissa, △) or 0.5µg valinomycin plus variable amounts of nigericin (bottom abscissa, ●) were added. The variable substrate was AcPyADP. S-13, 5-chloro-3-*t*-butyl-2'-chloro-4'-nitrosalicylanilide. For other details, see (19).

The significance of this question rests on the fact that the reaction shown in eq. (2) is at all times at thermodynamic equilibrium, because there is essentially no difference in the nature and concentration of substrates and products. Results showed that, indeed, the rate of reaction (2) as catalyzed by SMP-bound TH was increased severalfold upon membrane energization (18). Since no energy could be conserved in the products for the reasons just stated, energy utilization by this reaction had necessarily to result in considerable entropy increase, which we interpreted to mean that during catalysis the enzyme undergoes large energy-induced conformation changes (18). This interpretation agreed with other studies from our laboratory which showed that in the energy-promoted transhydrogenation reaction NADH → AcPyADP, there was a linear relationship between $\ln(V_{max}/K_m^{AcPyADP})$ and the degree of membrane energization (Figure 3), indicating that energization increased enzyme affinity for AcPyADP (19).

The second clue was provided by the results of others (20-22) and ourselves (10,23) showing that NADPH greatly increases the sensitivity of the bovine TH to several inhibitory protein modifiers, including trypsin. These data suggested that NADPH binding causes substantial changes in TH conformation. We have investigated the effect of substrates on cleavage of the enzyme by trypsin and its modification by *N*-ethylmaleimide, and the results are summarized below.

Effect of substrates on cleavage of TH by trypsin - As originally demonstrated by Ernster and coworkers (24), TH is highly sensitive to inhibition by trypsin. This sensitivity, as shown in Figure 4, is greatly increased in the presence of NADPH. NADP was also somewhat stimulatory, while NAD had no effect and NADH offered a slight protection. Study of NADPH concentration indicated a half-maximal effect at 34 μM NADPH, which is close to the apparent K_m of 20 μM for NADPH at pH = 7.5 (25). Figure 5 depicts on SDS-polyacrylamide gel the effects of the above substrates on fragmentation of TH by trypsin, and Figure 6 shows the cleavage points that resulted in the formation of large and analyzable fragments. The most sensitive bond was the K_{410}-T_{411} bond. Also, the presence of NADP(H) resulted in the formation of a 48 kDa fragment by cleavage of the R_{602}-L_{603} bond, which did not happen in the absence of NADP(H). These results, especially the distances of these locations from the NADP(H) binding domain near the C-terminus, suggested that NADPH >> NADP binding causes global conformation changes in the TH molecule.

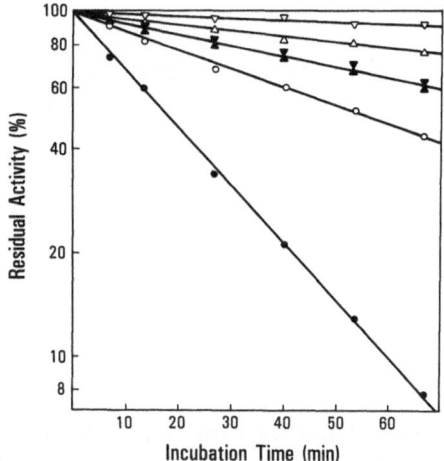

Figure 4. Effect of substrates on the inactivation of purified TH by trypsin. TH (2.0 mg/ml) was suspended in 50 mM Tris-acetate, pH 7.5, containing 0.001% potassium cholate, and treated at 23°C with trypsin (trypsin: TH = 1:400). Where indicated the incubation mixture contained 0.4 mM each of NADPH (●), NADP (○), NADH (△), NAD (▲), no substrate (▼), no substrate or tyrpsin (▽). Aliquots of the mixtures were removed at the intervals shown and assayed for NADPH to AcPyAD transhydrogenase activity. 100% activity was 35-40 μmol AcPyAD reduced per min per mg of protein at 37°C.

Effect of substrates on the inhibitory modification of TH by N-ethylmaleimide - As seen in Table I, the inhibition rate of TH by NEM was also greatly accelerated in the presence of NADPH. NAD and NADH had no effect, while NADP offered protection. The effects of NADPH and NADP were pH-dependent, as will be

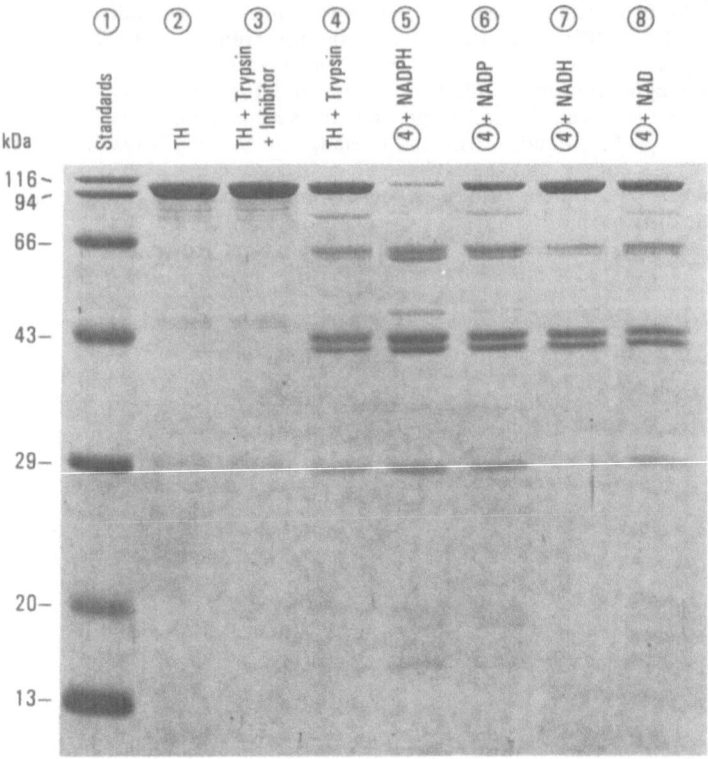

Figure 5. Sodium dodecyl sulfate-polyacrylamide gel electrophoresis of TH treated with trypsin in the absence and presence of substrates. Conditions were the same as in Figure 4. After 67 minutes of incubation, soybean trypsin inhibitor was added to the reaction mixtures, then aliquots were removed, denatured, and subjected to slab gel electrophoresis by the method of Laemmli (27). The faint bands seen in lanes 2 and 3 resulted from autolysis of TH while being concentrated after column purification.

seen below. The half-maximal effect of NADPH occurred at 13.4 μM NADPH at pH 7.5, which again was very close to the apparent K_m^{NADPH} at this pH (25), and suggested that the NADPH effect results from its binding at the catalytic site. Comparison of the extent of enzyme modification by [^{3}H]NEM in the absence and presence of NADPH, followed by cyanogen bromide fragmentation of the protein and examination of the fragments, indicated that Cys-893 was the residue whose sensitivity to NEM was increased in the presence of NADPH (26). As seen in Figure 6, Cys-893 is located 113 residues upstream from the tyrosyl residue modified by FSBA in the NADPH binding domain of TH. A second [^{3}H]NEM-modified cysteine residue was also identified at position 626 (see Figure 6), but NADPH had no effect on the extent of radioactivity incorporated in this position.

TABLE I

Effects of Substrates on the Inactivation Rate of Purified Nicotinamide
Nucleotide Transhydrogenase by NEM

Additions	$k-k_0(min)^{-1}$
None	0.253
NAD	0.253
NADH	0.200
NADP	0.115
NADPH	1.800
5'AMP	0.258
2'AMP	0.150
NMN	0.230
NMNH	0.253
2'AMP + NMN	0.152
2'AMP + NMNH	0.168

TH (10 μg of protein in 100μl of 50 mM Tris-acetate, pH7.5, containing 0.001%
potassium cholate) was incubated at 23°C with 0.4 mM *N*-ethylmaleimide in the
absence or the presence of 0.2 mM concentration of each nucleotide, then
sampled at intervals and assayed for NADPH to AcPyAD transhydrogenase
activity. The rate constants in the presence (k) and absence (k_0) of the inhibitor
were calculated from the slopes of semilogarithmic inhibition time course plots.
For other details, see (26).

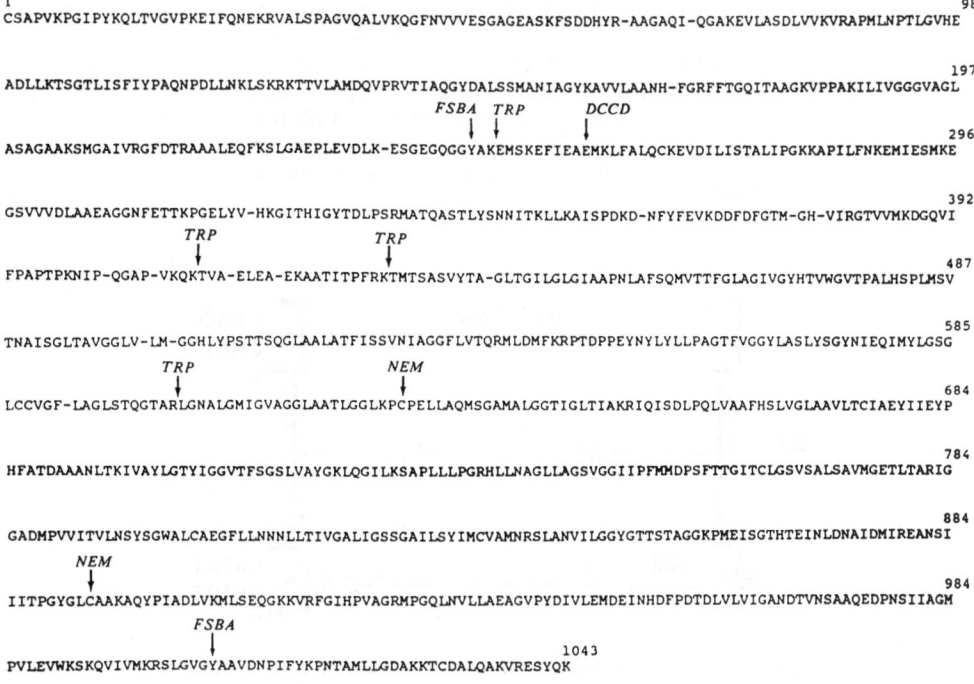

Figure 6. **Amino acid sequence of bovine TH showing the sites of modification by *p*-
fluorosulfonylbenzoyl-5'-adenosine (FSBA), *N,N*'-dicyclohexylcarbodiimide (DCCD) and
N-ethylmaleimide (NEM), and cleavage by trypsin (TRP).** The dashes (spacers) are
those used in Figure 1.

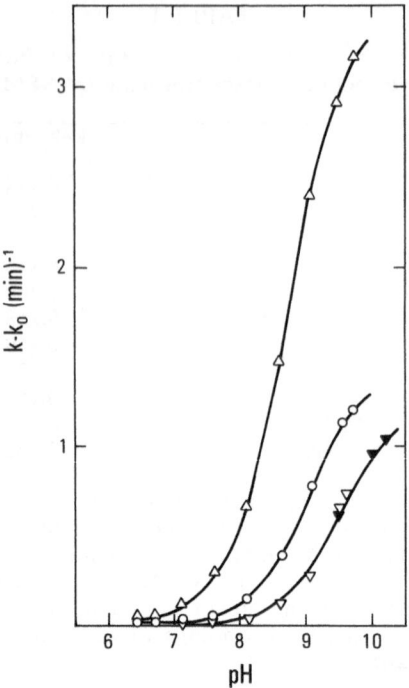

Figure 7. pH dependence of the inactivation rate constant (k) of TH by *N*-ethylmaleimide in the absence (○) and presence of 0.2 mM NADPH (△) or NADP (▽,▼). k and k_o, pseudofirst order rate constants, respectively, in the presence and absence of the inhibitor, were determined from semilogarithmic plots of the inactivation time course. For other details, see (26).

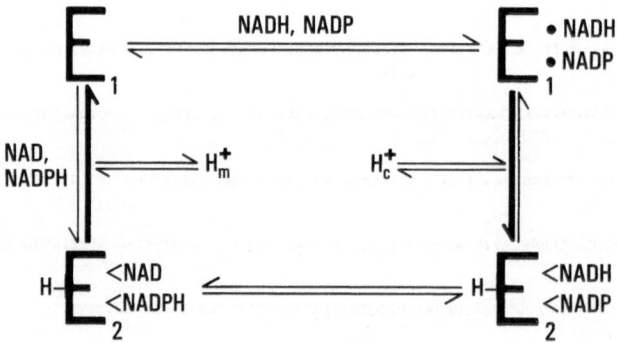

Figure 8. Hypothetical transhydrogenase reaction scheme showing proton uptake and release coupled to TH conformation change (E_1, E_2) and substrate affinity change (low affinity, ●; high affinity, <).

Particularly interesting was the pH dependency of Cys-893 modification by NEM. As seen in Figure 7, the estimated pK_a of Cys-893 was $pK_a \simeq 9.1$ in the absence of substrates, $pK_a \simeq 9.5$ in the presence of NADP, and $pK_a \simeq 8.7$ in the presence of NADPH. In other words, the pK_a of Cys-893 changed by 0.8 of a pH unit depending on whether the catalytic site contained NADP or NADPH. Other experiments showed that, unlike NEM which causes a near complete inhibition, modification of Cys-893 by methylmethanethiosulfonate results in no more than 75% inhibition (26). These results suggested that Cys-893 is not an essential residue, and that the extent of inhibition accompanying its modification is related to the size of the foreign mass introduced at this site. The non-essentiality of Cys-893 is important, because it indicates absence of an interaction with NADP or NADPH. Hence, the opposite effects of NADP and NADPH on the pK_a of Cys-893 could be attributed to the different ways in which these ligands affect the conformation of TH and the environment of Cys-893.

4. CONCLUSIONS

The results presented above suggest that in the forward transhydrogenation reaction, protonic energy is consumed to change the conformation of the enzyme (presumably by protonation of appropriate residues). This conformation change increases substrate binding energy and facilitates hydride ion transfer from NADH to NADP. In the reverse reaction (eq. 1), the energy for $m \rightarrow c$ proton translocation comes from the difference in the concentrations of the substrates (NADPH + NAD) and the products (NADP + NADH). As demonstrated above, NADPH and NADP alter TH conformation in different ways, and as shown for Cys-893, these conformation changes can have a substantial effect on the pK_a of prototropic residues of the protein. Thus, one can conceive of the possibility that such substrate/product-induced enzyme conformation changes might alter the pK_a of appropriate residues on opposite sides of the membrane, thereby resulting in proton uptake on one side and proton release on the other side. These considerations are depicted in the hypothetical scheme of Figure 8.

ACKNOWLEDGEMENTS

The work of the authors' laboratory was supported by The United States Public Health Service Grant GM24887. This is publication number 6071 MEM from the Research Institute of Scripps Clinic, La Jolla, California.

REFERENCES

1. Fisher, R. R., and Earle, S. R. (1982) in *The Pyridine Nucleotide Coenzymes* (Everse, J., Anderson, B., and You, K.-S., eds) pp. 279-324 Academic Press, New York.
2. Rydström, J. (1977) *Biochim. Biophys. Acta 463*, 155-184.
3. Earle, S. R., and Fisher, R. R. (1980) *J. Biol. Chem. 255*, 827-830.
4. Eytan, G. D., Persson, B., Ekebacke, A., and Rydström, J. (1987) *J. Biol. Chem. 262*, 5008-5014.
5. Yamaguchi, M., Hatefi, Y., Trach, K., and Hoch, J. A. (1988) *J. Biol. Chem. 263*, 2761-2767.
6. Clarke, D. M, Loo, T. W., Gillam, S., and Bragg, P. D. (1986) *Eur. J. Biochem. 158*, 647-653.
7. Yamaguchi, M., Hatefi, Y., Trach, K., and Hoch, J. A. (1988) *Biochem. Biophys. Res. Commn. 157*, 24-29.
8. Wakabayashi, S., and Hatefi, Y. (1987) *Biochem. Int. 15*, 915-924.
9. Phelps, D. C., and Hatefi, Y. (1985) *Biochemistry 24*, 3503-3507.
10. Phelps, D. C., and Hatefi, Y. (1984) *Biochemistry 23*, 4475-4480.

11. Wakabayashi, S., and Hatefi, Y. (1987) *Biochem. Int. 15*, 667–675.
12. Solioz, M. (1984) *Trends in Biochem. Sci. 9*, 309–312.
13. Olson, J. A., and Anfinsen, C. B. (1953) *J. Biol. Chem. 202*, 841–856.
14. Kaplan, N. O., Colowick, S. P., and Neufeld, E. F. (1953) *J. Biol. Chem. 205*, 1–15.
15. Galante, Y. M., Lee, Y., and Hatefi, Y. (1980) *J. Biol. Chem. 255*, 9641–9646.
16. Van de Stadt, R. J., Nieuwenhuis, F. J. R. M., and Van Dam, K. (1971) *Biochim. Biophys. Acta 234*, 173–176.
17. Phelps, D. C., Galante, Y. M., and Hatefi, Y. (1980) *J. Biol. Chem. 255*, 9647–9652.
18. Hatefi, Y., Phelps, D. C., and Galante, Y. M. (1980) *J. Biol. Chem. 255*, 9526–9529.
19. Hatefi, Y., Yagi, T., Phelps, D. C., Wong, S.-Y., Vik, S. B., and Galante, Y. M. (1982) *Proc. Natl. Acad. Sci U.S.A. 79*, 1756–1760.
20. Blazyk, J. F., Lam, D., and Fisher, R. R. (1976) *Biochemistry 15*, 2843–2848.
21. Earle, S. R., O'Neal, S. G., and Fisher, R. R. (1978) *Biochemistry 17*, 4683–4690.
22. Persson, B., Hartog, A. F., Rydström, J., and Berden, J. A. (1988) *Biochim. Biophys. Acta 953*, 241–248.
23. Yamaguchi, M., and Hatefi, Y. (1985) *Arch. Biochem. Biophys. 243*, 20–27.
24. Junti, K., Torndal, U. B., and Ernster L. (1970) in *Electron Transport and Energy Conservation* (Tager, J. M., Papa, S., Quagliariello, E., and Slater, E. S., eds) pp 257–271, Adriatica Editrice, Bari.
25. Teixeira da Cruz, A., Rydström, J., and Ernster, L. (1971) *Eur. J. Biochem. 23*, 203–211.
26. Yamaguchi, M., and Hatefi, Y. (1989) *Biochemistry 28*, 6050–6056.
27. Laemmli, U.K. (1970) *Nature (London) 277*, 680–685.

STRUCTURES OF BEEF MITOCHONDRIAL COMPLEX I SUBUNITS

Sadao Wakabayashi*, Ryoji Masui*, Hiroshi Matsubara*, and
Youssef Hatefi**

*Department of Biology
Osaka University
Toyonaka, Osaka 560, Japan
**Division of Biochemistry
Scripps Clinic and Research Foundation
La Jolla, CA 92037, U.S.A.

SUMMARY

Nine polypeptides were isolated from the iron-protein fragment of
beef heart mitochondrial complex I. Their apparent molecular weights
were estimated to be 75K, 49K, 30K, 20K, 18K, 15K, and 13K by SDS-
polyacrylamide gel electrophoresis. Among them both the 30K and 13K bands
each contained two distinct polypeptides. The amino acid analysis of each
subunit suggested that the 75K, 49K, two 30K, 20K, 15K and one of the
13K polypeptides had enough cysteines and histidines to chelate the iron-
sulfur clusters. The amino acid sequences of the 13K polypeptides were
determined to show no apparent structural similarity to other proteins.

INTRODUCTION

Mitochondrial complex I is the most intricate system in the
respiratory chain and catalyzes the reduction of ubiquinone by NADH linked
to the proton translocation across the inner membrane (1). It contains 22
-24 non-heme iron and 22-24 acid labile sulfur atoms relative to FMN, and
approximately 25 unlike polypeptides. These polypeptides can be divided
into three groups when treated with chaotropes (2-4). The flavoprotein
fragment (FP) contains an FMN and two iron-sulfur clusters and shows NADH
dehydrogenase activity. It consists of three polypeptides with Mr of 51K,
24K and 9K. The iron-protein fragment (IP) consists of six major polypeptides (75K, 49K, 30K, 18K, 15K, and 13K) and has three binuclear and
one tetranuclear clusters. The insoluble residue (P fraction) contains
the most of complex I polypeptides and has at least two iron-sulfur
clusters.

Understanding of the mechanism of complex I function is still
immature at present and this may to some extent come from the lack of
structural information of complex I components, while the primary
structures of most polypeptide components of the other respiratory
complexes have been elucidated (5,6). The nucleotide sequence of the
mammalian mitochondrial DNA revealed the presence of eight unidentified

reading frames, seven of which were later shown to encode the hydrophobic
components of complex I (7-10). The partial amino acid sequence and
complete cDNA sequence of 24K polypeptide have been determined (11-13) and
cDNA sequence of 49K polypeptide has also recently been determined (14).
In order to obtain further structural information of complex I components,
we resolved the IP fraction systematically, obtained nine polypeptides in
pure form and sequenced two of them.

MATERIALS AND METHODS

      Beef heart complex I was prepared and resolved into three fractions
by the treatment with chaotrope as described (15). The IP fragment was
treated with trichloroacetic acid and carboxymethylated (16). The
carboxymethylated IP was further citraconylated (17) and applied to a
Toyopearl HW-65 column (2 X 180 cm) equilibrated with 50 mM Tris-HCl
buffer, pH 7.5, containing 6 M urea. The first and second fractions
contained 75K and 49K polypeptides, respectively, and further purified by
rechromatography. The third fraction was separated by a Sephacryl S-200
(HR) column (2 X 120 cm) in the same solvent into three fractions. These
fractions were separately applied to a DEAE-cellulofine column. The
column was developed by a linear gradient of NaCl from 0 M to 1 M in 50 mM
Tris-HCl buffer, pH 8.0, or 50 mM to 1 M ammonium bicarbonate. Thus, two
30K, 20K, and two 13K polypeptides were obtained. The 18K and 15K
polypeptides were eluted together from the DEAE-cellulofine column and
separated by a CM-cellulose column in an ammonium acetate buffer, pH 5.3,
containing 8 M urea after decitraconylation with 10 % acetic acid. The
purity of each polypeptide was assessed by SDS-polyacrylamide gel electrophoresis (18) in 15 % gel containing 12 % glycerol.

      Amino acid analysis was performed essentially according to Spackmen
et al. (19) with an Irica model A-5500 amino acid analyzer. The amino-
terminal sequence was determined by manual Edman degradation method (20)
and the phenylthiohydantoin (PTH) derivatives were identified by HPLC
(21). The carboxyl-terminal sequence was analyzed by digestion of protein
or peptide with carboxypeptidase A or Y (Worthington, Oriental Yeast) and
detection of released amino acid on the analyzer. To obtain the complete
sequences of the two 13K polypeptides, they were digested with various
proteolytic enzymes (trypsin (Worthington), lysylendopeptidase (Wako),
staphylococcal V8 protease (Miles), and endoproteinase Asp-N (Behringer))
and the peptides were separated by reverse-phase HPLC (Shodex ODS-pak F-
511A column). The identification of amino-terminal blocking group was
performed by digestion of the amino-terminal peptides with acylamino-acid-
releasing enzyme (Takara) and comparison of retention time of the digest
with authentic N-acetyl-Ala or N-acetyl-Gly on HPLC (22).

RESULTS AND DISCUSSION

Separation of Component Polypeptides

      Initial attempt to resolve the IP fragment was carried out by Ragan
et al. (23). They purified three iron sulfur proteins with molecular
weight of 75K, 49K, and 30K + 13K. In order to determine the primary
structures of all the IP components, we had to prepare a relatively large
amount of each polypeptide and could not apply their method. Also it was
difficult to prepare the iron-sulfur proteins holding the intact forms of
iron-sulfur centers because of their lability to aerobic conditions.

Therefore, the IP fragment was first treated with trichloro-acetic acid and cysteine residues were converted to S-carboxymethylcysteines. The carboxymethylated IP fraction was insoluble in normal aqueous solutions, and was chromatographed on a gel filtration column in 6 M urea. Under these conditions some polypeptides aggregated and were eluted at the void volume of the column. Thus the sample was citraconylated. The introduction of negative charges to the polypeptides led them soluble in the slightly alkaline solution to be separated in a simple aqueous solution. However, even under these conditions, some polypeptides still aggregated. Separation of the citraconylated sample in 6 M urea solution was most effective. After the first separation on a Toyopearl HW-65 column in 6 M urea, each fraction was further purified by ion-exchange chromatography as described in Materials and Methods. In some cases the final purification was performed by a reverse-phase HPLC using a butyl column (Shodex RSpak D4-613) in 0.05 % triethylamine.

The amino acid compositions of the isolated polypeptides are shown in Table I. The composition of the 49K polypeptide deduced from the cDNA sequence (14) is also listed in the table and agreed well with the obtained value. All polypeptides except for the 18K and 13K-B polypeptides had enough cysteines and histidines to be the candidates for iron-sulfur proteins. The possible location of the iron-sulfur centers in IP fraction will be discussed later.

Table I. Amino Acid Compositions of IP Component Polypeptides

	75K	49K	30K-A	30K-B	20K	18K	15K	13K-A	13K-B
Asx	69	40(39)	28	26	15	20	11	10 (9)	5 (5)
Thr	42	20(19)	12	14	6	14	7	7 (7)	6 (6)
Ser	37	18(19)	15	14	8	10	8	4 (4)	3 (3)
Glx	73	47(45)	34	34	25	18	20	13(12)	20(20)
Pro	32	27(28)	20	18	13	7	7	3 (3)	8 (7)
Gly	53	33(31)	10	13	8	9	9	13(14)	6 (6)
Ala	66	31(28)	22	23	11	12	9	4 (3)	9 (8)
Cys	17	6 (6)	2	5	7	0	2	3 (3)	1 (1)
Val	57	32(30)	22	16	11	11	6	10(10)	7 (7)
Met	20	19(19)	1	3	2	3	2	0 (0)	2 (2)
Ile	39	21(24)	11	12	5	6	7	4 (4)	7 (8)
Leu	62	37(36)	24	24	18	9	8	4 (4)	15(15)
Tyr	15	20(20)	11	11	4	3	2	3 (4)	3 (3)
Phe	18	16(16)	16	13	8	6	6	3 (3)	0 (0)
Lys	31	18(21)	12	12	14	14	10	5 (5)	12(13)
His	10	11(13)	5	4	4	1	3	3 (4)	2 (2)
Arg	40	29(28)	20	20	13	11	11	7 (7)	4 (4)
Trp	nd	nd (8)	nd	nd	nd	nd	nd	nd (0)	nd (4)
total	681	(430)	265	262	172	154	128	(96)	(114)

Cysteines are determined as S-carboxymethylcysteine. Values in parentheses are deduced from the sequence. nd; not determined.

Sequence Studies of the 13K-A Polypeptide

The amino-terminal sequence of carboxymethylated 13K-A polypeptide was determined by manual Edman degradation up to the 6th residue. The carboxyl-terminal sequence of the polypeptide was analyzed by carboxy-peptidase A digestion. Histidine (1.7 mol/mol) and glutamine (1.8 mol/mol) were released for 1 h incubation, in conjunction with the sequence analysis of the carboxyl-terminal peptide showing the sequence of -Gln-Gln-His-His.

The citraconylated 13K-A polypeptide was first digested with trypsin, giving seven peptides (R-1 to R-7). A tripeptide corresponding to residues 25 to 27 could not be recovered. Manual Edman degradation of each peptide determined the amino acid sequence of about 70 % of the protein. In order to determine the amino acid sequence of the remaining parts and the order of these tryptic peptides, the 13K-A polypeptide was decitraconylated and digested with lysylendopeptidase. Six peptides (K-1 to K-6) were recovered and sequenced. These studies established the amino acid sequence of 13K-A polypeptide except for the region around the residues 20 to 25. Then, the polypeptide was digested with endoproteinase Asp-N and six peptides (D-1 to D-6) were obtained. The sequence studies of these peptides established the complete amino acid sequence of the 13K-A polypeptide as summarized in Fig. 1 (A). The total number of amino acid residues was 96 and the molecular weight was calculated to be 10,536.

Sequence Studies of the 13K-B Polypeptide

Edman degradation of the carboxymethylated 13K-B polypeptide gave no PTH-amino acid suggesting the amino-terminus of this protein to be blocked. The carboxyl-terminal sequence was determined to be -Pro-Ile by carboxy-peptidase Y digestion (Pro 0.14; Ile 0.40; for 5 min, Pro 0.62; Ile 0.89; for 20 min).

The 13K-B polypeptide was first digested with trypsin as in the case of the 13K-A polypeptide and five peptides (R-1 to R-5) could be obtained by HPLC after decitraconylation. Manual Edman degradation of these peptides (except for peptide R-1) revealed the amino acid sequence of about 80 % of the protein. Peptide R-1 gave no PTH-amino acid and was asssumed to be the amino-teminal peptide. It was further digested with lysylendopeptidase and the resulted peptide R1-K-3 was sequenced. In order to obtain overlaps of these tryptic peptides, the 13K-B polypeptide was decitraconylated and digested with lysylendopeptidase. The amino acid sequences of the resulted peptides (K-1 to K-15) were determined and the whole sequence of the 13K-B polypeptide except for the amino-terminal region was established. The peptides R1-K-1 and K-1 had the same amino acid compositions and blocked amino terminus. The carboxyl-terminal sequence of R1-K-1 was determined by carboxypeptidase Y digestion to be -Leu-Leu-Lys. Peptide K-1 was digested with acylamino-acid-releasing enzyme and N-acetyl-alanine and a peptide K1-2 were separated by HPLC. The amino terminus of K1-2 was determined to be glycine. These results together with the amino acid compositions of the peptides K-1 and K1-2 indicated that the structure of amino-terminal peptide was Ac-Ala-Gly-Leu-Leu-Lys. Thus the amino acid sequence of the 13K-B polypeptide was determined as in Fig. 1 (B). The only ambiguous identification was Trp-107 which was assigned by amino acid composition of peptide K-14. The total number of amino acid residues of the 13K-B polypeptide was 114 and the calculated molecular weight was 13,130 including the amino terminal acetyl group.

(A)

Fig. 1. Summary of the sequence studies of the 13K-A (A) and 13K-B (B) polypeptides. R-, K-, D-, and S- refer to the peptides obtained by digestion with trypsin, lysylendopeptidase, endoproteinase Asp-N, and staphylococcal protease, respectively. Arrows (⟶) and (⟵)mean that the residues above were identified by manual Edman degradation and carboxypeptidase digestion. Dotted arrows mean ambiguous identification.

Possible Location of Iron-Sulfur Centers

There are several classes of iron-sulfur proteins with different type of iron-sulfur clusters, all of which are supported by characteristic amino acid sequences, especially characteristic arrangements of cysteines and histidines.   In complex I at least eight iron-sulfur clusters are assumed to be present, six of which are detected by ESR (4).   The resolution of complex I and isolation of IP fragment showed 75K, 49K and 30K + 13K polypeptides to be the iron-sulfur proteins (23). However, the sequence of 49K polypeptide recently reported showed no homology to the amino acid sequence of any type of iron-sulfur proteins (14).   Therefore, the 49K polypeptide may be a new type of iron-sulfur protein or may not be an iron-sulfur protein. The 75K polypeptide has nearly four iron and sulfur atoms and was assumed to have two 2Fe-2S clusters. The amino acid analysis of isolated 75K polypeptide showed 17 cysteines per mol enough to chelate two clusters and possibly as many as four iron-sulfur clusters. The amino acid sequences of two 13K polypeptides determined here also showed no sequence similarity to any type of iron-sulfur proteins, nor any proteins whose amino acid sequences are known. This suggests that the possibility of these 13K polypeptides to be iron-sulfur proteins is low. Ozawa et al. reported that the 13K polypeptide of the IP fraction was the coenzyme Q binding protein (24). It is not clear that their 13K polypeptide corresponds to 13K-A or 13K-B polypeptide, but both have no sequence homology to the Q binding protein in complex III (25). Since the presence of one 2Fe-2S cluster in the complex of 13K + 30K polypeptides was suggested (23), the 30K polypeptide is probably an iron-sulfur protein. We could isolate two 30K polypeptides, whose amino acid compositions are very similar to each other, although the contents of some amino acids are definitely different. It is not sure at present whether these are really distinct polypeptides or not, and at least one of these (probably the 30K-B polypeptide) is a candidate for an iron-sulfur protein.   It should be noted, however, that the 20K polypeptide has a high content of cysteine although it is not the major component of IP fragment.

ACKNOWLEDGEMENTS

We thank C. Muñoz for preparing the mitochondria and extracts.   This work was supported in part by Grants-in-Aid for International Scientific Research Program (#63044089, #01044089), Priority Area of Bioenergetics (#01617002) and No.01470148 from the Ministry of Education, Science and Culture of Japan.

REFERENCES

1. Hatefi, Y., Haavik, A. G., and Griffiths, D. E. (1962) J. Biol. Chem. 237, 1676-1680
2. Hatefi, Y. (1985) Ann. Rev. Biochem. 54, 1015-1069
3. Hatefi, Y., Ragan, C. I., and Galante, Y. M. (1985) in The Enzymes of Biological Membranes (Martonosi, A. N., ed) Vol. 4, pp. 1-70, Plenum Press, New York
4. Ragan, C. I., Ohnishi, T., and Hatefi, Y. (1986) in Iron-Sulfur Protein Research (Matsubara, H., Katsube, Y., and Wada, K., eds) pp. 220-231, Japan Sci. Soc. Press, Tokyo / Springer-Verlag, Berlin
5. Buse, G., Steffens, G. C. M., Biewald, R., Bruch, B., and Hensel, S. (1987) in Cytochrome Systems (Papa, S., Chance, B., and Ernster, L. eds) pp. 261-270, Plenum Press, New York

6. Link, T. A., Schägger, H., and Von Jagow, G. (1987) in Cytochrome Systems (Papa, S., Chance, B., and Ernster, L. eds) pp. 289-301, Plenum Press, New York

7. Anderson, S., Bankier, A. T., Barrell, B. G., de Bruijn, M. H. L., Coulson, A. R., Drouin, J., Eperon, I. C., Nierlich, D. P., Roe, B. A., Sanger, F., Schreier, P. H., Smith, A. J. H., Staden, R., and Young, I. G. (1981) Nature 290, 457-465

8. Bibb, M. J., Van Etten, R. A., Wright, C. T., Walberg, M. W., and Clayton, D. A. (1981) Cell 26, 167-180

9. Chomyn, A., Mariottini, P., Cleeter, M. W. J., Ragan, C. I., Matsuno-Yagi, A., Hatefi, Y., Doolittle, R. F., and Attardi, G. (1985) Nature 314, 592-597

10. Chomyn, A., Cleeter, M. W. J., Ragan, C. I., Riley, M., Doolittle, R. F., and Attardi, G. (1986) Science 234, 614-618

11. Von Bahr-Lindström, H., Galante, Y. M., Persson, M., and Jörnvall, H. (1983) Eur. J. Biochem. 134, 145-150

12. Nishikimi, M., Hosokawa, Y., Toda, H., Suzuki, H., and Ozawa, T. (1988) Biochem. Biophys. Res. Commun. 157, 914-920

13. Pilkington, S. J., and Walker, J. E. (1989) Biochemistry 28, 3257-3265

14. Fearnley, I. M., Runswick, M. J., and Walker, J. E. (1989) EMBO J. 8, 665-672

15. Galante, Y. M., and Hatefi, Y. (1978) Methods Enzymol. 53, 15-21

16. Crestfield, A. M., Moore, S., and Stein, W. H. (1963) J. Biol. Chem. 238, 622-627

17. Dixon, H. B. F., and Perham, R. N. (1968) Biochem. J. 109, 312-314

18. Laemmli, U. K. (1970) Nature 227, 680-685

19. Spackman, D. H., Moore, S., and Stein, W. H. (1958) Anal. Chem. 30, 1190-1206

20. Blombäck, B., Blombäck, M., Edman, P., and Hessel, B. (1966) Biochim. Biophys. Acta 115, 371-396

21. Zimmerman, C. L., Apella, E., and Pisano, J. J. (1977) Anal. Biochem. 77, 569-573

22. Tunasawa, S., and Narita, K. (1982) J. Biochem. 92, 607-613

23. Ragan, C. I., Galante, Y. M., and Hatefi, Y. (1982) Biochemistry 21, 2518-2524

24. Ozawa, T., Nishikimi, M., Suzuki, H., Tanaka, M., and Shimomura, Y. (1987) in Bioenergetics (Ozawa, T. and Papa, S. eds) pp. 101-119, Japan Sci. Soc. Press, Tokyo / Springer-Verlag, Berlin

25. Wakabayashi, S., Takao, T., Shimonishi, Y., Kuramitsu, S., Matsubara, H., Wang, T.-Y., Zhang, Z.-P., and King, T. E. (1985) J. Biol. Chem. 260, 337-343

INHIBITION OF THE MITOCHONDRIAL COMPLEX I ACTIVITY

BY FATTY ACID DERIVATIVES OF VANILLYLAMIDE

Yoshiharu Shimomura,[*] Teruo Kawada,[+] Kazumi Tagami,[*] and
Masashige Suzuki[*]

[*]Laboratory of Biochemistry of Exercise and Nutrition
Institute of Health and Sport Sciences, University of
Tsukuba, Tsukuba 305, Japan; and [+]Laboratory of Nutritional
Chemistry, Department of Food Science and Technology
Faculty of Agriculture, Kyoto University, Kyoto 606, Japan

SUMMARY

Fatty acid derivatives of vanillylamide were found to inhibit the
mitochondrial NADH oxidase activity, but almost not the succinate oxidase
activity. These results suggest that the compounds are specific
inhibitors of the Complex I activity. A study using purified Complex I
demonstrated that the compounds inhibit NADH-coenzyme Q reductase
activity of the enzyme. However, these did not inhibit rotenone-
insensitive NADH-menadione reductase activity and electron transfer from
NADH through iron-sulfur centers in Complex I. Kinetic analyses with
double-reciprocal plots showed that these compounds were competitive
inhibitors with respect to coenzyme Q_1. These findings suggest that the
compounds bind to the coenzyme Q binding site of Complex I. The natures
of acyl groups greatly affected the inhibitory potencies of the
compounds, suggesting that the acyl group of compound is important for
the binding of compound to the binding site of Complex I.

INTRODUCTION

Fatty acid derivatives of vanillylamide are contained in hot peppers,
which are the most common spices used in foods in many parts of the world
(1). Especially, 8-methylnon-trans-enoyl vanillylamide (capsaicin) is
known as the pungent principle of red pepper species (Capsicum species)
(1). It has been reported that pharmacological doses of capsaicin causes
damage of mitochondria in certain spinal and preoptic neurons of rats
(2,3), and that capsaicin inhibits energy metabolism of mitochondria
prepared from rat liver (4,5).
We have examined in detail the inhibitory effects of fatty acid
derivatized vanillylamides on the mitochondrial electron transfer chain
(6). In this paper, we will report that the compounds inhibited the
Complex I activity in the electron transfer chain and that the nature of
fatty acid side chain of the compound had a great effect on the
inhibitory action.

EXPERIMENTAL PROCEDURES

Materials. Rat liver mitochondria were prepared by the method of Schnaitman and Greenawalt (7). Hypotonic-treatment of the mitochondria was performed according to the method of Chudapongse and Janthasoot (5). Complex I and Complex I-III were isolated from beef heart mitochondria by the methods of Hatefi (8) and Hatefi and Stiggall (9), respectively, and stored at -70°C until use. The Complex I preparation had enzymic activity of 8.5 μmol NADH oxidized/min/mg of protein at 30°C, and the Complex I-III preparation 15 μmol cytochrome c reduced/min/mg of protein at 30°C. 8-Methylnonanoyl vanillylamide (dihydrocapsaicin), heptanoyl vanillylamide (C7-VA), nonanoyl vanillylamide (C9-VA), decanoyl vanillyl-amide (C10-VA), undecanoyl vanillylamide (C11-VA), lauroyl vanillylamide (C12-VA), palmitoyl vanillylamide (C16-VA), stearoyl vanillylamide (C18-VA), and nonanoyl veratrylamide were synthesized by the method of Rangoonwala and Seitz (10). Those with more than 95% purity, as verified by gas-liquid chromatography and high performance liquid chromatography, were used in the experiment. Capsaicin (98% purity), cholic acid, and deoxycholic acid were purchased from Sigma (St. Louis). Coenzyme Q_1 was obtained as a gift from Eisai Co., Ltd., Tokyo. Other chemicals used were of reagent grade.

Analytical methods. Oxygen consumption of hypotonic-treated mitochondria was assayed at 30°C by the method of Chudapongse and Janthasoot (5) using NADH or succinate as an electron donor. Activities of Complex I and Complex I-III were assayed at 30°C by the method of Hatefi (8) and Hatefi and Stiggall (9), respectively, unless otherwise stated. Varing concentrations of fatty acid derivatized vanillylamid solutions in ethanol were prepared prior to use, and 10 μl of the solution was added into the reaction mixtures of oxygen consumption assays (final 3 ml) and enzyme activity assays (1 ml). EPR measurements (11) were performed with a JEOL Model JES-FE3XG spectrometer (JEOL, Ltd., Tokyo). The temperature of the sample for EPR measurements was controlled by means of a variable temperature cryostat (Air Products Model LTR-3-110). EPR operating conditions: microwave frequency, 9.07 GHz; modulation amplitude, 1 G; microwave power, 1.0 mW; time constant, 0.1 s; scanning time, 250 G/min; temperature, 5-24°K. Coenzyme Q_1 was estimated spectrophotometrically as described by Crane and Barr (12). Protein was assayed by the Lowry method as modified by Hartree (13) using bovine serum albumin as a standard.

RESULTS AND DISCUSSION

The NADH oxidase activity of mitochondria was inhibited in a dose dependent manner by fatty acid derivatives of vanillylamide. The typical data of inhibition at 50 μM compounds were given in Table 1. The degrees of inhibition were markedly different among the compounds: the inhibiory effect was high (approximately 70% inhibition) in dihydrocapsaicin, C11-VA, and C12-VA, but low (less than 10%) in C16-VA, and C18-VA. These results indicate that the fatty acid moieties of compounds have a great effect on the inhibitory potency. On the other hand, the succinate oxidase activity of mitochondria was little affected by all of the compounds at 50 μM (Table 1). These results indicate that the compounds retarded electron transfer from NADH to coenzyme Q in the mitochondrial electron transfer chain, suggesting that the compounds are specific inhibitors of the Complex I activity.

Table 1. Inhibitory effects of fatty acid derivatized vanillylamides on mitochondrial oxygen consumption. Assay conditions are described under Experimental Procedures.

Compound (50 µM)	Percent inhibition	
	NADH oxidase	Succinate oxidase
Capsaicin	33	3
Dihydrocapsaicin	70	5
C7–VA	58	4
C9–VA	35	1
C10–VA	13	3
C11–VA	71	4
C12–VA	71	4
C16–VA	6	0
C18–VA	7	0

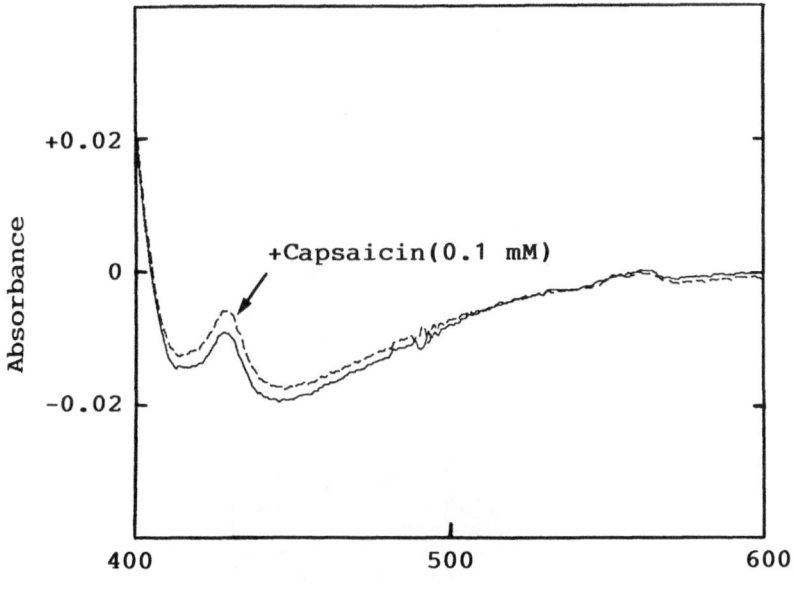

Fig. 1. Effect of capsaicin on the difference spectrum of NADH reduced minus oxidized form of Complex I. The Complex I preparation was suspended at 1.2 mg of protein/ml in 50 mM Tris–Cl, pH 8.0, containing 0.66 M sucrose, 1 mM histidine, 0.1% cholate, and 0.2% deoxycholate. The difference spectrum was recorded at room temperature after addition of 0.375 mM NADH into the sample cuvette. Capsaicin was added at 0.1 mM into the sample cuvette just prior to addition of NADH.

Table 2. I_{50} values for fatty acid derivatives of vanillylamides with isolated Complex I. The NADH–coenzyme Q reductase activity of Complex I was measured using 60 μM coenzyme Q_1. The compounds were added into the reaction mixture just prior to addition of the enzyme. Other conditions are as described under Experimental Procedures.

Compound	I_{50} (μM)
Capsaicin	48
Dihydrocapsaicin	21
C7–VA	18
C9–VA	45
C10–VA	3.5
C11–VA	3.0
C12–VA	2.5
C16–VA	60
C18–VA	95

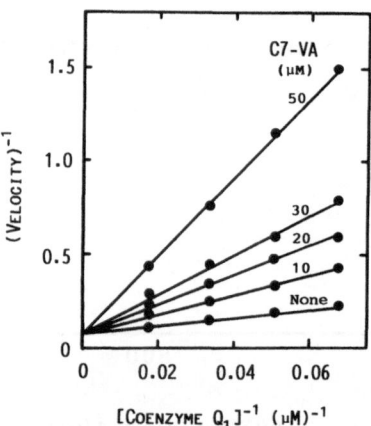

Fig. 2. Inhibition of NADH–coenzyme Q reductase activity of Complex I by C7–VA. The enzyme activities were assayed as described under Experimental Procedures, except that the coenzyme Q_1 concentration was varied from 15 to 60 μM. Velocity is given as micromoles of NADH oxidized per minute per milligram of protein.

The findings obtained in the mitochondrial experiments described above addressed us to clarify in detail the inhibitory effects of fatty acid derivatized vanillylamide on the Complex I activity using the purified Complex I preparation. It was shown in this experiment that the compounds inhibited electron transfer from NADH to coenzyme Q in Complex I, but not the rotenone-insensitive electron transfer from NADH to menadione. Fig. 1 shows the effect of capsaicin on the difference spectrum between 400 and 600 nm of the NADH reduced minus oxidized form of Complex I. The compound had almost no effect on the difference spectrum, suggesting that the compound did not inhibit NADH reduction of FMN and iron-sulfur groups in Complex I. Since a number of iron-sulfur centers have been identified by EPR spectroscopy at liquid helium temperatures (14), the effects of capsaicin, dihydrocapsaicin, and C10-VA on the EPR specra of Complex I were examined. All of the compounds tested had almost no effect on the spectra measured at 5-24°K, suggesting that the compounds did not interrupt electron transfer from NADH through iron-sulfur centers in Complex I. The natures of fatty acid derivatized vanillylamide inhibitions were further characterized by double reciprocal plots. Typical data for C7-VA are shown in Fig. 2. The compound inhibited the Complex I activity in a competitive manner with respect to coenzyme Q_1. All of the other compounds used in this study also inhibited the enzyme activity in the same manner. These results suggest that the compounds bind to the substrate (coenzyme Q_1) binding site of Complex I, due to the structural similarity between the compounds and coenzyme Q.

The effects of varing concentrations of compounds on the Complex I activity were examined, and I_{50} values (amount of the compound needed to inhbit 50% of the Complex I activity) were estimated (Table 2). The values of C10-VA, C11-VA, and C12-VA were markedly low (2.5-3.5 μM), but those of C16-VA, and C18-VA were high (60-95 μM), indicating that the former compounds are potent inhibitors of the Complex I activity, but the latter compounds are not. These results suggest that the acyl moiety affects the binding of compounds to the enzyme. Probably, acyl moieties of 10-12 carbons are appropriate for the binding.

The effect of nonanoyl veratrylamide, which has a methoxy group in place of a hydroxy group of vanillylamide, on the Complex I activity were also examined. This compound also inhibited the enzyme activity in a competitive manner with respect to coenzyme Q_1. The I_{50} values was 12 μM and was approximately one third compared with that of C9-VA (45 μM), suggesting that a fatty acid derivative of veratrylamide is a stronger inhibitor than a fatty acid derivative of vanillylamide. This may be due to the more similar structure between veratrylamide and quinone than between vanillylamide and quinone.

C10-VA was a potent inhibitor of the Complex I activity in the assay of purified enzyme, and the degree of inhibition was almost the same as that of C11-VA (Table 2). However, C10-VA inhibition of the mitochondrial NADH oxidase activity was markedly low compared with that of C11-VA (Table 1). To solve this problem, the inhibitory effects of C10-VA and C11-VA on the Complex I-III activity were examined using the isolated Complex I-III preparation, because it had been posturated that the purity of enzyme preparation might affect the accessibility of C10-VA, but not that of C11-VA, to the binding site of Complex I. Results (Fig. 2) showed that the rate of enzyme activity is linear after addition of C11-VA, whereas it decreased gradually after addition of C10-VA, showing a curved line. These results indicate that C11-VA showed the maximal inhibitory effect just after addition, whereas C10-VA required time to reach the maximal inhibition after addition, suggesting that accessibility of C10-VA to Complex I decreased in the presence of Complex III in the Complex I-III preparation. Probably, the binding of C10-VA to Complex I in mitochondria was also inhibited by other proteins (and phospholipids) in mitochondria, resulting in the low inhibitory effect of compound on the mitochondrial NADH oxidase activity.

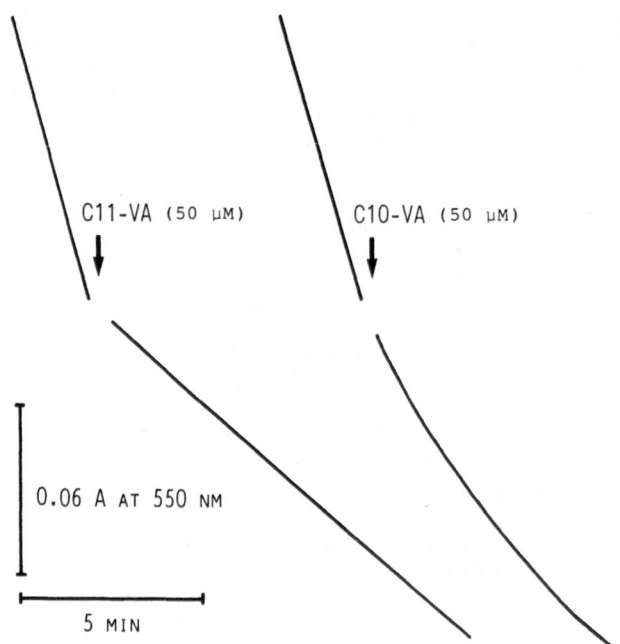

C11-VA (50 μM)

C10-VA (50 μM)

0.06 A AT 550 NM

5 MIN

Fig. 3. Inhibitory effect of C10-VA and C11-VA on the Complex I-III activity. The NADH-cytochrome c reductase activity of Complex I-III were assayed using the isolated Complex I-III preparation as described under Experimental Procedures.

In this study, it has been clearly demonstrated that fatty acid derivatives of vanillylamide inhibit the mitochondrial Complex I activity. It is known that pharmacological doses of capsaicin causes mitochondrial damage (2,3,15), and that capsaicin inhibits the mitochondrial energy metabolism (4,5). These may, at least partly, result from the direct inhibitory effect of capsaicin on Complex I in the mitochondrial electron transfer chain.

Many inhibitors have been used to elucidate catalytic mechanisms of enzymes. Fatty acid derivatives of vanillylamide and veratrylamide are new inhibitors of the Complex I activity. These compounds might be useful as tools to investigate the catalytic mechanism of Complex I.

REFERENCES

1. Suzuki, T., and Iwai, K., (1984) in The Alkaloids (Brossi, A., Ed.), Vol. **23**, pp. 227–299, Academic Press, New York.
2. Joo, F., Szolcsanyi, J., and Jancso-Gabor, A. (1969) Life Sci. **8**, 621–626.
3. Szolcsanyi, J., Joo, F., and Jancso-Gabor, A. (1971) Nature (London) **229**, 116–117.
4. Chudapongse, P., and Janthasoot, W. (1976) Toxicol. Appl. Pharmacol. **37**, 263–270.
5. Chudapongse, P., and Janthasoot, W. (1981) Biochem. Pharmacol. **30**, 735–740.
6. Shimomura, Y., Kawada, T., and Suzuki, M. (1989) Arch. Biochem. Biophys. **270**, 573–577.
7. Schnaitman, C., and Greenawalt, J. W. (1968) J. Cell Biol. **38**, 158–175.
8. Hatefi, Y. (1978) Methods Enzymol. **53**, 11–14.
9. Hatefi, Y., and Stiggall, D. L. (1978) Methods Enzymol. **53**, 5–10.
10. Rangoonwala, R., and Seitz, G. (1970) Dtsche. Apoth. Ztg. **110**, 1946–1949.
11. Ohnishi, T., and Leigh, J. S. (1974) Biochem. Biophys. Res. Commun. **56**, 775–782.
12. Crane, F. L., and Barr, R. (1971) Methods Enzymol. **18**, 137–165.
13. Hartree, E. F. (1972) Anal. Biochem. **48**, 422–427.
14. Ragan, C. I. (1976) Biochim. Biophys. Acta **456**, 249–290.
15. Nopanitaya, W., and Nye, S. W. (1974) Toxicol. Appl. Phamacol. **30**, 149–161.

New 4-Hydroxypyridine and 4-Hydroxyquinoline Derivatives as Inhibitors of NADH-Ubiquinone Oxidoreductase in the Respiratory Chain II*

Kun Hoe Chung*1, Kwang Yun Cho*1, Yasuko Asami*2, Nobutaka Takahashi*2 and Shigeo Yoshida*3

*1 Korea Research Institute of Chemical Technology, P. O. Box 9, Daedeog Danji, Taejon, Korea 302-343
*2 Department of Agricultural Chemistry, The University of Tokyo, Bunkyo-ku, Tokyo 113, Japan. *3 Chemical Regulation of Biomechanisms Lab., The Institute of Chemical and Physical Research (RIKEN), Wako-shi Saitama 351-01, Japan

SUMMARY

For elucidation of structural requisites of piericidin-like compounds for inhibitors to NADH-UQ oxidoreductase biomimetic derivatives of 4-hydroxypyridine and 4-hydroxyquinoline were systematically designed, synthesized and examined with mitochondria and submitochondria. The modification of arylalkyl sidechains were revealed to show high inhibitory activity of compounds. Also, sidechains of alkoxy-ω-phenylalkyl and alkyl-ω–phenylalkyl groups were regarded as good templates for examination of inhibitors to NADH-ubiquinone oxidoreductase. Among synthetic compounds 6-phenylhexyl derivatives of hydroxypyridine and hydroxyquinoline

----

*: Part I is ref. 17.   Abbreviations: RET, respiratory electron transport; ETP, electron transport particles; NADH-UQ oxidoreductase, NADH-ubiquinone oxidoreductase;   Tris, Tris(hydroxymethyl)amino-methane;   EDTA, ethylenediaminetetraacetic acid.

showed highest inhibitory activity as much as ubicidins.   The modification of hydrophilic part did not increase the activity well, but combination of methoxy, ethoxy and propoxy groups maintained their activity like dimethoxy compound.   Especially inhibition activity of 6-substituted hydroxyquinolines was diminished dramatically.

INTRODUCTION

Most inhibitors of RET system were discovered as naturally occurring substances, but a few series of synthetic compounds have been found from random screening researches.   Examples of the former are piericidins (1)[1], rotenone (2)[2], myxalamid (3)[3], stigmatellin (4)[4] and antimycin (5)[5], and the examples of the latter are benzimidazoles (6)[6], amytal (7)[7], fenaminosulf (8)[8] and carboxin (9)[9]. Among these compounds, piericidins, rotenone, myxalamid, benzimidazoles and amytal have been classified as inhibitors of NADH-UQ oxidoreductase in site I of the respiratory chain (Fig.1). Although rotenone is commonly used as a standard to compare the inhibitory activity of various compounds, piericidin A is the most potent inhibitors of NADH-UQ oxidoreductase site.   Earlier studies[1, 10] have suggested that the structural similarity between piericidins and ubiquinone (UQ, 10) could be correlated with the competitive behavior of the inhibitors at the same site of NADH-UQ oxidoreductase, where ubiquinones act as electron carriers.   The relationships between piericidins and rotenone were clearly identified by means of extensive biochemical studies using radioactive compounds.   The results again suggested that piericidins should have stronger binding force with the

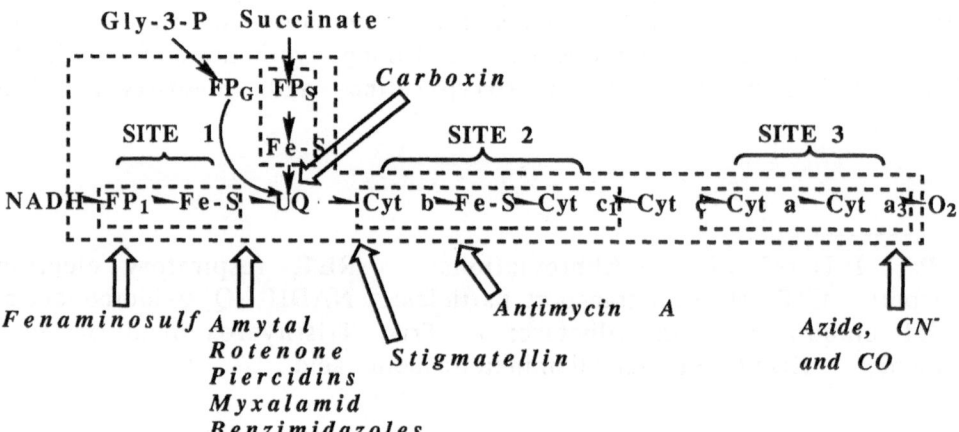

Figure 1.   Respiratory Electron Transport (RET) System & Inhibitors

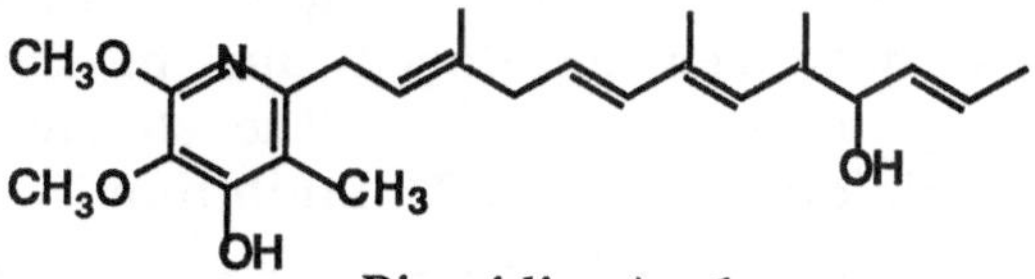

Piercidin A, **1**

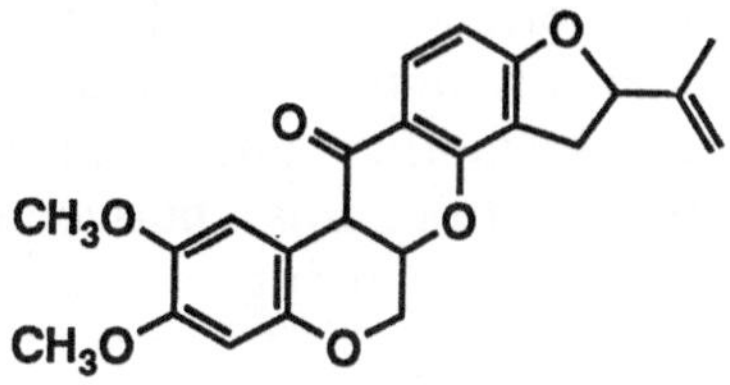

Rotenone, **2**

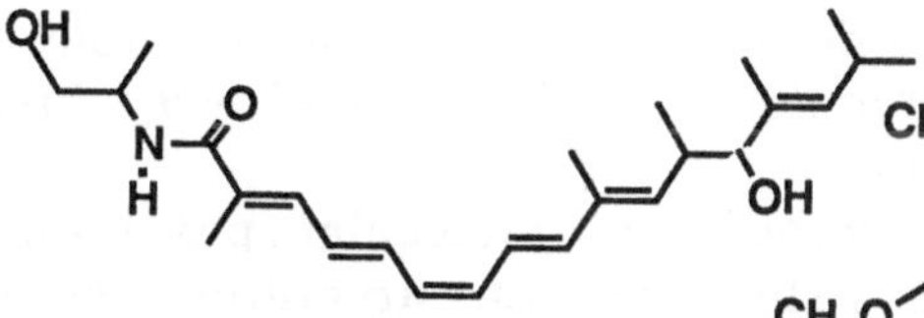

Myxalamid B, **3**

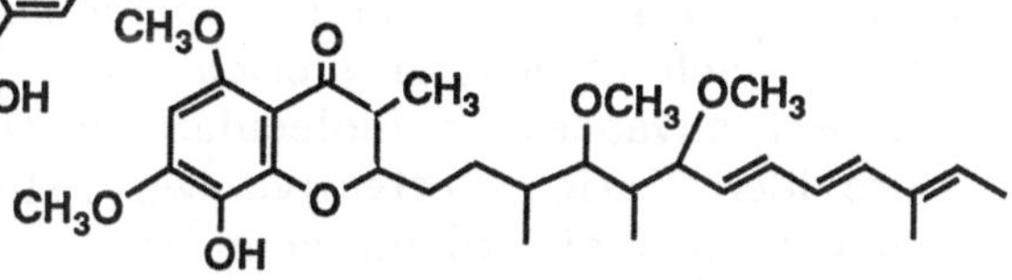

Stigmastelin A, **4**

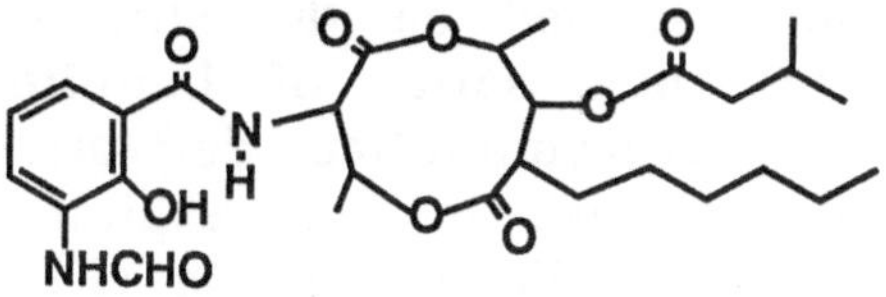

Antimycin A, **5**

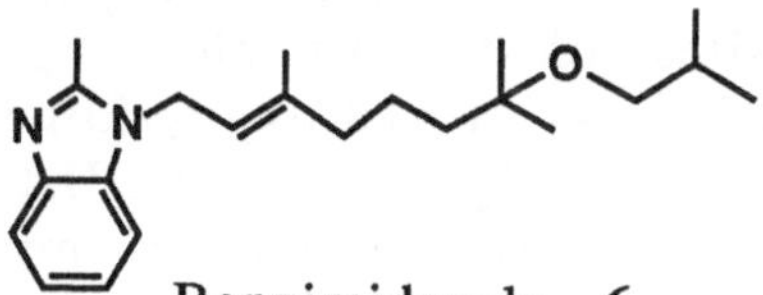

Benzimidazole, **6**

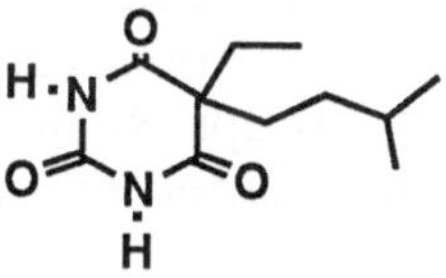

Amytal, **7**

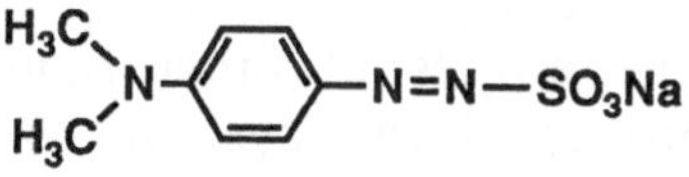

Fenaminosulf, **8**

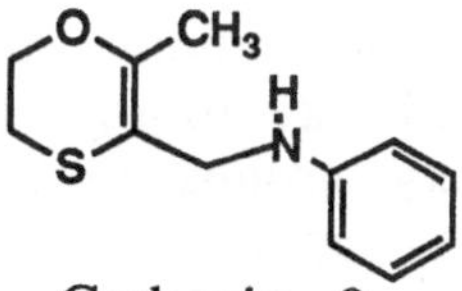

Carboxin, **9**

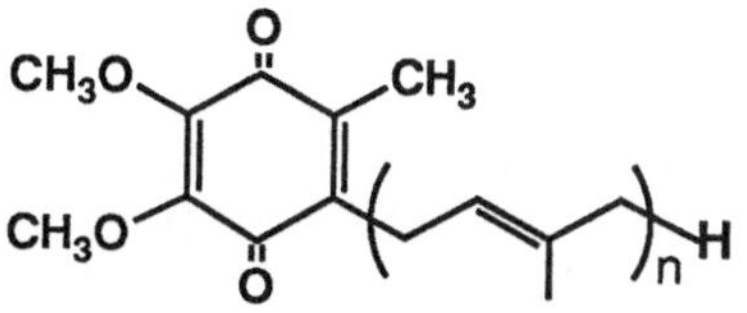

Ubiquinone, **10** (n; 6-10)

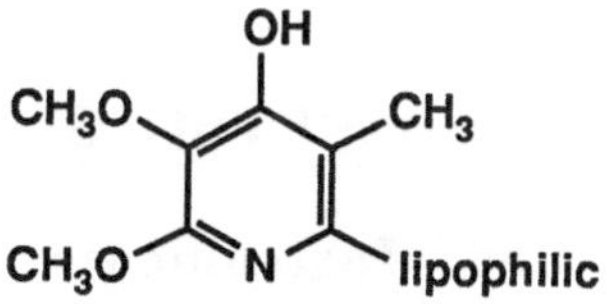

Type A, **11**

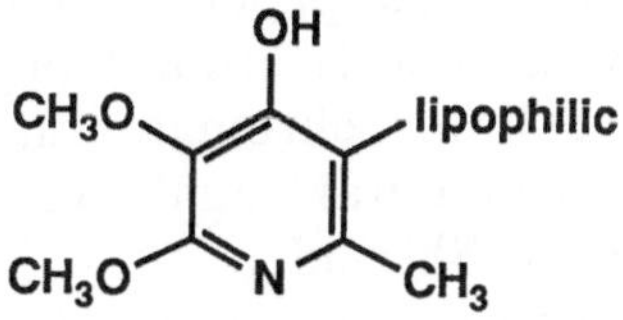

Type B, **12**

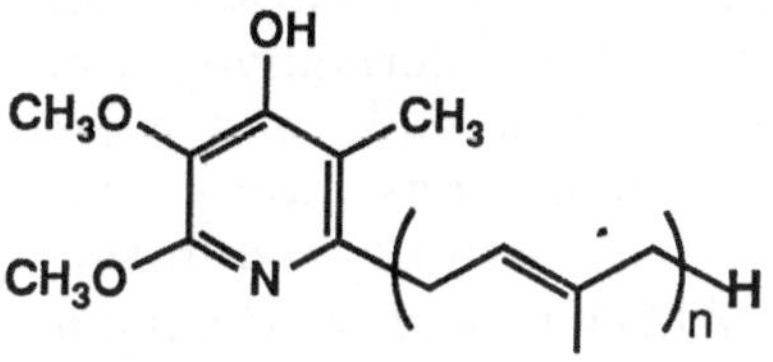

Ubicidin, **13** (n; 1-10)

ubiquinone reducing site than rotenone[11, 12].   Although piericidins are extremely effective in killing many kinds of insects[13, 14], applications of this type of natural product have not been developed due to their chemical instability and high toxicity to mammals[1, 10].   The structure-activity relationships between piericidins and RET inhibition was investigated by using some derivatives of piericidin homologs and synthetic analogs[15, 16], and the essential structures for the inhibitory activity were proposed as illustrated by Types **A** (**11**) and **B** (**12**).   In fact piericidin-like compounds of ubicidins (**13**), which are totally synthesized and possess simple isoprenoid chains[16], showed very high activity as much as natural piericidins.

Recent advances in molecular biology and molecular physiology have provided various effective ways to interpret structures and/or the characteristics of protein complex.   If there is a good molecular probe to determine the UQ-binding site, the sequence of UQ-receptor protein, such as NADH-UQ oxidoreductase, will be easily revealed.   In this sense, in a previous paper[17], we described on the structural requisites of the lipophilic part in two new series of inhibitors classified as 4-hydroxypyridine and 4-hydroxyquinoline derivatives for the inhibition of NADH-UQ oxidoreductase in the respiratory electron transport system of mitochondria and submitochondria (ETP). The manner of this study might be described as a biorational approach where all the objects should be considered with consciousness of the biological parameters of target inhibitors reflecting binding affinity to the receptor, mobility and stability in the living entity.   In this paper we would like to report the effects of other functional groups such as dimethoxy groups in hydroxypyridine system, as complements of the previous paper.

MATERIALS AND METHODS

Chemicals   and   Instruments

Compounds of 2,3-dialkoxy-4-hydroxypyridine systems were synthesized by a modified method[18] of the previous paper[15], and 4-hydroxyquinoline derivatives were obtained by a modification[18] of the classical methods[19, 20], and were purified by column chromatography. All the compounds were identified by instrumental analyses as follows: IR, NMR and mass spectrum with a Shimadzu IR-435, a Bruker AM-400 and a Finnigan INCOS-50 spectrometer, respectively. Biochemicals were purchased from Sigma Chemical Company. Inhibitory activity of compounds was measured by an oxygen electrode system of Rank Brothers.

Isolation of rat liver mitochondria

Suspensions of rat liver mitochondria were prepared in usual way as follows[21]. A male albino rat (5 week old) was decapitated to remove the liver which were immersed in 50 mL of ice-cold (0~4 °C) 0.25 M sucrose-0.1 mM EDTA in a tared beaker. The liver (about 10 g) were rinsed with the buffer, sliced into several pieces with a pair of scissors, and homogenized smoothly by hand with a 20 mL glass-Teflon homogenizer. The homogenate was centrifuged at 900 g for 10 minutes. The supernatant was decanted into a flask and the pellet was homogenized and centrifuged as for the first run. Combined supernatants were recentrifuged at 8,000 g for 20 minutes. The upper layer was discarded and the pellet was suspended to repeat the same centrifugation. Finally the pellet was resuspended by gentle swirling with 0.25 M sucrose-10 mM Tris-0.1 mM EDTA (pH 7.4, 5 mL). The protein content was determined by the modified Lorry method[22].

Isolation of bovine heart submitochondria

ETP (electron transport particles: submitochondria) were obtained from bovine heart mitochondria which was prepared by an established method[23]. The sucrose mitochondrial suspension was adjusted to pH 8.5 at 0~4 °C with 0.1 N KOH and maintained at this pH value for thirty minutes by addition of alkali solution. The suspension was then treated with a 20 mL glass-Teflon homogenizer at 0 °C. The treated suspension was centrifuged at 19,000 g for 7 minutes under cooling conditions (0~4 °C). The supernatant was collected and the pellet was again homogenated by the above procedure. The combined supernatants were centrifuged for 30 minutes at 80,000 g. The final pellet was taken up in 0.25 M sucrose, and the pH was adjusted to 7.5 with acetic acid, and stored in liquid nitrogen.

Bioassay method

Respiratory inhibition of compounds was measured by an oxygen electrode of Clark type at 25 °C in a total volume of 2 ml of a medium (pH 7.4) consisting of a mitochondrial suspension (0.2 mL), the phosphate buffer (1.8 mL), $MgCl_2$ (10 mmoles), ADP (0.5 mmoles) and L-glutamic acid (10 mmoles), or in the presence of a ETP suspension (0.2 mL), the phosphate buffer (1.8 mL), $MgCl_2$ (10 mmoles), cytochrome C (0.03 mmoles), ADP (0.5 mmoles) and NADH (2 mmoles). The inhibitory activity of compounds is expressed as a pI_{50} value, the negative logarithm of inhibitor amount (moles/mg-protein) at 50 % inhibition.

RESULTS AND DISCUSSION

In a previous study[9, 15-17] on the structure-activity correlation of piericidins, effects of their functionality on activity were summarized in terms of the following facts: 1) hydrogenation on all double bonds in the sidechains of piericidins caused remarkable loss of activity in the mitochondria bioassay, but it was clear that these reduced compounds showed fairly good activity on NADH-UQ oxidoreductase; 2) piericidin analogs holding three isoprenoid units (farnesyl) as a sidechain demonstrated high activity as natural piericidins and ω-[{(w/o) alkyl or alkoxy} phenyl] alkyl side chain ones;   3) the inhibitory activity was high when the side chain was in alpha position (A type) and the length of sidechain in piericidin was 13- 14 carbons; however, optimal aralkyl sidechain length of synthetic compounds was 9-11 carbon numbers (length of phenyl ring was counted as 3 carbon numbers);   4) 4-quinolinol derivatives were found as a novel class of inhibitors at the NADH-UQ reductase site;   5) β-methyl group in 4-hydroxypyridine and 4-hydroxyquinoline derivatives was the best group for the inhibitory activity;   6) a free phenolic hydroxy group on the pyridine ring was necessary to maintain the high level of activity.

Based on those results, we intended to get more information about the inhibitory activity of piericidin-like inhibitors synthesized from the structural relation of the native substrates at the binding sites (of ubiquinones and/or menaquinones).   The results of the structure-activity relationships are summarized in Table 1-3.

A) Effect of lipophilic sidechain

As shown in Table 1, the derivatives carrying isoprenoid sidechain on the hydroxypyridine and hydroxyquinoline system (14~16, 24~25) showed higher activity than saturated linear ones (22~23) and especially, A type compounds (14, 24) were 10 times more active than B type ones (15, 25).   These results are consistent with those of previous paper[17].   Among the compounds attached with ω-phenylalkyl group sidechains compounds (17~21, 26~29) and compounds of 19, 27 and 28 showed optimally high level of inhibition.   As mentioned in the previous paper, phenyl group in the sidechain was regarded as a good template to examine sterical and functional limits for the inhibitors at the receptor site of the NADH-UQ oxidoreductase because of its bulkiness and π-electron clouds in the phenyl part as well as the case of photosynthetic inhibitors[24-26]. Also, β-alkyl group effect in the A type compounds was examined exclusively[17] and, here, we confirmed that the methyl group was the

Table 1. Inhibitory Effect of Lipophilic Sidechain in Hydroxypyridine and Hydroxyquinoline System against NADH-UQ
Oxidoreductase of Intact Mitochondria and ETP.

Compound		Sidechain		pI_{50}
No.	Structure	R_1	R_2	ETP
14 A		$-CH_3$	$-CH_2$-Geranyl	11.0
15 B		-Geranyl	$-CH_3$	10.0
16 B	OH	-Farnesyl	$-CH_3$	10.3
17 A	CH_3O ... R_1	$-CH_3$	$-(CH_2)_3$-Ph	9.5(8.4)
18 A		$-CH_3$	$-(CH_2)_4$-Ph	10.6(10.0)
19 A	CH_3O N R_2	$-CH_3$	$-(CH_2)_6$-Ph	10.8(10.4)
20 A		$-CH_3$	$-(CH_2)_8$-Ph	10.4(10.2)
21 A		$-CH_3$	$-(CH_2)_{10}$-Ph	10.2(10.2)
1	Piericidin A			11.4(11.0)
2	Rotenone			10.8(10.8)

Compound		Sidechain		pI_{50}
No.	Structure	R_1	R_2	ETP
22 A		$-CH_3$	$-(CH_2)_9$-H	9.4
23 A		$-CH_3$	$-(CH_2)_3$-H	9.3
24 A		$-CH_3$	$-CH_2$-Geranyl	11.1
25 B		-Geranyl	$-CH_3$	10.1
26 A		$-CH_3$	$-(CH_2)_4$-Ph	10.1
27 A	OH	$-CH_3$	$-(CH_2)_6$-Ph	10.7(8.4)
28 A	R_1	$-CH_3$	$-(CH_2)_8$-Ph	10.7(8.4)
29 A		$-CH_3$	$-(CH_2)_{10}$-Ph	10.5(8.2)
30 A	N R_2	$-CH_3$	$-(CH_2)_2$-Ph-OC_4H_9	10.4(9.2)
31 -		$-C_2H_5$	$-(CH_2)_2$-Ph-OC_4H_9	9.4
32 -		$-C_3H_7$	$-(CH_2)_2$-Ph-OC_4H_9	8.8
33 -		-H	$-(CH_2)_7$-H	8.2
34 -		-H	$-(CH_2)_9$-H	8.6
35 -		-H	$-(CH_2)_{13}$-H	8.2

value in parenthesis: mitochondrial bioassay

Table 2. Relation between Inhibitory Activity and Steric and Electronic Effect of the Sidechain in Hydroxypyridine and Hydroxyquinoline System

Compound No.	Structure	Sidechain R	pI_{50} ETP
36 P		ortho-	10.5
37 P		meta-	10.5
38 P		para- $-(CH_2)_4-$ OCH_3	10.5
39 Q		ortho-	10.2
40 Q		meta-	10.1
41 Q		para-	10.5
18 P		$-(CH_2)_2-CH_2-CH_2-Ph$	10.6
42 P		$-(CH_2)_2-CH(CH_3)-CH_2-Ph$	10.4
26 Q		$-(CH_2)_2-CH_2-CH_2-Ph$	10.1
43 Q		$-(CH_2)_2-CH(CH_3)-CH_2-Ph$	10.3
44 P		$-(CH_2)_2-Ph-OC_6H_9$	9.6
45 P		$-(CH_2)_2-Ph-OPh$	10.5
46 Q		$-(CH_2)_2-Ph-OC_6H_9$	10.4
47 Q		$-(CH_2)_2-Ph-OPh$	10.2

Compound No.	Structure	pI_{50} ETP
48		< 7.5
49		8.2
50		10.0

Table 2. (continued)

No.	Compound Structure	Sidechain R_1	R_2	pI_{50} ETP
51		-H	-H	< 7.5
52		-t-Butyl	-H	< 7.5
53		-H	-$C_{10}H_{21}$	< 7.5
54		-H	-Citronellyl	< 7.5
55				7.6

most effective one in inhibiting the NADH-UQ oxidoreductase by comparing the activity patterns of compounds (**22~23** vs **33~35; 30** vs **31~32**) in 4-hydroxyquinoline system.

Because the active natural piericidins and synthetic ubicidins had 3'-methyl group in the sidechain and the function of the nearest methyl group was regarded important, we examined the steric and electronic effect of the sidechain as shown in Table 2. The effect of geometrical isomeric sidechains in hydroxypyridine derivatives (**36~38**) appeared to be the same but among hydroxyquinoline compounds (**39~41**) the para-isomer (**41**) showed higher activity than meta- and ortho-ones. The effect of 3'-methyl group showed opposite tendency in pyridine and quinoline derivatives (**18** vs **42; 26** vs **43**) as well as in the case of terminal hexyloxy and phenoxy groups (**44** vs **45; 46** vs **47**). In the case of quinolines, the inhibitory effect might be dependent on their lipophilicity. On the other hand, the steric and electronic effect far from the nuclei in the long chain might be more effective than that of near one, because the compounds **18~19, 27~28** and **45** showed high activity. Here we are trying to find out the size of a partial binding niche and we expect that it is possible to get much information by an application of the knowledge of photosynthetic electron transport system, which may evolutionarily relate to the mitochondrial system.

The effect of hydroxy group in the sidechain is shown in the compounds **48~50**. The hydroxy group in the side chain has large effect to inhibition according to its position. 2'-Hydroxy group in compounds of **48** and **49** caused to reduce the activity compared to their precursors, respectively. In the compound **50**, the terminal hydroxy group also has an effect of lowering inhibitory activity, however piericidin A which has a terminal hydroxy group showed higher activity than piericidin B, which has terminal methoxy group instead of hydroxy one[15].

The compounds **51~55** which were synthesized as simple rotenone-type showed almost no activity, even though rotenone is a highly active polycyclic compound. This fact is similar to the effect of β-alkyl group in the A type compounds because the extension of the methyl group into an ethyl and a propyl group reduced the activity gradually (**30** vs **31** vs **32**)[17]. It is considered that the size of binding niche around β-methyl group is not so wide that large alkyl group in β-position interferes with strong binding of compound to NADH-UQ oxidoreductase. In this sense, it may be necessary to design new type of sidechains by the combination of steric factor and double bond character for the highest inhibition on the UQ-receptor.

B) Effect of other functional groups instead of dimethoxy groups of pyridine derivatives

Although the effect of alkoxy groups to the inhibition has not been investigated, it was regarded that methoxy groups were essential part to show inhibitory activity. Compounds which have various alkoxy groups instead of dimethoxy groups (**56~60** in Table 3) showed higher level of inhibition than the dimethoxy derivative **56**. Rotenone and piericidin show the highest inhibition in the mitochondria as well as in ETP, however, inhibition level of compounds **56~60** in the mitochondrial bioassay is 10 times lower than that in ETP. It may indicate that some other steric factors are required in the lipophilic sidechain of piericidin-like inhibitors to penetrate from outer membrane to inner membrane of mitochondria. Quinoline derivatives, which were attached with electron donating methoxy or electron withdrawing nitro group on the nuclei showed dramatically low activity (**24** vs **61**; **62** vs **63**) and reduced quinoline ones (cyclohexenopyridines; **65**, **67** and **69**) showed slightly low activity. It is considered that even hydrogenated geranyl sidechain is fairly fitted to lipophilic part of binding niche regardless of hydrophilic part. As mentioned in the previous paper, quinoline derivatives as a new group of inhibitors showed good activity to NADH-UQ oxidoreductase[17].

According to the inhibitory tendency of alkoxypyridine compounds (**56~60**) and substituted quinoline ones (**61**, **63**), it is clear that the hydrophilic part of the binding niche is small. This is, also, supported by the effect of alkyl group on β-position in the nuclei (**30~32** in Table 1).

Also, the last part of Table 3 shows the activity of compounds of which nucleus is modified. **HQNO** (2-heptyl-4-quinolinol-N-oxide) is known to be an inhibitor of RET system[27, 28]. Therefore, in the

Table 3. Inhibitory Effect of Modified Nuclei Instead of Dimethoxyhydroxypyridine System against NADH-UQ Oxidoreductase of Intact Mitochondria and ETP.

Compound No.	R_1	R_1	pI_{50} Mitochondria	ETP
56	$-OCH_3$	$-OCH_3$	9.6	10.6
57	$-OCH_3$	$-OC_2H_5$	10.0	11.1
58	$-OCH_3$	$-OC_3H_7$	10.0	10.9
59	$-OC_2H_5$	$-OCH_3$	10.0	10.8
60	$-OC_2H_5$	$-OC_2H_5$	10.0	10.9

Compound No., Type		B*	X	R_1	R_1	pI_{50} ETP
24	A	aro	-H	$-CH_3$	$-CH_2$-Geranyl	11.1
61	A	aro	$-OCH_3$	$-CH_3$	$-CH_2$-Geranyl	10.0
62	B	aro	-H	-Farnesyl	$-CH_3$	10.2
63	B	aro	$-NO_2$	-Farnesyl	$-CH_3$	<7.5
64	A	aro	-H	$-CH_3$	$-(CH_2)_6$-Ph	10.7
65	A	sat	-H	$-CH_3$	$-(CH_2)_6$-Ph	10.0
66	B	aro	-H	-4H-Geranyl	$-CH_3$	10.2
67	B	sat	-H	-4H-Geranyl	$-CH_3$	10.0
68	B	aro	-H	-4H-Farnesyl	$-CH_3$	10.0
69	B	sat	-H	-4H-Farnesyl	$-CH_3$	10.2

*; aro: aromatic,   sat: saturated

Table 3. (continued)

Compound No.	Structure	PI$_{50}$ ETP
70		9.5
HQNO		10.1
71		9.8
72		8.5
73		8.3

beginning, it was predicted that N-oxide compounds should show more potent activity than their precursors. One typical example of compound **70** showed lower inhibitory activity than its precursor. In the previous paper a piericidin of which phenolic hydroxy group was protected by acyl group showed lowered activity[15]. As shown, the activity of compound **72** and **73** whose hydroxy groups are protected with carboethoxy, was decreased dramatically. However, because the inhibitory effect depends on the characteristics of the functional groups, this result can be understood by considering hydrophilic binding sites of these piericidin-like compounds. Therefore, further investigation is necessary to elucidate the clear effect of substituents on the nucleus for the potent inhibition to NADH-UQ oxidoreductase.

In addition, we also observed a gap of inhibitory effect between mitochondrial and ETP bioassay (**17~21, 27~30** in Table 1 and **56~60** in Table 3), which was explained by the barrier effect of outer membrane in the previous paper[17]. Similar effect was observed between chloroplast and thylakoid bioassay of inhibitors in the photosynthetic electron transport system[29].

Although a bound form of ubiquinone in the photoreaction center of photosynthetic bacteria has been revealed well by studies using X-ray crystallography[30-33], there is in fact very little information on the binding site of ubiquinone in NADH-UQ oxidoreductase. The various structural requisites for effective inhibition may be more easily explained by allowing the introduction of bacterial knowledge into a model of the ubiquinone binding domain of NADH-UQ oxidoreductase, considering evolutionary relationships of photosynthetic and mitochondrial system[34] (Fig. 2). The size of the niche is presumably limited by the length of the sequential polypeptide which may involve various peptide units which bind with a particular part of the inhibitor, as in the case of the photoreaction center[35]. In this model, piericidin A1 and rotenone suggest a suitable basic frame for the lipophilic part of the inhibitor (A and B in Figure 2), and a proper fit of inhibitors is assumed to depend on the flexibility of the sidechain. In this sense, this is a new approach to the study of binding at the NADH-UQ oxidoreductase site, although it is recognized that topological analyses employing such lipophilic sidechains of piericidin-like inhibitors will require much more variation of those structures.

The structure-activity relationships of piericidin-like compounds as inhibitors on the ubiquinone-NADH oxidoreductase region can be summarized as follow: 1) the inhibitory activity is high when the side chain in C-2 position (A type) and optimal length of side chain is 9-11 carbon numbers (length of phenyl ring was counted as 3 carbon numbers), 2) C-3 methyl group in pyridines and quinolines is the best group for the activity and the activity decreases in the order of methyl> ethyl> propyl, 3) hydroxy group in the side chain diminishes the inhibitory activity, 4) combination of methoxy, ethoxy and propoxy groups instead of dimethoxy ones in the pyridine derivatives showed almost same activity as the dimethoxy compound, 5) 6-methoxy and nitro groups in the quinolinols decreased the activity dramatically, 6) Structures of free hydroxy group and nitrogen on the nucleus is essential to high inhibitory activity.

Synthetic piericidin-like compounds with an aralkyl sidechain were shown to inhibit NADH-UQ oxidoreductase, however they were less effective against the intact mitochondria than against the ETP (submitochondria). This result suggests that structure-activity studies on inhibitors of NADH-UQ oxidoreductase should be directed at ETP assays in order to avoid confusion about sidechain effects. That is to say, one factor for inhibition, such as affinity with the binding site, might be confused with another like permeability at the outer membrane.

Inhibition of NADH-UQ oxidoreductase due to piericidin-like compounds was mainly affected by the length of the lipophilic part, regardless of bulkiness (location of a phenyl group and methyl one) in

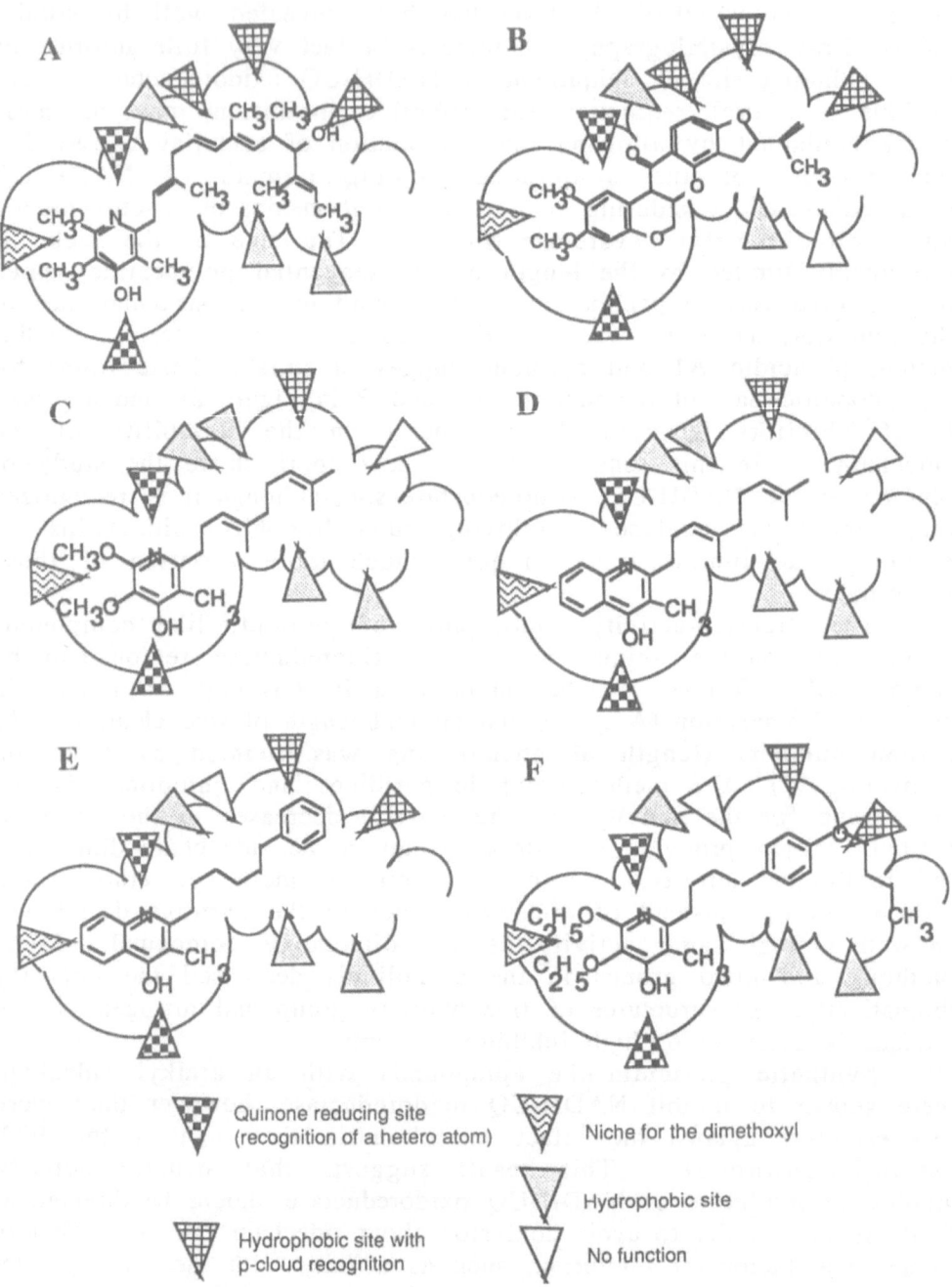

Figure 2. Hypothetical Binding Manners of Piercidin-like Inhibitors at the Ubiquinone Binding Site of NADH-UQ Oxidoreductase Site

the sidechain. The vicinal dimethoxy functionality of the 4-hydroxy-pyridine ring system is replaceable with a fused benzene to form the 4-hydroxyquinolines as a novel class of inhibitors at the site but substituted benzenoid diminished the activity. The methyl group was demonstrated to be the optimal functionality in that area of the nuclei of the pyridine derivatives so that either deletion or insertion of a methylene on the group eliminated its activity[15, 17]. A methyl group on the nuclei of inhibitors belonging to the class of 4-hydroxyquinoline was also shown to be a functionality which optimized activity, but the sterical limit of the binding niche around the group should be more tolerant than in the case of the pyridine derivatives. In addition, even cyclohexeno-4-pyridinol (reduced quinolinol) derivatives showed fairly high inhibition to NADH-UQ oxidoreductase.

The various structures for effective inhibition was more easily explained by the introduction of photosynthetic bacterial knowledge into a model of the ubiquinone binding domain of NADH-UQ oxidoreductase, because the structure of photoreaction center of photosynthetic bacteria, which may be evolutionarily related to the mitochondrial system, has been revealed well. In this way, a proper fit of a inhibitor to the binding niche is assumed to depend on the flexibility of the sidechain. This may be a new approach to the study of binding at the NADH-UQ oxidoreductase site, although much topological analyses of inhibitors would be required by the variation of those structures.

The finding of a large variation for synthetic inhibitors of NADH-UQ oxidoreductase is valuable for further investigations on the mitochondrial respiration system because these compounds are readily modified to introduce a functionality to probe the binding site; for example a radio-isotope, a photo-affinity label and/or a chiral center etc. It will also be very interesting to confirm the hydrophilic part and sidechain effects in other classes of inhibitors of NADH-UQ oxidoreductase such as benzimidazoles, myxalamid and amytal.

ACKNOWLEDGEMENTS

We thank Yasuaki Hariya of Yashima Chemical Industry Ltd. for skilled assistance in biological assays. This research was supported in part by Grants in Aid of the Ministry of Education, Science and Culture, Japan and in part by the Korean Ministry of Science & Technology. We are indebted to the Japan Society for the Promotion of Science and the Korea Science and Engineering Foundation for international collaborative funding.

REFERENCES

1. Hall, C., Wu, M., Crane, F. L., Takahashi, N., Tamura, S. and Folkers, K. (1966) Biochem. Biophys. Res. Commun., **25**, 373-377.
2. Fukami, H. and Nakajima, M. in Naturally Occurring Insecticides (ed. by Jacobson, M. and Crosby, D. G. Marcel), p. 71, Dekker, New York, 1971.
3. Gerth, K., Jansen, R., Reifenstahl, G., Hofle, G., Irschik, H., Kunze, B., Reichenbach, H. and Thierbach, G. (1983) J. Antibiotics, **36**, 1150.
4. Thierback, G., Kunze, B., Reichenbach, H. and Hofle, G. (1984) Biochem. Biophys. Acta, **765**, 227.
5. Slater, E. C. (1973) Biochem. Biophys. Acta, **301**, 105.
6. Nakagawa, Y., Kuwano, E., Eto, M. and Fujita, T. (1985) Agric. Biol. Chem., **49**, 3569.
7. Horgan, D. J. and Singer, T. P. (1968) J. Biol. Chem., **243**, 834-843.
8. Schewe, T., Heibsch, C. and Halangk, W. (1975) Acta. Biol. Med. Ger., **34**, 1767.
9. Ramsey, R., Ackrell, B., Coles, C., Singer, T., White, G. and Thorn, G. (1981) Proc. Natl. Acad. Sci. USA, **78**, 825.
10. Jeng, M., Hall, C., Crane, F. L., Takahashi, N., Tamura, S. and Folkers, K. (1968) Biochemistry, **7**, 1311.
11. Singer, T. P. and Gutman, M. (1971) Adv. Enzymol., **34**, 79.
12. Gutman, M., Singer, T. P. and Casida, J. E. (1979) J. Biol. Chem., **245**, 1992.
13. Mitsui, T., Fukami, J., Fukunaga, K., Sagawa, T., Takahashi, N. and Tamura, S. (1969) Botyu-Kagaku, **34**, 126-134.
14. Yoshida, S. and Takahashi, N. (1978) Heterocycles, **10**, 425.
15. Yoshida, S., Nagao, Y., Watanabe, A. and Takahashi, N. (1980) Agric. Biol. Chem., **44**, 2921-2924.
16. Gutman, M. and Kliatchko, S. (1976) FEBS Lett., **67**, 348-353.
17. Chung, K. H., Cho, K. Y., Asami, Y., Takahashi, N. and Yoshida, S. (1989) Z. Naturforsch., **44c**, 609.
18. Chung, K. H., Cho, K. Y., Asami, Y., Takahashi, N. and Yoshida, S. (1989) Agric. Biol. Chem., submitted.
19. Leonard, N. J., Herbrandson, H. F. and Heyningen, E. M. V. (1946) J. Am. Chem. Soc., **68**, 1279-1281.
20. Price, C. C. and Jackson, W. G. (1946) J. Am. Chem. Soc., **68**, 1282.
21. Fleischer, S., McIntyre, J. O. and Vidal, J. C., in Methods in Enzymology (eds by Fleischer, S. and Packer, L.), Vol. **55**, pp. 32-39, Academic Press, New York, San Francisco, London 1979.
22. Miller, G. L. (1959) Anal. Chem., 31, 964.
23. Crane, F. L., Glenn, J. L. and Green, D. E. (1956) Biochem. Biophys. Acta, **22**, 475-487.
24. Huppatz, J. L. and Phillips, J. N. (1987) Z. Naturforsch., **42c**, 679.
25. Kirino, O. and Takayama, C. (1987) J. Syn. Org. Chem. Japan, **45**, 1107-1118.

26. Huppatz, J. L. and Phillips, J. N. (1987) Z. Naturforsh., **42c**, 674.

27. Brandon, J. R., Brocklehurst, J. R. and Lee, C. P. (1972) Biochemistry, **11**, 1150-1154.

28. Lightbown, J. W. and Jackson, F. L. (1956) Biochem. J., **63**, 130.

29. Hauska, G., in Photosynthesis I, Encyclopedia of Plant Physiology, New series (eds by Trebst, A. and Avron, M.), Vol **5**, pp. 253-265, Springer-Verlag, Berlin, Heiderberg, New York, 1977; cf. Asami, T., Takahashi, N. and Yoshida, S. (1986) Z. Naturforsch., **41c**, 751-757.

30. Deisenhofer, J., Epp, O., Miki, K., Huber, R. and Michel, H. (1985) Nature, **318**, 618-623.

31. Allen, J. P., Feher, G., Yeates, T. O., Komiya, H. and Rees, D. C. (1987) Proc. Natl. Acad. Sci. USA, **84**, 5730-5734.

32. Allen, J. P., Feher, G., Yeates, T. O., Komiya, H. and Rees, D. C. (1987) Proc. Natl. Acad. Sci. USA, **84**, 6162-6166.

33. Yeates, T. O., Komiya, H., Rees, D. C., Allen, J. P. and Feher, G. (1987) Proc. Natl. Acad. Sci. USA, **84**, 6438-6442.

34. Alberts, B., Bray, D., Lewis, J., Raff, M., Roberts, K. and Watson, J. D., in Molecular Biology of the Cell, Garland Publisher Inc., New York, 1983.

35. Michel, H., Epp, O. and Deisenhofer, J.(1986) EMBO J. **5**, 2445.

EFFECTS OF REDOX STATE FOR UBIQUINONE-10 ON THE

MEMBRANE FLUIDITY OF CARDIOLIPIN-CONTAINING LIPOSOMES

Bo-ji Cheng, Zhen-hong Yang*, Yu-zhi Dong,
Kun-rong Cheng and Ke-chun Lin

Dept. of Biophysics, Beijing Medical University
* National Research Laboratories of Natural and Biomimetic
Drugs, Beijing Medical University, Beijing, China

SUMMARY

The effect of redox state for ubiquinone-10 on the membrane fluidity of cardiolipin-containing liposomes has been studied with spin labels 5- and 16-doxyl stearate derivatives. Incorporation of ubiquinone-10 or its reduced form to the membrane significantly increases the membrane fluidity at the depth near membrane surface. The reduced form has more significant effect than oxidized form has. In the core of lipid bilayer, however, rotational correlation time of spin label 16-NS increases upon incorporation of ubiquinone-10 or its reduced form. After binding of adriamycin to cardiolipin, the fluidizing effect of ubiquinone-10 is abolished. It is suggested that cardiolipin plays important role in determination of dynamic characteristics of the membrane via its interaction with ubiquinone-10.

INTRODUCTION

Ubiquinone-10 is a mobile electron carrier which is one of the components for respiratory chain. Chance pointed out earlier that the motion of carrier was important for electron transfer along the chain[1]. Inner mitochondrial membrane is a fluid membrane which provides an environment suitble for the motion of respiratory chain components[2,3]. However, the mechanism for regulation of components' motion is still not understood completely.

In model systems and natural membranes, the physico-chemical properties of ubiquinone-10 and its interaction with some kinds of phospholipid have been investigated with different approaches[4-9]. However, study of the interaction between cardiolipin and ubiquinone-10 is still rare. Cardiolipin, a phospholipid specific to mitochondrial inner membrane, is present in the range of about 20% of total lipids for this membrane[10], which makes this membrane more fluid[8,11]. In order to understand the mechanism to regulate the motion of ubiquinone-10 or other components in the membrane, it is necessary to characterize the change in dynamic property induced by ubiquinone-10 or others on cardiolipin-containing membrane. It is well known that some of lipophilic compounds can induce the perturbation on the lipid bilayer, for example the change in membrane fluidiy. We report here the effect of redox state for ubiquinone-10 on the cardiolipin-containing membrane fluidity and the

influence of adriamycin on the effect induced by ubiquinone-10, studied
by the use of two spin labels 5-and 16-doxyl stearate derivatives, which
probe the surface and the core of the membrane, respectively.

EXPERIMENTAL PROCEDURES

Preparation of cardiolipin-containing liposome membrane

Cardiolipin and dipalmitoylphosphatidylcholine (DPPC) were purchased
from Sigma Co.(USA) and used without further purification. Cardiolipin and
DPPC (1:1 wt/wt) were mixed and dissolved in absolute ethanol. After
ethanol had been evaporated by using rotatory evaporator, phospholipids
were hydrated with Tris-HCl buffer (20 mM Tris, 1 mM EDTA, pH 7.4).
Cardiolipin-containing liposomes were then prepared by ultrasonic
homogenizer model us-150 (Japan). The phospholipids concentration for the
liposomes was 4 mg/ml.

Incorporation of ubiquinone-10 into the membrane

Ubiquinone-10 was supplied by Dr. Kozo Utsumi and Eisai Co. (Japan).
Absolute ethanol was used to dissolve ubiquinone-10. Reduced ubiquinone-10
was prepared by sodium dithionite according to the method of Rieske[12].
The reduced form of ubiquinone-10 was extracted by cyclohexane and the
solvent was evaporated. After that, the ubiquinone-10 ethanolic solution
was made at concentration of 5.8 mM.   With Beckman DU-7 ubiquinone-10
reduction was monitored, verifying that the sample contained totally
reduced ubiquinone-10.

Incorporation of both ubiquinone-10 forms into the membrane was
achieved according to[13]. $5\mu l$ of 5.8 mM ethanolic solution of
ubiquinone-10 or its reduced form was added to the liposomes suspension
and the mixture was vigorously stirred with a vortex mixter for 5 minutes.

ESR study

Spin labels used in the experiment were Sigma production. $5\mu l$ of 6.5
mM ethanolic solution of 5- or 16- (N-oxyl-4',4'-dimethyl-oxazolidine)
derivatives of stearic acid (5- or 16- NS) was added to the liposome
membrane. Into which both ubiquinone-10 had been incorporated by vigorous
stir for 5 minutes. The molar ratio of phospholipid to spin label was
about 80:1.

Electron Spin Resonance (ESR) spectra were recorded with ESP 300 ESR
spectrometer (Bruker). From the spectra recorded for 5-NS, the distances
of the outer and inner extremes of hyperfine splitting (denoted by 2A∥
and 2A⊥) were measured, order parameter (S) was then calculated in terms
of equation[14].

$$S=0.568*(A_{\parallel} -A_{\perp})/a' \qquad a'=(A_{\parallel} +2A_{\perp})/3$$

The freedom of 16-NS in the membrane was calculated as rotational
correlation time (τc) in terms of equation[15]

$$\tau c=6.5*1010*\omega o[(ho/h-1)-1]$$

where ωo is the linewidth of medium field line, ho and h-1 are the heights
of the medium and high field lines, respectively.

RESULTS

In order to investigate the effects of both forms of ubiquinone-10 on the membrane fluidity of cardiolipin-containing liposomes, spin labels 5- and 16-NS were used to measure the fluidity at different depth in the lipid bilayer. For comparison under the same condition, 5 μl of absolute ethanol

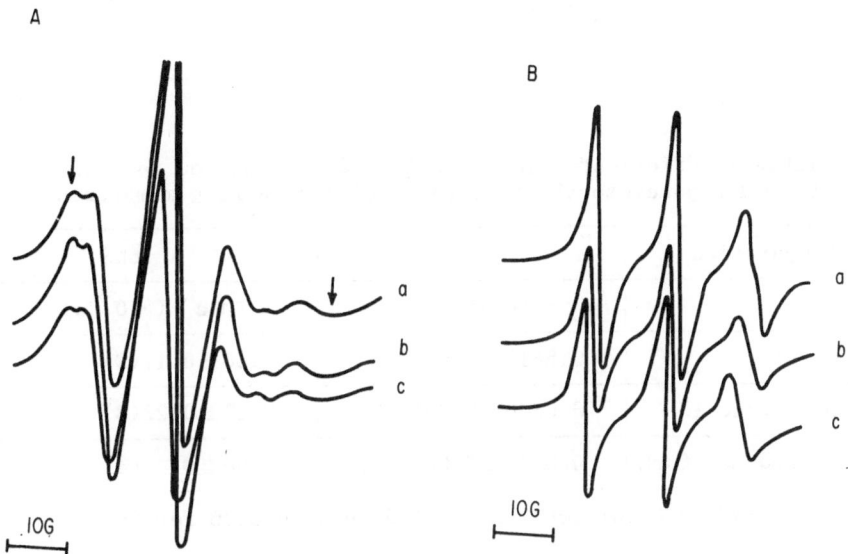

Fig.1. ESR spectra of cardiolipin-containing membranes
labeled with 5-NS (A) and 16-NS (B)
a.control, b.+ubiquinone-10, c.+its reduced form.

was added to control membrane. Fig.1 shows three representative ESR spectra of the membranes with ubiquinone-10 and its reduced form as well as control membrane.

The ESR spectra for 5-NS clearly exhibit some ordering effect by the spin label on the membrane. From the spectra recorded, the outer and inner extremes of hyperfine splitting were measured and order parameters (S) were calculated. The results obtained are listed in Table 1. Either ubiquinone-10 or its reduced form can induce significant decrease of order parameter, indicating that both forms of ubiquinone-10 can fluidize the cardiolipin-containing membrane at the depth near membrane surface. It is found that the reduced form of ubiquinone-10 can induce more obvious fluidizing effect on the membrane in comparison with the oxidized form, i.e. ubiquinone-10 itself. This result suggests that the interaction of the reduced form of ubiquinone-10 with phospholipids is different from that of the oxidized form.

   Spin label 16-NS is located in an almost disordered envirnoment, thus
its ESR spectrum shows an isotropic feature (Fig. 1B). From the linewidth
of the medium field line (ωo) and the height of medium field (ho) and the
high field line (h-1), the rotational correlation time (τc) can be
calculated (see Table 1). The value of τc for the control membrane is the
smallest among the three membranes, indicating that freedom of motion for
the end of acyl chain of lipid is not restricted in the core of lipid
bilayer compared with those membranes incorporated by both form of
ubiquinone-10. It is interesting that the restricted degree for the motion
induced by the reduced form is less than that induced by ubiquinone-10
itself.

Table 1. Effect of ubiquinone-10 and its reduced form on
the order parameter(S) of cardiolipin-containing membranes

Preparation	5-NS	16NS
	S (mean $\pm$ Sx)	τc (*10^{-10} s)
1. control	0.5617 ± 0.0082(5)	7.60 ± 0.17(3)
2. +ubiquinone-10	0.5340 ± 0.0017(3)	8.47 ± 0.23(3)
3. +its reduced form	0.5245 ± 0.0061(4)	9.00 ± 0 (4)

* For 5-NS, $P<0.001$ for 1-2 and 1-3, $0.02<P<0.05$ for 2-3.

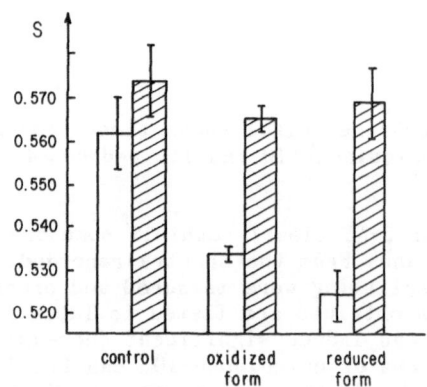

Fig.2. Effects of ubiquinone-10 on the order parameter (S)
of cardiolipin-containing membrane before (□)
and after (▨) adriamycin binding.

    Adriamycin, an antitumor drug, can bind to cadiolipin. This drug was
used to reveal the role of cardiolipin in the fluidizing effects induced by
ubiquinone-10 and its reduced form on the membrane. Prior to incorporation
of both forms of ubiquinone-10 and spin label 5-NS, 2μl of 10mM adriamycin
was added to the liposome suspension and mixed well. After binding of
adriamycin to cardiolipin, the order parameters for membranes labeled with
5-NS become increased in comparison with those membranes without adriamycin
binding (Fig. 2). This result suggests that adriamycin makes the membrane
more order due to its binding to cardiolipin. Moreover, this also indicates
that the fluidizing effect induced by both forms of ubiquinone-10 result
from their interaction with cardiolipin.

DISCUSSION

    It is well known that some lipophilic compounds could be inserted
into lipid bilayer to alter the membrane fluidity, for example, cholesterol
and vitamins[14,18,19]. Ubiquinone-10 is a lipidic component of the
respiratory chain in the inner mitochondrial membrane, and a mobile
component between complexes. That this compound enhances membrane fluidity
in both model system and mitochondrial membrane has been reported by Lenaz
and his group[8,13,17]. Our results show that ubiquinone-10 and its reduced
form behave in opposite way at different regions of cardiolipin-containing
membrane. At the depth near membrane surface both forms of this compound
increase the membrane fluidity, decrease it in the core of membrane. To
explain the difference mentioned above depends on the knowledge on the
location of ubiquinone-10 within membrane. Crane suggested that this
compound was intecalated between the two monolayers[20]. But it seems
unlikely because in this case the ordering of lipid bilayer near surface
would not be affected. Based on the present study, it may be implied that
the quinone ring is near the membrane surface and the hydrophobic tail is
accommomdated in the core. Recently, the concept that ubiquinone-10 forms
a seperate phase in mixed lipid-ubiquinone-10 dispersion has been proposed
by many laboratories[21,22,23]. Chatelier and Sawyer proposed a two-site
model for the location of ubiquinone-10, one of the sites was in the
interior of the membrane and the other near the surface[24].

    As an electron carrier, ubiquinone-10 undergoes the transition of
redox state during electron transfer. It is found that the reduced form of
this compound fluidizes the membrane more obviously near the surface, and
restricts the motion freedom of the end of acyl chain to a smaller extent
than ubiquinone-10 itself does. One of the differences between the two
forms is the increased polarity for the reduced form[8]. Therefore, one
possibility is that at the surface the reduced form experiences exclusion
to disorder the lipid bilayer. With inner mitochondrial membrane, we
reported that the membrane fluidity was affected by the redox state of the
compounds in the respiratory chain[25]. In the present study, it is also
proved in model system. The fluidizing effect induced by both forms of
ubiquinone-10 near the membrane surface is abolished after adriamycin
binding to cardiolipin. Cardiolipin itself is one of factors to fluidize
the membrane[11]. Moreover, the complexes of respiratory chain including
complex I, III, IV require cardiolipin for their activity(for a review see
Ref.10). It is suggested that cardiolipin may play an important role in
determination of the dynamic characteristics of the membrane. According to
the random collision model proposed by Hackenbrock et al[26,27], the

present study throws light on the mechanism to regulate the lateral diffusion and collision of components in the inner mitochondrial membrane. It is
also suggested that the interaction of redox state of components with phospholipids play a role to regulate their motion.

ACKNOWLEDGEMENT

        This work were supported by the National Natural Science Foundation
of China and the Grant from National Research Laboratories of Natural and
Biomimetic Drugs. We thank Dr. Kozo Utsumi and Eisai Co. for supplyment of
ubiquinone-10.

REFERENCE

 1. Chance, B. (1974) Ann. N. Y. Acad. Sci. 277, 613-625
 2. Vanderkooi, J. M. and Chance, B. (1972) FEBS Lett. 22, 23-26
 3. Hochli, M. and Hackenbrock, C. R. (1976) Proc. Natl. Acad. Sci. USA 73,
    1636-1640
 4. Chance, B., Erecinska, M. and Radda, G. (1975)   Eur. J. Biochem. 54,
    521-529
 5. Degli Esposti, M., Ferri, E. and Lenaz, G. (1981) Ital. J. Biochem. 30,
    437-452
 6. Kingsley, P. B. and Feigenson, G. W. (1981) Biochim. Biophys. Acta 635,
    602-618
 7. Katsikas, H. and Quinn, P. J. (1982) Biochim. Biophys. Acta 689, 363-369
 8. Fato, R., Bertoli, E., Parenti, Castelli, G. and Lenaz, G (1985) FEBS
    Lett. 172, 6-10
 9. Cornell, B. A.,   Keniry, M. A.,   Robertson, R. N.,   Weir, L. E.   and
    Westerman, P. W. (1987) Biochemistry 26, 7702-7707
10. Daum,G.(1985) Biochim. Biophys. Acta 822, 1-42
11. Stuhrie-Sekalek, L. and Stanacev, N. E. (1977)   Can. J. Biochem. 55,
    186-204
12. Rieske, J. S. (1976) Biochim. Biophys. Acta 456, 195-237
13. Spisni, A., Masotti, L., Lenaz, G., Bertoli, E., Pedulli, G. F. and
    Zannon, C. (1978) Arch. Biochem. Biophys. 190, 454-458
14. Lenaz, G. (1977) in Membrane Protein and Their Interaction with Lipids
    (Capoldi, R. A. ed.) pp. 47-150, Marcel Dekker, New York
15. Yang, F., Guo, B. and Huang, Y. (1983)   Biochim. Biophys. Acta 724,
    104-110
16. Goormaghtigh, E. and Ruysschaert, J. M. (1984) Biochim. Biophys. Acta
    779, 271
17. Spishi, A., Sartor, G., Lenaz, G. and Massotti, L. (1981) Membr.
    Biochem. 4, 149-157
18. Meeks, R. G., Zaharevitz, D. and Chen, R. F. (1981) Arch. Biochem.
    Biophys. 207, 141-147
19. Quinn, P. J. (1981) Prog. Biophys. Molec. Biol. 38, 1-104
20. Crane, F. L. (1977) Ann. Rev. Biochem. 46, 439-469
21. Katsikas, H. and Quinn, P. J. (1981) FEBS Lett. 133, 230-234
22. Stidham, M.A.,MacIntosh, T. J. and Siedow, J. N.(1984) Biochim.
    Biophys. Acta 767, 423-431
23. Fato, R., Battino, M., Degli Esposti, M., Parenti Castelli, G. and
    Lenaz, G. (1986) Biochemistry 25, 3378-3390
24. Chatelier. R. C. and Sawyer, W. H. (1985) Eur. Biochem. J. 11, 179-185
25. Cheng, B. J., Feng, Y. Y. and Lin, K. C. (1986) Chinese Biochem. J. 2,
    37-43 (Eng. Abstr.)
26. Hackenbrock, C. R. (1981) Trens Biochem. Sci. 6, 151-154
27. Hackenbrock, C. R., Chazatte, B. and Gupte, S. S. (1986) J. Bioenerg.
    Biomembr. 18, 331-368

FEATURES OF ASSEMBLY AND MECHANISM OF YEAST MITOCHONDRIAL

UBIQUINOL:CYTOCHROME C OXIDOREDUCTASE

J.A. Berden, P.J. Schoppink, W. Hemrika and P. Nieboer

E.C. Slater Institute for Biochemical Research
University of Amsterdam, Plantage Muidergracht 12
1018 TV Amsterdam, The Netherlands

SUMMARY

Several deletion mutants of S. cerevisiae, deficient in one of the subunits of the yeast mitochondrial ubiquinol: cytochrome c oxidoreductase, contain subnormal amounts of various subunits, due to proteolytic breakdown as consequence of improper assembly. Only deletion of the 17 kDa subunit or the Fe-S protein does not have any significant effect on assembly. It is concluded that the complex is assembled from various subcomplexes of which the subcomplex consisting of cytochrome b, the 11 kDa subunit and the 14 kDa subunit is sensitive to proteolytic breakdown in the absence of any of the two core proteins. The Fe-S protein has to be integrated into this subcomplex to become proteinase-resistant. Cytochrome b is absent in the 11 and 14 kDa^0 mutants, and present at very low amounts in the 40 and 44 kDa^0 mutants.

Using digitonin for selective disruption of the outer and both outer and inner membrane of the yeast mitochondria the topology of the various subunits is investigated. It is concluded that the 11 kDa subunit can be digested from the intermembrane space, while the 14 kDa can be disrupted only from the matrix. The core proteins, especially the 40 kDa subunit, protect this subunit from proteolytic breakdown.

Studies with embellin, fluorescamin and dicyclo-hexylcarbodiimide show that one of the two antimycin-binding sites in the dimeric complex can be occupied without any effect on the rate of electron transfer from exogenous quinol to cytochrome c, while the reduction of cytochrome b by exogenous quinol via centre i is fully blocked. It is concluded that in steady-state electron transfer the reaction between cytochrome b and exogenous quinone (at centre i) is not involved. The relation of these phenomena with the known effect of e.g. dicyclohexylcarbodiimide on the proton translocation is briefly discussed.

INTRODUCTION

The research on ubiquinol:cytochrome c oxidoreductase has

a long-standing and at the same time an out-standing history.
For most studies on the mitochondrial enzyme bovine heart was
the source of the enzyme, both as part of the mitochondrial
inner membrane and as isolated enzyme. A good second was
Neurospora crassa. The studies on the mitochondrial ubiquinol:
cytochrome   c oxidoreductase have been very fruitful, and
both on the mechanism of electron transfer and the topology
of most of the subunits a great deal of knowledge has been
obtained (see e.g. refs 1-4). The molecular details of the
enzyme structure, however, and the mechanism of proton
translocation remain in the dark, although many proposals for
the mechanism of proton translocation, combined with or
contrary to the Q-cycle mechanism of electron transfer, have
been proposed.

Some years ago we decided to search for the role of the
subunits without prosthetic group in assembly and function of
the complex and yeast seemed to be the most suitable organism
for this type of studies, due to the possibilities of a
molecular genetic approach and its capability to grow in the
absence of a functional respiratory chain. The first step was
the construction of deletion mutants and the results of this
approach are summarized in this contribution. To construct,
on the basis of the obtained data, a suitable model for
assembly we also studied the topology of the various subunits
of the enzyme, part of which could be anticipated from the
analogy with the enzyme from bovine heart and Neurospora
crassa. On the role of the various subunits in the
functioning of the enzyme, relevant data were only obtained
for the 17 kDa protein. Further detailed knowledge will have
to come from more subtle mutations.

Accidentally the yeast enzyme appeared to be very suitable
for the study of the overall functioning of the enzyme due to
the differentiation between the two antimycin-binding sites
in the dimeric complex which was revealed during the study of
the inhibition of the reduction of cytochrome b in the
presence of an inhibitor of centre o. Occupation of one of
the two sites in the dimer appeared not to influence the rate
of steady-state electron transfer, although the reduction of
cytochrome b via centre i was fully inhibited.

METHODS

Preparations

Commercially grown yeast cells, kindly supplied by Gist-
Brocades b.v. (Delft, The Netherlands), were broken
mechanically with a Dynomill type KD-L (5) and mitochondria
were isolated by differential centrifugation. Submitochondrial particles were prepared by sonication (6) and used to
study the effect of inhibitors of Complex III. Complex II-III
was isolated essentially as described (7) except that the 60%
ammonium sulphate pellet containing complex II-III was
resuspended in 100 mM potassium phosphate pH 7.4, 1 mM EDTA
and 0.5 mM PMSF to which 0.5% (w/v) lauryl maltoside had been
added. After dialysis overnight at 4° C against the same
buffer containing 0.015% (w/v) lauryl maltoside, the

preparation was concentrated using a Diaflo filter size PM
30. As the last step FPLC chromatography on a Superose 6
size-exclusion column (Pharmacia) was used to separate
complex III. The elution buffer employed contained 100 mM
potassium phosphate pH 7.4, 1 mM EDTRA, 0.5 mM PMSF and 0.1%
(w/v) lauryl maltoside. The complex III preparation was
finally concentrated with a Diaflo filter size PM 30 and
stored in liquid nitrogen.

Strains and media

Eschericia coli JM 101 and JM 107 (8) or EC 490 (9) were
used to propagate recombinant DNA constructs. E. coli
transformants were grown in 2 YT medium in the presence or
absence of 50 µg/ml ampicillin.
Saccharomyces cerevisiae strain HR2 (a, his 4, leu 2, trp 1)
was used to obtain the mutants lacking the 11 kDa (subunit
VIII) gene (10), the 14 kDa (subunit VII) gene (11), the 17
kDa (subunit VI) gene (12) or the 40 kDa (subunit II) gene
(13). Mutants lacking the 44 kDa (subunit I) gene or the 24
kDa Rieske Fe-S protein (subunit V) gene were provided by Dr.
A. Tzagoloff, Columbia University, New York and Dr. B.L.
Trumpower, Dartmouth Medical School, Hanover, respectively.
Transformation of yeast was carried out as described (14).
Transformants were seleted on minimal media containing 0.67%
(w/v) yeast nitrogen base, 2% (w/v) glucose, 2% (w/v) agar
and supplemented with the appropriate amino acids (0.02
mg/ml).

Northern blot analysis

Isolation of RNA was done essentially as described (15).
RNA was resuspended in 0.1 x SSC and, after addition of an
equal volume of layermix (7 M urea, 0.1 x SSC, 10% (w/v)
glucose, 0.05% (w/v) orange G) was size-fractionated on a
1.25% (w/v) agarose gel in Tris/borate/EDTA and capillary
blotted to Hybond-N (Amersham) using 10 x SSC as buffer.
Further treatment was as recommended by the manufacturer.

Cell lysates, isolation of mitochondria and immunoblotting

Cell lysates and mitochondria were prepared as described
(12,16), using zymolyase to prepare spheroplasts. The
mitochondria were used for functional studies as well as for
topology experiments and immunoblotting. Protein
concentrations were determined by the Lowry method (17).
Total cell and mitochondrial proteins were separated on SDS-
polyacrylamide slab gels (18). The protein transfer to
nitrocellulose and the immunoreactions were carried out
essentially as desribed (19) except that visualization of the
antigen-antibody complex was mostly performed using the
horseradish peroxidase colour assay as an alternative to [125]I-
labelled protein A (20).

Digitonin/Proteinase K treatment

Mitochondria were diluted to 5 mg/ml im 0.6 M sorbitol, 25
mM potassium phosphate pH 7.4 and incubated for 1 minute in
the absence of digitonin, with a concentration of digitonin
able to disrupt the outer membrane or with a concentration of
digitonin able to disrupt both the outer and inner membrane.
After incubation the samples were diluted 6-8 fold with the

same buffer and control samples were taken to measure cytochrome b_2 and fumarase activities. The rest of each sample
was divided into three equal fractions: one fraction was used
as a control without proteinase K, one as a control for the
proteinase K activity, containing both proteinase K (0.25
mg/ml) and 0.5% Triton X-100 and to the third fraction only
proteinase K was added. After 30 minutes incubation at 22°C
2 mM PMSF was added to stop protease activity and protein was
precipitated with 5% TCA. The pellets were dissolved in
Laemli sample buffer and used for electrophoresis and
immunoblotting.

Miscellaneous

Published procedures were employed for manipulation of DNA
(21), the assys of QH_2:cytochrome c oxidoreductase and
cytochrome oxidase activities (12,16) and spectral analysis
(12). Restriction and other enzymes used in DNA manipulation
were purchased from Boehringer, Biolabs and Sigma and used as
recommended by the manufacturers. Radioactive chemicals were
obtained from New England Nuclear and Amersham. $Q_{2H}2$ was
synthesized in our laboratory by A.F. Hartog. Embelline was
purchased from ICN, dicyclohexylcarbodiimide and florescamine
from Sigma and antimycin A and myxothiazol were obtained from
Boehringer. All other chemicals used were of the highest
purity available. The antisera used were raised in rabbits or
mice.

RESULTS AND DISCUSSION

I. ASSEMBLY AND TOPOLOGY OF YEAST COMPLEX III

Construction of deletion mutants

Fig. 1 shows the polyacrylamide gel electrophoresis
pattern of isolated yeast complex III with the nomenclature
of the different bands as used in this paper. The band with
molecular weight of about 20 kDa is as yet unidentified. It
is not impossible that this band represents CPB4, described
by Tzagoloff et al (22) as a 20 kDa protein whose expression
is needed for the synthesis/assembly of Complex III.
The genes for four subunits of the yeast ubiquinol:cyto-
chrome c oxidoreductase have been cloned in our laboratory by
the group of Grivell (23,24). These are the genes for
subunits II (40 kDa), VI (17 kDa), VII (14 kDa) and VIII (11
kDa). Using a one step gene-disruption technique (25) we have
constructed deletion mutants deficient in the expression of
each. The essential features of the procedure are illustrated
in Fig. 2 for the construction of a mutant carrying a
deletion in the gene for the 14 kDa subunit: The structural
gene is excised from the plasmid, replaced by the LEU2 gene
and the resulting DNA fragment, consisting of the marker gene
flanked by respectively 700 and 105 bp of the 3' and 5'
flanking sequences of the 14 kDa gene, is offered to
competent yeast cells. The wild-type strain is leu⁻ and
selection of integrants is made via growth in the absence of
leucine. Since integration is the result of homologous
recombination, insertion of the transformimg DNA may occur at

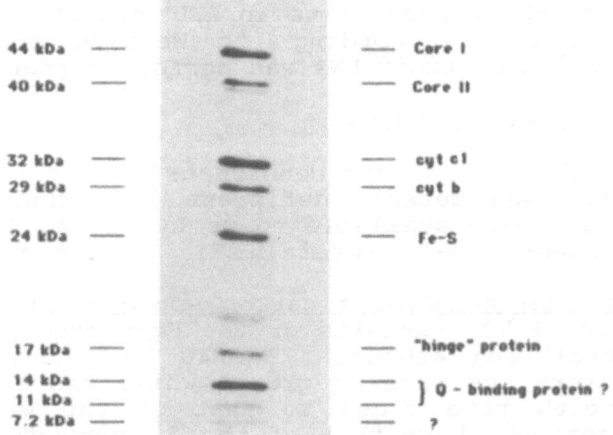

Fig. 1. Yeast complex III.   The complex was isolated
as described in **METHODS** and analyzed by electrophoresis
through a 15% polyacrylamide slab gel in the presence of
SDS. The gel was stained with Coomassie Brilliant Blue.

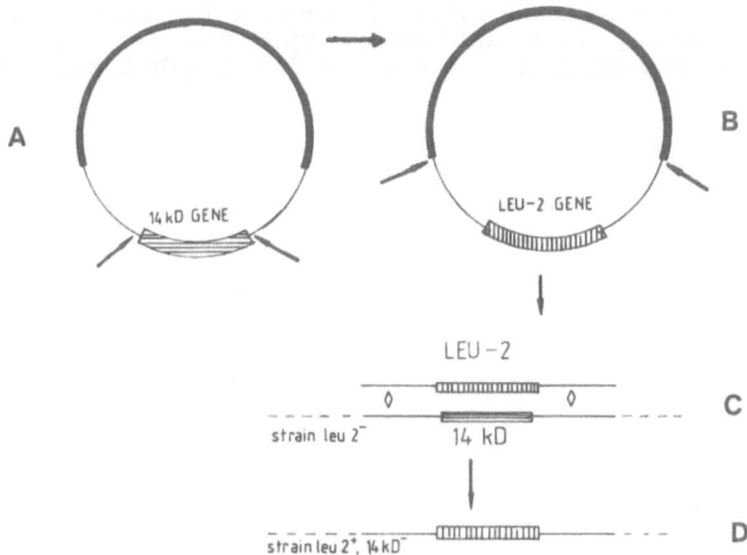

**Fig. 2. Schematic representation of the strategy for
the deletion of a gene.** The deletion of the 14 kDa
gene of the HR2 wild-type strain of <u>S. cerevisiae</u> is
represented.

either the leu2 locus or in the 14 kDa gene. Western blotting
of cell lysates of LEU⁺ colonies was therefore used to select
appropriate transformations for further analysis. Especially
in the case of the 14 kDA gene one of the flanking sequences
is very short (105 bp) and as a consequence most integration
events occur at the leu2 locus. Those cells still produce the

14 kDa protein.   As a control for the integration of the
marker gene at the correct site in LEU⁺ transformants lacking
the 14 kDa protein according to Western blot analysis,
Southern blotting of their DNA was performed (not shown).

Analysis of deletion effects

    Together with the mutants constructed in our laboratory we
have also analysed mutants deficient in either the 44 kDa
subunit I (26), made available to us by dr. A. Tzagoloff, or
the 24 kDa Rieske Fe-S protein (27), a gift from dr. B.L.
Trumpower.
    Growth on non-fermentable carbon sources. Mutants lacking
the 44, 14 or 11 kDa subunits or the Fe-S protein were found
to be glycerol⁻ and ethanol⁻. The 40 kDa⁰ mutant, however,
displayed a very low rate of growth while the 17 kDa⁰ mutant
showed a growth rate close to that of the wild type. It
should be mentioned here that of the disruptants lacking the
17 kDa subunit about 25% exhibited a glycerol⁻ phenotype.
Further analysis revealed that the latter cells contained a
secondary mutation, responsible for a lack of cytochrome b
and cytochrome c oxidase, and probably located in the mt DNA
since crossing with petite cells did not abolish the
glycerol⁻ phenotype.
    Western blot analysis. The effects of the deletions on the
level of the various subunits of complex III were analysed by
Western blotting with antibodies specific for one or more
subunits of the complex. After we had established that all

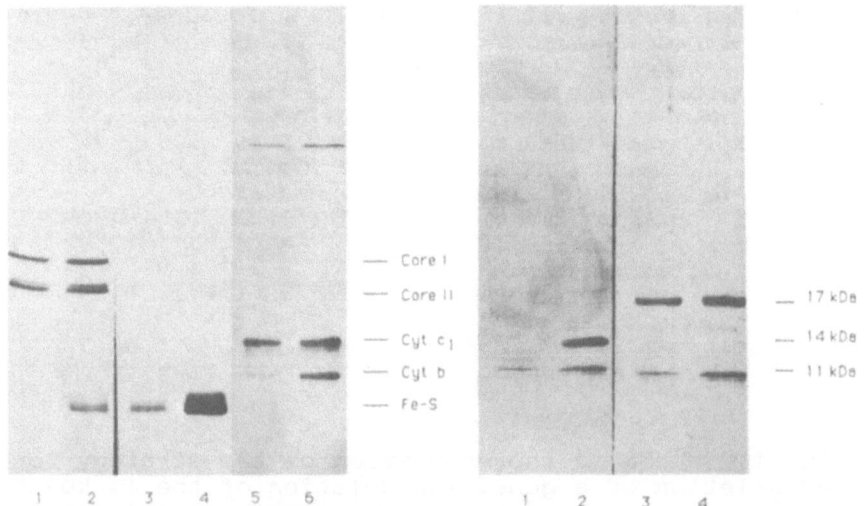

Fig. 3. Western blot analysis of the 14 kDa⁰ mutant.
50 μg mitochondrial protein from HR2 wild type cells
(even numbers) or 14 kDa mutant cells (odd numbers)
were electrophoresed as in Fig.1. After blotting to
nitrocellulose, the different pairs of lanes were
incubated with antibodies against each of the complex
III subunits, as indicated in the figure. The GARPO
system was used for visualization of bands.

subunits accumulated in the mitochondria, we used mitochondria isolated from cells grown on maltose (to avoid glucose repression) for SDS poplyacrylamide gel electrophoresis and immunoblotting. The mutants appeared to be of various types as described below.

A first class of mutants consists of the 11 kDa⁰ and 14 kDa⁰ mutants. Results with the 14 kDa⁰ mutant are shown in Fig. 3. The blots are characterized by trace amounts of apocytochrome *b* and low amounts of the 11 kDa polypeptide and the Fe-S protein, while the levels of the two core proteins, cytochrome *c₁* and the 17 kDa subunit are not affected. The response of our antbodies towards the 7.2 kDa subunit is such that no conclusion about the level of this subunit can be made. Results with the 11 kDa⁰ mutant are nearly identical: only the numbers 14 and 11 have to be exchanged. In neither mutant can cytochrome *b* be detected spectrally (not shown) and in some blots also no apocytochrome *b* is detectable. We cannot distinguish, therefore, whether the glycerol⁻ phenotype of these mutants is due to the lack of cytochrome *b* (as consequence of interrupted assembly) or to the lack of the 11 or 14 kDa subunit. Subsequent experiments, in which the 11 kDa⁰ mutant was complemented with truncated forms of the 11 kDa gene (28) showed that eleven amino acids at the C-terminus are not essential either for assembly, or for functioning of the complex, but removal of 44 amino acids from the C-terminal part resulted in partly deficient assembly. The residual electron transport activity from ubiquinol to cytochrome *c*, however, was proportional with the amount of cytochrome *b* and the conclusion must be that the C-terminal part of the 11 kDa polypeptide, although relevant for a correct assembly, is not directly involved in electron transfer.

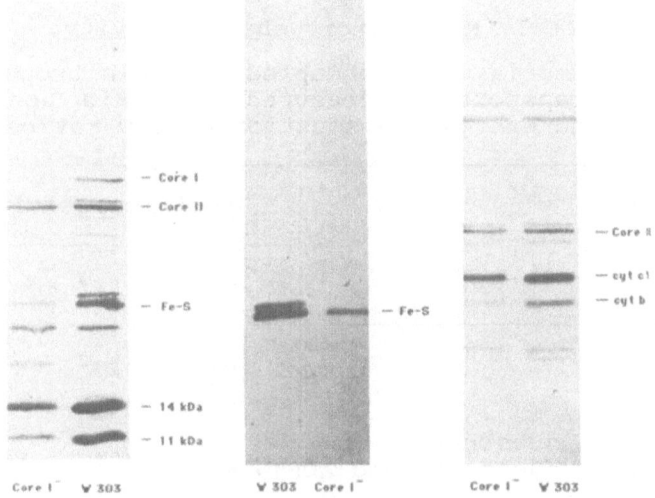

Fig. 4. Western blot analysis of the 44 kDa⁰ mutant. See legend to Fig. 3 for details of electrophoresis and immuno-decoration. The complex III subunits recognized by each set of antibodies are indicated.

The 40 and 44 kDa⁰ disruptants fall into the second class of mutants. In these the Fe-S protein, the 14 kDa subunit and the 11 kDa polypeptide are present at a low level, while cytochrome c_1, the 17 kDa subunit, the 7.2 kDa subunit (not easily visible in the figure) and the one remaining core protein are all present at normal levels (Fig. 4). Apocytochrome c is always clearly visible and also spectrally some cytochrome b can be detected (not shown). In agreement with these results could some ubiquinol:cytochrome c oxidoreductase activity (sensitive to both antimycin and myxothiazol) be detected in the mutants. In the 40 kDa⁰ mutant the activity reached 5% of the wild-type level, enough for a very low rate of growth on glycerol, while in the 44 kDa⁰ mutant this activity was only about 2% of wild-type level, apparently not enough to sustain growth on glycerol. Further analysis of the residual electron transfer activities revealed that in both mutants the complex III activity was proportional with the spectrally detectable amount of cytochrome b, in other words the Turnover Number, calculated in relation with the cytochrome b content, was the same as found for the wild-type strain. The conclusion has to be drawn, then, that both core proteins are not directly involved in electron transfer, although they are quite essential for a proper assembly. Additionally, they may be relevant for proton translocation.

A third class of mutants, as far as assembly is concerned, is formed by the 17 kDa⁰ and Fe-S⁰ mutants. In both mutants the absence of one subunit has no effect on the presence of all the other subunits. As may be expected from the role of the Fe-S protein in electron transfer, the Fe-S⁰ mutant shows a glycerol⁻ phenotype, while the 17 kDa⁰ mutant is glycerol+. From the finding that in all mutants of the first two classes the level of the Fe-S protein is severely reduced, a certain influence of the absence of the Fe-S protein on the levels

TABLE I. Respiratory chain activity.

The mitochondria were uncoupled with dinitrophenol. Electron transport was measured as NADH:O_2 specific activity and energy transduction as P/O ratios.

Conditions	Specific activity of	
	wild type	17 kDa⁰ mutant
	nmol. min^{-1}. mg^{-1}	
State 4 (with substrate)	200	170
State 3 (with substrate and ADP)	460	390
Uncoupled	910	500
P/O ratio	1.8	1.8

of cytochrome b, the core proteins and the 14 and 11 kDa
protein could be expected, but apparently the relationship
between these subunits and the Fe-S protein for assembly is
only unidirectional.

Since the 17 kDa^0 mutant showed proper assembly of complex
III and nearly normal growth on non-fermentable carbon
sources, the electron transport activity was further
analysed. The state 3 respiration of mitochondria from the

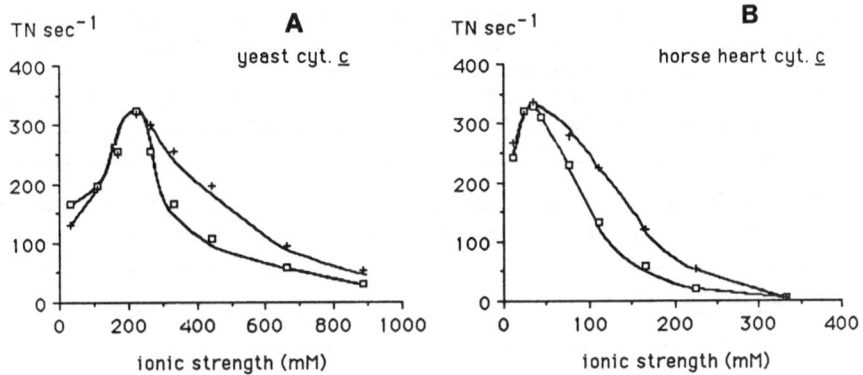

Fig. 5. Ionic strength dependency of the rate of reduction of cytochrome c by submitochondrial particles with
25 µM Q$_2$H$_2$, using yeast (panel A) or horse-heart (panel
B) cytochrome c. The medium contained EDTA and KCN and
the ionic strength was varied with potassium phosphate,
pH 7.5. The concentration of cytochrome c was 15 µM.
TN, turnover rate; +, HR2 wild type; o, 17 kDa^0 mutant.

mutant cells was only slightly diminished relative to that of
mitochondria from the wild-type strain and the P/O ratio was
not affected (Table I). Uncoupled respiration, however, was
clearly diminished and therefore the ubiquinol:cytochrome c
oxidoreductase activity of submitochondrial particles from
the mutant was investigated, with both horse-heart and yeast
cytochrome c as acceptor (29). It appeared that at ionic
strength values below the optimal value (30 mM for horse-
heart cytochrome c as acceptor and 220 mM for yeast
cytochrome c as acceptor) the electron transfer activity
(both Vmax and Km) was not affected by the absence of the 17
kDa polypeptide. At higher values of ionic strength, however,
a substantial inhibitory effect of the mutation was observed
(Fig. 5). Since at low ionic strength binding of cytochrome c
to complex III is quite tight and dissociation of the bc$_1$-
cytochrome c complex is highly rate-limiting, while at high
ionic strength the association becomes more rate limiting,
the most simple interpretation of the data is the assumption
that the 17 kDa subunit increases the affinity between
cytochrome c and complex III, in agreement with its very
strong negative charge. An effect of the absence or presence
of the 17 kDa protein, therefore, is only detected at high

ionic strength. To test this interpretation experiments were carried out in the presence of ATP or ADP. These nucleotides bind to oxidized cytochrome c (30) and decrease the positive charge of this protein. As a consequence the affinity between cytochrome c and complex III should be diminished and possibly an effect of the absence or presence of the 17 kDa

TABLE II. <u>Inhibition of cytochrome c reduction by Complex III by ATP as measured in yeast mitochondria</u>.

Mitochondrial ubiquinol:cytochrome c oxidoreductase activity was measured in 100 mM potassium phosphate (I=225 mM) at pH 7.5. Both horse-heart and yeast cytochrome c were used at a concentration of 15 µM. The substrate was 25 µM Q_2H_2. ATP and ADP were added at a concentration of 5 mM.

	Horse-heart cytochrome c (activity is expressed			Yeast cytochrome c in µmol. min^{-1}. mg^{-1})		
	no addition	+ATP	+ADP	no addition	+ATP	+ADP
HR2	1.20	0.81	1.13	2.26	2.17	2.17
17 kDa^0	0.26	0.15	0.23	1.94	1.29	1.66
40 kDa^0	--	--	--	0.20	0.18	0.18

protein could be observed under conditions of low or optimal ionic strength. The data of Table II show that this is indeed the case. With the wild-type particles ATP has an effect on the quinol:cytochrome c oxidoreductase activity only at high ionic strength, but the combination of ATP and the mutation affects this activity also at low ionic strength. The effect of ATP is stronger than that of ADP, confirming the relevance of the charge of ATP for the effect.

   Northern blot analysis. To be sure that the effects of the deletion of a gene for one subunit on the level of one of the other subunits was not caused by an effect of the deletion on the transcription, we compared mutant and wild-type strains with respect to the steady-state levels of various mRNAs. The Northern blots of Fig. 6 show the level of the mRNAs for the polypeptides that are reduced in level in the 14 kDa^0 mutant, actin mRNA being used as an internal control for the amount of RNA present. It can be seen that in the mutant the mRNAs for cytochrome b, the Rieske Fe-S protein and the 11 kDa subunit are present at the same level as in the wild type, while the mRNA for the 14 kDa protein itself is not detectable. We conclude that the synthesis of the subunits is not affected by the deletion and, since the subunits are recovered in the mitochondria, import into mitochondria is also not affected. The most probable explanation for the reduction in subunit levels is thus deficiency in assembly, with consequent degradation by proteolysis.

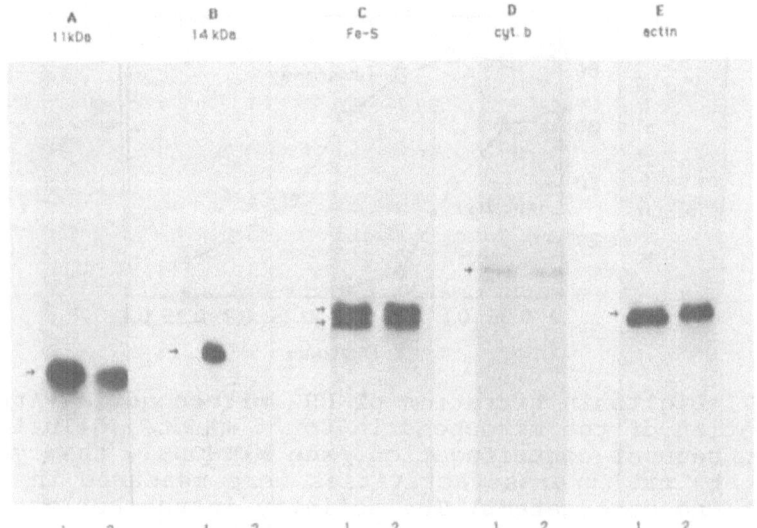

Fig. 6. Northern blot analysis of mRNAs in the 14 kDa⁰ mutant. Total RNA was isolated from both wild-type and 14 kDa⁰ mutant cells and separated on agarose gels. After transfer to Hybond-N, RNAs were hybridized with [^{32}P]-labelled DNA probes for the genes encoding the 11 kDa, 14 kDa or Fe-S subunit, as well as cytochrome b or actin.

Topology of the subunits of complex III

The effects of the various deletions on the levels of the different subunits of complex III, as decribed above, indicate a requirement for certain interactions between subunits for a proper assembly of the complex. The formulation of a model for assembly, however, also requires some knowledge about the topology of the subunits in the complex. The available data, obtained with the bovine-heart (31) or <u>Neurospora crassa</u> (32) enzyme indicate by analogy that the two core proteins extend from the membrane into the matrix, that cytochrome <u>b</u> is inside the core of the membrane and that cytochrome c_1, the Fe-S protein and the 17 kDa protein largely extend into the inter-membrane space. About the topology of the 11 and 14 kDa subunits not much is known, although for the bovine-heart analogues of both subunits a role in binding of ubiquinone has been postulated (33,34). For the bovine-heart analogue of the 11 kDa subunit a role in binding of antimycin (35) and in centre o activity (36) has been claimed. Japa et al. (37) conclude that the 14 kDa subunit faces the mitochondrial matrix, but their conclusion has been challenged (38).

To answer the questions on topology we used digitonin treatment of mitochondria, followed by proteinase K treatment, to see under which conditions the various subunits could be digested by the protease. Fig. 7 shows that digitonin could be used to make specifically the outer membrane permeable for proteins. At a digitonin concentration

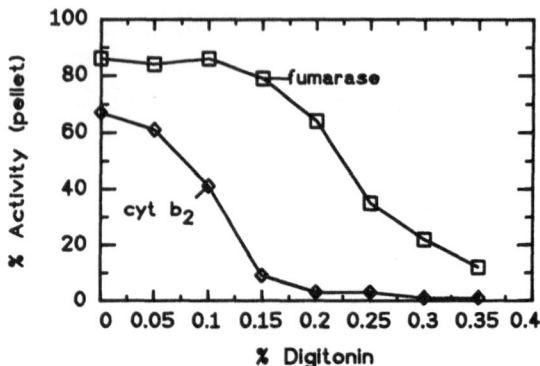

Fig. 7. Digitonin titration of HR2 mitochondria. After incubation of the mitochondria for 1 minute, dilution and subsequent centrifugation (see METHODS), the cytochrome b_2 and fumarase activities were measured in both pellet and supernatant. The activity in the pellet is given as percentage of the total activity.

of 0.14 mg/ml cytochrome b_2 activity is recovered in the supernatant, while the fumarase activity is still largely recovered in the pellet, indicative for intactness of the inner mitochondrial membrane. When the experiment, as described in the METHODS section, was carried out with the wild-type strain HR2, the blot of Fig. 8 was obtained. The control lanes show that proteinase K is active and that no subunit of complex III is broken down in the absence of proteinase K. Panel A shows that even in the absence of digitonin some breakdown occurs, indicating that in about 20% of the mitochondria the outer membrane is not intact any more. This same phenomenon could be seen in Fig. 7. At 0.14 mg/ml digitonin the 11 kDa subunit has disappeared, as well as the 17 kDa subunit. Cytochrome b and cytochrome c_1 are still partly intact, while the Fe-S protein seems not affected at all. At higher concentrations of digitonin, where also the inner membrane has been disrupted, the two cytochromes have completely disappeared and also core 1 is largely broken down. The core 2 and Fe-S protein and the 14 kDa subunit are still present at normal concentrations, so it is not clear from which side of the membrane they can be attacked. Since the core 2 protein, just as the core 1 protein, should be located at the matrix side of the membrane, this protein is apparently very resistent against proteolytic breakdown. Since we have seen above that the complex of cytochrome b, the 14 kDa protein and the 11 kDa protein, possibly together with the Fe-S protein, is destabilized in the absence of one of the core proteins, it is quite well possible that the proteinase resistence of the

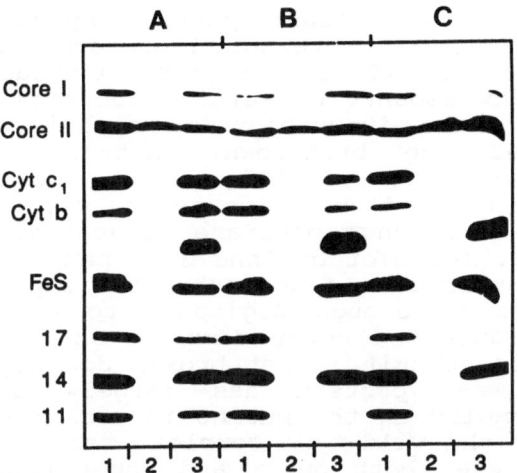

Fig. 8. Digitonin/proteinase K treatment of HR2 mito-chondria. After incubation of mitochondria with the apprpriate amounts of digitonin, samples were further incubated with no proteinase or Triton X-100 (lanes 1), with 0.25 mg/ml proteinase K + 0.5% Triton X-100 (lanes 2) or with 0.25 mg/ml proteinase K (lanes 3). Subsequent electrophoresis and immunoblotting was carried out as described in METHODS.
A: no digitonin; B: 0.14% digitonin; C: 0.6% digitonin.

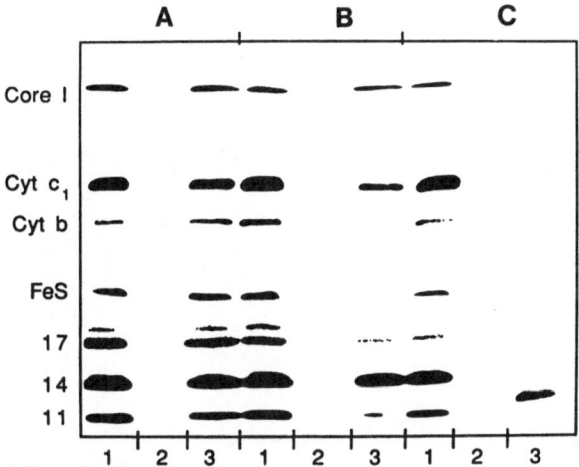

Fig. 9. Digitonin/proteinase K treatment of 40 kDa⁰ mitochondria. The experimental details are similar to those of Fig. 7. The only difference is that in B 0.125% digitonin is used for specific degradation of the outer membrane in stead of 0.14%.

14 kDa subunit is at least partly dependent on to the presence of the core proteins. Therefore we repeated the experiment with the 40 kDa° mutant. Although this mutant contains only low amounts of several subunits, including the 14 kDa subunit, the immunoblotting technique is sensitive enough to detect any breakdown. Control experiments (not shown) revealed that slightly lower concentrations of digitonin were required, relative to the wild-type strain, to make the outer and inner membrane permeable. The result is shown in Fig. 8. The blot of lane 3 of panel C clearly shows that in this mutant the 14 kDa subunit disappears completely in the presence of enough digitonin to disrupt the inner membrane. When only the outer membrane is disrupted (lane 3 of panel B) this subunit is not broken down by proteinase K, although the Fe-S protein has largely disappeared. So although the complex in the mutant is less resistent against proteinase than the wild-type complex, the 14 kDa subunit is only disrupted when both outer and inner membrane are made permeable. We may, then, safely conclude that the 14 kDa subunit, contrary to the 11 kDa subunit, is not extending into the inter-membrane space, but faces the matrix, in agreement with the conclusion of Japa et al. (37). The protection of the 14 kDa protein by the core 2 protein is in agreement with the finding of Lorusso et al. (39) that also in isolated complex III from bovine heart the analogous subunit is very resistent against proteolytic breakdown. Contrary to the results with the isolated enzyme (39) we found that reduction of the mitochondria with dithionite even increased the resistence against proteinase K treatment.

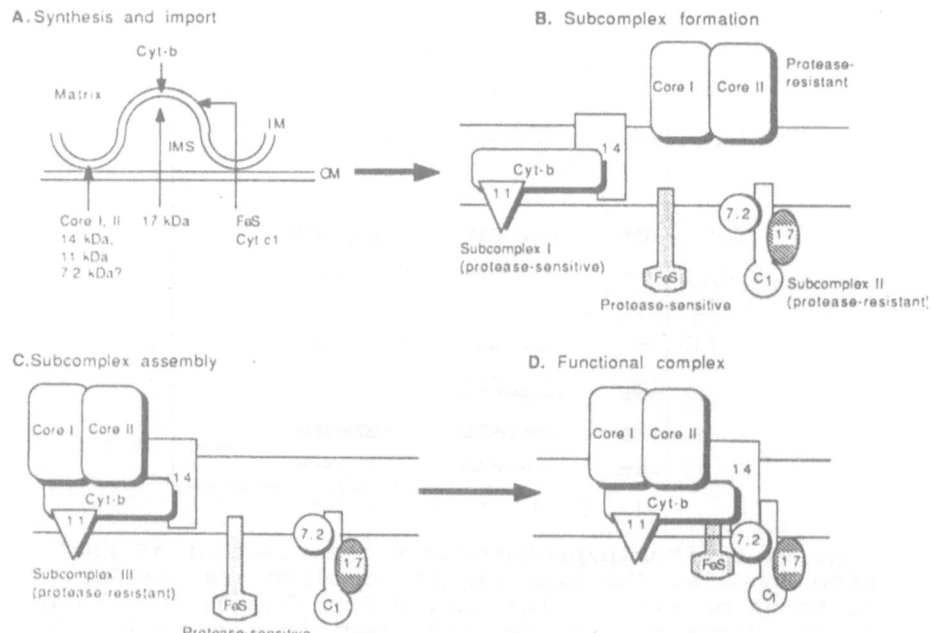

Fig. 10. Model for the assembly of yeast complex III.

A model for assembly

On the basis of the data presented above we have developed
a model for the assembly of complex III in yeast (Fig. 10).
Contrary to the proposal of Crivellone et al. (40) we propose
the intermediary formation of subcomplexes. The core of the
complex is formed by a subcomplex consisting of cytochrome b,
the 11 kDa and the 14 kDa subunit. Each of the three
components is absolutely required for the formation of this
complex. This also fits with previous data (36) showing that
in the absence of cytochrome b the 11 and 14 kDa subunits are
also missing. This subcomplex, however, is still sensitive to
proteolytic breakdown, unless it is protected by the
attachment of the two core proteins. The complex of the two
core proteins, on the other hand, is very resistent
against proteolytic breakdown, even when it is not attached
to the cytochrome b subcomplex. The Fe-S protein forms its
own group. Integration of the hydrophobic domain of this
protein into the cytochrome b subcomplex has no effect on the
stability of the latter complex but is essential for the
protease resistance of the Fe-S protein itself. The
cytochrome c_1 subcomplex, finally, contains cytochrome c_1 and
at least the 17 kDa protein and possibly the 7.2 kDa protein.
In mutants in which both the cytochrome b subcomplex and the
17 kDa protein are absent, also cytochrome c_1 is strongly
diminished, but introduction of the gene for the 17 kDa
protein restores not only the 17 kDa protein, but also the
cytochrome c_1 (18). These proteins, therefore, have a strong
interaction, the 17 kDa protein stabilizing cytochrome c_1. We
do not know whether this stabilization is reciprocal.
Since our antibodies do not recognize the 7.2 kDa subunit
very well, we cannot draw any firm conclusion about the
localisation and subunitinteractions of this polypeptide. We
suppose, however, that it is associated with the cytochrome
c_1 subcomplex at the inner-membrane side of the complex and
plays no important role in assembly, in agreement with
unpublished data of dr. Trumpower concerning the properties
of a 7.2 kDa^0 mutant.

II. DIFFERENTIATION OF TWO ANTIMYCIN-BINDING SITES

Reduction of cytochrome b via centre i of complex III

Since cytochrome b in complex III can be reduced *via* two
pathways, centres i and o according to the Q-cycle hypothesis
(41), the reduction of cytochrome b *via* centre i can only be
studied in the presence of an inhibitor of centre o, like
myxothiazol. Since antimycin in stoicheiometric amounts (1
antimycin/cytochrome c_1) inhibits both the reduction of
cytochrome b in the presence of myxothiazol and the steady-
state electron transfer from quinol to cytochrome c, it is
generally assumed that the redox reaction between quinone/
quinol and cytochrome b is part of the steady state electron
transfer pathway. Using embellin (2,5-dihydroxy-3-undecyl-
1,4-benzoquinone) as inhibitory Q-analogue it was found,
however, that the reduction of cytochrome b was inhibited at
concentrations (4-5 μM) that only slightly affected the

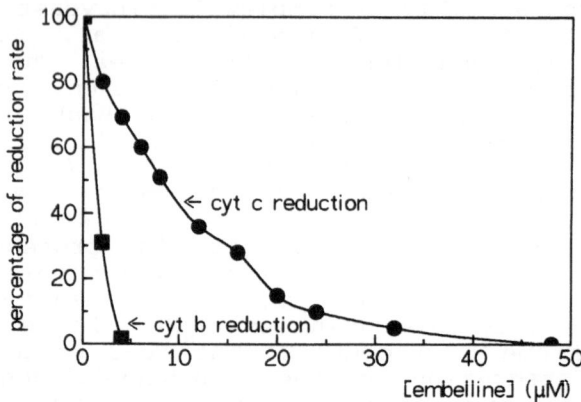

Fig. 11. Inhibition of the reduction of cytochrome b and cytochrome c by embellin. Bovine-heart submito-chondrial particles (3 mg/ml) were suspended in a phosphate buffer at pH 7.4 containing 0.5 mM KCN and the reduction of cytochrome b (in the presence of 3 μM myxothiazol) and cytochrome c by 25 μM ubiquinol-2 was measured at the apprpriate wavelength pairs.

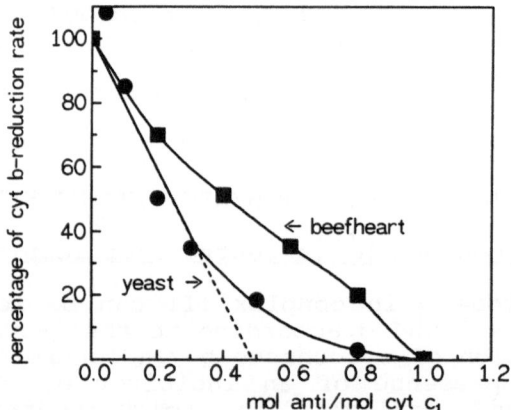

Fig. 12. Inhibition of cytochrome b reduction by antimycin. Submitochondrial particles from yeast (5 mg protein/ml) or bovine heart (3 mg protein/ml) were suspended in a sorbitol/phosphate buffer at pH 7.4 in the presence of 0.5 mM KCN, 5 μM myxothiazol and vary-ing amounts of antimycin. The reduction of cytochrome b by 40 μM duroquinol was measured spectrophotometrically.

steady-state electron transfer (Fig. 11). At concentrations
of about 40-50 µM reduction of cytochrome c was inhibited,
but at this concentration centre o was inhibited as well, as
evidenced by the inhibition of electron transfer from the Fe-
S cluster to cytochrome c_1 (not shown). The reduction of
cytochrome b in the presence of myxothiazol could also be
inhibited with DCCD and fluorescamim at concentrations of
about 100 µM. At this concentration the steady state electron
transfer was not inhibited at all and with DCCD in fact 100
times as much was needed to inhibit steady-state electron
transfer. These data suggested that the reaction step between
quinol and cytochrome b is not part of the electron pathway
from quinol to cytochrome c.

Differentiation between antimycin-binding sites

Studying the inhibition of the reduction of cytochrome b
(in the presence of myxothiazol) by antimycin, we found that
in yeast submitochondrial particles, contrary to the results
with bovine-heart particles, the inhibition curve was
hyperbolic, indicating the presence of two different
antimycin-binding sites, the occupation of one of them being
responsible for the inhibition (Fig. 12, cf ref. 42).
On the basis of earlier data (43 - 45), indicating that DCCD
lowers the number of antimycin-binding sites, we studied the
effects of DCCD and fluorescamin on the inhibition of
electron transfer by antimycin. Fig. 13 shows that in the
presence of 100 µM DCCD or fluorescamin, under which
conditions the rate of electron transfer to cytochrome c is
not affected, only slightly more than 1 antimycin per 2
cytochrome c_1 are required to inhibit the electron transfer
from succinate to oxygen completely. Also the reduction of
cytochrome c (in the presence of KCN) is inhibited by the
same concentration of antimycin (not shown). Using increasing
concentrations of DCCD or fluorescamin the subsequent titer

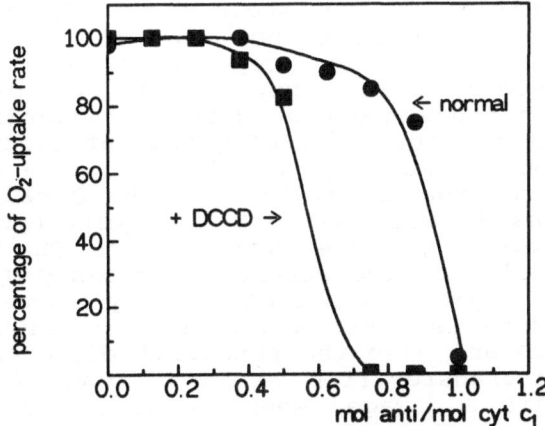

Fig. 13. **Inhibition of electron transfer by antimycin in
the presence of DCCD.** Yeast submitochondrial particles were
suspended in a sorbitol/phosphate buffer (pH 7.4) and the
rate of oxygen consumption in the presence of 10 mM succinate and varying amounts of antimycin was measured in an
oxygraph vessel in the presence or absence of 100 µM DCCD.

for antimycin did not fall below the value of 1 antimycin per
2 cytochrome c_1, that is 1 antimycin per complex III dimer.
Addition of DCCD and fluorescamin together, however, resulted
in a further decrease of residual antimycin-binding sites,
accompanied with a partial inhibition of electron transfer
(not shown). It may be concluded, then, that occupation of
one of the 2 antimycin-binding sites in the dimeric complex
(we do not know whether this is just one of the two
principally identical sites or specifically one site with
specific properties and localisation) inhibits the reaction
between exogenous quinone/quinol and cytochrome b, but that
both sites have to be occupied to inhibit electron transfer
to cytochrome c.

Electron transfer and proton translocation

The studies on the effect of DCCD on complex III have
clearly shown (see ref. 44) that both with the bovine-heart
and the yeast enzyme binding of DCCD inhibits the detection
of proton translocation. According to the data of Brand et
al. (46), however, electron transfer still generates a
potential difference betweeen the quinone/quinol couple and
cytochrome c of about 200 mV. These latter authors concluded
that the thermodynamic data were so convincing that the,
supposedly contradictory, kinetic experiments had to be
dismissed. This seems not very scientific, since no arguments
were given for the non-reliability of the kinetic data.
Similar data have been reported recently for complex I (47).
So what can be the explanation? It is clear from our data
that one antimycin-binding site in the dimeric complex is
only involved in electron transfer between cytochrome b and
exogenous quinone/quinol. Upon blockade of this one site also
the formation of semiquinone is 50% inhibited (P. Nieboer and
J. Berden, unpublished experiments) indicating that one Q-
binding site is not available any more. Weiss et al. (48)
have shown that for electron transfer only one Q per dimeric
complex is required. We may suggest that by DCCD the two-
electron transfer between quinol and cytochrome b (C.A.M.
Marres and S. de Vries, in preparation) is inhibited, but
one-electron steps still may occur, probably between
cytochrome b and the endogenous quinone. According to the Q-
cycle hypothesis the formation of quinol at centre i requires
two protons from the matrix, but if we suppose that DCCD and
fluorescamin block the interaction of the Q reduction centre
with the protons as well as with the quinone/quinol in the
matrix, the protons for the formation of QH_2 have to come
from the other side of the membrane, probably via protonated
groups in the enzyme. This proton relay, however, will be
slow and therefore is state 4 respiration maintained in the
presence of DCCD and also the potential difference generated
by the electron transfer from centre o to the Q at centre i.
In particles with a damaged membrane or in isolated complex
the protonation of the reduced quinone at centre i will be
more rapid, despite of the blockade of the normal pathway, so
no or little inhibition of electron transfer is observed.
Only at very high concentrations of DCCD also the proton
transfer from outside is blocked and electron transfer
becomes inhibited. The maintenance of a potential difference

in the thermodynamic measurements of Brand et al. (46) is therefore explained by the fact that electron transfer is much faster than the compensating proton transfer.

CONCLUSIONS

1. Complex III is assembled from subcomplexes, two of which are proteinase resistent (the core proteins and the cytochrome c_1 subcomplex) and two of which are proteinase sensitive (the cytochrome b subcomplex and the Fe-S protein). The former is protected against proteolysis by the attachment of the core proteins and the latter by insertion into the cytochrome b subcomplex.

2. The 11 kDa subunit extends into the inner membrane space (most likely with the C-terminus) and the 14 kDa subunit faces the matrix. The latter, however, is certainly transmembrane, since it is closely associated with centre o.

3. For none of the subunits without prosthetic group an essential function in electron transfer has been established, although most of them are required for proper assembly. The term "proper assembly" may be understood as making the polypeptides with a prosthetic group proteinase resistant.

4. One of the two antimycin-binding (Q-binding) sites in the dimeric complex III is not required for electron transfer from quinol to cytochrome c. This site is, however, essential for the two-electron transfer between exogenous quinol/quinone and cytochrome b and for the uptake of protons from the matrix space.

ACKNOWLEDGEMENTS

We like to thank prof. dr. L.A. Grivell for his continuous interest and advices. This research was supported in part by grants from the Netherlands Organization for Scientific Research (NWO) under the auspices of the Netherlands Foundation for Chemical Research (SON).

REFERENCES

1. Rich, P.R. (1984) Biochim. Biophys. Acta 768, 53-79.
2. De Vries, S. (1985) J. Bioenerg. Biomembr. 18, 195-224.
3. Perkins, S.J. and Weiss, H. (1983) J. Mol. Biol. 168, 847-866.
4. Crofts, A.R. and Wraight, C.A. (1983) Biochim. Biophys. Acta 726, 149-186.
5. Deters, D., Muller, U. and Homberger, H. (1976) Anal. Biochem. 70, 263-267.
6. Mason, T.L., Poyton, R.O., Wharton, D.C. and Schatz, G. (1973) J. Biol. Chem. 248, 1346-1354.
7. Palmer, G. (1978) Methods in Enzymology 53, 113-121.
8. Yannish-Perron, C., Vieira, J. and Messing, J. (1984) Gene 33, 103-119.
9. Hollenberg, C.P., Degelmann, A., Kustermann-Kuhn B. and Roger, H.D. (1976) Proc. Natl. Acad. Sci. USA 73, 2072-2076.

10. Maarse,A.C., De Haan, M., Schoppink, P.J., Berden, J.A. and Grivell, L.A. (1988) Eur. J. Biochem. 172, 179-184.
11. Schoppink, P.J., Berden, J.A. and Grivell, L.A. (1989) Eur. J. Biochem. 181, 475-483.
12. Schoppink, P.J., Hemrika, W., Reynen, J.M., Grivell, L.A. and Berden, J.A. (1988) Eur. J. Biochem. 173, 115-122.
13. Oudshoorn, P., van Steeg, H., Swinkels, B.W., Schoppink, P.J. and Grivell, L.A. (1987) Eur. J. Biochem. 163, 97-103.
14. Klebe, R.J., Harries, J.V., Sharp, D. and Douglas, M.D. (1983) Gene, 25, 333-341.
15. Losson, R. and Lacroute, F. (1979) Proc. Natl. Acad. Sci. USA 76, 5134-5137.
16. Schoppink, P.J., Grivell, L.A. and Berden, J.A. (1987) in: Advances in Membrane Biochemistry and Bioenergetics (Kim, C.H., Tedeschi, H., Diwan, J.J. and Salerno, J.C., eds) Plenum Publ. Corp., New York, pp. 129-140.
17. Lowry, O.H., Rosebrough, N.J., Farr, A.L. and Randall, R.J. (1951) J. Biol. Chem. 193, 265-275.
18. Laemmli, U.K. (1970) Nature (London) 227, 680-685.
19. Vaessen, R.T.M.J., Kreike, J. and Groot, G.S.P. (1981) FEBS Lett. 124, 193-196.
20. Avrameas, S. and Ternynck, T. (1971) Immunochemistry 8, 1175-1185
21. Maniatis, T., Fritsch, E.F. and Sambrook, J. (1982) Molecular cloning: A laboratory manual, CSH Laboratory Press, Cold Spring Harbor, New York.
22. Tzagoloff, A., Crivellone, M.D., Gampel, A., Muroff, I., Nishikimi, M. and Wu, M. (1988) Phil. Trans. R. Soc. Lond. B 319, 107-120.
23. Van Loon, A.P.G.M., De Groot, R.J., van Eijk, E., van der Horst, G.T.J. and Grivell. L.A. (1982) Gene 20, 323-337.
24. Van Loon, A.P.G.M., Maarse, A.C., Riezman, H. and Grivell, L.A. (1983) Gene 26, 261-272.
25. Rothstein, R.J. (1983) Methods in Enzymology 101, 202-211
26. Tzagoloff, A., Wu, M. and Crivellone, M. (1986) J. Biol. Chem. 261, 17163-17169.
27. Ljundahl, P. (1987) Ph.D. Thesis, Dartmouth Medical School, Hanover, N.H., U.S.A.
28. Schoppink, P.J., De Jong, M., Berden, J.A. and Grivell, L.A. (1989) Eur. J. Biochem. 181, 681-687.
29. Schoppink, P.J., Hemrika, W. and Berden, J.A. (1989) Biochim. Biophys. Acta 974, 192-201.
30. Corthesy, B.A. and Wallace C.J.A. (1988) Biochem. J. 252, 349-355.
31. Kim, C.H. and King, T.E. (1983) J. Biol. Chem. 258, 13543-13551.
32. Karlsson, Hovmoller, S., Weiss, H. and Leonard, K. (1983) J. Mol. Biol. 169, 287-302.
33. Wang, T.Y. and King, T.E. (1982) Biochem. Biophys. Res. Commun. 104, 591-596.
34. Berden, J.A., Van Hoek, A.N., De Vries, S. and Schoppink, P.J. (1988) in: Cytochrome Systems, Molecular Biology and Bioenergetics (Papa, S., Chance, B. and Ernster, L., eds) Plenum Press, New York, pp. 523-531.
35. Van Keulen, M.A. and Berden, J.A. (1985) Biochim. Biophys. Acta 808, 32-38.

36. Marres, C.A.M., De Vries, S. and Slater, E.C. (1982) Biochim. Biophys. Acta 682, 160-167.
37. Japa, S., Zhu, Q.S. and Beattie, D.S. (1987) J. Biol. Chem. 262, 5441-5444.
38. De Vries, S. and Marres, C.A.M. (1987) Biochim. Biophys. Acta 895, 205-239.
39. Lorusso, M., Cocco, T., Boffoli, D., Gatti, D., Meinhardt, S., Ohnishi, T. and Papa, S. (1989) Eur. J. Biochem. 179, 535-540.
40. Crivellone, M.D., Wu, M. and Tzagoloff, A. (1988) J. Biol. Chem. 263, 14323-14333.
41. Mitchell, P. (1976) J. Theor. Biol. 62, 327-367.
42. Bechmann, G., Zweck, A. and Weiss, H. (1988) EBEC Reports 5, p. 82.
43. Clejan, L. and Beattie, D.S. (1983) J. Biol. Chem. 258, 14271-14275.
44. Degli Esposti, M. and Lenaz, G. (1985) J. Bioenerg. Biomemb. 17, 109-121.
45. Van Hoek, A.N. (1989) Ph.D. Thesis, University of Amsterdam, Amsterdam, The Netherlands.
46. Brand, M.D., Al-Shawi, M.K., Brown, G.G. and Price, B.D. (1985) Biochem. J. 225, 407-411.
47. Vuokila, P.T. and Hassinen, I.E. (1989) Biochim. Biophys. Acta 974, 219-222.
48. Linke, P., Bechmann, G., Gothe, A. and Weiss, H. (1986) Eur. J. Biochem. 158, 615-621.

STRUCTURAL STUDIES OF *EUGLENA* CYTOCHROME c_1[*]

Hiroshi Matsubara, Kuniaki Mukai[**], and Sadao
Wakabayashi

Department of Biology, Faculty of Science
Osaka University, Toyonaka, Osaka 560, Japan

SUMMARY

Euglena gracilis complex III prepared from its submitochondrial
particles showed an atypical difference absorption spectrum for
cytochrome c_1 with α-peak at 561 nm. Pyridine ferrohemochrome of
cytochrome c_1 showed its α-peak at 553 nm. The amino-terminal sequence of
cytochrome c_1 suggested that its heme moiety was covalently linked to
cysteine-39 through a single thioether bond. Phenylalanine occupied
position 36 usually occupied by cysteine affording the other thioether
bond. The total amino acid sequence of mature cytochrome c1 was
determined by sequencing of its cDNA. The cDNA clone consisted of 872
base pairs encoding the mature protein with 243 amino acids. The putative
sequence contained an unusual heme binding sequence, -F-A-P-C-H-, instead
of the typical sequence, -C-X-Y-C-H-, found in other eukaryotic C-type
cytochromes. Sequence comparison, gene manipulation, and physical studies
indicated the heme ligands and a binding site to cytochrome *c*.

INTRODUCTION

Mitochondrial respiratory electron transfer chain has two distinct
C-type cytochromes, cytochrome *c* and c_1, and electrons are transferred
from Rieske iron-sulfur protein successively to cytochrome c_1, cytochrome
c, and cytochrome oxidase. Cytochromes *c* and c_1 have a common
heme-binding feature, that is, the heme c_1 is covavlently bound to the
polypeptide chain at two cysteine residues through thioether bonds.[1]
However, cytochrome c from *Euglena* and *Crithidia*, protozoa, have heme c
bound to their polypeptide chains through a single thioether bond with
atypical absorption spectra, α-peaks being at the longer wavelengths,
557-558 nm, than those of other eukaryotic cytochromes *c*.[2]

Protozoa occupy a unique evolutionary position and their
mitochondria have several distinctive properties, although the

---

[*]This work was supported in part by Grants-in-Aid for Scientific Research
on Priority Area of Bioenergetics and on Nos. 01470148 and 62470148 from
the Ministry of Education, Science and Culture of Japan.
[**]Present Address: Mitsubishi-Kasei Institute of Life Science, Machida,
Tokyo 194, Japan.

respiratory chain seems to have typical complexes I-IV.[3] Cytochrome c_1 is
one of the components of ubiquinol-cytochrome c reductase (alternatively,
cytochrome bc_1 complex or complex III) and has been a major subject of
research in our laboratory in terms of structure-function and
evolutionary relationship.[4-14] Since *Euglena* cytochrome c showed such a
unique structural feature as mentioned above, we suspected that
cytochrome c_1 might also show an unusual structural feature.

The main purpose of the present paper is to describe the isolation
and characterization of *Euglena* cytochrome c_1 with special reference to
its heme-binding and its amino acid sequence deduced from the nucleotide
sequences of cDNA encoding the mature protein. Several sequences of
cytochrome c_1 will be compared and discussions will be carried out on
ligand residues to the heme iron and a region responsible for interacting
with cytochrome c with the aid of supplemental evidence of gene
manipulation of yeast cytochrome c_1 and a physical experiment on beef
cytochrome c_1.

EXPERIMENTAL PROCEDURES

Detailed experimental procedures for preparing *Euglena* complex III
and analyzing the heme-binding structure were previously described,[12,13]
and only brief outlines of these procedures will be given below.

Euglena gracilis SM-ZK, a mutant lacking chloroplasts, was cultured
aerobically in the dark and harvested at an early stationary phase. Cells
stored in a frozen state were thawed and treated with trypsin in 50 mM
Na-EDTA. The trypsin-treated cells were homogenized and differentially
centrifuged to collect mitochondrial membranes. The membranes were
sonicated and again differentially centrifuged. Submitochondrial
particles were precipitated at 100,000 xg and treated with Triton X-100
and 0.2 M NaCl to solubilize complex III. The complex was chromatographed
on a DEAE-Toyopearl 650M column developed with a concentration gradient
of NaCl. Complex III was further applied to a hydroxylapatite column
developed with Na-phosphate buffer and guanidine-HCl to separate
cytochrome c_1 from other subunits. The following chromatographies on
Toyopearl HW-55F and Sephadex G-150 columns developed with Na-phosphate
buffer containing SDS and 2-mercaptoethanol gave a protein-chemically
pure preparation of cytochrome c_1.

Cytochrome c_1 thus obtained was treated with nitrophenylsulfenyl
chloride in 70% formic acid to remove covalently bound heme c. The
apocytochrome c_1 was treated with iodoacetic acid to modify cysteine
residues for the sequence analysis. Carboxymethyl-derivative was digested
with trypsin and lysyl-endopeptidase, and the digest was applied to an
octadecylsilane column developed with a linear concentration gradient of
acetonitrile in trifluoroacetic acid to isolate a peptide contributing to
the heme-binding. Amino acid composition and sequence analysis of
proteins and peptides were performed as described.[6,13]

Total RNA was extracted from *Euglena* cells with the guanidine
thiocyanate method.[15] Poly(A)[+]RNA was prepared by chromatography on an
oligo(dT)-cellulose column.[16] A cDNA library in λgt11 was prepared by the
method of Gubler and Hoffman[17] and Huynh et al.[18] An antiserum was raised
against *Euglene* cytochrome c_1 in a rabbit. The cDNA library was screened
according to Huynh et al.[18] The antiserum was diluted 200-fold and
treated with an extract of *E. coli* Y1090 before use. Bound antibodies

were visualized using an alkaline phosphatase-conjugated goat anti-rabbit immunoglobulin antibody.[19] Affinity-preparation of antibodies cross-reacting with expressed fusion protein was performed as described.[20] Immunoblot analysis was carried out with the antibodies. DNA sequence analysis was performed by the dideoxy method.[21,22]

RESULTS

Preparation and Characterization of Euglena Complex III

Fresh *Euglena* mitochondrial membrane preparation showed the antimycin-sensitive ubiquinol-cytochrome *c* reductase activity, but it became insensitive by storing the cells at -20°C for several weeks. Myxothiazol was less effective on the activity in contrast to antimycin. For the subsequent work the stored cells were used.

The reductase was solubilized with Triton X-100 and approximately two-thirds of proteins were removed by the first extraction. The complex III remaining in the precipitate was solubilized by the second Triton X-100 extraction. The extract was purified on a DEAE-Toyopearl column. Complex III was adsorbed on the column and eluted in the middle of the salt gradient as shown in Fig. 1A. SDS-polyacrylamide gel electrophoresis of the major fraction showing the absorbance at 415 nm revealed 10 polypeptides. About 22-fold purification was achieved on the basis of cytochromn c_1 content. Further purification on hydroxylapatite and Toyopearl HW-65F chromatographies did not show any improvement. The subunit composition of *Euglena* complex III was similar to those of other eukaryotic complexes III. Subunits III and IV corresponded to cytochromes *b* and *c*, respectively, identified by heme staining, molecular mass, and amino-terminal sequence analysis as given in Fig. 1B. The amino-terminal sequence of band IV was identical with that of cytochrome c_1 of *Euglena* given later. That of band III was blocked and treatment with acid-methanol revealed the terminal-sequence to be M-K-I-S-L-Y-N-H-Y-H. Hydrophobic nature of band III judged from the amino acid composition supported it to be cytochrome *b*.

The relative molecular mass of the complex III in the Triton X-100 micelle was about 570 kDa estimated by a gel filtration method, indicating that the complex was in a dimeric form. The molar ratio between cytochrome *b* and c_1 was approximately one to one instead of two to one usually observed for other cytochrome bc_1 complexes.[24] A certain amount of acid-labile sulfide was detected, indicating the presence of Rieske iron-sulfur protein.[24]

Difference specctrum of the fully reduced minus fully oxidized *Euglena* complex III (Fig. 2, top) showed a maximum at 562 nm similar to other complexes III, but no shoulder at 552-553 nm due to the typical cytochromes c_1 was observed. That of the ascorbate reduced minus ferricyanide oxidized sample (Fig. 2, bottom) showed a maximum at 561 nm unusual for the typical cytochrome c_1, being a shift of the wavelength by 8 nm to the red.[25] The presence of cytochrome b was indicated by the absorption maximum at 563 nm observed by the difference spectrum of dithionite reduced minus ascorbate reduced complex (Fig. 2, middle).

Pyridine Ferrohemochromes of Euglena Complex III Preparation

Table 1 gives the α-peaks of pyridine ferrohemochromes of *Euglena*

Matsubara et al.

complex III and its acid-acetone treated preparation, comparing with
those prepared from hemin, horse heart cytochrome *c*, bovine heart
cytochrome c_1, and *Euglena* cytochrome *c* (*c*-558). The pyridine
ferrohemochrome of *Euglena* complex III showed the α-peak at 554.8 nm, but
cytochrome c_1 remaining in the precipitate after extracting heme b by

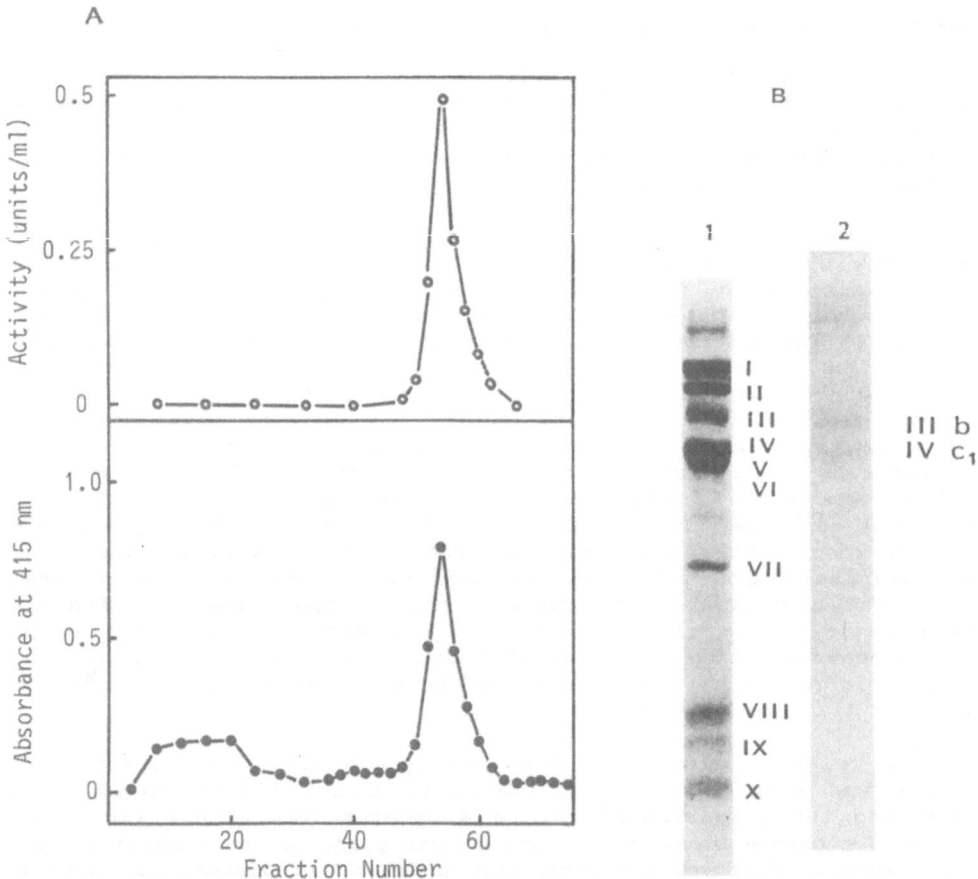

Fig. 1. Elution profile of *Euglena* complex III from a DEAE-
Toyopearl column and SDS-urea-polyacrylamide gel electrophoretic patterns of the complex. A) After the first Triton
X-100 extraction, the submitochondrial particles, about 970
mg protein, was further extracted with Triton (1.9 mg/mg
protein) and applied to a 3.6 x 14 cm DEAE-Toyopearl column.
The column was washed with 150 ml of a buffer consisting of 50
mM Tris buffer, pH 7.5, 0.1% Triton X-100, 0.1 M NaCl, and 0.5
mM EDTA. Complex III was eluted with 600 ml of a linear
gradient of 0.1 to 0.3 M NaCl in the above buffer. Fractions (
15 ml each) were monitored by the absorbance at 415 nm and the
ubiquinol-cytochrome *c* reductase activity. B) Comolex III was
analyzed on an 18% gel containing 6 M urea using SDS-
polyacrylamide gel electrophoresis.[23] The gels were stained
with Coomassie brilliant blue (1) and peroxidase activity (2).

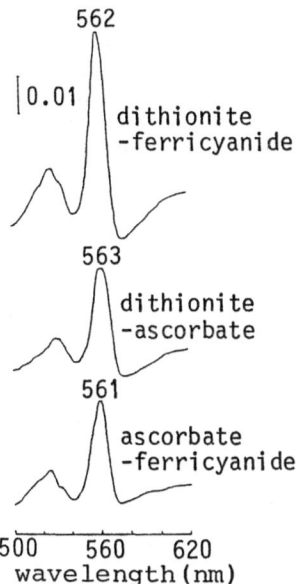

Fig. 2. Difference absorption spectra of *Euglena* complex III. The figures from top to bottom are self-evident. The top figure is for reduced minus oxidized, the middle for reduced minus reduced, and the bottom for reduced minus oxidized with various redox agents. The protein concentration was 0.5 mg/ml.

acid-acetone treatment of the complex showed the α-peak at 553 nm, indicating that the heme c covalently bound to the polypeptide chain was similar to that of *Euglena* cytochrome c (c-558).[2]

Amino Acid Composition and Amino-Terminal Sequence of Euglena Cytochrome c_1

Euglena cytochrome c_1 was purified in a denatured form by chromatographies on hydroxylapatite, Toyopearl HW-55F, and Sephadex G-150 in the presence of SDS and guanidine-HCl. About 12 nmoles of cytochrome c_1 was purified from 60 mg of complex III. Cytochrome c_1 thus obtained was colorless and showed no peroxidase activity on SDS gels. Carboxy-methyl-cytochrome c_1 was prepared from the complex III and purified as mentioned above to yield 90 nmoles of the derivative from 120 mg of the complex.

Table 1. Comparison of Various Pyridine Ferrohemochromes.

Sample	α-peak (nm)
Hemin	555.9
Horse heart cytochrome c	550.0
Bovine heart cytochrome c_1	550.0
Euglena cytocrome c (c-558)	553.1
Euglena complex III	554.8
Acid-acetone treated complex III	
Precipitate	553.0
Supernatant	555.6

Pyridine ferrohemochrome was prepared as described.[28] The complex III was mixed with ice-cold acetone containing 2% conc. HCl to give a 90% acetone mixture. After centrifugation, the supernatant and precipitate were prepared.

Matsubara et al.

Amino acid compositions of these preparations are given in Table 2 and they were found to be nearly identical with each other. These values agreed well with that deduced from cDNA sequence as mentioned later. The amino-terminal sequence of cytochrome c_1 gave the first 38 amino acid up to proline as given below. The residue at position 36 corresponding to the first cysteine residue responsible for heme linkage in other cytochromes c_1[4,6,30-35] was replaced by phenylalanine. Further sequence analysis with this sample was not attempted.

```
1              10                 20                 30
G-V-D-S-H-P-P-A-L-P-W-P-H-F-Q-W-F-Q-G-L-D-W-R-S-V-R-R-G-K-E-
              40
V-Y-E-Q-V-F-A-P-X-H-S-L-S-F-I-K
```

A heptadecapeptide (residues 30-46) was purified from a proteolytic digest of carboxymethyl-cytochrome c_1 with an yield of 28%. Its composition is given in Table 2 together with that deduced from the sequence given above. This sequence overlapped with the amino-terminal sequence directly determined with cytochrome c_1 at residues 30-38,

Table 2. Amino Acid Compositions of *Euglena* Cytochrome c_1, Carboxymethyl-Cytochrome c_1, and Heme-Peptide.

Amino acid	Cytochrome c_1	Carboxymethyl-cytochrome c_1[a]	Heme-peptide[b]
Carboxylmethly-cysteine	n.d.	3 (3)	
Aspartic acid	} 24	} 24 (14)	
Asparagine		(9)	
Threonine	10	11 (10)	0.27 (0)
Serine	15	14 (15)	2.0 (2)
Glutamic acid	} 25	} 24 (14)	} 3.1 (2)
Glutamine		(8)	(1)
Proline	21	20 (20)	1.4 (1)
Glycine	21	20 (19)	0.64 (0)
Alanine	19	19 (17)	1.2 (1)
Valine	14	15 (16)	1.7 (2)
Methionine	2	5 (5)	0.19 (0)
Isoleucine	8	7 (8)	0.85 (1)
Leucine	18	18 (16)	1.1 (1)
Tyrosine	11	13 (14)	0.98 (1)
Phenylalanine	13	13 (13)	1.9 (2)
Lysine	12	12 (12)	0.94 (1)
Histidine	7	5 (6)	0.96 (1)
Arginine	16	17 (16)	0.10 (0)
Tryptophan	n.d.	n.d. (8)	n.d.
unknown residues			(1)
Total number of residues	236	240 (243)	(17)

Samples were hydrolyzed with 6 N HCl at 110°C for 21 h with a small amount of thioglycolic acid. The hydrolysates were analyzed with an amino acid analyzer (Irica model A-5500).[29] Number of the total amino acids was assumed to be about 240 excluding tryptophan not determined (n.d.). Those in parentheses were deduced from the sequence analyses of cDNA[a] and peptide[b].

extending the amino-terminal sequence to a total of 46 residues. However, the residues 39 was unidentified.

Histidine-40, the probable fifth ligand to heme iron in cytochrome c_1, was conserved together with several other residues when the sequence was aligned with those of other cytochrome c_1.[4,6,30-35]

These results indicated that the amino-terminal sequence of *Euglena* cytochrome c_1 included the heme-binding region and probably the unidentified residue was a derivative of cysteine, to which the heme was originally attached. The reason why the amino acid residue at position 39 could not be identified is not clear at present, although it was later identified as cysteine from the cDNA sequence analysis.

Isolation of cDNA Clones Encoding Euglena Cytochrome c$_1$ and the Nucleotide Sequences

A λgt11 cDNA library was screened with rabbit polyclonal antibodies raised against *Euglena* cytochrome c_1. Twenty-four positive clones were isolated from 68,000 recombinant clones. The affinity-prepared antibodies recognized a subunit corresponding to cytochrome c_1. The insert size of the positive clones varied from 0.4 to 1.8 kilobase pairs (kbp) and 22 out of 24 clones were 0.8 to 0.9 kbp in size. One of the 0.9 kbp inserts was subcloned into a phage vector M13mp18 and sequenced.

Figure 3 shows the restriction map and the strategy for its nucleotide sequence analysis of the insert. Nucleotide sequence of the 872-bp insert was determined as given in Fig. 4 with the deduced amino acid sequence. A coding region was flanked by a 3'-untranslational region followed by poly(A). The longest 1.8-kbp insert was also subcloned and sequenced. This insert started at the forth nucleotide of the 872-bp insert and the identical sequence of the 846 nucleotides was followed by poly(A) and an unrelated sequence. The 1.8-kbp insert resulted probably from a blunt-ent ligation during the construction of the cDNA library.

The deduced amino acid sequence included that of the mature *Euglena* cytochrome c_1. The amino-terminal sequence of the cytochrome started from the 21st nucleotide. Amino acid sequences of four peptides purified from a digest of carboxymethyl-cytochromr c_1 were also assigned in the deduced sequence (Fig. 4). Thus the deduced amino acid sequence was the one of the mature protein of *Euglena* cytochrome c_1. The protein was composed of 243 amino acid residues, giving a molecular weight of 27,855 excluding the heme. This work will appear in *J. Biochem.* 106(No.3)(1989).

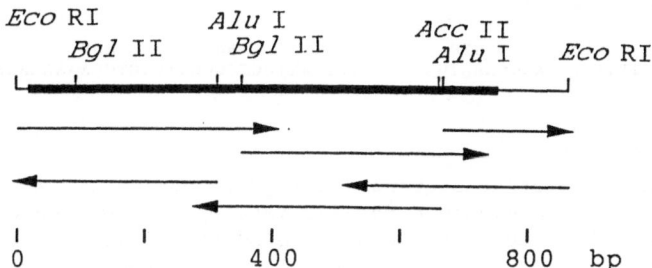

Fig. 3. Restriction map and sequence strategy for a cDNA encoding *Euglena* cytochrome c_1. Encoding region of the mature protein is shown by the solid line. Arrows show direction and extent of the sequencing.

```
                                    ↓        10        20
                              TTTTTTTTTTCGGAAACATG

           30        40        50        60        70        80
    GGAGTTGATTCCCATCCTCCTGCTCTTCCATGGCCTCACTTCCAGTGGTTTCAGGGCCTT
  1 G  V  D  S  H  P  P  A  L  P  W  P  H  F  Q  W  F  Q  G  L   20

           90       100       110       120       130       140
    GATTGGAGATCTGTGCGCCGTGGGAAAGAGGTCTACGAACAAGTTTTTGCGCCATGCCAT
 21 D  W  R  S  V  R  R  G  K  E  V  Y  E  Q  V  F  A  P  C  H   40

          150       160       170       180       190       200
    TCACTGAGTTTCATTAAATACCGCCATTTCGAGGCTTTCATGTCGAAGGAAGAGGTCAAA
 41 S  L  S  F  I  K  Y  R  H  F  E  A  F  M  S  K  E  E  V  K   60

          210       220       230       240       250       260
    AACATGGCAGCAAGTTTTGAAGTGGACGATGATCCAGATGAAAAGGGAGAGGCAAGGAAG
 61 N  M  A  A  S  F  E  V  D  D  D  P  D  E  K  G  E  A  R  K   80

          270       280       290       300       310       320
    CGCCCAGGGAAGCGCTTTGATACTGTTGTCCAGCCATACAAAAATGAGCAAGAAGCTCGA
 81 R  P  G  K  R  F  D  T  V  V  Q  P  Y  K  N  E  Q  E  A  R  100

          330       340       350       360       370       380
    TATGCCAACAATGGTGCGCTGCCACCAGATCTCAGTGTTATCACCAATGCACGACATGGT
101 Y  A  N  N  G  A  L  P  P  D  L  S  V  I  T  N  A  R  H  G  120

          390       400       410       420       430       440
    GGGGTGGACTACATTTATGCCCTTCTGACGGGTTATGGAAGGCCAGTTCCTGGGGGAGTG
121 G  V  D  Y  I  Y  A  L  L  T  G  Y  G  R  P  V  P  G  G  V  140

          450       460       470       480       490       500
    CAGTTGTCAACCACTCAATGGTACAACCCATATTTCCACGGTGGTATCATTGGAATGCCT
141 Q  L  S  T  T  Q  W  Y  N  P  Y  F  H  G  G  I  I  G  M  P  160

          510       520       530       540       550       560
    CCTCCTCTCACCGATGACATGATTGAGTATGAGGACGGAACGCCAGCAAGTGTTCCTCAA
161 P  P  L  T  D  D  M  I  E  Y  E  D  G  T  P  A  S  V  P  Q  180

          570       580       590       600       610       620
    ATGGCGAAGGATGTTACATGTTTTCTGGAGTGGTGCTCAAACCCCTGGTGGGATGAGAGG
181 M  A  K  D  V  T  C  F  L  E  W  C  S  N  P  W  W  D  E  R  200

          630       640       650       660       670       680
    AAGTTGCTCGGCTACAAGACCATCGCCACGCTGGCTGTGATCGCGGTCAGCTCTGGGTAT
201 K  L  L  G  Y  K  T  I  A  T  L  A  V  I  A  V  S  S  G  Y  220

          690       700       710       720       730       740
    TACAATCGGTTCCTCTCGGGTCTGTGGCGATCCCGCCGCCTTGCCTTCCGGCCGTTCAAC
221 Y  N  R  F  L  S  G  L  W  R  S  R  R  L  A  F  R  P  F  N  240

          750       760       770       780       790       800
    TACTCCAAATGATTCTCCCGACATGGTTGATGACCATCAATTTGGTGTCCCACCTATGTT
241 Y  S  K  *                                                  243

          810       820       830       840       850       860
    TGAGGTGTTTTCAACCAAATGTACACTTGTCATTGGTCCCGTTCTCTGCAAAAAAAAAAA

          870
    AAAAAAAAAAA
```

Fig. 4. Nucleotide sequence of *Euglena* cytochrome c_1 cDNA and deduced amino acid sequence for the mature protein. Linker sequence at 5'- and 3'-ends of the clone are omitted. Amino acid sequence is numbered from the amino-terminus of the mature protein. Underlines show the regions identified by the sequence analyses on the protein and peptide fragments from carboxymethyl-cytochrome c_1. An arrow indicates the 5'-end of another insert.

DISCUSSION

The most remarkable difference between the complexes III of *Euglena* and other organisms was the spectrophotometric properties of cytochrome c_1. Absorption maximum of the slightly asymmetric α-band of *Euglena* cytochrome c_1 was at 561 nm, being shifted by 8 nm to the red, if compared with those of other cytochromes c_1 at 552-553 nm.[25] This peculiar phenomenon was also found in *Euglena* cytochrome *c* (*c*-558),[2,26,27] which was the electron acceptor of the complex III and bound the heme *c* at only one cysteine residue through a single thioether bond, giving a confidence that *Euglena* cytochrome c_1 had also an atypical heme-binding mode as evidenced later.

The isolated *Euglena* complex III was composed of 10 polypeptide chains similar to those of other eukaryotes.[24] Two subunits, III and IV, were identified as the heme-binding subunits, cytochromes *b* and c_1, respectively, both of which were positive in peroxidase heme-staining on the electrophoretic gel, molecular size, 37 and 32.5 kDa, and the amino-terminal sequences.

Antimycin insensitive nature of the complex III after storing was regarded as the electron transfer catalyzed by only the iron-sulfur protein and cytochrome c_1, and cytochrome *b* was not involved.[2] The content of cytochrome *b* was lower, one mole per mole of cytocheome c_1, than expected, and this may be compatible to the present observation.

The amino-terminal sequence analysis of *Euglena* cytochrome c_1 indicated that the heme was linked through only a single thioether bond at cysteine-39, although no concrete evidence was obtained at this stage. The indication was solely based on the homologies of the *Euglena* sequence to those of other cytochromes c_1 in this region. A particular important residue, histidine-40, was conserved to be probably the fifth ligand to the heme iron. Therefore, the position 39 was probably cysteine affording the thioether bond.

Measurements of the mass of the heptadecapeptide and the retention times of the unknown residue at position 39 on an amino acid analyzer and on HPLC of its phenylthiohydantoin derivative suggested it to be S-(2-hydroxyethyl)cysteine, but a mechanism deriving this compound from heme *c* linkage was speculated with skeptical conclusions.[13]

Amino acid sequence of *Euglena* cytochrome c_1 is compared with those of others in Fig. 5 showing the cysteines linked to the heme (⌐) in the shaded boxes. They are apparently homologous with insertions or deletions at certain regions. An atypical heme-binding sequence in *Euglena* cytochrome c_1 was confirmed by establishing the total sequence of the mature protein. *Euglena* cytochrome c_1 binds the heme through a single thioether bond at cysteine-39 in a similar manner found in the electron transfer partner, *Euglena* cytochrome *c* (*c*-558).

Euglena cytochrome *c* and c_1, both of which are the electron transfer componets in mitochondria, showed the same mode of heme-binding as mentioned above in contrast with cytochrome *c*-552 in *Euglena* chloroplasts, which binds the heme through two thioether bonds as found in common.[36] Mitochondrial cytocheomes of C-type may have a common heme attachment mechanism in protozoa which may be different from those of the C-type cytochromes in *Euglena* chloroplasts and in other organisms.

Matsubara et al.

```
                                                         ┌─┐↓
              1         .        20         .        40              .
              Δ       Δ  Δ       *      Δ**   **    *  **Δ            *
(E)        GVDSHPPALPWPHFQWFQGLDWRSVRRGKEVYEQVFAPCHSLSFIKYRHF
(Y)        MTAAEHGLHAPAYAWSHNGPFETFDHASIRRGYQVYREVCAACHSLDRVAWRTL
(N)        MTPAEEGLHATKYPWVHEQWLKTFDHQALRRGFQVYREVCASCHSLSRVPYRAL
(B)        SDLELHPPSYPWSHRGLLSSLDHTSIRRGFQVYKQVCSSCHSMDYVAYRHL
(H)        SDLELHPPSYPWSHRGLLSSLDHTSIRRGFQVYKQVCASCHSMDFVAYRHL
(Rs)       NSNVQDHAFSFEGIFGKFDQAQLRRGFQVYSEVCSTCHGMKFVPIRTL
(Rc)       NSNVPDHAFSFEGIFGKYDQAQLRRGFQVYNEVCSACHGMKFVPIRTL
(P)       ·AAAQEAGDSHAAAHIEDISFSFEGPFGKFDQHQLQRGLQVYTEVCSACHGLRYVPLRTL

              60       Δ      Δ      .     Δ       Δ     *ΔΔΔ    *     ΔΔ Δ    100   ΔΔ   ΔΔ ΔΔΔΔ*
              Δ          Δ         Δ
(E)        EAF·MSKEEVKNMAASFEVDDDPDEKGEARKRPGKRFDTVVQPYKNEQEARYANNGALPP
(Y)        VGVSHTNEEVRNMAEEFEYDDEPDEQGNPKKRPGKLSDYIPGPYPNEQAARAANQGALPP
(N)        VGTILTVDEAKALAEENEYDTEPNDQGEIEKRPGKLSDYLPDPYKNDEAARFANNGALPP
(B)        VGVCYTEDEAKALAEEVEVQDGPNEDGEMFMRPGKLSDYFPKPYPNPEAARAANNGALPP
(H)        VGVCYTEDEAKELAAEVEVQDGPNEDGEMFMRPGKLFDYFPKPYPNSEAARAANNGALPP
(Rs)       SDDGGPQLDPTFVREYAAGLDTIIDKDSGEERDRKETDMFPTRVGDGMG·········P
(Rc)       ADDGGPQLDPTFVREYAAGLDTIIDKDSGEERDRKETDMFPTRVGDGMG·········P
(P)        ADEGGPQLPEDQVRAYAANFD·ITDPETEEDRPRVPTDHFPTVSGEGMG·········P

                           .              120         .         140          .
              ***  Δ  **Δ                 **  Δ*    ΔΔ**Δ       Δ  Δ        *Δ
(E)        DLSVITNARH·············GGVDYIYALLTGYGRPVPGGVQLST·TQWYNPY
(Y)        DLSLIVKARH·············GGCDYIFSLLTGYPDEPPAGVALPP·GSNYNPY
(N)        DLSLIVKARH·············GGCDYIFSLLTGYPDEPPAGASVGA·GLNFNPY
(B)        DLSYIVRARH·············GGEDYVFSLLTGYC·EPPTGVSLRE·GLYFNPY
(H)        DLSYIVRARH·············GGEDYVFSLLTGYC·EPPTGVSLRE·GLYFNPY
(Rs)       DLSVMAKARAGFSGPAGSGMNQLFKGIGGPEYIYRYVTGFPEENPACAPEGIDGYYYNEV
(Rc)       DLSVMAKARAGFSGPAGSGMNQLFKGMGGPEYIYNYVIGF·EENPACAPEGIDGYYYNKT
(P)        DLSLMAKARAGFHGPYGTGLSQLFNGIGGPEYIHAVLTGYDGEEKEEAGA···VLYHNAA

                                         ↓
                                        160         .         180          .
              *  Δ   Δ                   *               *****    *  ΔΔ**   ** *
(E)        FHGGII················GMPPPLTDDMIEYEDGTPASVPQMAKDVTCFLEWCS
(Y)        FPGGSI················AMARVLFDDMVEYEDGTPATTSQMAKDVTTFLNWCA
(N)        FPGTGI················AMARVLYDGLVDYEDGTPASTSQMAKDVVEFLNWAA
(B)        FPGQAI················GMAPPIYNEVLEFDDGTPATMSQVAKDVCTFLRWAA
(H)        FPGQAI················AMAPPIYTDVLEFDDGTPATMSQIAKDVCTFLRWAS
(Rs)       FQVGGVPDTCKDAAGIKTTHGSWAQMPPALFDDLVTYEDGTPATVDQMGQDVASFLMWAA
(Rc)       FQIGGVPDTCKDAAGVKITHGSWARMPPPLVDDQVTYEDGTPATVDQMAQDVSAFLMWAA
(P)        FAGNWI················QMAAPLSDDQVTYEDGTPATVDQMATDVAAFLMWTA

              200         .         220         .         240
              *  Δ  **  * Δ                          Δ          Δ
(E)        NPWWDERKLLGYKTIATLAVIAVSSGYYNRFLSGLWRSRRLAFRPFNYSK
(Y)        EPEHDERKRLGLKTVIILSSLYLLSIWVKKFKWAGIKTRKFVFNPPKPRK
(N)        EPEMDDRKRMGMKVLVVTSVLFALSVYVKRYKWAWLKSRKIVYDPPKRPPPATNLALPQQRAKS
(B)        EPEHDHRKRMGLKMLLMMGLLLPLVYAMKRHKWSVLKSRKLAYRPPK
(H)        EPEHDHRKRMGLKMLMMMALLVPLVYTIKRHKWSVLKSRKLAYRPPK
(Rs)       EPKLVARKQMGLVAVVMLGLLSVMLYLTNKRLWAPYKRQKA
(Rc)       EPKLVARKQMGLVAMVMLGLLSVMLYLTNKRLWAPYKGHKA
(P)        EPKMMDRKQVGFVSVIFLIVLAALLYLTNKKLWQPIKHPRKPE
```

Fig. 5. Sequence comparison of various cytochromes c_1. They are aligned to give a homology most probable by inserting gaps, ·. *, Identical residues at the same positions in all cytochromes c_1; Δ, those only in eukaryotic cytochromes. Numbering is for *Euglena* protein. Symbols and boxed regions are explained in the text. (E), *Euglena* (present paper); (Y), yeast[30]; (N), *Neurospora*[31]; (H), human[35]; (Rs), *Rhodobacter sphaeroides*[32]; (Rc) *R. capsulatus*[33]; (P), *Paracoccus*[34].

The most probable heme ligands have been supposed to be an imidazole nitrogen of histidine and a sulfur of methionine[4,37,38] and the comparison of all cytochromes c_1 sequences fulfils the conditions with histidine-40 and methionine-159 (*Euglena*'s numbering) indicated by arrows in Fig. 5. However, an EPR study suggested the sulfur atoms of a cysteine and a methionine to be responsible for the axial ligands contrary to the general idea.[39]

Proton NMR spectral studies[40] on beef ferri- and ferro-cytochromes c_1 showed that the spectral pattern for cytochrome c_1 was similar to that for *Pseudomonas* cytochrome c-551 whose coordination structure was different from that of cytochrome c, the orientation of the methionine residue with respect to the heme plane in cytochrome c_1 was apparently similar to that of cytochrome c-551, and the fifth ligand of the heme iron was not provided by a cysteine but by a histidine.

We have independently replaced histidine-44 (yeast numbering, Fig. 5, (Y)) to phenylalanine or tyrosine and methionine-164 to leucine or lysine, respectively, in yeast cytochrome c_1 by applying the site directed mutagenesis (Nakai, M., et al., unpublished results). The three mutants with phenylalanine, tyrosine, or leucine showed poor growths of yeast cells in a glycerol medium with no existence of cytochrome c_1 in the cells. In contrast to these, the one with lysine in place of methionine showed an apparent existence of cytochrome c_1 or a hemoprotein similar to it with a poor growth of the cells in a glycerol medium, indicating the incorporation of the heme into the protein moiety to be assembled in complex III. These results strongly suggest that the histidine-44 (histidine-41 in beef cytochrome c_1) and methionine-164 (methionine-160 in beef) are the axial ligands to the heme iron.

Cytochrome c_1 is inserted into the cytoplasmic side of the mitochondrial inner membrane as a component of the complex III.[41] We have demonstrated by deletion experiments with yeast cytochrome c_1 that the region between the carboxyl-terminal 17 and 71 resides is necessary for the function of cytochrome c_1.[9] This agrees with our earlier proposal that hydrophobic region near the carboxyl-terminus (Fig. 5, a region at around residue 210 in a shaded box) associates with, or is buried in, the membrane.[4,6]

We have also pointed out that beef cytochrome c_1 has two strongly acidic regions at around residue 70 (Fig. 5, in a box) and 170 (Fig. 5 in a shaded box) which are the candidates for interacting with cytochrome c.[4,6] These regions were chemically modified and a binding site in cytochrome c_1 for cytochrome c was postulated to be either the region at around residue 170[42] or that at around residue 70.[43] A deletion experiment suggested that the region at around residue 70 did not affect on the cell growth of yeast even after deletion, indicating that this region was unnecessary for the function of cytochrome c_1 (Nakai, M. et al., unpublished results).

ACKNOWLEDGEMENTS

We express our deep appreciations to Drs. Y. Takahashi and K. Saeki for their interests and discussions throughout these works and to Messrs. Y. Yao, H. Toyosaki, and Y. Yoshida for their technical assistances. A generous gift of a *Euglena* culture and helpful suggestions of Drs. S. Kitaoka and Y. Nakano are also greatly appreciated. Unpublished data on gene manipulation experiments with yeast cytochrome c_1 were kindly informed by Mr. M. Nakai.

REFERENCES

1. Dickerson, R. E., and Timkovich, R. (1975) in *The Enzymes 11* (Boyer, P D., ed.) pp.397-547, Academic Press, New York
2. Pettigrew, G. W., Leaver, J. L., Meyer, T. E., and Ryle, A. P. (1975) *Biochem. J.* **147**, 291-302
3. Scarpless, T. K., and Butow, R. A. (1970) *J. Biol. Chem.* **245**, 50-57
4. Wakabayashi, S., Matsubara, H., Kim, C. H., Kawai, K., and King, T. E. (1980) *Biochem. Biophys. Res. Commun.* **97**, 1548-1554
5. Wakabayashi, S., Takeda, H., Matsubara, H., Kim, C. H., and King, T. E. (1982) *J. Biochem.* **91**, 2077-2085
6. Wakabayashi, S., Matsubara, H., Kim, C. H., and King T. E. (1982) *J. Biol. Chem.* **257**, 9335-9344
7. Mukai, K., Miyazaki, T., Wakabayashi, S., Kuramitsu, S., and Matsubara, H. (1985) *J. Biochem.* **98**, 1417-1425
8. Mukai, K., and Matsubara, H. (1986) *J. Biochem.* **100**, 1165-1173
9. Hase, T., Harabayashi, M., Kawai, K., and Matsubara, H. (1987) *J. Biochem.* **102**, 401-410
10. Hase, T., Harabayashi, M., Kawai, K., and Matsubara, H. (1987) *J. Biochem.* **102**, 411-419
11. Mukai, K., and Matsubara, H. (1987) in *Adv. in Membrane Biochemistry and Bioenergetics* (Kim, C. H., Tedeschi, H., Diwan, J. J., and Salerno, J. C., eds.) pp.179-184, Plenum Press, New York
12. Mukai, K., Yoshida, M., Yao, Y., Wakabayashi, S., and Matsubara, H. (1988) *Proc. Japan Acad.* **64**, *Ser. B*, 41-44
13. Mukai, K., Yoshida, M., Toyosaki, H., Yao, Y., Wakabayashi, S., and Matsubara, H. (1989) *Eur. J. Biochem.* **178**, 649-656
14. Nakai, M., Harabayashi, M., Hase, T., and Matsubara, H. (1989) *J. Biochem.* **106**, 181-187
15. Chirgwin, J. J., Przbyla, A. E., Mac Donald, R. J., and Rutter, W. J. (1979) *Biochemistry* **18**, 5294-5299
16. Aviv, H., and Leder, P. (1972) *Proc. Natl. Acad. Sci. U.S.A.* **69**, 1408-1412
17. Gubler, U., and Hoffman, B. J. (1983) *Gene (Amst.)* **25**, 263-269
18. Huynh, T. V., Young, R. A., and Davis, R. W. (1985) in *DNA Cloning, A Practical Approach* (Glover, D., ed.) Vol. I, pp.49-78, IRL Press, Oxford
19. Mierendorf, R. C., Percy, C., and Young, R. A. (1987) *Methods in Enzymol.* **152**, 458-469
20. Weinberger, C., Hollengerg, S. M., Thompson, E. B., Harmon, J. M., Brower, S. T., Cidlowski, J., Thompson, E. B., Rosenfeld, M. G., and Evance, R. M. (1985) *Science* **228**, 740-742
21. Sanger, F., Nicklen, S., and Coulson, A. R. (1977) *Proc. Natl. Acad. Sci. U.S.A.* **74**, 5463-5467
22. Tabor, S., and Richardson, C. C. (1987) *Proc. Natl. Acad. Sci. U.S.A.* **84**, 4767-4771
23. Kadenback, B., Jarausch, J., Hartmann, R., and Merle, P. (1983) *Anal. Biochem.* **129**, 517-521
24. Hatefi, Y. (1985) *Ann. Rev. Biochem.* **54**, 1015-1069
25. Yu, C. A., Yu, L., and King, T. E. (1972) *J. Biol. Chem.* **247** 1012-1019
26. Pettigrew, G. W. (1973) *Nature* **241**, 531-533
27. Lin, D. K., Niece, R. L., and Fitch, W. M. (1973) *Nature* **241**, 533-535

28. Paul, K. G., Theorell, H., and Akeson, A. (1953) *Acta Chem. Scand.* **7**, 1284-1287

29. Spackman, D. H., Moore, S., and Stein, W. H. (1958) *Anal. Chem.* **30**, 1190-1206

30. Sadler, I., Suda, K., Schatz, G., Kaudewitz, F., and Haid, A. (1984) *EMBO J.* **3**, 2137-2143

31. Römisch, J., Tropschug, M., Sebald, W., and Weiss, H. (1987) *Eur. J. Biochem.* **164**, 111-115

32. Gabellini, N., and Sebald, W. (1986) *Eur. J. Biochem.* **154**, 569-579

33. Davidson, E., and Daldal, F. (1987) *J. Mol. Biol.* **195**, 13-24

34. Kurowski, B., and Ludwig, B. (1987) *J. Biol. Chem.* **262**, 13805-13811

35. Nishikimi, M., Ohta, S., Suzuki, H., Tanaka, T., Kikkawa, F., Tanaka, M., Kagawa, Y., and Ozawa, T. (1988) *Nucleic Acid Res.* **16**, 3577

36. Pettigrew, G. W. (1974) *Biochem. J.* **139**, 449-459

37. Kaminsky, L. S., Chiang, Y.-L., and King, T. E. (1975) *J. Biol. Chem.* **250**, 7280-7287

38. Meyer, T. E., and Kamen, M. D. (1982) *Adv. Prot. Chem.* **35**, 105-212

39. Tervoort, M. J., and Van Gelder, B. F. (1983) *Biochim. Biophys. Acta* **722**, 137-143

40. Funahashi, T. (1987) *Some Structural Studies of Hemo-proteins in Mitochondrial Respiratory Chain*, Doctoral thesis, No. 960, Faculty of Engineering, Kyoto University

41. Bell, R. L., Sweetland, J., Ludwig, B., and Capaldi, R. A. (1979) *Proc. Natl. Acad. Sci. U.S.A.* **76**, 741-745

42. Broger, C., Salardi, S., and Azzi, A. (1983) *Eur. J. Biochem.* **131**, 349-352

43. Stonehuerner, J., O'Brient, P., Geren, L., Millett, F., Steidl, J., Yu, L., and Yu, C.-A. (1985) *J. Biol. Chem.* **260**, 5392-5398

HOW WELL IS CYTOCHROME *c* ENGINEERED?

Emanuel Margoliash[*], Abel Schejter[†], Thomas I. Koshy[*], Thomas L. Luntz[*], and Eric A. E. Garber[*]

[*]Department of Biochemistry, Molecular Biology and Cell Biology
Northwestern University, Evanston, Illinois 60208, USA and
[†]Sackler Institute of Molecular Biology, Sackler Faculty of Medicine
Tel Aviv University, Tel Aviv, Israel 69978

SUMMARY

For a protein as old as cytochrome *c*, which shows exquisite adaptation to its particular environment in the species carrying it, and which has been extensively studied, the properties of site-directed mutants should be readily predictable and fall within our understanding of how the evolutionary process has operated on the protein. Among the relatively few mutants at evolutionarily constant residues studied so far, some yield cytochromes *c* of apparently normal functionality and dramatically greater stability than the wild-type protein. This poses the question of how close to structural perfection have the present day cytochromes *c* evolved?

Examples are discussed of such stabilization mutants at tyrosine 67, asparagine 52 and threonine 78, all three residues being hydrogen bonded to a molecule of water held in the interior of the protein. In contrast, mutants at the invariant proline 30 show the expected deterioration in function. To explain this situation, it is suggested that the mutations of this series which impart improved stability, are nevertheless potentially deleterious in that they may allow mutations at other sites, which would otherwise be of minor consequence, to result in a grossly disturbed protein. Such an evolutionarily important phenomenon may be termed *second site deficiency*. The significant energetic cost of maintaining the internal water molecule structure of the native protein, which appears to be the function held in common by residues 67, 52 and 78, may be used to preclude such a drastic eventuality, thereby keeping open a variety of evolutionary pathways to the protein that would otherwise be forbidden.

INTRODUCTION

Examining the relation between structure, functional activities and evolution in cytochrome c

Mitochondrial cytochrome *c* has probably been studied more extensively than any other protein (1-4). Just to list in useful detail all the work done on it, or in which it was used as a model globular protein for numerous studies, would exceed the space available for this presentation.   Of those more directly pertinent to the present argument are: the amino acid sequences of over one hundred eukaryotic cytochromes c (5,6), some unpublished, which have given particularly good estimates of the acceptable variability of each residue; high resolution crystallographic studies of the cytochrome c of the tuna (7,8), rice (9) and of baker's yeast (10), from crystals grown in near-saturated ammonium sulfate, and more recently, with crystals grown at the lower ionic strength (11), at which the structure more closely resembles that which exists under physiological conditions; identification of the whole 2D-NOESY nmr spectrum of the protein (12,13), allowing the examination of fine structural changes in solution; studies of the properties of a remarkably large number of chemically-modified cytochromes c (14-17), those at the surface-located lysines having served to identify the relatively large area on the 'front' of the protein which includes the solvent-accessible edge of the heme and serves as the interaction domain with its mitochondrial reaction partners (18); extensive examinations of the properties of the heme and its surroundings; numerous studies of the electron exchange properties of the protein, in terms of kinetics and binding, with both inorganic redox compounds and its physiological protein reaction partners (3,4,19), including cytochrome oxidase (cytochrome aa_3), cytochrome c reductase (cytochrome $bc_{1 \text{ complex}}$), yeast cytochrome c peroxidase, cytochrome b_5 and sulfite oxidase, mostly as ordinary solution kinetics, except for the peroxidase with which it is also possible to measure the rate of electron transfer in the preformed complex of enzyme and cytochrome c (20-22); and most recently, the cloning of nine cytochrome c genes, including some from mammals, birds, insects, coelenterates and fungi (23-32), their directed mutagenesis and expression in cultures of *Saccharomyces cerevisiae*.

With such an extensive background, one may have expected that the properties of site-directed mutants would be readily understood, so that given a determination of the spatial conformation of such a mutant by nmr or x-ray crystallography, it should not be difficult to interpret the relation between such a structure and the known evolutionary variations of cytochrome c over the widest of taxonomic scales.   Indeed, there are several examples indicating that cytochrome c is well adapted to its physiological environment in a given species, and in particular to the cytochrome oxidase which is clearly part of the selective environment of the protein.   Thus, the simian cytochromes c, those of apes, Old World and New World monkeys, react normally with higher primate cytochrome oxidases, but bind too tightly to non-primate mammalian oxidases leading to an unusually low steady state turnover rate (33). Conversely, non-primate mammalian cytochromes c bind weakly to primate oxidases and show the corresponding electron transfer kinetic parameters.   Clearly, primate cytochromes c and oxidases have co-evolved, so as to maintain an overall binding affinity compatible with an appropriate turnover, the cytochrome c moving towards tighter binding, the oxidases towards weaker.   These changes are probably related to the substitution of several residues in the enzymic interaction domain of the cytochrome c for more hydrophobic ones, and the decrease in the number of hydrophobic and negatively charged residues in the segment of subunit II of the higher primate oxidases considered to represent the binding site for cytochrome c (34). Interestingly, the prosimian primate proteins, from the lemur and the tree shrew, react either in an intermediate fashion or like the non-primate mammalian cytochromes c.

Another such example is provided by the fungal cytochromes c from baker's yeast, *S. cerevisiae*, iso-1 and iso-2, from *Candida krusei* (35) and from the smut fungus, *Ustilago sphaerogena* (72).   All of these proteins react normally with baker's yeast cytochrome oxidase, but bind too tightly to beef or horse cytochrome oxidase, leading to a decreased turnover rate.   Conversely, the horse protein binds weakly to the yeast oxidase.   Here again, the contribution of the cytochrome c to the overall binding affinity appears to have increased in the fungal proteins, while that of the cytochrome oxidase appears to have decreased, maintaining a balance leading to 'normal' kinetic activity.

A possibly important complication in attempting to relate amino acid sequence changes to function, and functional changes to evolutionary development, by studying the properties of natural and site-directed mutants of the holoprotein, is introduced by the complex life cycle of the protein. The apoprotein is made on free cytoplasmic ribosomes, it penetrates into the intermembrane space of mitochondria, by unknown means and by an unknown pathway, and then has the heme prosthetic group covalently bound to cysteines 14 and 17 *via* thioether bonds by the heme lyase. The protein then presumably folds or refolds into its final structure. The holoprotein is no longer capable of traversing the outer mitochondrial membrane and remains confined for the rest of its rather long life in its functional location, the intermembrane space of the mitochondrion (36-38). The structural requirements underlying this biosynthetic process need to be understood as thoroughly as those affecting the functional activities of the holoprotein. Otherwise, it may not be possible to judge, for example, whether a site-directed mutant at an evolutionarily invariant residue lacking any detected effect on the holoprotein, nevertheless represents an unacceptable structure because it has a deleterious effect on the biosynthetic pathway. This is likely to be particularly significant under natural conditions different from those for a laboratory grown strain. That this requirement may not be as onerous as implied, has recently been indicated by studies showing that the carboxyl-terminal half of the amino acid sequence is not required for the protein to follow successfully the biosynthetic pathway, both *in vivo* (R. Scarpulla, personal communication) and *in vitro* (38,39).

Thus, with its complex biosynthesis, its long life cycle and clear-cut examples of exquisite adaptation of the protein to its specific environment, it could be expected that each cytochrome *c* has reached a high level of adaptation, including appropriate structural stability and functional capacity. Evolutionarily constant residues should not be capable of substitution without some loss of a significant attribute of the protein, while mutations at variable residues should produce cytochromes *c* in which defects of structure and/or function may in some cases be more difficult to detect. This may require careful testing with physiological interaction partners from the species of origin. Most of the site-directed mutants examined to date have been of yeast cytochromes *c* grown in cultures of yeast (40-54). More recently, some vertebrate and invertebrate cytochromes *c* have also been expressed in yeast, and mutants of these proteins described (55-60). Though the expected relation between evolutionary invariance and stringency of structural requirements is often observed, this is by no means always the case, at least by the relatively crude tests of functional attributes applied to date, *in vitro* and *in vivo*. With the exception of the set of mutants we term the 'water mutants' described below, in which the proteins appear to be improved in comparison to the wild-type (48,59), the unusual situations have been cases in which changes in invariant residues have apparently not led to the expected deterioration of properties (see, for example, Refs. 41,44,47,48).

Here we also discuss site-directed mutants of rat and *Drosophila melanogaster* cytochromes *c* at proline 30 (56,58), an invariant residue situated on the 'right lower front side' of the protein, which cause a destabilization of the bond between the sulfur of methionine 80 and the heme iron, namely a weakening of the functional closed form of the heme crevice. These are contrasted to site-directed mutants of rat and *S. cerevisiae* cytochromes *c* at tyrosine 67, taken to phenylalanine (Y67F)[1], at asparagine 52, taken to isoleucine (N52I) and at threonine 78, taken to valine (T78V). In their wild-type state, these three residues have their side chains hydrogen bonded to a single water molecule, held in the inside of the protein on the 'left side' near the lower edge of the heme prosthetic group. The non-hydrogen bonding side chains introduced into these mutants apparently result either in the expulsion of the internal water

---

[1]Abbreviations used: Y67F, the mutant in which tyrosine 67 was taken to phenylalanine; N52I, asparagine 52 to isoleucine; T78V, threonine 78 to valine; T78S, threonine 78 to serine; P30A, proline 30 to alanine; P30V, proline 30 to valine; DMc, *Drosophila melanogaster* cytochrome *c*; RNc, rat cytochrome *c*; CYC1, the gene encoding the iso-1-cytochrome *c* gene; YPL, rich lactate medium; YPD, rich dextrose medium.

molecule, or other changes of this hydrogen bonded water molecule network. These are the proteins we term 'water mutants.' All three sustain the growth of yeast cultures at normal rates, all are more stable than the wild-type proteins with regard to the methionine 80 sulfur-heme iron bond, and the phenylalanine 67 and isoleucine 52 mutants are as or more active than the wild-type in their reaction with cytochrome oxidase *in vitro*. The valine 78 mutant displays a large change in the heme environment, which occurs as it is extracted from the yeast. This structural rearrangement results in a non-functional auto-oxidizable protein with a much lower redox potential, a catastrophic change which occurs despite the fact that the yeast have grown at a normal rate utilizing the cytochrome *c* for the operation of the terminal respiratory chain.

A hypothesis is offered to explain the evolutionary invariance of the destabilizing hydrogen bonded internal water molecule structure, even though appropriate mutants of the three hydrogen bonding residues yield cytochromes *c* that in several cases are better than the wild-type proteins in terms of stability and, in two cases examined so far, for electron transfer function.

EXPERIMENTAL PROCEDURES

Expression of recombinant cytochromes c

Expression of the rat cytochrome *c* pseudogene, RC9, coding for the natural rat cytochrome *c*, and its mutants in cultures of *S. cerevisiae* was as described (59). In brief, the coding sequences and 50 bases of 5' nontranslated region of the gene were cloned behind the yeast CYC1 promoter and this 1.2 kb fragment cloned into the yeast episomal plasmid YEp13 (61). Alternatively, the yeast-rat chimeric gene, the 2 micron DNA and the Leu2 marker from YEp13 were cloned into the pGEM-7Zf(+) vector (Promega Biotec) so that the same construction could be used as a shuttle vector for mutagenesis and expression in yeast.

The gene which expresses the previously cloned (29,30) *D. melanogaster* cytochrome *c* in yeast consists of the complete CYC1 promoter, including the 5' nontranslated region of the gene, behind which is ligated the drosophila cytochrome *c* gene coding sequence. To this end, the start codon of the CYC1 gene was mutated to an ATC and the 1.9 kb Bam HI-Eco RI fragment used as the promoter. The *D. melanogaster* cytochrome *c* gene was digested with Xmn I which cleaves the gene 40 bases downstream from the start codon. The 5' end of the gene was reconstructed using a pair of complementary oligonucleotides which regenerate the 40 bases of the gene with an Eco RI site immediately upstream from the start codon. This fragment was cloned between the modified CYC1 promoter and a 350 bp fragment of the 3' end of the gene, the latter deemed necessary to ensure proper termination of transcription. The resulting Bam HI-Hind III 2.5 kb gene was cloned into YEp13 for expression in yeast. The CYC1 gene was cloned as a Bam HI-Hind III fragment into YEp13 (Fig. 1).

Site-directed mutagenesis was performed in M13 by the procedure of Zoller and Smith (62) using, as template, DNA isolated from bacterial strain JM105 infected with the appropriate phage. Alternatively, the protocols of Kunkel et al. (62) were used. This procedure utilizes a strain of *Escherichia coli*, CJ236, a bacterium deficient in two enzymes: dUTPase (*dut*), producing elevated levels of dUTP in the cell which is incorporated into the DNA, and uracil N-glycosylase (*ung*), which results in the presence of uracil in the DNA to remain uncorrected. Mutagenesis of phage DNA which has been passaged through (*dut, ung*) *E. coli*, when transformed into cells wild-type for these two enzymes, results in a mutation frequency of 50-80%. This eliminates the need to identify putative mutants by plaque lifts and hybridization experiments. Transformants are simply sequenced directly to find an appropriate mutant.

Mutant genes, once cloned behind the appropriate promoters and into the shuttle

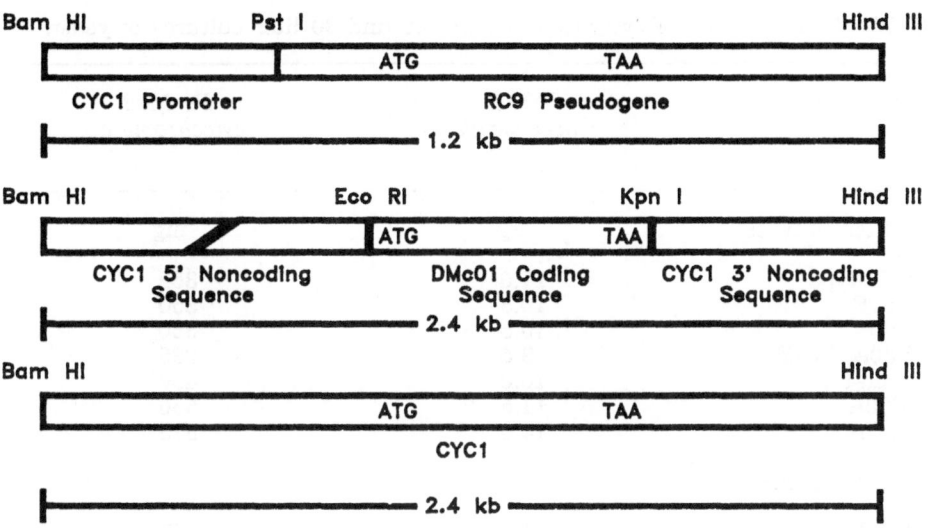

Figure 1. Construction of genes for the expression of the rat, *Drosophila melanogaster* and yeast cytochromes c. The Bam H1-Hind III fragments were cloned into the shuttle vector YEp13.

vector, were transformed into the yeast strain GM-3C-2 [trp1-1, leu2-2, leu2-112, his4-519, CYC1-1, cyp3-1, (cyc7⁻), gal⁻] (64). This strain does not express either of the two forms of yeast cytochromes c, iso-1 and iso-2. Cells were plated on simple dextrose medium supplemented with tryptophan and histidine to select for transformants. These were then replica gridded on media containing a non-fermentable carbon source (lactate or glycerol) supplemented with the same amino acids to determine whether or not the cytochrome c introduced could complement the yeast electron transport chain.

Large scale production and purification of recombinant cytochromes c

Growth of large (30 or 240 liter) cultures was accomplished using modifications of the protocol of Stewart *et al.* (65). Cultures of 5 ml were started in a rich lactate medium (YPL medium: 1% yeast extract, 2% Bacto peptone, 1% lactate neutralized to pH 5 with potassium hydroxide) at 30 °C. Once these cultures had reached stationary phase they were used to inoculate three flasks containing the same YPL medium to a minimum density of 1×10^5 cells/ml. These cultures were incubated at 30 °C with vigorous aeration until they had reached stationary phase, and were used to inoculate 12 liters of rich sucrose medium (2.7% yeast extract, 2.7% Bacto-peptone, 4% sucrose), supplemented with 10 ml of 98% ethanol, 50,000 units of penicillin and 50 mg of streptomycin, per liter, in a Braun 30D Bioreactor. Vigorous aeration and stirring were maintained. When the culture reached a density of 2-3×10^8 cells/ml, 15 more liters of the supplemented rich medium were added over a period of 10 to 12 hours. After the medium feed was complete, the vessel was allowed to incubate for another 4 to 6 hours to ensure the culture had reached stationary phase. Cells could be harvested at this point and frozen at -20 °C as centrifuged cakes of yeast. Alternatively, the 30 liter culture was used to inoculate 150 liters of supplemented rich media in a Braun 300D Bioreactor. After 4 hours, another 30 liters of medium were added and this repeated once more. The 240 liter culture was allowed to grow overnight and harvested the next morning. This strategy takes advantage of the observation that the greatest levels of cytochrome c are produced some time after the cells reach stationary phase (see Fig. 2D). By maintaining thick cultures and increasing their volumes slowly, much larger amounts of cells and cytochrome c were obtained than when, for example, 3 liters of inoculum were added to 27 liters of

Table I. Production of cytochromes c in 240 and 30 liter cultures of yeast.

Cytochromes c	Yeast (wet weight)	Purified cytochrome c
In 240 liter cultures:	kg	mg
RNc, wild-type	11.0	352
RNc, Y67F	10.5	350
RNc, P30A	10.5	360
RNc, P30A, Y67F	9.5	285
DMc, wild-type	10.0	450
DMc, P30A	11.5	550
DMc, A47S	14.0	875
In 30 liter cultures:		
RNc, N52I	1.40	75
RNc, P30V	0.98	62
CYC1, N52I	0.75	75
CYC1, Y67F	0.87	80
CYC1, T78S	1.22	102

medium and the culture grown overnight. Cell and cytochrome c yields obtained from these protocols are shown in Table I.

Cytochrome c was extracted from the yeast using a modification of the method of Sherman *et al.* (66). All steps were carried out at 4 °C. Yeast (1-2 kg, centrifuged wet weight) was stirred with 500 ml of ethyl acetate and 250 ml of 1M NaCl/kg overnight, the extract centrifuged (4000g, 30 min), the supernatant clarified by two filtrations through Hy-Flo Supercell (Fisher Scientific), diluted 20-fold with 10 mM phosphate buffer, pH 7, and passed through an 8.5 x 2.5 cm bed of carboxymethyl cellulose (Whatman) (CMC). The resin was washed with 1 l of phosphate buffer, repacked, and the cytochrome c eluted with 0.1 M phosphate, pH 7.0, containing 0.5M NaCl. The colored eluate was brought to 65% saturation with ammonium sulfate for 1 hour, the precipitate removed by filtration and the filtrate dialyzed against 5 to 6 changes of 10 mM phosphate, pH 7, over 48 hours, the first changes of buffer containing a few drops of concentrated ammonium hydroxide. After dialysis the cytochrome c was concentrated on CMC. The rat and yeast cytochromes c were eluted as before and dialyzed against the 10 mM phosphate buffer to prepare them for HPLC purification. The drosophila cytochrome c, when concentrated after dialysis, was displaced from the top of the resin, apparently by some other proteins, from which it could be separated by developing the column with 10 mM phosphate, pH 7, containing 90 mM NaCl. The cytochrome c was diluted, concentrated on CMC and prepared for HPLC.

Yeast and rat cytochromes c were purified on a sulfopropyl preparative HPLC column (Waters SP 5PW, 21.5 mm x 15 cm) operated at a flow rate of 2 ml/min. The column was equilibrated in the A buffer (10 mM phosphate, pH 7.0, 2 mM dithiothreitol) and developed with an elution buffer, B (Buffer A containing 250 mM NaCl for the rat cytochromes c, and 500 mM NaCl for the yeast cytochromes c), after loading 25 to 30 mg of protein. The yeast cytochromes c were developed by washing the column with buffer A for 5 min. increasing the proportion of buffer B to 75% over 25 min., maintaining this value for 10 min., and then increasing it to 100% over 50 min. The rat cytochromes *c* were chromatographed by bringing buffer B to 50% over 10 minutes and increasing it to 100% over the next 200 minutes. The drosophila cytochromes *c* were chromatographed on a carboxymethyl silica column (SynChropak CM300, 250 x 21.2 mm) employing the same buffers as those used for the rat cytochromes *c*, operated at a flow rate of 10 ml/min. Cytochrome *c* (10 mg) was loaded, washed for 5 min. the percent of buffer B increased to 30% over 15 min. and then increased another 10% over 2 hours. Fractions were collected, pooled and rechromatographed if necessary. All chromatographies were monitored at 410 or 526 nm, isosbestic points for ferric and ferrous cytochrome *c*.

Analytical methods

Visible spectra were recorded on a Hitachi model 557 double wavelength double beam spectrophotometer. Spectra of the 695 nm band were corrected as described by Kaminsky *et al.* (67). Alkaline and acid titrations of the 695 nm band were performed on samples of approximately 0.1 mM ferricytochrome *c* in 0.1M KCl. The pH was changed by touching to the solution a sealed pipette which had been dipped in 0.1 to 1 N HCl or KOH and allowed to partially dry, and was monitored in the cuvette using a microelectrode. Thermal titrations of ferricytochrome *c* were carried out in 0.1 M phosphate, pH 6.5 employing a variable temperature water bath connected to the cuvette assembly of the spectrophotometer. Temperatures were recorded in the cuvette using a Cole-Parmer 8508-45 Type K thermocouple thermometer (Cole-Parmer Instrument Co.). The temperature was increased in less than 5 °C increments and the system allowed to equilibrate completely before spectra were recorded. Urea titrations were carried out in 0.1 M phosphate, pH 7.0, by dissolving the protein in a high concentration of urea and monitoring the reappearance of the 695 nm band as the sample was diluted with buffer. Appropriate volume corrections were employed in determining the final urea concentrations.

The cytochrome *c* content of yeast cells was measured using a modification of the method of Williams *et al.* (68). From cultures of yeast grown in 1 liter of a rich dextrose medium (YPD medium: 2% Bacto peptone, 1% yeast extract, 2% dextrose), aliquots were removed, the cells centrifuged (4000g, 20 min.), weighed and frozen. The samples were resuspended at a concentration of 0.5 gm wet weight yeast/ml in 1 M sorbitol, 0.1 M sodium citrate, pH 7.0, 0.06 M EDTA containing 1 mg/ml zymolyase 20T (Miles Laboratories) and gently mixed at room temperature for 40 min. Yeast suspension (0.5 ml), 0.1 ml of 0.1 M sodium phosphate, pH 6.5, and 0.1 ml of 5% deoxycholate were placed in each of two 1 ml cuvettes which were balanced against each other between 500 nm and 650 nm. The material in one cuvette was reduced by the addition of 0.1 ml of 50 mM sodium ascorbate and a few grains of dithionite, while 0.1 ml of 50 mM potassium ferricyanide was added to the other cuvette to oxidize the reference sample. The samples were well mixed with a pipetting device and by repeated inversion. After 2 minutes the difference spectrum was recorded and the amount of cytochrome *c* per gram of yeast was determined using the extinction coefficients given by Williams (68).

RESULTS

Growth of yeast strains carrying mutant cytochromes c

Employing the yeast strain GM-3C-2, which does not express either iso-1 or iso-2 cytochrome *c*, the genes being deleted and carrying a nonsense mutation, respectively (64), and inoculating cells transformed with either the wild-type or various mutant cytochrome *c* genes carried by the multi-copy plasmid, YEp13, into a rich sucrose medium it became evident that, under strong aeration, the mutant strains behaved identically to that carrying the corresponding wild-type (Fig. 2). The cells grew rapidly to a density of 1 to 5 x 10^8 cells/ml, the cytochrome *c* content started rising rapidly near the end of the log phase of growth, peaked some 3 to 6 hours into the stationary phase and dropped irregularly after that (see Fig. 2D). The growth of the cultures bearing mutant cytochromes *c* were also the same as those with the wild-type protein when energy could only be derived by respiration, as in a 1% lactate medium, YPL (see Fig. 2A for rat cytochromes *c* and Fig. 2B for yeast iso-1-cytochromes *c*). Thus, the growth curves and cytochrome *c* contents were indistinguishable for the strains carrying the tyrosine 67 to phenylalanine mutant rat cytochrome *c* and the wild-type rat protein, the proline 30 to alanine drosophila cytochrome *c* and the wild-type drosophila protein, and the yeast tyrosine 67 to phenylalanine, asparagine 52 to isoleucine and threonine 78 to serine iso-1 cytochromes *c* and the wild-type yeast iso-1 cytochrome *c*. However, it was clear that the yeast cytochrome *c* caused the cells to grow about twice as fast as the rat protein (Fig. 2A).

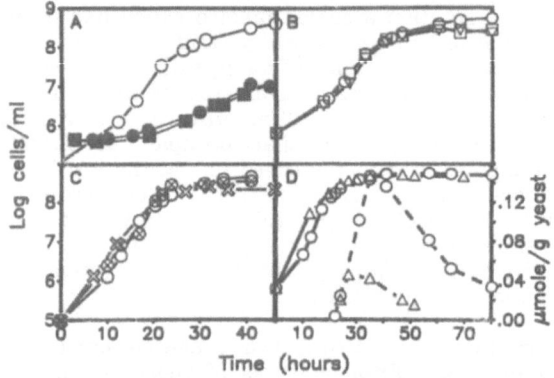

Figure 2. Growth curves of yeast (GM-3C-2) transformed with various cytochrome *c* genes in YEp13. **A.** CYC1, wild-type (open circles); RNc, wild-type (closed squares), grown in YPL. **B.** CYC1, wild-type (open circles); CYC1, N52I (open inverted triangles); CYC1, Y67F (open squares), grown in YPL. **C.** CYC1, wild-type (open circles); DMc, wild-type (crossed circles); DMc, P30A (open crosses), grown in YPD. **D.** CYC1, wild-type (open circles); CYC1, T78V (open triangles), grown in YPD; the cytochrome *c* content of these cultures (μmoles/gram wet weight of yeast are shown with the same symbols and dashed lines.

The only exception to these types of relation in the growth of mutants and corresponding wild-type proteins observed so far, occurred in the case of the threonine 78 to valine mutant of the yeast protein. Here, the mutant protein supported growth as well as the wild-type protein in the 2% dextrose medium (YPD), but the cytochrome *c* content remained much lower with the mutant (Fig. 2D). Because once prepared, the threonine 78 to valine iso-1 yeast cytochrome *c* spontaneously changes to a functionless protein (see below), it is probable that the strain carrying such an unstable protein managed to grow by continuing to make cytochrome *c* at a rate sufficient to replace that which had deteriorated. It is also possible that intracellular conditions tend to slow down the process occurring upon purification. Nevertheless, it is remarkable that no change in growth could be detected, even though growth on a medium containing glucose presumably requires cytochrome c function only in its later stages when the sugar is depleted. Indeed, even in lactate medium (YPL), this mutant strain grew as well as that carrying the wild-type protein. Furthermore, it is difficult to argue that under the conditions employed cytochrome c function was unnecessary at all stages of the growth curve, since the yeast strains carrying the rat protein grew distinctly slower than those carrying the yeast cytochrome c in the lactate medium (Fig. 2A). It is also interesting to note that in the glucose medium the strain carrying the drosophila protein grew indistinguishably from that carrying yeast iso-1 cytochrome c (Fig. 2C).

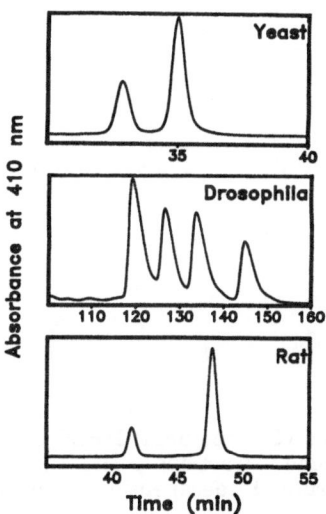

Figure 3. Chromatography of recombinant ferrous cytochromes c on HPLC cation exchange resins (see Experimental Procedures). The two fractions of yeast and rat cytochromes c are both trimethylated at lysine 72, but differ in that the front fraction is blocked at the N-terminus. The four drosophila cytochromes c are all unblocked and differ by being tri-, di-, mono-, and unmethylated at lysine 72 in order of elution.

Identity of the recombinant cytochromes c prepared in cultures of S. cerevisiae

When the cytochromes c prepared by the recombinant procedures described above were examined by cation exchange HPLC all appeared to consist of more than one fraction. Thus, the rat cytochrome c showed two peaks (Fig. 3). Upon being subjected to Edman degradation on an Applied Biosystems model 4775 protein sequencer, the first fraction, Fraction I, appeared to be blocked, while Fraction II sequenced normally, yielding the N-terminal amino acid sequence of rat cytochrome c for 10 residues.

Furthermore, the chymotryptic N-terminal decapeptide of the recombinant Fraction I material co-chromatographed on an HPLC C-18 reversed phase column under a gradient from 0 to 45% acetonitrile, containing throughout 0.1 M K_2HPO_4, pH 7.2 (70), precisely with the corresponding chymotryptic peptide of the natural protein prepared from rat hearts (Acetyl-G-D-V-E-K-G-K-K-I-F) (5), while the corresponding peptide of Fraction II, chromatographed in front of that from Fraction I (Fig. 4). This indicated that the yeast cells could N-terminally acetylate the rat cytochrome c. For both the wild-type and mutant recombinant rat cytochromes c prepared under the growth conditions described above, about a quarter of the protein was obtained in the N-terminally acetylated form, as earlier reported (59). However, these observations did not complete the identification of the recombinant products, because Fraction I did not co-chromatograph with natural rat cytochrome c by HPLC cation exchange

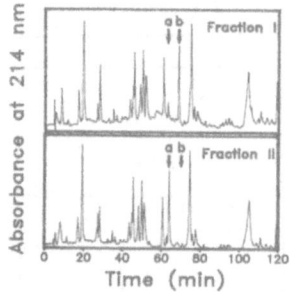

Figure 4. Reverse phase (C-18 column) HPLC of chymotryptic digests of rat wild-type cytochrome c fractions I and II. a and b indicate the positions of the N-terminal decapeptide, unblocked and blocked respectively.

chromatography, but migrated slightly before it. The problem was solved when it was found that the yeast cells also caused extensive, but not always complete, trimethylation of lysine 72 in all the recombinant cytochromes c. This secondary modification could best be quantified by nmr, determining the number of protons corresponding to the resonance at 3.35 ppm of the ferricytochrome c. In this way it could be shown that both Fraction I and Fraction II of the rat protein were fully trimethylated, the 3.35 ppm resonance yielding approximately 9 protons per molecule of cytochrome c, respectively, while the natural rat protein showed no trimethyl-lysine. It could also be shown that the chymotryptic peptide containing lysine 72 and prepared from either Fraction I or II (L-E-N-P-K-K-Y, Residues 68 to 74), chromatographed on reverse phase HPLC slightly in front of the corresponding peptide from the natural protein.

In the case of the recombinant drosophila cytochrome c the situation was somewhat different. A major peak that did not appear to be homogeneous was obtained by cation exchanger HPLC. This material yielded a normal drosophila

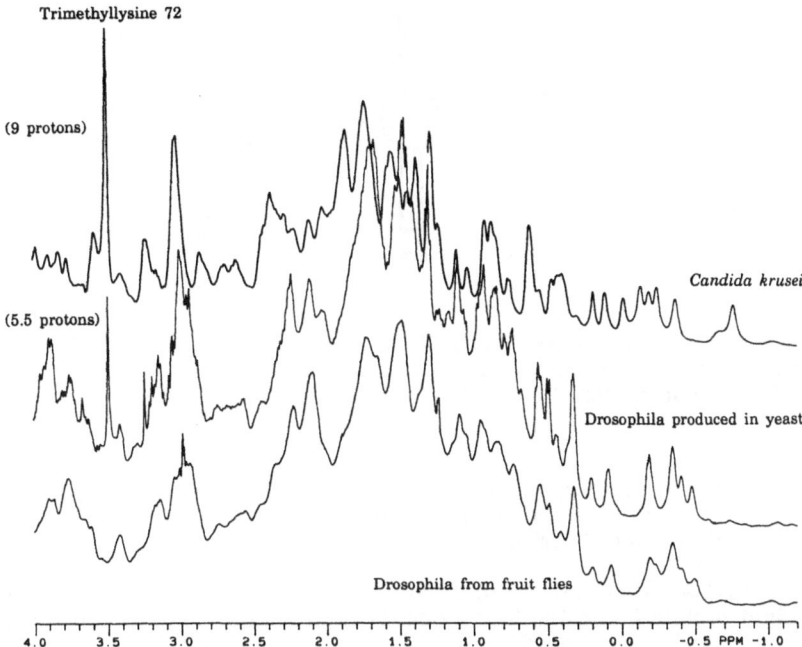

Figure 5. ¹H-nmr spectra of the aliphatic regions of *Candida krusei*, recombinant drosophila and *Drosophila melanogaster* cytochromes *c*. Data were collected on a Varian XL-400 with a cycling time of 0.75 sec. and a flip angle of 80° as described previously [J. D. Rush et al. (1988) *J. Biol. Chem.* **263** 7514-7520]. The three spectra are scaled to the same areas for their respective leucine 68 and thioether 2 methyl resonances at approximately -2.6 and -2.5 ppm.

cytochrome *c* Edman degradation pattern for 10 residues (G-V-P-A-G-D-V-E-K-G), indicating that the cytochrome *c* was not blocked at the N-terminal residue. By nmr, heterogeneity was apparent in the multiple resonances observed for the heme methyls-3 and -8 centered at 33.5 ppm and 35.1 ppm, respectively. It also became evident that the resonance at 3.35 ppm for trimethyllysine 72 corresponded only to 5.7 protons, as compared to 9.0 protons for the trimethylated cytochrome *c* of another yeast, *Candida krusei*, or the absence of the resonance seen for the non-methylated natural drosophila cytochrome *c* (Fig. 5). Following careful exploration of chromatographic conditions, it became possible to separate the recombinant drosophila protein into 4 peaks, which correspond to tri-, di-, mono- and unmethylated drosophila cytochromes *c*, from the lower to the higher ionic strength range of the chromatographic gradient (Fig. 3). This is a difficult separation to achieve and has been obtained satisfactorily only on an analytical scale. For preparative purposes, each peak had to be extensively rechromatographed to homogeneity.

Mutants at proline 30

When considering which mutant proteins could best be employed to study the properties of the central inorganic complex of cytochrome *c* consisting of the heme and its two axial ligands, the sulfur atom of methionine 80 and the imidazole of histidine 18, it was thought that changing these ligands themselves would likely produce such profound alterations in properties that the resultant mutant proteins would be of little relevance to the understanding of cytochrome *c*. More readily interpretable results should be obtained by modifying the second tier of residues which interact with the heme ligands, but are not heme ligands themselves. Two such

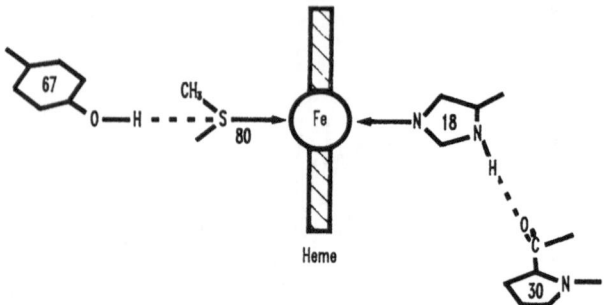

Figure 6. The spatial sequence of residues which involve the axial ligands of cytochrome *c*; tyrosine 67, methionine 80, the heme iron, histidine 18 and proline 30.

residues are tyrosine 67 and proline 30. These are part of a phylogenetically invariant spatial sequence consisting of proline 30, followed by histidine 18, the heme iron, methionine 80 and tyrosine 67 (Fig. 6).

Thus, the first set of mutants of drosophila and rat cytochromes *c* were generated at the invariant proline 30 on the 'right side' of the protein. The backbone carbonyl of this proline is hydrogen bonded to the nitrogen of the imidazole of histidine 18. This hydrogen bond fixes the orientation of the plane of this imidazole with respect to the heme. It is kept perpendicular to the heme plane, bisecting it into two identical rectangles. Proline 30 is just carboxyl-terminal to a sharp γ turn of the peptide chain, occurring at residues 27, 28 and 29 (8), lysine, valine and glycine in drosophila cytochrome *c*, and lysine, threonine and glycine in rat cytochrome *c*. This γ turn is also unusual in that it is held by one hydrogen bond only, that between the amino group of the first residue to the carboxyl of the last, forming an 11-atom ring, as compared to the more usual hydrogen bond between the carboxyl-group of the first residue and the amino group of the last, forming a 7-atom ring (see Fig. 9B). It was expected that changing the rather rigid proline side chain to others that would permit greater flexibility in this area, might destabilize the hydrogen bond to the heme iron-liganded imidazole, allowing more mobility to the imidazole, thus weakening the closed form of the heme crevice and leading to a deterioration in function. If this occurred it would probably be accompanied by a repacking of the region of the γ turn into a more stable structure.

Proline 30 was replaced by alanine in drosophila cytochrome *c* and by alanine or valine in rat cytochrome *c* (58-60), and the properties of the mutant proteins were indeed essentially as expected. It was found that the 695 nm band of the ferric cytochrome *c*, associated with the ligation of a low spin ferric heme iron by the bivalent sulfur of methionine 80, behaved in a manner indicating a weakening of the iron-sulfur bond. The equilibrium considered to represent this system is:

$$\text{(Met) S-Fe}^{3+}\text{-Imidazole} \rightleftharpoons \text{X-Fe}^{3+}\text{-Imidazole (His)} \qquad \text{Eq. 1}$$

The 695 nm band is present as long as the sulfur is bonded to the heme iron, but disappears when the methionine is displaced by an as yet undetermined non-methionine ligand, denoted as X. A large number of processes can lead to this displacement, such as extremes of pH to the alkaline and the acid side, increased temperature, appropriate concentrations of urea and any other conditions capable of sufficiently disturbing the native conformation of the protein (69).

It was found (60) that the alkaline pK_a of the 695 nm band for both the drosophila and rat mutant proteins, in which proline 30 was replaced by alanine or valine, was 0.7 to 0.8 pH units below that of the wild-type cytochromes c (Fig. 7 and Table II), that the temperature for half disappearance of the 695 nm band was similarly decreased by 21 °C for the drosophila protein, and 9 °C or 12 °C for the rat protein (Table II), while the concentration of urea for half disappearance of the 695 nm band went from 5.5 M for the drosophila wild-type cytochrome c to 0.3 M for the mutant, and similarly from 7.2 M for the rat wild-type protein to 3.8 for the mutant (Table II). That the changes in protein conformation resulting from the mutations at residue 30 were largely local, became evident when it was found that measurements of circular dichroism spectra showed that concentrations of urea that resulted in the disappearance of the 695 nm band had very little effect on the ellipticity of the molecule as a whole. All the changes in these mutant cytochromes c are indicative of decreased stability of the heme iron-methionine sulfur bond.

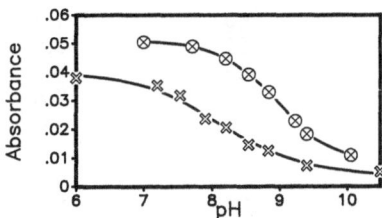

Figure 7. pH titration of the 695 nm band of the following cytochromes c: DMc, wild-type (crossed circles); and DMc, P30A (open crosses).

The interesting features of the effects of temperature on these mutants are depicted in Fig. 8, in which is plotted ΔG° for the thermal titration of the 695 nm band, calculated from $\Delta G^\circ = -RT \ln K$, where K is the equilibrium constant for Equation 1 above, as a function of temperature. The linear portion of the dependence represents the shifting of equilibrium as the temperature increases, while the sudden curving downwards of the line corresponds to a large increase in the heat capacity of the system which occurs as generalized denaturation of the protein structure takes place (71). Remarkably, the wild-type proteins show the overall heat denaturation phenomenon within the temperature range tested, while the proline 30 mutants do not, the temperature dependence remaining linear throughout, even though the mutants cross the $\Delta G = 0$ line, namely the half point for the disappearance of the 695 nm band, at temperatures clearly lower than the points at which the wild-type proteins do.

This behavior, together with that elicited by the urea and alkaline titrations, lead to the following conclusions: (i) the proline to alanine or valine mutants have decreased stability of the heme iron to methionine 80 sulfur bond, even though the

Table II. Stability of the methionine sulfur-heme iron bond of recombinant cytochromes *c* (see Ref. 60)

Cytochromes *c*	$T_{1/2}$	[Urea]$_{1/2}$	pK_a
	°C	M	pH
Drosophila			
DM*c*, wild-type	67	5.5	9.0
DM*c*, P30A	46	0.3	8.2
Rat			
RN*c*, wild-type	60	7.7	9.5
RN*c*, P30A	51	3.8	8.8
RN*c*, P30V	48	---	8.8
RN*c*, Y67F	90	7.3	10.7
RN*c*, N52I	88	---	---
Yeast			
CYC1, wild-type	---	---	8.3
CYC1, Y67F	---	---	9.4
CYC1, N52I	---	---	9.2
CYC1, T78V	---	---	9.3

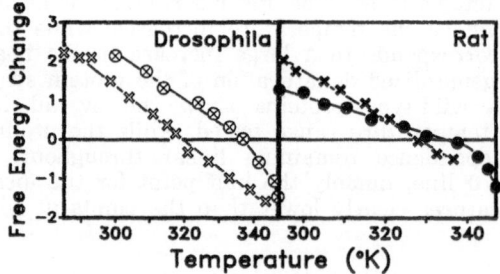

Figure 8. Thermal titrations of wild-type and P30A mutants of drosophila and rat cytochromes *c*. **Left panel:** DM*c*, wild-type (crossed circles); DM*c*, P30A (open crosses). **Right panel:** RN*c*, wild-type (closed circles), RN*c*, P30A (closed crosses).

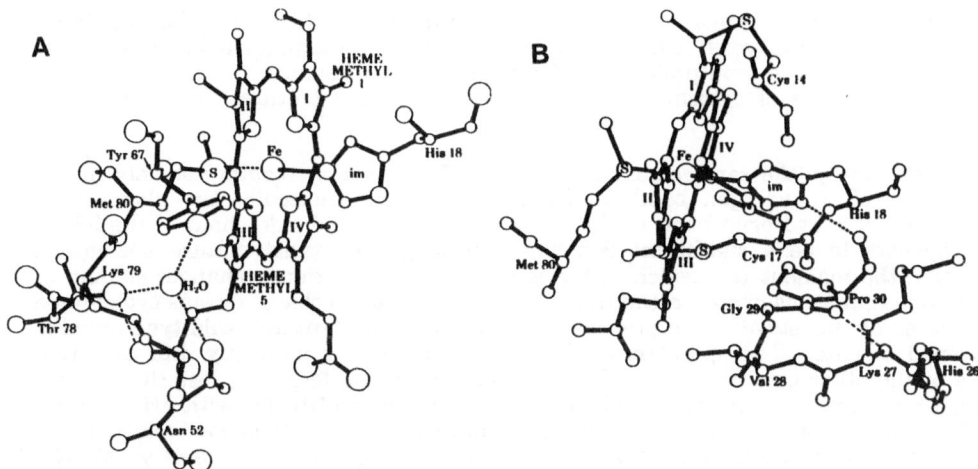

Figure 9. Diagram of portions of the structure of cytochrome *c* to illustrate: **A.** The internal water molecule hydrogen bonded to residues 67 (tyrosine), 78 (threonine), and 52 (asparagine) on the 'left' side of the heme. Note that the threonine side chain hydroxyl also hydrogen bonds to the anterior propionyl of the heme. **B.** The hydrogen bond between the carbonyl of proline 30 and the heme-liganded imidazole of histidine 18 on the 'right' side of the heme. Note also the γ turn of residues 27 (lysine), 28 (valine), and 29 (glycine) immediately preceding the proline, and lacking one of the two hydrogen bonds commonly found in such bends. The coordinates for tuna ferricytochrome *c* were employed (8).

mutation is at a considerable distance from that bond and on the other side of the heme plane; (ii) these effects are entirely local, since the mutant cytochromes *c* are as stable, or even possibly more stable than the wild-type proteins with regard to general denaturation (60). This stabilization could conceivably result from the removal of the proline 30 side chain, allowing a more usual γ turn to form at residues 27, 28 and 29, as mentioned above. It should be noted that not all residues are acceptable instead of proline 30, since mutants carrying a leucine at that position could not sustain the growth of a yeast culture at a rate sufficient to allow the isolation of the mutant protein. A similar observation was made by Wood *et al.* (46) who found that the yeast iso-2 cytochrome *c* mutant in which proline 30 was replaced by threonine was similarly deficient.

Finally, all the mutants at proline 30 that have been prepared show only slight decreases in their reduction potential, that are unlikely to cause any functional difficulties. Indeed as already noted, they sustain the growth of cultures of yeast in a normal fashion (60).

The water mutants

The second set of mutants examined were those of residues on the 'left' side of heme plane which are apparently involved in maintaining a single water molecule in the hydrophobic interior of the protein, near the 'lower' edge of the heme plane. The three residue side chains holding this water molecule in place are those of tyrosine 67, at the 'top left', asparagine 52, at the 'bottom left' and threonine 78 at the 'lower front' of the heme crevice (Fig. 9). Tyrosine 67 to phenylalanine mutant cytochromes *c* were obtained for the rat and yeast proteins, as were asparagine 52 to isoleucine cytochromes c, and a threonine 78 to valine mutant in the yeast protein. All these mutants allowed the yeast cultures to grow at rates indistinguishable from those sustained by the wild-type proteins, and all except the threonine 78 mutant resulted in normal amounts of cytochrome c to accumulate in the cultures, as noted above (Table I).

Interestingly, the replacement of threonine 78 by valine results not only in the elimination of a hydrogen bond to the internal water molecule, but also in the elimination of a hydrogen bond to the anterior propionic acid side chain of the heme. This appears to be a significant fact since, although the yeast culture has grown normally, the moment the cytochrome c is collected on a cation exchanger column of CM-cellulose, it turns brown, rather than retaining the pink color of ferrous cytochrome c exhibited at this stage of the preparation by all other mutant and wild-type proteins examined to date. The spectrum of this mutant protein is distinctly different from those of the other cytochromes c. In particular, the ferric form has lost the 695 nm band, which appears to be replaced by one at 650 nm which titrates with pH similarly to the ordinary 695 nm band. This ferric protein is also no longer reducible by ascorbate, only by dithionite. How such a protein can allow a culture of yeast cells to grown can only be surmised at this time. Possibly some intracellular factor or condition retards it from undergoing the catastrophic change observed *in vitro*, and the low level of cytochrome c observed reflects a more rapid turnover than normal, being maintained by an increased rate of synthesis. These and related possibilities are being examined.

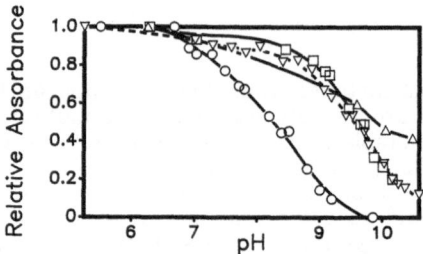

Figure 10. pH titrations of the 695 nm or 650 nm band of the following cytochromes c: CYC1, wild-type (open circles); CYC1, N52I (open inverted triangles); CYC1, Y67F (open squares); CYC1, T78V (open triangles).

What is remarkable is that all the 'water' mutants, including the threonine 78 to valine mutant, display increased stabilization of the heme iron to methionine 80 sulfur bond. The pH titrations of the 695 nm band of the tyrosine 67 to phenylalanine and asparagine 52 to isoleucine mutants show increased apparent pK_a values of about 1 pH unit, as does the titration of the 650 nm band of the threonine 78 to valine yeast mutant protein (Fig. 10). Furthermore, the temperature titrations of these mutants all show dramatic increases in the stability of the 695 nm band of the ferric forms. The temperatures for the half disappearance of these bands are about 30 °C

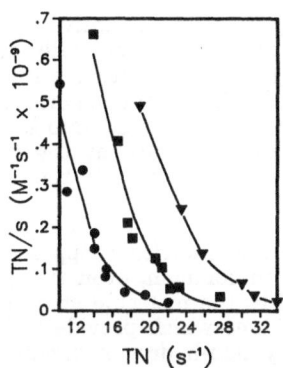

Figure 11. Eadie-Hofstee representation of the polarographic kinetics of the cytochrome oxidase reaction employing a Keilin-Hartree beef heart preparation as source of enzyme, and the following recombinant cytochromes *c*: Rnc, wild-type (solid circles); RNc, Y67F (solid squares); and RNc, N52I (solid triangles). See references 14 and 35 for experimental procedures.

higher indicating considerable increases in the stability of the functional closed form of the heme crevice (see Table II).

Finally, the electron transfer activities with cytochrome oxidase of the tyrosine 67 to phenylalanine and the asparagine 52 to isoleucine mutants of the rat protein were shown to be higher than that of the corresponding wild-type rat protein (Fig. 11). Since these measurements were by the polarographic assay, and no change in the apparent K_m of the high affinity phase of the kinetics was observed, only an increase in V_{max}, the observed changes may result from an increased rate of reduction of the cytochrome *c* in its cytochrome oxidase complex by the reductant employed, N,N,N',N'-tetramethylphenylenediamine (72). Nevertheless, it is clear that the mutant proteins are either as good as or better than the wild-type protein in the cytochrome oxidase reaction.

Like the proline 30 mutants, the 'water' mutants, except that at threonine 78, show minor decreases in reduction potentials, the E_0 values being decreased by 25 mV or less. As already discussed, they also maintain the growth of yeast cultures at normal rates.

DISCUSSION

The contrast between the properties of the cytochrome *c* mutants at proline 30 of the drosophila and rat proteins, and those of the 'water mutants' at tyrosine 67, asparagine 52 and threonine 78 of the yeast iso-1 and rat proteins is dramatic. As described above, the proline mutants probably allow for increased lability or mobility of a hydrogen bond which tends to fix the orientation, with respect to the heme plane of the second heme axial ligand, the imidazole of histidine 18. This, in turn, *via* a trans-effect decreases the strength of binding to the heme iron of the first axial ligand, the sulfur of methionine 80. This leads to a weakening of the closed functional form of the heme crevice, as observed by various effects on the 695 nm band of ferric cytochrome *c*, a spectral transition characteristic of this heme iron-sulfur bond. Thus, a disturbance of a hydrogen bond present in the native structure leads to a weakening of the conformation in which the protein is functionally active. Moreover, this effect is only the result of a local change in conformation, while there are indications that the overall result may be a strengthening of the general structure of the protein. As it were, a change which allows the protein to fall into a slightly more stable state is accompanied by a weakening of the structure at the functional center, the heme prosthetic group. That such an event is not fixed in the course of evolution is entirely as expected.

The water mutants behave in a diametrically opposed fashion. All such mutants have a fundamental characteristic in common, even though the three residues which define them are widely distributed in the molecule at positions 67, 52 and 78. Thus, when the side chains of these residues which provide hydrogen bonds to an internal water molecule, are substituted by non-hydrogen bonding moieties, it is the water structure that must either disappear or be strongly modified. Paradoxically, the common characteristic is a dramatic increase in the stability of the closed functional form of the heme crevice, as quantified by measuring the effects of pH, heat and urea on the 695 nm band of the ferric cytochrome c. Furthermore, this is accompanied by either no change or an improvement in the electron transfer activity of two of these mutants with their major physiological interaction partner, cytochrome oxidase.

If the water mutants are indeed as good or better functionally than the wild-type protein, and clearly more stable, why are the residues involved in the hydrogen bonded water structure unchanged, not allowing such apparently superior variants to be fixed in the course of evolution? It is of course possible that a more detailed examination of these mutants will uncover aspects of their functional activities which are less well adapted to mitochondrial respiratory chain physiology than are those of the corresponding wild-type cytochromes c. The hidden deficiencies may also be in other aspects of the protein's biology, such as rates of intracellular breakdown and/or synthesis, or even in reactions with physiological partners other than respiratory chain components. such as the sulfite oxidase of mammalian liver mitochondria, the peroxidase of yeast mitochondria, etc. . . At this time it is not possible to judge the likelihood that these or other such parameters are the crucial influences maintaining the evolutionary invariance of the triad of water binding side chains under consideration.

However, there are indications that another type of problem may underlie the evolutionary invariance of this water bonded structure in cytochrome c. Thus, threonine 78 has its side chain hydrogen bonded both to the water molecule and the anterior propionyl side chain of the heme (7-10). When it is replaced by a valine in yeast iso-1 cytochrome c, both hydrogen bonds are lost, and there is a dramatic change in the properties of the protein. There are obvious spectral changes, the protein becomes rapidly auto-oxidizable and cannot be reduced with ascorbate, only with dithionite, all indicating large structural changes in the environment of the prosthetic group that have rendered the cytochrome c functionless. That such a protein can nevertheless manage to sustain the growth of a culture of yeast cells is remarkable and not understood. Whatever the mechanism of the *in vivo* functionality of this mutant, the isolated product represents a catastrophic change.

Interestingly, similar spectral changes, of lesser magnitude, have been observed with other mutants in which the conserved water structure is changed and the closed structure of the heme crevice is destabilized (unpublished). An example of these is a rat cytochrome c double mutant in which tyrosine 67 is changed to phenylalanine and proline 30 is changed to alanine. Thus, there would appear to be a set of residues in which certain mutations result in a functional protein, or which lead to a minimal disturbance. However, when these mutations are coupled with a second mutation which modifies the hydrogen bonded water structure, which by itself would produce a protein of enhanced stability and functionality, this leads to a catastrophic change in the structure of the environment of the heme prosthetic group. It is as if the water structure is maintained because in its absence a relatively large set of other changes would be disastrous, notwithstanding that the water mutants themselves may be proteins of excellent functionality. Thus, the hydrogen bonded water structure allows much more leeway to cytochrome c in phylogenetic descent, since it is likely that relatively numerous evolutionary pathways become passable that would otherwise be forbidden.

This secondary set of residues and mutants are being identified and examined, and this appears to represent the first case of such a phenomenon with cytochrome c. We would like to propose it be termed *second site deficiency*.

Acknowledgments

The authors are grateful to Dr. B. D. Hall for the yeast iso-1 cytochrome c gene, Dr. R. Zitomer for the GM-3C-2 yeast strain, Dr. K. L. Ngai of the Northwestern University Biotechnology Facility for the protein sequencing, the synthesis of oligonucleotides and the operation of fermentors and Ms. Cita R. Orendain for producing the manuscript.

REFERENCES

1. Margoliash, E., and Schejter, A. (1966) *Protein Chemistry* **21**, 113-286.
2. Dickerson R. E., and Timkovich, R. (1975). In *The Enzymes* Vol. **11**, 3rd edition (P.D. Boyer, ed.), Academic Press, NY, 397-547.
3. Ferguson-Miller, S., Brautigan, D. L., and Margoliash, E. (1979). In *The Porphyrins* (D. Dolphin, ed.), Academic Press, NY, 149-240.
4. Pettigrew, G. W., and Moore, G. R. (1987) *Cytochromes c - biological aspects*, Springer-Verlag, Berlin, Heidelberg, NY, London, Paris, Tokyo.
5. Borden, D., and Margoliash, E. (1976). In *Handbook of Biochemistry and Molecular Biology* Vol. **III**, 3rd edition, Chemical Rubber Co., Cleveland, OH, 268-279.
6. Hampsey, D. M., Das, G., and Sherman, F. (1986) *J. Biol. Chem.* **261**, 3259-3271.
7. Takano, T., and Dickerson, R. E. (1981) *J. Mol. Biol.* **153**, 79-94.
8. Takano, T., and Dickerson, R. E. (1981) *J. Mol. Biol.* **153**, 95-115.
9. Ochi, H., Hata, Y., Tanaka, N., Kakudo, M., Sakurai, T., Aihara, S., and Morita, Y. (1983) *J. Mol. Biol.* **116**, 407-418.
10. Louie, G. V., Hutchinson, W. L. B., and Brayer, G. D. (1989) *J. Mol. Biol.* **199**, 295-314.
11. Walter, M. H., Westbrook, E. M., Tykodi, S., Uhm, A. M., and Margoliash, E. (1990) *J. Biol. Chem.* **265**, 4177-4180.
12. Wand, A. J., and Stefano, D. L. (1989) *Biochemistry* **28**, 186-194.
13. Feng, Y., Roder, H., Englander, S. W., Wand, A. J., and Stefano D. L. (1989) *Biochemistry* **28**, 195-203.
14. Ferguson-Miller, S., Brautigan, D. L., and Margoliash, E. (1978) *J. Biol. Chem.* **253**, 149-159.

15. Brautigan, D. L., Ferguson-Miller, S., and Margoliash, E. (1978) *Methods in Ezymol.* **53**, 128-164.
16. Osheroff, N., Brautigan, D. L., and Margoliash, E. (1960) *J. Biol. Chem.* **255**, 8245-8251.
17. Margoliash, E. and Bosshard, H. R. (1983) *TIBS* **8**, 316-320.
18. Koppenol, W. H. and Margoliash, E. (1982) *J. Biol. Chem.* **257**, 4426-4437.
19. Marcus, R. A. and Sutin, N. (1985) *Biochim. Biophys. Acta* **811**, 265-322.
20. Ho, P. S., Sutoris, C., Liang, N., Margoliash, E., and Hoffman, B. M. (1985) *J. Am. Chem. Soc.* **107**, 1070-1071.
21. Liang, N., Kang, C. H., Ho, P. S., Margoliash, E., and Hoffman, B. M. (1986) *J. Am. Chem. Soc.* **108**, 4665-4666.
22. Nocek, J. M., Liang, N., Wallin, S. A., Mauk, A. G., and Hoffman, B. M. (1990) *J. Am. Chem. Soc.* **112**, 1623-1625.
23. Montgomery, D. L., Hall, B. D., Gillam, S., and Smith, M. (1978) *Cell* **14**, 673-680.
24. Smith, M., Leung, D. W., Gillam, S., Astell, C. R., Montgomery, D. L., and Hall, B. D. (1979) *Cell* **16**, 753-761.
25. Montgomery, D. L., Leung, D. W., Smith, M., Shalit, P. R., Faye, G., and Hall, B. D. (1980) *Proc. Natl. Acad. Sci. USA* **77**, 541-545.
26. Scarpulla, R. C., Agne, K. M., and Wu, R. (1981) *J. Biol. Chem.* **239**, 3109-3112.
27. Russell, P. R. and Hall, B. D. (1982) *Mol. Cell Biol.* **2**, 106-116.
28. Limbach, K. J. and Wu, R. (1983) *Nucleic Acids Res.* **11**, 8931-8950.
29. Swanson, M. S., Zieminn, S. M., Miller, D. D., Garber, E. A. E., and Margoliash, E. (1985) *Proc. Natl. Acad. Sci. USA* **82**, 1964-1968.
30. Limbach, K. J. and Wu, R. (1985) *Nucleic Acids Res.* **13**, 617-630.
31. Limbach, K. J. and Wu, R. (1985) *Nucleic Acids Res.* **13**, 631-644.
32. Miller, D. D. (1988) Ph.D. Dissertation, Northwestern University, Evanston, IL.
33. Osheroff, N., Speck, S. H., Margoliash, E., Veerman, E. C. I., Wilms, J., König, B. H., and Muijsers, A. O. (1983) *J. Biol. Chem.* **258**, 5731-5738.
34. Luntz, T. L. and Margoliash, E. (1988). In *Cytochrome Systems: Molecular Biology and Bioenergetics* (S. Papa, B. Chance and L. Ernster, eds.) Plenum Press, NY and London, 271-279.
35. Dethmers, J. K., Ferguson-Miller, S., and Margoliash, E. (1979) *J. Biol. Chem.* **254**, 11973-11981.
36. Matsuura, S., Arpin, M., Hannum, C., Margoliash, E., Sabatini, D. D., and Morimoto, T. (1981) *Proc. Natl. Acad. Sci. USA* **78**, 4368-4372.
37. Hartle, F., Pfanner, N., Nicholson, D. W., and Neupert, W. (1989) *Biochim. Biophys. Acta* **988**, 1-45.
38. Sprinkle, J. R., Hakvoort, T. B. M., Koshy, T. I., and Margoliash, E. (1990). In the Proceedings for the Symposium "New Trends in Biological Chemistry" in honour of Professor Kunio Yagi, Nagoya, Aug. 22-23, 1989 (T. Ozawa, ed.) Japan Scientific Society Press.
39. Sprinkle, J. R., Hakvoort, T. B. M., Koshy, T. I., Miller, D. D., and Margoliash, E. (1990) *Proc. Natl. Acad. Sci. USA*, in press.
40. Pielak, G. J., Mauk, A. G., and Smith, M. (1985) *Nature* **313**, 152-153.
41. Ernst, J. E., Hampsey, D. M., Stewart, J. W., Rakovsky, S., Goldstein, D., and Sherman, F. (1985) *J. Biol. Chem.* **260**, 13225-13236.
42. Ramdas, L., Sherman, F., and Nall, B. T. (1986) *Biochemistry* **25**, 6952-6958.
43. Ramdas, L. and Nall, B. T. (1986) *Biochemistry* **25**, 6959-6964.
44. Holzschu, D., Principio, L., Conklin, K. T., Hickey, D. R., Short, J., Rao, R., McLendon, G., and Sherman, F. (1987) *J. Biol. Chem.* **262**, 7125-7131.
45. Hazzard, J. T., McLendon, G., Cusanovich, M. A., Das, G., Sherman, F., and Tollin, G. (1988) *Biochemistry* **27**, 4445-4451.
46. Wood, L. C., Muthukrishnan, K., White, T. B., Ramdas, L., and Nall, B. T. (1988) *Biochemistry* **27**, 8554-8561.
47. Das., G., Hickey, D. R., Principio, L., Conklin, K. T., Short, J., Miller, J. R., McLendon, G. and Sherman, F. (1988) *J. Biol. Chem.* **263**, 18290-18297.
48. Hickey, D. R., McLendon, G., and Sherman, F. (1988) *J. Biol. Chem.* **263**, 18298-18305.
49. Das, G., Hickey, D. R., McLendon, D., McLendon, G., and Sherman, F. (1989) *Proc. Natl. Acad. Sci. USA* **86**, 496-499.

50. Michel, B., Mauk, A. G., and Bosshard, H. R. (1989) *FEB* **243**, 149-152.
51. Sorrell, T. N. and Martin, P. K. (1989) *J. Am. Chem. Soc.* **111**, 766-767.
52. Cutler, R. L., Davies, A. M., Creighton, S., Warshel, A., Moore, G. R., Smith, M., and Mauk, A. G. (1989) *Biochemistry* **28**, 3188-3197.
53. Pearce, L. P., Gärtner, A. L., Smith, M., and Mauk, A. G. (1989) *Biochemistry* **28**, 3152-3156.
54. Nall, B. T., Zuniga, E. H., White, T. B., Wood, L. D., and Ramdas, L. (1989) *Biochemistry* **28**, 9834-9839.
55. Tanaka, T., Ashikari, T., Shibano, Y., Amachi, T., Yoshizumi, H., and Matsubara, H. (1988) *J. Biochem.* **103**, 954-961.
56. Koshy, T. I., Luntz, T. L., Schejter, A., Garber, E. A. E., and Margoliash, E. (1988) *Faseb J.* **2**, Abstract #1676.
57. Tanaka, Y., Ashikari, A., Shibano, Y., Amachi, T., Yoshizumi, H., and Matsubara, H. (1988) *J. Biochem.* **104**, 477-480.
58. Koshy, T. I. and Margoliash, E. (1988) *J. Cell Biol.* **107**, Abstract #3541.
59. Luntz, T. L., Schejter, A., Garber, E. A. E., and Margoliash, E. (1989) *Proc. Natl. Acad. Sci. USA* **86**, 3524-3538.
60. Koshy, T. I., Luntz, T. L., Schejter, A., and Margoliash, E. (1990) *Proc. Natl. Acad. Sci. USA*, in press.
61. Broach, J., Strathern, J., and Hicks, J. (1979) *Gene* **8**, 121-133.
62. Zoller, M. and Smith, M. (1984) *DNA* **3**, 479-488.
63. Kunkel, T. A., Roberts, J. D., and Zakour, R. A. (1987) *Methods in Enzymol.* **154**, 367-382.
64. Faye, G., Leung, D. W., Tatchell, K., Hall, B. D., and Smith, M. (1981) *Proc. Natl. Acad. Sci. USA* **78**, 2258-2262.
65. Stewart, J. W., Sherman, F., Shipman, N., and Jackson, M. (1971) *J. Biol. Chem.* **246**, 7429-7445.
66. Sherman, F., Stewart, J. W., Parker, J. H., Inhaber, E., Shipman, N., Putterman, G. J., Gardisky, R. L., and Margoliash, E. (1968) *J. Biol. Chem.* **243**, 5446-5456.
67. Kaminsky, L. S., Miller, V. J., and Davison, A. J. (1973) *Biochemistry* **12**, 2215-2221.
68. Williams, Jr., J. N. (1964) *Arch. Biochem. Biophys.* **107**, 537-543.
69. Davis, L. A., Schejter, A., and Hess, G. P. (1974) *J. Biol. Chem.* **240**, 2624-2632.
70. Vensel, W. H., Fujita, V. S., Tarr, G. E., Margoliash, E., and Kayser, H. (1983) *J. Chromatogr.* **266**, 491-500.
71. Privalov, P. L. and Khechinashvili, N. N. (1974) *J. Mol. Biol.* **86**, 665-684.
72. Garber, E. A. E. and Margoliash, E. (1990) *Biochim. Biophys. Acta.* **1015**, 279-287.

SUBUNIT I IS THE CATALYTIC CENTER OF P. DENITRIFICANS CYTOCHROME C OXIDASE

Michele Müller and Angelo Azzi

Institut für Biochemie und Molekularbiologie
Universität Bern, Bühlstrasse 28, 3012 Bern
Switzerland

INTRODUCTION

Cytochrome c oxidases of either mammalian or bacterial origin are membrane bound, multisubunit complexes responsible for catalyzing the electron transfer from ferrocytochrome c to dioxygen.[1,2] In most cases, a vectorial electrogenic proton translocation has been shown to be coupled to this process. The number of subunits present in eukaryotes is larger (up to 13 in mammals) than in prokaryotes that contain two to three subunits only.[3,4] The three subunits present in several bacteria are highly homologous to the three largest subunits in eukaryotes. Bacteria have been therefore used as a simple systems to understand the role of quaternary structure on the function of the enzyme. The oxidases of some bacteria, such as *P. denitrificans*, although they have at least three genes coding for the three mentioned polypeptides,[5] are normally isolated with two subunits only.[6,7] Such a characteristics has simplified their study and has permitted to establish that electron and proton transfer properties of the two subunits oxidase of *P. denitrificans* are very similar to those of the more complex mammalian enzyme.[8,9,10]

The active centers of cytochrome c oxidase consist of two porphyrin irons and two copper atoms whose location[11,12] and mechanism of participation in the catalytic process are still not fully established. Since the three largest subunits of all cytochrome c oxidases are homologous and they have almost identical spectral and functional properties it has been concluded that the limited number of evolutionary conserved residues must be responsible for the characteristics of the active centers.

The recent discovery that histidine 24 of bovine heart enzyme is not conserved in the oxidases of *Leishmania tarentolae*, *Trypanosoma brucei* and *Crithidia*

Bioenergetics, Edited by C. H. Kim and
T. Ozawa, Plenum Press, New York, 1990

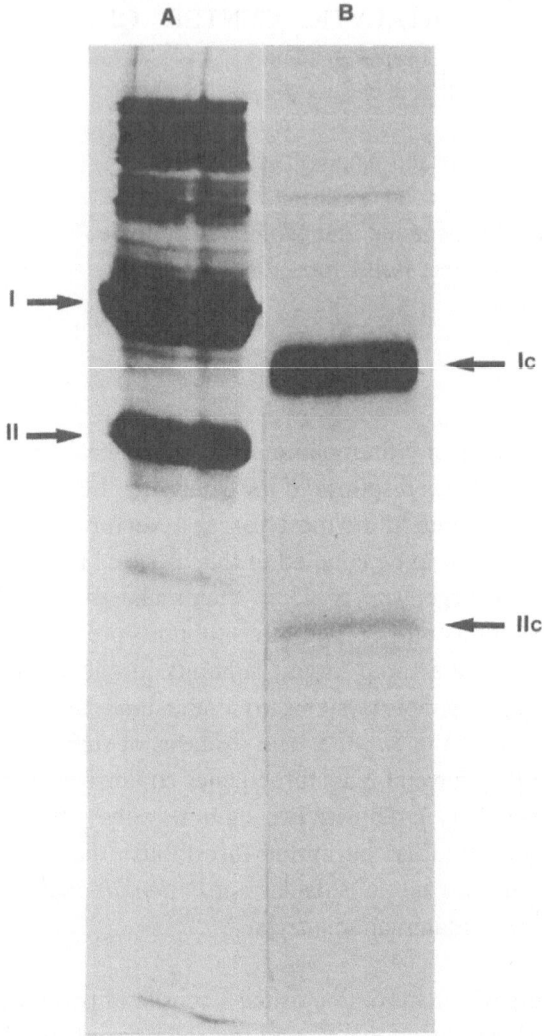

Fig. 1.

Sodium dodecylsulfate-polyacrylamide gel electrophoresis of *P. denitrificans* cytochrome c oxidase.
Lane A) native enzyme; Lane B) Enzyme treated with chymotrypsin and *S. aureus V8* protease. I: undigested subunit I; Ic: subunit I digested by chymotrypsin and by *S. aureus V8* protease whose apparent molecular weight was 43,000. II: undigested subunit II; IIc: subunit II digested by chymotrypsin which apparent molecular weight was 24,000.

fasciculata is not compatible with subunit II being a heme binding site.[13] In fact the attribution of precise ligands to the heme of cytochrome a (two histidyl residues) and Cu_A (two histidyl and one or two cysteinyl residues), obtained by electron nuclear double resonance (ENDOR) and extended X-ray absorption fine structure (EXAFS) in bovine heart oxidase,[14,15] made it impossible to find in subunit II of *P. denitrificans* oxidase enough conserved residues, capable of coordinating both metal centers.

Recently the metal analysis of cytochrome c oxidase was revisited by using inductively coupled plasma atomic emission spectrometry,[16,17] proton-induced X-ray emission[18] and by energy dispersive X-ray fluorescence spectrometry.[19] The results of these analyses indicate the presence of 3 copper and 2 iron atoms per monomeric molecule of cytochrome c oxidase. This implies that subunits I and II must contain the ligands for these 5 metal centers since the two subunit cytochrome c oxidase of *P. denitrificans* has spectral characteristics not distinguishable from the eukaryotic enzymes.

LOCALIZATION OF THE HEMES A AND COPPER B

The first experimental evidence that only subunit I contains cytochrome a (Fe_A), cytochrome a_3 (Fe_B) and copper B was given by studies on proteolytic cleavage of *P. denitrificans* enzyme.[20-23]

A first partial proteolytic degradation of *P. denitrificans* cytochrome c oxidase was obtained by chymotrypsin. The two resulting fragments (not shown) had lower molecular weight as compared with the native protein (Fig. 1; lane A). An N-terminus sequence analysis indicated that subunit II was digested at the residue phenylalanine 72 as shown in the model of Fig. 2. The consequence of this is that the remaining larger fragments implicated in copper binding, contrary to the native polypeptide, does not have a blocked N-terminus. Further digestion of the chymotrypsin treated enzyme with *Staphylococcus aureus V8* protease resulted in specific degradation of subunit II and in no further degradation of subunit I (Fig. 1, lane B).

As it appears from the gel, contaminating proteins were present in low amount. A further indication that purification of subunit I had occurred is given by the data of UV relative to Soret band absorption of Table 1.

The purified subunit I showed visible spectral properties not distinguishable from the native enzyme (Fig. 3). The (CO-dithionite reduced) *minus* (reduced) spectrum (not shown) indicated that the heme a to heme a_3 ratio remained 1:1 after protease digestion.

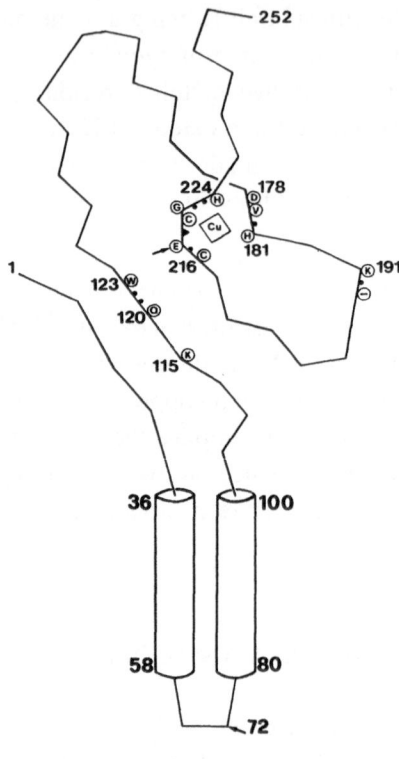

Fig. 2.

Schematic model of *P. denitrificans* cytochrome c oxidase subunit II.
The numbers in the diagram correspond to the sequence of *P.
denitrificans* oxidase. The open circles represent evolutionarily conserved
amino acids (one-letter code) or charges (+ and -). Filled triangles
represent conserved position without any apparent special function. The
arrow represents the conserved glutamyl residue, which was protected by
cytochrome c against modification by water-soluble carbodiimide. The
proposed ligands for copper were histidyl-residues 181, 224 and cysteinyl-
residue 216. An additional ligand (aspartyl 178?) may also coordinate
this copper atom. The cylinders represent two evolutionarily conserved
α-helices spanning the hydrophobic region of a membrane. Chymotrypsin cleaved at position 72.

The above data confirm experimentally the existence of two hemes within
subunit I, as it was inferred from the analysis of evolutionary conserved residues
needed for liganding the Fe and the Cu. Subunit I of *P. denitrificans* as purified
above, possessed cytochrome c oxidase activity (K_m = 2.8 μM; V_{max} 168 s^{-1}) and salt
dependence similar to the native enzyme.

Table 1. Absorption ratios of *P. denitrificans* cytochrome c oxidase and of its purified subunit I.

	$\dfrac{A_{280-310nm}}{A_{424-480nm}}$	$\dfrac{A_{424-480nm}}{A_{820-700nm}}$
Bovine heart	2.8-3.0	145[a]
P. denitrificans	1.4-1.6	116-148
P. denitrificans subunit I	1.1-1.2	107-122

[a] Data calculated from Gelles and Chan[29]

WHERE ARE COPPER A, COPPER B AND COPPER X?

The presence of two hemes and of activity associated with subunit I poses the question of the role and of the location of the copper. It has been confirmed by several laboratories that the enzyme contains three copper atoms and that for one of them enough ligands are present in subunit II. It is clear that subunit I should coordinate two more copper atom with amino acid side chains presumably conserved in all sequences analyzed till now. Since there is no cysteine conserved in subunit I, this ligand can be excluded for the second and third copper atom. Beside histidine, however, also tyrosine was found as copper ligand in the galactose oxidase[24] whereas methionine was found to be a very weak and a not obligatory ligand of copper in azurin.[25]

At the moment different experimental evidences indicate that copper A is localized in subunit II.[14,15,26-28] At variance with them the actual work on *P. denitrificans* gives indication that copper A may be located on subunit I. In Fig 3 a spectrum of the one-subunit preparation shows the near-infrared absorption of copper A. The absorption disappeared upon addition of dithionite as known to occur in the intact two-subunit enzyme. The preparation analyzed till now had different amounts of contamination by subunit II up to the 25% of its original content. Nevertheless, the absorption ratio between the heme a α-band and copper A remained unchanged. To further support these indications, EPR experiments are presently carried out in our laboratory.

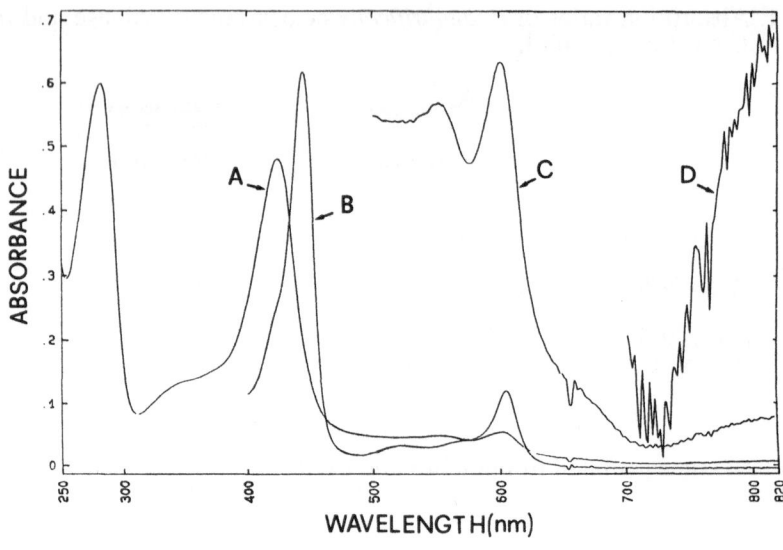

Fig. 3.

Spectral analysis of *P. denitrificans* oxidase after digestion with chymotrypsin and *S. aureus V8* protease.
A.: Air-oxidized spectrum showing maxima at 280, 424, 600 and 820 nm. The α-band and the near-infrared regions are shown also by the enlarged traces (C and D).
B: Dithionite reduced spectrum showing maxima at 446 and 606 nm. The absorption of the near-infrared band was bleached by the reduction with dithionite. The values of the maxima of these spectra are identical to the native enzyme.

ACKNOWLEDGEMENTS

Financial support of the Swiss National Science Foundation (No. 3.525.086 and 3100-27745.89) is gratefully acknowledged.

REFERENCES

1. A. Azzi, K. Bill, R. Bolli, R. P. Casey, K. A. Nałęcz, and P. O'Shea, Molecular aspects of the structure-function relationship in cytochrome c oxidase, in: "Structure and Properties of Cell Membranes," G. Benga, ed., Vol. 2, CRC Press Inc., Boca Raton, FL (1985).
2. M. Wikström, M. Saraste, and T. Penttilä, Relationship between structure and function in cytochrome oxidase, in: "The Enzymes of Biological Membranes," A.N. Martonosi, ed., Vol. 4, Plenum Press, New York, NY (1985).

3. J. A. Fee, D. Kuila, M. W. Mather, and T. Yoshida, Respiratory proteins from extremely thermophilic, aerobic bacteria, <u>Biochem. Biophys. Acta</u> 853:153 (1986).

4. B. Ludwig, Cytochrome c oxidase in prokaryotes, <u>FEMS Microbiol. Review</u> 46:41 (1987).

5. M. Raitio, T. Jalli, and M. Saraste, Isolation and analysis of the genes for cytochrome c oxidase in *Paracoccus denitrificans,* <u>EMBO J.</u> 6:2825 (1987).

6. B. Ludwig, and G. Schatz, A two-subunit cytochrome c oxidase (cytochrome aa_3) from *Paracoccus denitrificans*, <u>Proc. Natl. Acad. Sci. USA</u> 77:196 (1980).

7. B. Ludwig, Cytochrome c oxidase from *Paracoccus denitrificans*, <u>Methods Enzymol.</u> 126:153 (1986).

8. J. K. V. Reichhardt, and Q. H. Gibson, Turnover of cytochrome c oxidase from *Paracoccus denitrificans,* <u>J. Biol. Chem.</u> 258:1504 (1983).

9. M. Solioz, E. Carafoli, and B. Ludwig, The cytochrome c oxidase of *Paracoccus denitrificans* pumps proton in a reconstituted system, <u>J. Bioenerg. Biomembr.</u> 257:1579 (1982).

10. R. Bolli, K. A. Nałęcz, and A. Azzi, Cytochrome c oxidase from *Paracoccus denitrificans* in Triton X-100: aggregation state and kinetics, <u>J. Bioenerg. Biomembr.</u> 18:277 (1986).

11. D. B. Winter, W. J. Bruynickx, F. G. Foulke, N. P. Grinich, and H. S. Mason, Location of heme a on subunits I and II and copper on subunit II of cytochrome c oxidase, <u>J. Biol. Chem.</u> 255:11408 (1980).

12. M. Corbley, and A. Azzi, Resolution of bovine heart cytochrome c oxidase into smaller complexes by controlled subunit denaturation, <u>Eur. J. Biochem.</u> 139:535 (1984).

13. R. Benne, J. Van Den Burg, J. P. J. Brakenhoff, P. Sloof, J. H. VanBoom, and M. C. Tromp, Major transcript of the frameshifted coxII gene from trypanosome mitochondria contains four nucleotides that are not encoded in the DNA, <u>Cell</u> 46:819 (1986).

14. T. H. Stevens, C. T. Martin, H. Wang, G. W. Brudvig, C. P. Scholes, and S. I. Chan, The nature of Cu_A in cytochrome c oxidase. <u>J. Biol. Chem.</u> 257:12106 (1982).

15. P. M. Li, J. Gelles, S. I. Chan, R. J. Sullivan, and R. A. Scott, Extended X-ray absorption fine structure of copper in Cu_A-depleted, p-(hydroxymercuri)-benzoate-modified, and native cytochrome c oxidase, <u>Biochemistry</u> 26:2091 (1987).

16. O. Einarsdóttir, and W. S. Caughey, Bovine heart cytochrome c oxidase preparations contain high affinity binding sites for magnesium as well for zinc, copper, and heme iron, <u>Biochem. Biophys. Res. Comm.</u> 129:840 (1985).

17. G. C. M. Steffens, R. Biewald, and G. Buse, Cytochrome c oxidase is a three-copper, two-heme-A protein, <u>Eur. J. Biochem.</u> 164:295 (1987).

18. E. Bombelka, F.-W. Richter, A. Stroh, and B. Kadenbach, Analysis of the Cu, Fe and Zn contents in cytochrome c oxidases from different species and tissues by proton-induced X-ray emission (PIXE), Biochem. Biophys. Res. Comm. 140:1007 (1986).

19. M. Oeblad, E. Selin, B. Malmström, L. Strid, R. Aasa, and B.G. Malmström, Analytical characterization of cytochrome oxidase preparations with regard to metal and phospholipids contents, peptide composition and catalytic activity, Biochim. Biophys. Acta 975:267 (1989).

20. M. Müller, N. Labonia, B. Schläpfer, and A. Azzi, Cytochrome c oxidase: past, present and future, in: Cytochrome Systems: Molecular Biology and Bioenergetics, S. Papa, B. Chance, and L. Ernster, eds., Plenum Press, New York (1987).

21. M. Müller, B. Schläpfer, and A. Azzi, Preparation of a one-subunit cytochrome oxidase from *Paracoccus denitrificans*: spectral analysis and enzymatic activity, Biochemistry 27:7546 (1988).

22. M. Müller, B. Schläpfer, and A. Azzi, Cytochrome c oxidase from *Paracoccus denitrificans*: both hemes are located in subunit one, Proc. Natl. Acad. Sci. USA 85:6647 (1988).

23. M. Müller, and A. Azzi, Subunit I is the catalytic center of *Paracoccus denitrificans* cytochrome c oxidase, Ann. N.Y. Acad. Sci. 550:13 (1988).

24. M. M. Whittaker, V. L. De Vito, S. A. Asher, and J. W. Whittaker, Resonance Raman evidence for tyrosine involvement in the radical site of galactose oxidase, J. Biol. Chem. 264:7104 (1988).

25. B. G. Karlsson, R. Aasa, B. G. Malmström, and L. G. Lundberg, Rack-induced bonding in blue copper proteins: spectroscopic properties and reduction potential of the azurin mutant Met-121→Leu, FEBS Lett. 253:99 (1989).

26. J. Hall, A. Moubarak, P. O'Brien, L. P. Pan, I. Cho, and F. Millet, Topological studies of monomeric and dimeric cytochrome c oxidase and identification of the copper A site using a fluorescence probe, J. Biol. Chem. 263:8142 (1988).

27. C. T. Martin, C. P. Scholes, and S. I. Chan, On the nature of cysteine coordination to Cu_A in cytochrome c oxidase, J. Biol. Chem. 263:8420 (1988)

28. R. A. Scott, W. G. Zumft, C. L. Coyle, and D.M. Dodey, *Pseudomonas stutzeri* N_2O reductase contains Cu_A-type sites, Proc. Natl. Acad. Sci. USA 86:4082 (1989).

29. J. Gelles, and S. I. Chan, Chemical modification of the CuA center in cytochrome c oxidase by sodium p-(hydroxymercuri)benzoate, Biochemistry 24:39634 (1985).

THE SEPARATION BETWEEN CYTOCHROME $\underline{A}$ AND CYTOCHROME $\underline{A}_3$ IN THE ABSOLUTE SPECTRUM

Taketomo Fujiwara, Yoshihiro Fukumori and
Tateo Yamanaka

Department of Life Science, Faculty of Science
Tokyo Institute of Technology, O-okayama, Meguro-
ku, Tokyo 152, Japan

ABSTRACT

$\underline{aa}_3$-Type cytochrome was purified from <u>Halobacterium halobium</u> (1). The cytochrome contained two heme $\underline{a}$ molecules per molecule but no copper. It did not show cytochrome $\underline{c}$ oxidase activity. One of the two heme $\underline{a}$ molecules in the cytochrome was reduced with ascorbate + TMPD, while the other was not reduced with this reducing reagents. The heme $\underline{a}$ molecule reducible with ascorbate + TMPD did not react with CO, while the heme $\underline{a}$ molecule reducible only with $Na_2S_2O_4$ reacted with CO. Therefore, cytochrome $\underline{a}$ or heme $\underline{a}_A$ in the cytochrome was separated from cytochrome $\underline{a}_3$ or heme $\underline{a}_B$ on the reduction with ascorbate + TMPD; the γ peaks of ferrocytochrome $\underline{a}$ and ferricytochrome $\underline{a}_3$ were observed spectrophotometrically in the absolute spectrum. As Cu_A is known to be unnecessary for cytochrome $\underline{aa}_3$ to oxidize ferrocytochrome $\underline{c}$ (2), these results mentioned above show that copper atom, Cu_B mediate electrons between heme $\underline{a}_A$ and heme $\underline{a}_B$.

INTRODUCTION

From a study on the effect of CO on the absorption bands of cytochromes in the mitochondria, Keilin and Hartree (3) concluded that cytochrome $\underline{c}$ oxidase consisted of two components, $\underline{a}$ and $\underline{a}_3$. Since then, the oxidase was called cytochrome $\underline{a}$ + $\underline{a}_3$. The mitochondrial oxidase shows two absorption bands at 445 and 605 nm, respectively. They postulated that the band at 445 nm was attributable mostly to the $\underline{a}_3$ component and the band at 605 nm to the $\underline{a}$ component, and that the $\underline{a}_3$ component was reactive with CO while the $\underline{a}$ component was not.

Okunuki and co-workers (4,5) insisted that as cytochrome $\underline{c}$ oxidase preparation was not separated into the two components by any purification methods which they tried, the enzyme molecule was composed of only cytochrome $\underline{a}$ and this cytochrome was reactive with CO (6). Yonetani (7) separated spectrophotometrically $\underline{a}$ and $\underline{a}_3$ components from each other in the difference spectra and showed that about half of the absorbance

---

Abbreviation: TMPD, $\underline{N},\underline{N},\underline{N}',\underline{N}'$,-tetramethylphenylenediamine

Bioenergetics, Edited by C. H. Kim and
T. Ozawa, Plenum Press, New York, 1990

at 445 nm was attributable to the a_3 component while more than 70% of the absorbance at 605 nm was attributable to the a component. However, he could not show these results in the absolute absorption spectra.

Previously, it was thought that each of cytochromes a and a_3 contained one heme a molecule per molecule (6,8). Nowadays, it has been established that cytochrome c oxidase contains two heme a molecules per molecule, consequently it is cytochrome aa_3 different from cytochrome a + a_3 (9-11), although Thiobacillus novellus cytochrome c oxidase contains only one heme a molecule per minimal structural unit which is composed of one molecule each of subunits (12). Therefore, it is easily recognized that the different microenvironments around the two heme a molecules will cause the occurrence of the a and a_3 components or heme a_A and heme a_B. Namely, at the present time, it is easily understandable that although cytochrome c oxidase preparation is not divided into the two proteins, cytochromes a and a_3 are separated spectrophotometrically from each other in the difference spectra.

Recently, we have succeeded in purifying from Halobacterium halobium a cytochrome which resembles greatly cytochrome aa_3 in the spectral properties; although it contains two heme a molecules per molecule it does not have copper at all (1). Consequently, it is inactive and does not show the 830 nm-peak. So, we call it "cytochrome aa_3". With the "cytochrome aa_3" which lacks both Cu_A and Cu_B, we have separated spectrophotometrically cytochrome a from cytochrome a_3 in the absolute spectrum (1); the γ peaks of ferricytochrome a_3 and ferrocytochrome a have been separately observed in the absolute spectrum.

EXPERIMENTAL PROCEDURES

Large-scale cultivation of H. halobium L-33 (kindly supplied by Prof. J.K. Lanyi, University of California, Irvine, USA) was performed as described in (13). The "cytochrome aa_3" of the bacterium was purified by the method as previously described in (1). Spectrophotometric measurements were performed with a Shimadzu MPS-2000 spectrophotometer using cuvettes of 1 cm light path. The content of heme a was determined assuming ε_{mM} at 587 nm for pyridine ferrohemochrome a to be 29.2. Copper content was determined with a Varian AA-875 atomic absorption spectrophotometer.

RESULTS AND DISCUSSION

When ascorbate + TMPD was added to H. halobium "cytochrome aa_3", its γ peak was half reduced while its α peak was approx. 80% reduced (Fig. 1). The partially reduced "cytochrome aa_3" did not react with CO. Therefore, it seemsed that cytochrome a was reduced on addition of ascorbate + TMPD while cytochrome a_3 was not. $Na_2S_2O_4$-reduced "cytochrome aa_3" reacted with CO. Thus, the difference spectrum of the cytochrome, $Na_2S_2O_4$-reduced + CO minus (ascorbate + TMPD)-reduced, showed the difference spectrum, a_3^{2+} + CO minus a_3^{3+}, as reported by Yonetani (7) (Fig.2). The

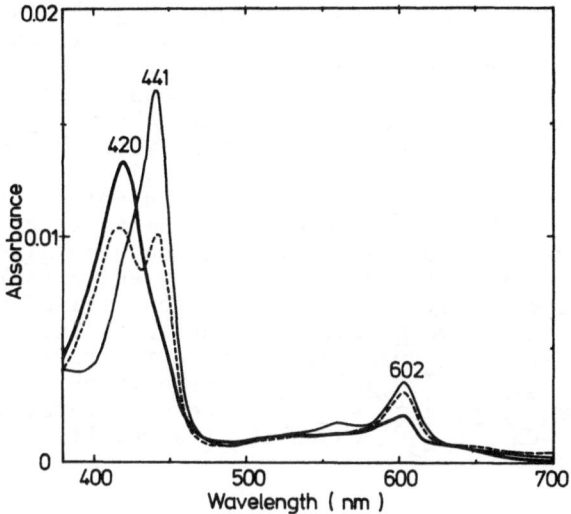

Fig. 1 Reduction of "cytochrome aa_3" with ascorbate+TMPD.
----, Resting state; ----, reduced with $Na_2S_2O_4$; ----, reduced with ascorbate + TMPD. Concentration of heme $\underline{a}$ was 0.16 µM.

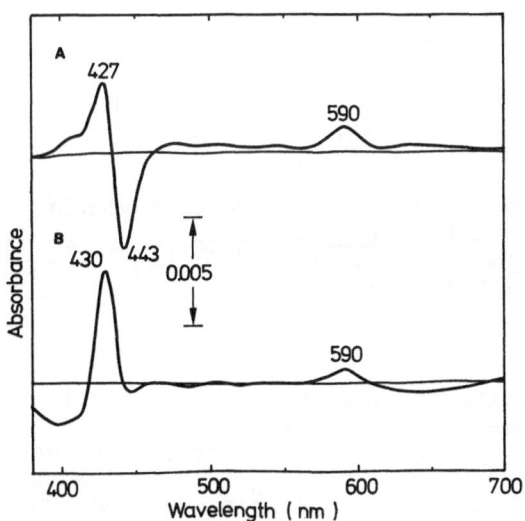

Fig. 2 The CO-difference spectra of "cytochrome aa_3".
A: +$Na_2S_2O_4$ + CO <u>minus</u> +$Na_2S_2O_4$ [$(\underline{a}^{2+}+\underline{a}_3^{2+} \cdot CO) - (\underline{a}^{2+}+\underline{a}_3^{2+})$].
B: +$Na_2S_2O_4$ + CO <u>minus</u> +(ascorbate + TMPD) [$(\underline{a}^{2+}+\underline{a}_3^{2+} \cdot CO) - (\underline{a}^{2+}+\underline{a}_3^{3+})$]. Concentration of heme $\underline{a}$ was 0.16 µM.

finding that cytochrome $\underline{a}$ of the "cytochrome $\underline{aa}_3$" was reduced with ascorbate + TMPD but cytochrome $\underline{a}_3$ was not, showed that electrons were not transferred from cytochrome $\underline{a}$ to cytochrome $\underline{a}_3$ or from heme $\underline{a}_A$ to heme $\underline{a}_B$ as Cu (Cu_B as described below) was absent. The results obtained with the "cytochrome $\underline{aa}_3$" were in good agreement with those shown by Yonetani with the difference spectra (7); about 66% of the absorbance at 441 nm (in the case of $\underline{H}.\ \underline{halobium}$ "cytochrome $\underline{aa}_3$") was attributable to cytochrome $\underline{a}$, while about 88% of the absorbance at 602 nm is attributable to cytochrome $\underline{a}$. Ferricytochrome $\underline{a}_3$ in the "cytochrome $\underline{aa}_3$" showed γ peak at 420 nm and ferrocytochrome $\underline{a}$ showed γ and α peaks at 441 and 602 nm, respectively. As the electron transfer between cytochrome $\underline{a}$ and cytochrome $\underline{a}_3$ was interrupted in the "cytochrome $\underline{aa}_3$", the difference spectra, ferrocytochrome $\underline{a}$ $\underline{minus}$ ferricytochrome $\underline{a}$ ($\underline{i.e.}$ $\underline{a}^{2+} - \underline{a}^{3+}$) and ferrocytochrome $\underline{a}_3$ $\underline{minus}$ ferricytochrome $\underline{a}_3$ ($\underline{i.e.}$ $\underline{a}_3^{2+} - \underline{a}_3^{3+}$) were determined easily without using $\underline{a}_3^{2+}$-CN and $\underline{a}_3^{3+}$-CN unlike the results obtained by Yonetani (7) (Fig.3).

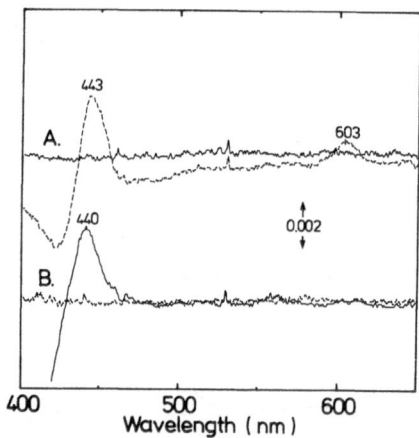

Fig. 3 The redox difference spectra of "cytochrome $\underline{aa}_3$".
A: +(ascorbate + TMPD) $\underline{minus}$ oxidized [$\underline{a}^{2+} - \underline{a}^{3+}$].
B: +$Na_2S_2O_4$ $\underline{minus}$ +(ascorbate + TMPD) [$\underline{a}_3^{2+} - \underline{a}_3^{3+}$].
Concentration of heme $\underline{a}$ was 0.16 µM.

In Table 1, some of the spectral properties of cytochromes $\underline{a}$ and $\underline{a}_3$ in the "cytochrome $\underline{aa}_3$" are shown.

From $\underline{Nitrosomonas\ europaea}$, cytochrome $\underline{c}$ oxidase has been obtained which lacks Cu_A but has Cu_B (2). The oxidase is almost as active as the enzyme which has both Cu_A and Cu_B. With the oxidase, the separation between cytochrome $\underline{a}$ and cytochrome $\underline{a}_3$ is not seen in the absolute spectrum on reduction by ascorbate + TMPD. Namely, electrons are transferred from the $\underline{a}$ component (or heme $\underline{a}_A$) to the $\underline{a}_3$ component (or heme $\underline{a}_B$) in the presence of Cu_B. Therefore, Cu_B of $\underline{aa}_3$-type cytochrome $\underline{c}$ oxidase is essential to its activity, while its Cu_A is not always necessary to its ferrocytochrome $\underline{c}$-oxidizing activity.

Table 1   Some of Spectral Properties of Cytochromes $\underline{a}$ and $\underline{a}_3$ in the "Cytochrome $\underline{aa}_3$".

Redox states of Components	Peaks (nm)	$\underline{A}_\gamma$ / heme $\underline{a}$ (mM^{-1})	$\underline{A}_\gamma/\underline{A}_\alpha$
$\underline{a}^{2+}\underline{a}_3^{2+} - \underline{a}^{3+}\underline{a}_3^{3+}$	441, 602	43	6.0
$\underline{a}^{2+} - \underline{a}^{3+}$	443, 603	48	4.4
$\underline{a}_3^{2+} - \underline{a}_3^{3+}$	440, 600	47	($\sim$10)
$\underline{a}_3^{2+}\cdot CO - \underline{a}_3^{3+}$	430, 590	62	7.2

REFERENCES

1. Fujiwara, T., Fukumori, Y. and Yamanaka, T. (1989) J. Biochem. 105, 287-292.
2. Numata, M., Yamazaki, T., Fukumori, Y. and Yamanaka, T. (1989) J. Biochem. 105, 245-248.
3. Keilin, D. and Hartree, E.F. (1939) Pros. Roy. Soc. London, B127, 167-191.
4. Okunuki, K., Sekuzu, I., Yonetani, T. and Takemori, S. (1958) J. Biochem. 45, 847-854.
5. Sekuzu, I., Takemori, S., Yonetani, T. and Okunuki, K. (1959) J. Biochem. 46, 43-49.
6. Okunuki, K. (1972) in Aspects of Cellular and Molecular Physiology (ed. by Hamaguchi, K.) University of Tokyo Press, Tokyo, pp. 57-73
7. Yonetani, T. (1960) J. Biol. Chem. 235, 845-852.
8. King, T.E. (1965) in Oxidases and Related Redox Systems (ed. by King, T.E., Mason, H.S. and Morrison, M.) Wiley, New York, p. 539.
9. Yamanaka, T., Kamita, Y. and Fukumori, Y. (1981) J. Biochem. 89, 265-273.
10. Ludwig, B. (1987) FEMS Microbiol. Rev. 46, 41-56.
11. Wikstrom, M., Saraste, M. and Penttila, T. (1985) in The Enzymes of Biological Membranes (ed. by Martonosi, A.N.) Plenum, New York, vol. 4, pp. 111-148.
12. Yamanaka, T., Fukumori, Y., Yamazaki, T., Kato, H. and Nakayama, K. (1985) J. Inorg. Biochem. 23, 273-277.
13. Fujiwara, T., Fukumori, Y. and Yamanaka, T. (1987) Plant Cell Physiol. 28, 29-36

Part I. 2. Energy Coupling and Ion Transport

WHICH ELECTRON-TRANSFERRING REACTIONS IN THE RESPIRATORY CHAIN CONTRIBUTE TO THE ENERGY CONSERVATION?

E.C. Slater

Department of Biochemistry
University of Southampton
Bassett Crescent East
Southampton SO9 3TU
England

SUMMARY

As starting point for this essay, it is assumed that the energy conserving act coupled with electron transfer in the mitochondrial respiratory chain is the transfer of negative charges from a specific site near the outer face of the inner membrane to a site near the inner face, and/or the transfer of positive charges in the opposite direction, resulting in a charge separation, negative inside. Little is known about which reactions are involved in Site-1 phosphorylation. Present evidence is in favour of the view that the only energy-conserving reactions in Site-2 phosphorylation is the transfer of electrons from ubisemiquinone, bound near the outer membrane, through cytochrome b to ubiquinone bound to the inner face. There is no electrogenic transfer of protons. Wikström's data indicates that in Site 3, about 20% of the charge transfer is due to electron transfer from ferrocytochrome c bound to the outer face to a binuclear centre in the middle of the membrane and 30% to the transfer of protons, required for the reduction of O_2 to H_2O, from the inner compartment to the centre. The remaining 50% is accounted for by the so-called "proton pump", the nature of which is still obscure.

INTRODUCTION

From work in the 1940's and 1950's, it became generally accepted that, coupled with the oxidation of NADH in the respiratory chain, there are three places or sites at which phosphorylation takes place[1]. Since the total yield of phosphorylation, expressed as the P:O or P:2e ratio, was also thought to be equal to 3, it was generally assumed that each site contributes equally, with a P:2e ratio of 1. By the mid 1960's, however, it was already known from measurements of the phosphate potential and the redox state of respiratory-chain components in State-4 mitochondria[2] that it is energetically an impossibly tight squeeze to maintain this for Site 2.

The concept of discrete phosphorylation sites received strong physical support by the isolation and characterization by Green, Hatefi and their colleagues of three separate large proteins[3], each composed of several subunits, that specifically catalyse the three electron-transferring steps that are linked, via a fourth large protein (the ATPase complex[4]), with the three phosphorylation sites.

Each of these large proteins contains a number of different electron-transferring centres, sometimes on different subunits, sometimes on the same. It is the transfer of electrons from one centre to a second of higher redox potential that provides the free energy that, in many cases, is conserved in a form that is utilized to make ATP from ADP and inorganic phosphate.

In this essay, I shall consider which of these reactions are involved in energy conservation and to what extent. Considerable progress has been made recently in our knowledge of the pathway of electron transfer within ubiquinol : cytochrome c oxidoreductase and cytochrome c oxidase. Little is known about NADH : ubiquinone oxidoreductase and I shall not deal with this enzyme. Since I no longer have a laboratory, I have no new data to put before you - only my perception of the significance of the work of others.

I shall not concern myself with the mechanism of oxidative phosphorylation. If I use, as I shall, the language of chemiosmotic theory, it does not necessarily mean that I believe that, in steady-state oxidative phosphorylation, protons are really pumped right across the membrane and sucked back in again through the ATPase. However, I do agree that charge separation across the mitochondrial inner membrane (or rather across the proteins directly involved) is the essential act of energy conservation in oxidative phosphorylation. My approach is mechanistic only insofar as the significance of the electron-transferring steps is concerned, not for the energy-conserving mechanism.

SITE 2

There is now wide support for the essential features of the Q cycle, originally proposed by Mitchell[5], as a description of the pathway of electron transfer in the oxidation of ubiquinol by cytochrome c . The variation of the Q cycle which is now favoured by most, if not all, workers in the field is shown in Fig. 1. Ubiquinol is oxidized by the high-potential Rieske Fe-S protein to the semiquinone anion, the two H atoms being expelled as protons. The Fe-S protein then passes its electrons to cytochrome c_1, another subunit of the reductase, and finally to cytochrome c bound in a collision complex to the reductase. The semiquinone is oxidised by yet another subunit - the two-haem cytochrome b - which is in turn oxidised by ubiquinone, bound to the inside of the membrane, the Q being reduced to the semiquinone. A second turn of the cycle delivers an electron to this semiquinone, reducing it to the quinol in a reaction in which two protons are taken up on the inside.

The net result of two turns of the cycle is the oxidation of *one* molecule of QH_2 (one is used up in the each turn of the cycle, one is reformed in the second turn) by two molecules of cytochrome c. No less than 14 individual single-electron transfers are involved in a net 2-electron transfer. Six of these (3 in each cycle) take place near the outer surface of the inner membrane, and the overall reaction, ubiquinol + ferricytochrome c —> ubisemiquinone + ferrocytochrome c, probably provides little free energy. The electrogenic reaction is the transfer of electrons from the semiquinone through the two haems of cytochrome b to Q, which, in two steps, is reduced to ubiquinol.

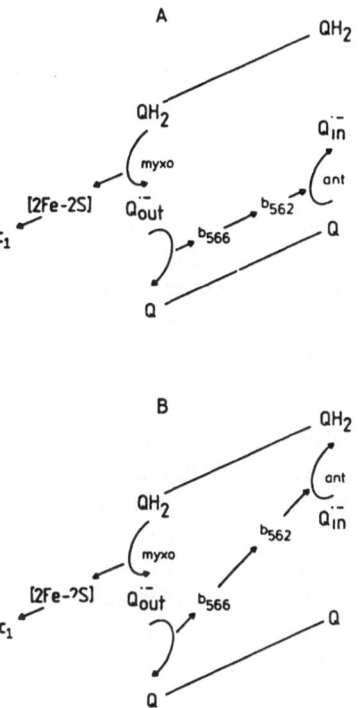

Fig. 1. The Q cycle. Two variants based on de Vries[6], with kind permisson of author and publisher. Straight full arrows represent electron transfer, curved arrows chemical equations. Protons are omitted. Q and QH_2 are freely diffusible. In this essay, it is assumed that variants A and B represent successive turns of the cycle.

Indeed, I have chosen this particular representation of the Q cycle since the protons are omitted from it and Dutton[7] and Rich[8] have shown that the entire electrogenic event (and therefore the energy conservation) is accounted for by electron transfer from the low-potential haem of cytochrome *b* to Q bound at the inner surface of the membrane, without either the proton ejection during oxidation of QH_2 or the proton uptake during the reduction of Q contributing to the electrogenicity. There is, it is true, a net translocation of two protons, but, as I see it, this is *caused by and not the cause of* the charge separation. This is, in my view, an important clarification and simplification. The whole Q cycle adds up to the electron-transferring arm of a Mitchell loop.[9]

SITE 3

Paradoxically, an analogous half-loop was originally proposed by Mitchell[9] to explain oxidative phosphorylation associated with the oxidation of ferrocytochrome *c* by cytochrome oxidase which was thought of as an electric wire conducting electrons from cytochrome *c* on one side of the membrane to the oxygen-reacting site on the other side, where protons are consumed in the reduction of oxygen to water. According to this concept, the oxidation of cytochrome *c* leads to a charge separation across the membrane of one unit per molecule of cytochrome c oxidized. Since it was also believed that two charges are required for the synthesis of one molecule of ATP in mitochondria, this agreed with the stoicheiometry of oxidative phosphorylation at this site, namely P:e =0.5 or P:O = 1, established by Lehninger[10] (see also ref. 11).

It is now widely believed, however, although I would not say that it is completely established, that a charge separation of 4 units is necesary for the synthesis of one molecule of ATP by mitochondria, 3 for the operation of the ATP synthase[12] and 1 for the electrogenic exchange of ADP for ATP[13]. At the same time, Wikström[14] established that, in the cytochrome c oxidase reaction, the charge separation is twice that envisaged in the Mitchell scheme, that is 2 per electron or 8 per molecule of oxygen reduced. Thus, the experimentally determined P:O ratio for this reaction is now thought to be explained by the ratio 4 /4, rather than 2/2!

The first attempts to establish which electron-transferring reactions in Site 3 are energy-conserving was made by Chance and Williams[15] on the basis of a "cross-over" between cytochromes c and a, that is to say cytochrome c becomes more reduced and cytochrome a more oxidized on transition from steady-state oxidative phosphorylation (State 3) to quasi-equiibrium State 4 (or static head). According to the crossover theorem[16], this locates a phoshorylation site. Muraoka and Slater[17], however, found conditions in which the cross-over lies between cytochrome a_3 and oxygen and located the energy-conserving step in this thermodynamically more attractive step.

Cytochrome c oxidase contains 4 electron acceptors, one (Cu_A) bound to one polypeptide and the other 3 - two haem a molecules and a second copper atom (Cu_B) - bound to a second subunit. One of the haem a molecules (Keilin and Hartree's[18] cytochrome a_3) forms a binuclear centre with Cu_B, where binding of oxygen and reduction to water takes place. Electrons from cytochrome c reach this binuclear centre via the other haem a molecule (cytochrome a), as indicated in Fig. 2, taken from a paper by Wikström[19], or Cu_A or both - this is still uncertain.

Starting from the pioneering work of Chance[20], four states of the binuclear centre are now clearly distinguishable spectroscopically - fully oxidized, fully reduced, a peroxy intermediate and a ferryl intermediate[19,21]. We can now write the following sequence of reactions in the reduction of O_2

(1)　　$Fe^{3+}\text{-}OH^- + 2e + O_2 + H^+$　$\longrightarrow$　$Fe^{3+}\text{-}O^- + H_2O$
　　　　Cu^{2+}　　　　　　　　　　　　　　　　　　　　　　$Cu^{2+}\text{-}O^-$
　　　　(O)　　　　　　　　　　　　　　　　　　　　　　　　　(P)

(2)　　$Fe^{3+}\text{-}O^- + e + 2H^+$　$\longrightarrow$　$Fe^{4+}{=}O^{2-}$
　　　　$Cu^{2+}\text{-}O^-$　　　　　　　　　　　　　　　　$Cu^{2+}\text{-}OH_2$
　　　　(P)　　　　　　　　　　　　　　　　　　　　　　　(F)

(3)　　$Fe^{4+}{=}O^{2-} + e + H^+$　$\longrightarrow$　$Fe^{3+}\text{-}OH^- + H_2O$
　　　　$Cu^{2+}\text{-}OH_2$　　　　　　　　　　　　　Cu^{2+}

The uptake of two electrons by the oxidised centre and binding of oxygen leads to the formation of a peroxy intermediate which is reduced to a ferryl intermediate and then back to oxidized form. The protonation steps shown here are appropriate for pH 7.2.

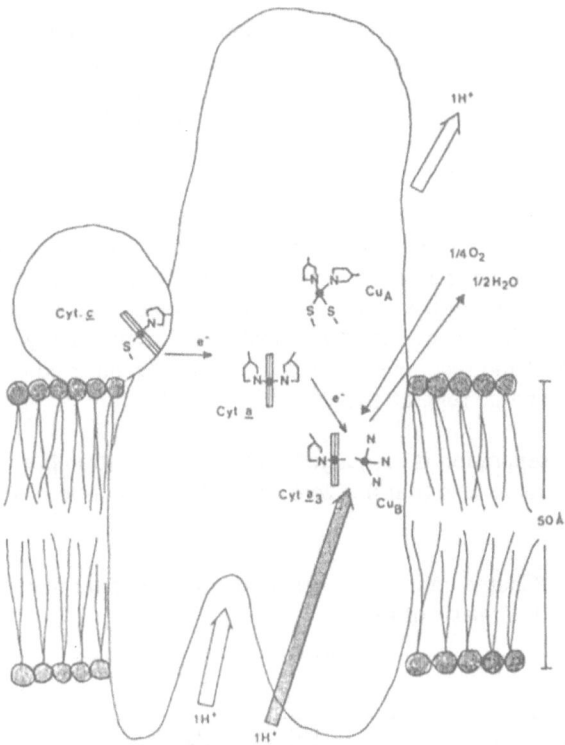

Fig. 2. Cytochrome oxidase. Reproduced from Wikström[19], with kind permisson of author and publisher.

As illustrated in Fig.2, the electrons move from cytochrome *c* on the surface of the oxidase to the binuclear centre buried within it, whereas the protons required for the formation of water travel from the other side. According to the effect of membrane potential on the redox potential of cytochrome *a*, measured by Hinkle and Mitchell[22], the binuclear centre is situated about half-way across the membrane. On the basis of the original Mitchell half loop, we may calculate the total transfer of charges for the three reactions given in the previous slide as shown in Table 1.

Table 1. Charge transfers across mitochondrial inner membrane, at pH 7.2, calculated on basis of a Mitchell half-loop

Reaction	charge transfer		
$[O] + 2e + H^+ \longrightarrow [P]$	$2 \times 0.5 + 0.5$	$=$	1.5
$[P] + e + 2H^+ \longrightarrow [F]$	$0.5 + 2 \times 0.5$	$=$	1.5
$[F] + e + H^+ \longrightarrow [O]$	$0.5 + 0.5$	$=$	$\underline{1.0}$
Sum : $4e + 4q^+$			$\underline{4.0}$

Thus, each of the electron-transferring steps contributes to the charge separation, that is to say to energy conservation. Actually, the relative contributions are pH-dependent, since the O state has two protonable groups, with similar pK values near 7.2. The question now arises: in which steps do the other 4 charges arise? Wikström[23] has supplied the answer to this question.

He has shown that reactions (2) and (3) are reversible, that is to say by putting a charge across the membrane, positive inside, electrons are driven from oxidized oxidase (O) or the ferryl intermediate (F) to the peroxy intermediate (P). What is more, using ATP as the energy source and assuming that the hydrolysis of one molecule of ATP is coupled with the movement of 4 charges, Wikström has shown that 2 additional charges are moved across the membrane in each of the reactions (2) and (3). Thus we must add 2 to both reactions (2) and (3) to yield the charge separations shown in Table 2.

Most of the charge separation is associated with the demonstratably reversible reactions (2) and (3) and considerably less (but not zero) with the oxygen-peroxy reaction, which is

Table 2. Charge transfers across mitochondrial inner membrane, calculated on basis of Fig.2

Reaction		charge transfer (q^+)	
$[O] + 2e + H^+$	$\longrightarrow [P]$	$2 \times 0.5 + 0.5$	$= 1.5$
$[P] + e + 2H^+ + 2q^+$	$\longrightarrow [F]$	$0.5 + 2 \times 0.5 + 2$	$= 3.5$
$[F] + e + H^+ + 2q^+$	$\longrightarrow [O]$	$0.5 + 0.5 + 2$	$= 3.0$
Sum $4e + 4H^+ + 4q^+$			8.0

Table 3. ΔG_o values for the cytochrome oxidase reaction

Reaction	ΔE_m (V)	n	ΔG_o(kcal)	
			[no q^+][a]	[+ q^+][b]
$[O] + 2c^2 + O_2 \longrightarrow [P] + 2c^{3+}$	0.29	2	-13.4	-7.2
$[P] + c^{2+} \longrightarrow [F] + c^{3+}$	0.86	1	-19.8	-5.3
$[F] + c^{2+} \longrightarrow [O] + c^{3+}$	0.72	1	-16.6	-4.1
Sum $4c^{2+} + O_2 \longrightarrow 4c^{3+}$	0.54	4	-49.8	-16.6

[a]Calculated without taking into account movements of charge across the membrane.
[b]Calculated on the basis of the charge transfers calculated in Table 2 moving against a charge potential of 180 mV, positive outside.

essentially irreversible. This is clearly shown by the ΔG_o values listed in Table 3, in which two calculations of the ΔG_o have been made, one without taking account of the charge separation and one taking this into account assuming the movement of charge against a potential of 180 mV. The pH is taken as 7.0 and the mid-point potentials of cytochrome c and oxygen as 260 and 800 mV, respectively. The redox potentials of P and F are from Wikström (personal communication) and that appropriate for Reaction (1) calculated by difference. Reactions (2) and (3) are sufficiently poised for Wikström[23] to be able to demonstrate the oxidation by cytochrome c , coupled to the hydrolysis of ATP, of O to F (at pH 7.2) and of F to P (at pH 8.3), with P:2e ratios of 1. However, Reaction (1) is essentially irreversible, accounting for earlier failures experiment to put sufficient energy into the system to bring about the oxidation of water to oxygen.

If the H^+:ATP ratio were 3 instead of 4, q values for the second and third reactions would each be reduced by 1.0 and ΔG_o values by 4.1 kcal, which would make them much less likely to be easily reversible. Moreover, the stoicheiometry of the F—>O reaction has been confirmed by an independent measurement not dependent on the hydrolysis of ATP[23]. All this gives good support for the assumption of 4 for the H^+:ATP ratio.

P:O RATIOS

There is one final point that I would like to make. If it is correct that energy conservation in Sites 2 and 3 leads to the separation of 2 and 4 charges per pair of electrons, respectively, and that the synthesis of one molecule of ATP requires a charge separation of 4 units, it follows that the mechanistic P:2e ratio for Sites 2 and 3 together, as in the oxidation of succinate, is 6/4 =1.5, not 2.0, as we were told in the 50's and 60's we should find experimentally if we were skillful enough to make well-coupled mitochondria. I do not mind: we[24] could never find more than about 1.4. Hinkle[25], in a careful study, also found a P:O ratio equal to 1.4.

Of course a low P:O ratio can always be explained away (and has been) by "leaks" or "slips" which cause the putative intermediate of oxidative phosphorylation either not to react with the ATP-synthesizing system or to do so abortively. It is most desirable, then, to find direct experimental methods of determining the mechanistic P:O ratio independent of "leaks" and "slips". Some years ago, Westerhoff *et al.*[26] devised such a method that took account of "leaks" and found a value of 1.46.

Beavis[27] and Lemasters[28], however, have reported higher values, namely 1.75 and 2.00, respectively.

P:O ratios are still controversial.

ACKNOWLEDGEMENT

I thank Professor M.Wikström for valuable discussions and for communicating unpublished material.

REFERENCES

1. Slater, E.C.(1966) *in* Comprehensive Biochemistry, Vol.14, M. Florkin and E.M.Stotz, ed., Elsevier, pp. 327-396.
2. Slater, E.C.(1970) *in* Electron Transport and Energy Conservation, J.M.Tager, S. Papa, E. Quagliariello and E.C. Slater, ed., Adriatica Editrice, Bari, pp.363-369.
3. Hatefi, Y.(1966) *in* Comprehensive Biochemistry, Vol.14, M. Florkin and E.M.Stotz, ed., Elsevier, pp. 199-231.
4. Kagawa, Y. and Racker, E.F.(1966) *J. Biol. Chem.* **241,** 2461-2466.
5. Mitchell, P.(1976) *J. Theor. Biol.* **62,** 327-367.
6. De Vries, S. (1986) *J. Bioenerg. Biomembr.* **18,** 195-224.
7. Robertson, D.E. and Dutton, P.L.(1988) *Biochim. Biophys. Acta* **935,** 273-291.
8. Hope, A.B. and Rich, P.R.(1989) *Biochim. Biophys. Acta* **975,** 96-103.
9. Mitchell, P.(1966), Chemiosmotic Coupling in Oxidative Phosphorylation, Glynn Research Ltd., Bodmin.
10. Nielsen, S.O. and Lehninger, A.L.(1954) *J. Amer. Chem. Soc.* **76,** 3860.
11. Chamalaun, R.A.F.M. and Tager, J.M.(1969) *Biochim. Biophys. Acta* **180,** 204-206.
12. Kagawa, Y.(1984) *in* Bioenergetics, New Comprehensive Biochemistry, Vol. 9, L. Ernster, ed., Elsevier, pp. 149-186.
13. LaNoue, K.F., Mizani, S.M. and Klingenberg, M.(1978) *J. Biol. Chem.* **253,** 191-198.
14. Wikström, M.(1977) *Nature* **266,** 271-273.
15. Chance, B. and Williams, G.R.(1956) *Adv. Enzymol.* **17,** 65-134.
16. Chance, B., Holmes, W., Higgins, J. and Connelly, J.M.(1958) *Nature* **182,** 1190-1193.
17. Muraoka,S. and Slater, E.C.(1969) *Biochim. Biophys. Acta* **180,** 227-236.
18. Keilin, D. and Hartree, E.F.(1939) *Proc. Roy. Soc. B* **127,** 167-191.
19. Wikström, M.(1987) *Chemica Scripta* **27B,** 53-58.
20. Chance, B., Saranio, C. and Leigh, J.S., Jr.(1975) *J. Biol. Chem.* **250,** 9226-9237.
21. Wikström, M.(1988) *Chemica Scripta* **28A,** 71-74.
22. Hinkle, P.C. and Mitchell, P.(1970) *J. Bioenerg.* **1,** 45-60.
23. Wikström, M.(1989) *Nature* **338,** 776-778.
24. Greengard, P., Minnaert, K., Slater, E.C. and Betel, I.(1959) *Biochem. J.* **73,** 637-646.
25. Hinkle, P.C. and Yu, M.L.(1979) *J.Biol.Chem.* **254,** 2450-2455.
26 van Dam, K., Westerhoff, H.V., Krab, K. and Arents, J.C.(1980) *Biochim. Biophys. Acta* **591,** 240-250.
27. Beavis, A.D. and Lehninger, A.L.(1986) *Eur. J. Biochem.* **158,** 315-322.
28. Lemasters, J.J.(1984) *J. Biol. Chem.* **259,** 13123-13130.

CYTOCHROME OXIDASE AND PEROXIDE METABOLISM

Yutaka Orii

Department of Public Health, Faculty of Medicine
Kyoto University
Kyoto 606, Japan

SUMMARY

Hydrogen peroxide accelerated the cytochrome oxidase-catalyzed oxidation of ferrocytochrome c whereas in the polarographic assay it depressed the oxygen consumption even though the extent was small. Calculations based on parallel determination of both cytochrome c oxidase and peroxidase activities revealed that 8-10% of the total electron flux was directed to the peroxide when nearly equal oxidizing equivalents of both oxidants were present in the reaction system. The depressed oxygen uptake suggests that hydrogen peroxide acts as an electron acceptor by drawing a portion of the electron flux which otherwise is directed to molecular oxygen. On the contrary, the accelerated oxidation of ferrocytochrome c suggests that the total electron flux through cytochrome oxidase increases possibly by activation of a reserve capacity of the enzyme to utilize hydrogen peroxide in addition to oxygen. The acceleration effect became pronounced as the oxygen concentration was lowered to 1.5-2.1 μM.

INTRODUCTION

Cytochrome oxidase catalyzes not only reduction of molecular oxygen to water utilizing electrons provided by reduced cytochrome c (1), but also peroxidatic oxidation of ferrocytochrome c under strictly anaerobic conditions (2,3). The latter function is not unexpected since it is conceivable that the peroxide metabolism is involved in the reduction process of oxygen to water (4), and early studies implicated a close relationship between the hydrogen peroxide complexes and so-called "oxygenated" compounds of cytochrome oxidase (5-7). Chance and his coworkers have proposed that peroxidase-like intermediates are formed in the reaction of cytochrome oxidase with oxygen at cryogenic temperatures and that a peroxy-type intermediate catalyzes oxidation of ferrocytochrome c and other donors (8,9). The experimental demonstration of the cytochrome c peroxidase activity of this enzyme at room temperatures has prompted me to propose that the reaction of half-reduced cytochrome oxidase with oxygen gives the same product as that obtained by reaction of the fully oxidized oxidase with hydrogen peroxide (10). Since then, many efforts have been made to analyze the reaction of cytochrome oxidase with hydrogen peroxide in great detail and to identify the reaction intermediates and products by optical and EPR spectroscopy at cryogenic and room temperatures as well (11-18).

Bioenergetics, Edited by C. H. Kim and
T. Ozawa, Plenum Press, New York, 1990

The achievements of these studies constituted basis for identification of both "peroxy" and "ferryl" forms of cytochrome oxidase as active intermediates in the cytochrome oxidase reaction (19, 20).

The cytochrome c peroxidase activity *per se*, on the other hand, was shown under anaerobic conditions to couple to the proton pumping as well as to generation of the membrane potential across the membrane of phospholipid vesicles reconstituted with cytochrome oxidase (21, 22). If these reactions have some physiological significance, at least the peroxidase activity should be recognized in the presence of oxygen, and the preliminary studies apparently supported this possibility (2, 10). In the present study, I have measured both oxidase and peroxidase activities in the presence of both molecular oxygen and hydrogen peroxide, and observed that a portion of the electron flux through cytochrome oxidase is directed to hydrogen peroxide.

MATERIALS AND METHODS

Cytochrome oxidase was purified from bovine heart muscle according to the procedure reported previously (23), dissolved in 0.05 M sodium phosphate buffer (pH 7.4) - 0.25% (v/v) Emasol 1130, and stored in liquid nitrogen until used. The catalytic activity of the present preparation was as described previously (2). The concentrations of cytochrome oxidase (in terms of a unit containing two heme a molecules) and cytochrome c were determined spectrophotometrically based on millimolar extinction coefficient differences as cited previously (23). Bovine liver catalase (2x crystallized, 32,000 Sigma units/mg protein) and horse heart cytochrome c were products of Sigma. Hydrogen peroxide was determined spectrophotometrically according to the method of Allen et al. (24).

Absorption spectra and time-dependent absorbance changes were recorded on either a Shimadzu dual wavelength spectrophotometer UV-300 or on a single beam type spectrophotometer (UNISOKU Co. Ltd. Hirakata, Osaka). An OXYGRAPH (Gilson model 5/6) was used to follow the oxygen consumption during the cytochrome oxidase reaction in the presence and absence of hydrogen peroxide. An anaerobic cell of a 9-ml capacity was made of a ceramic block and designed to allow simultaneous recording of an absorbance change and the oxygen concentration. Teflon tubings were fitted to the cell through the ports at the top cover for inlet and outlet of gas. Nitrogen gas was supplied to a Toray Oxygen Pump to yield a gas mixture containing a specified concentration of oxygen, and this was led to the reaction vessel through a bubbler to humidify the gas mixture. When the gas flow rate was around 35 ml/min, a 15-min bubbling was sufficient to equilibrate a solution in the vessel with oxygen. Then the tubing was pulled up above the solution surface and a gas flushing was continued throughout the measurement. Additions of reagents to the solution were made though the outlet port with a Hamilton microsyringe. All of the measurements were carried out at 25°C unless otherwise stated. To facilitate comparison of the enzyme activities as assayed spectrophotometrically and polarographically, the compositions of the reaction mixtures were maintained the same except for different total volumes. Although nonenzymic decomposition of hydrogen peroxide as well as autoxidation of ascorbate was diminished by EDTA, the three reagents altogether greatly enhanced the consumption of oxygen; that is, the rate of oxygen uptake of 9.6 mM ascorbate in a mixture of 0.05 M sodium phosphate (pH 7.4) - 0.25% (v/v) Emasol 1130 - 10 μM EDTA was 0.044 μM O_2/s, and this increased to 0.234μM O_2/s on addition of 1.4 mM hydrogen peroxide. If EDTA had been removed from the reaction mixture no such enhancement was observed. Thus, in order to eliminate such complexity, EDTA was omitted from the reaction mixtures unless otherwise described.

RESULTS AND DISCUSSION

Distinction between Oxidase and Peroxidase Reactions of Cytochrome Oxidase- Fig. 1 illustrates traces of the reduction level of cytochrome c under various reaction conditions. While the ordinary cytochrome oxidase reaction was in progress, the oxidation-reduction state of the cytochrome c usually remained at a constant level (Fig. 1A), which was determined by the balance of the reducing power of ascorbate and the oxidizing power of the cytochrome oxidase-oxygen system. In the present

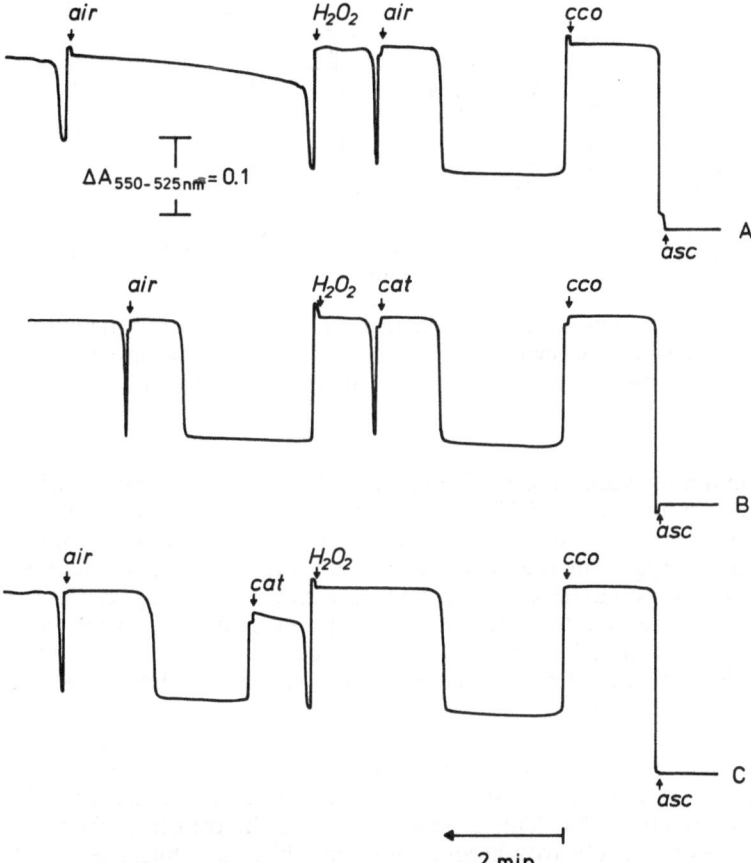

Fig. 1. Spectrophotometric monitoring of the reduction level of cytochrome c during the cytochrome oxidase and peroxidase reactions. An optical cuvette (1 cm light path) contained 2.0 ml of 0.05 M sodium phosphate buffer (pH 7.4)-0.25% (v/v) Emasol 1130 and 20 μl of 1.65 mM cytochrome c. Absorbance changes on addition of following reagents in a 20-μl portion each were followed on a Shimadzu dual wavelength spectrophotometer set at a wavelength pair of 550-525 nm. The reagents added were 1.0 M sodium ascorbate, 70 μM cytochrome oxidase and 47.2 mM hydrogen peroxide. Air was introduced to a reaction mixture at the time indicated by an arrow by moving an adder-mixer up and down two times in the solution.

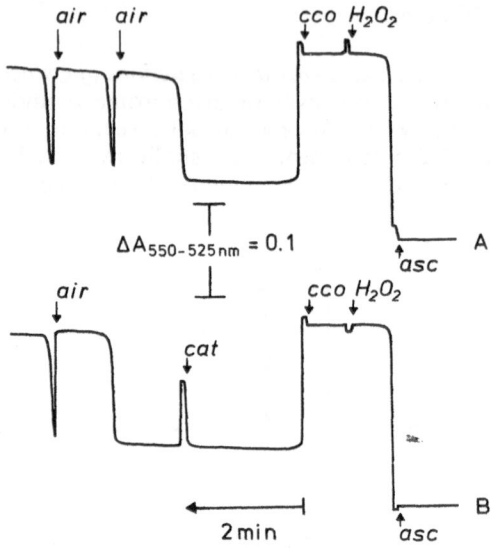

Fig. 2. Simultaneous utilization of oxygen and hydrogen peroxide by cytochrome oxidase. The experimental procedures were as described in the legend to Fig. 1.

study the conditions were chosen to maintain the reduction level considerably low (32% reduced in the case of Fig.1A), so that the absorbance change therefrom responding to lowering of the oxidizing power was easily manifested. Upon exhaustion of the dissolved oxygen in the reaction mixture the absorbance increased rapidly approaching the maximal level. The concentration of the oxygen divided by the duration time of the steady state gave an apparent rate of oxygen consumption. When the anaerobic reaction mixture was agitated by an adder-mixer up and down two times, the reduction level deflected downward responding to a concomitant introduction of oxygen, but reverted to the original level as soon as the oxygen was used up by the oxidase reaction. When hydrogen peroxide was added by the adder-mixer in the same way cytochrome c was not fully reduced even after exhaustion of the contaminated oxygen. A rapid absorbance increase ceased halfway, followed by a gradual increase (Fig. 1A). This change represents the transition from the reaction which utilizes oxygen as electron acceptor to that which uses hydrogen peroxide. The initial absorbance increase, or elevation of the reduction level of cytochrome c, indicates that the oxidizing power of cytochrome oxidase is higher with oxygen than with hydrogen peroxide. The size of the initial burst became small as the hydrogen peroxide concentration was increased. When bovine liver catalase had been added in advance, the cytochrome oxidase reaction was resumed immediately upon addition of hydrogen peroxide (Fig. 1B). Catalase added while an absorbance was increasing gradually in the presence of hydrogen peroxide but in the absence of oxygen also revived the oxidase reaction (Fig. 1C). Thus it is apparent that catalase/H_2O_2 instantaneously produces enough oxygen to initiate the oxidase reaction. The gradual absorbance increase is a manifestation of the susceptibility of the peroxidase activity on the peroxide concentration, and this would be due to the fact that the K_m value of cytochrome oxidase towards hydrogen peroxide, 0.18 mM (2), is in the range of the substrate concentration employed.

Simultaneous Action of Hydrogen Peroxide and Oxygen on Cytochrome Oxidase- After complete reduction of cytochrome c with ascorbate under the air, hydrogen peroxide and cytochrome oxidase were added to initiate the reaction (Fig. 2A). The reduction level of cytochrome c during the steady state was almost the same as that attained in the absence of hydrogen peroxide. After exhaustion of the dissolved oxygen the absorbance increased biphasically as observed previously (Fig. 1A). When catalase was added to the reaction mixture during the absorbance increase, hydrogen peroxide remaining was decomposed immediately and the ordinary cytochrome oxidase reaction was resumed (Fig. 2B). The duration of the steady state thus established may serve to estimate the concentration of hydrogen peroxide at the time of the catalase addition. For calibration different concentrations of hydrogen peroxide up to 565 μM were added to each of the reaction mixtures which had attained anaerobiosis after termination of the oxidase reaction. The duration of the steady state, t in s, was linearly correlated with the peroxide concentration, $[H_2O_2]$ in μM, by the equation t = 5.73 + 0.262 $[H_2O_2]$ (correlation coefficient=0.998; n=5), as illustrated in Fig. 3 (top), and this can be used to determine the concentration of hydrogen peroxide remaining when the oxidase reaction was terminated. Since the peroxide concentration prior to the oxidase reaction is known, the amount of peroxidase consumed during the reaction can be calculated. In one measurement as depicted in Fig. 2B cytochrome oxidase was shown to utilize hydrogen peroxide and oxygen simultaneously at the rates of 0.77 μM/s and 1.85 μM/s when the initial concentrations of both electron acceptors were 453 μM and 293 μM, respectively. After exhaustion of oxygen, hydrogen peroxide decreased as shown in Fig. 3 (bottom) giving the consumption rate of 0.89 μM H_2O_2/s. After correction for the nonenzymic reduction of hydrogen peroxide by ascorbate, which was determined to be 0.5 μM/s in parallel polarographic measurements, the consumption rates of hydrogen peroxide in the presence and absence of oxygen became 0.27 and 0.39 μM/s, respectively.

Polarographic Determination of Hydrogen Peroxide Utilization by Cytochrome Oxidase in the Presence and Absence of Oxygen- For the polarographic determination of the cytochrome oxidase and peroxidase activities some preparatory measures were undertaken as follows. Hydrogen peroxide is decomposed stoichiometrically to oxygen by catalase, and in the polarographic measurement the signal amplitude was linearly correlated with the peroxide concentration (μM) at least up to 600 μM. Contaminating heavy metal ions in solutions, if any, catalyzed decomposition of hydrogen peroxide in the absence of metal ion chelators like EDTA. In Emasol buffer 5.8 mM hydrogen peroxide evolved oxygen at a rate of 0.11 μM O_2/s but neither 0.4 nM cytochrome oxidase nor 13 μM ferrocytochrome c showed accelerating effects. When the peroxide was below 500 μM as employed in the present studies the oxygen evolution was almost negligible nor was cytochrome oxidase damaged. On the other hand, when hydrogen peroxide (460 μM) was incubated with ascorbate (9.8 mM) in the Emasol buffer the peroxide decreased with time. The initial consumption rate was 1 μM H_2O_2/s but soon it became 0.5 μM H_2O_2/s and the reaction proceeded almost linearly thereafter. In the polarographic assay, as ascorbate and hydrogen peroxide were allowed to react at least for 30 s before the initiation of the enzyme-catalyzed reactions, the latter value was used to correct for the nonenzymic reduction of hydrogen peroxide by ascorbate (see MATERIALS AND METHODS).

The cytochrome oxidase activities were examined in the presence and absence of 452 μM hydrogen peroxide and the results are summarized in TABLE I. The rate of oxygen uptake was inhibited 6% by the peroxide. When hydrogen peroxide was added while the oxidase reaction was in progress, the slope of a trace on the OXYGRAPH became less steep. After complete decomposition of hydrogen peroxide

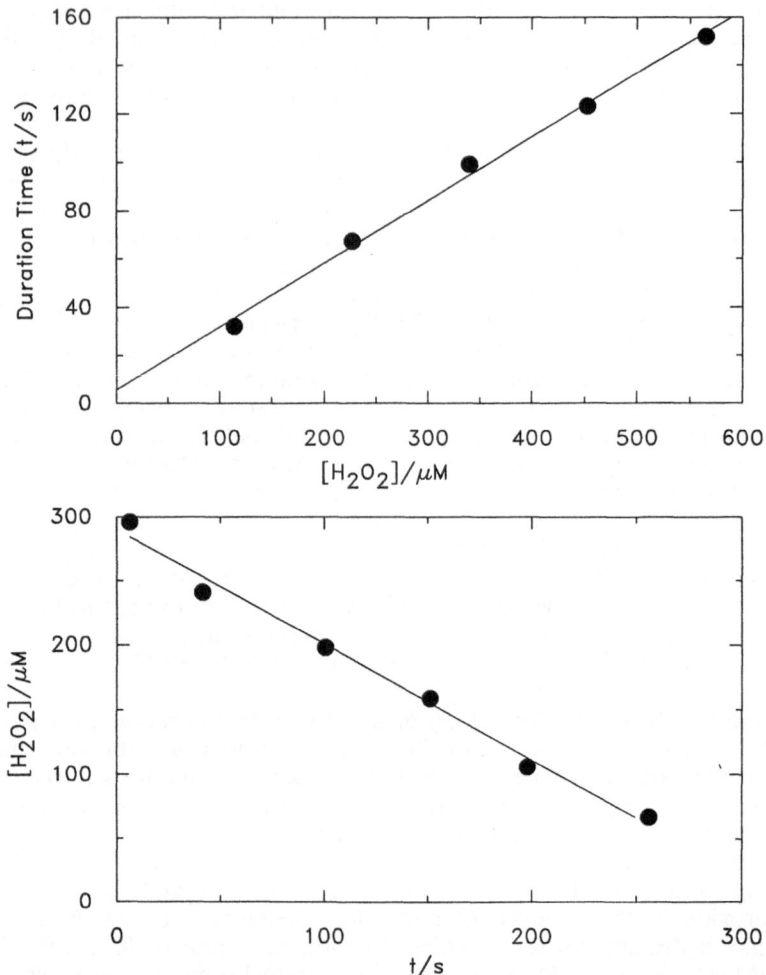

Fig. 3. (top) A calibration curve. (bottom) An apparent consumption of hydrogen peroxide by cytochrome oxidase under anaerobic conditions. Cytochrome c (1.58 mM), ascorbate (1.0 M), hydrogen peroxide (47.2 mM) and cytochrome oxidase (70 μM) were added in 20-μl portions to 2.0 ml of 0.05 M sodium phosphate buffer (pH 7.4)-0.25% (v/v) Emasol 1130 in this order as shown in Fig. 2B. After exhaustion of the dissolved oxygen the reaction which utilized hydrogen peroxide as an electron acceptor was allowed to proceed. Then at appropriate reaction times, catalase (125 μg in 5 μl) was added to the reaction mixture and the duration time of the steady state of the resumed cytochrome oxidase reaction was measured. The concentrations of hydrogen peroxide thus determined were plotted against the reaction time.

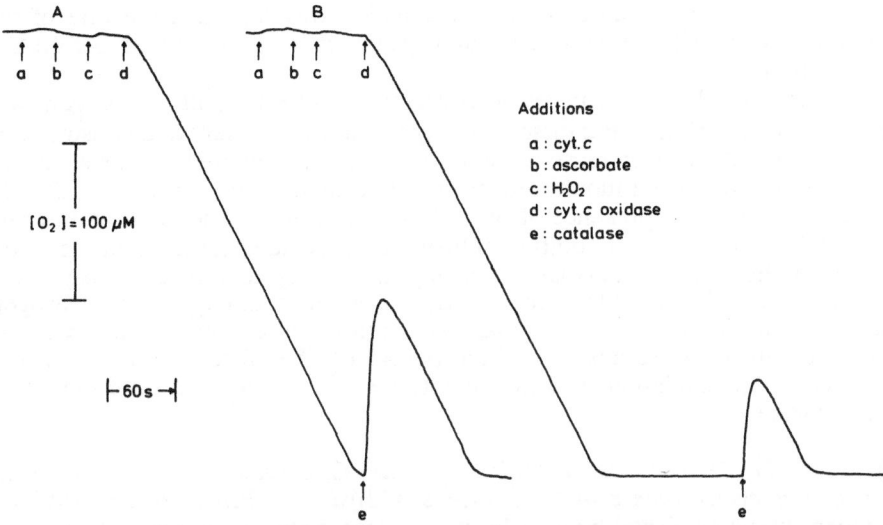

Fig. 4. Utilization of oxygen and hydrogen peroxide by cytochrome oxidase. To 1.6 ml of 0.05 M sodium phosphate buffer (pH 7.4)-0.25% (v/v) Emasol 1130 in a reaction vessel were added 16-μl portions of 1.58 mM cytochrome *c*, 1.0 M sodium ascorbate, 47.2 mM hydrogen peroxide and 140 μM cytochrome oxidase st every 20 s. As soon as the dissolved oxygen was exhausted or after the anaerobic utilization of the peroxide was allowed to proceed for about 120 s, 125 μg of catalase (5 μl) was added to decompose the peroxide remaining.

TABLE I. Utilization of Oxygen and Hydrogen Peroxide. The experimental details are as described in legend to Fig. 4.

| | Electron acceptor | | Consumption rate (μM/s) | |
	O_2	H_2O_2	O_2	H_2O_2
Expt. 1	+	+	1.22[a]	0.27
	+	$-$[c]	1.30[b]	
	$-$	+[d]		0.44
Expt. 2	+	+	1.41[a]	0.25
	+	$-$[c]	1.51[a]	
	$-$	+[d]		0.48

[a]Average of two determinations
[b]Average of three determinations
[c]The oxygen consumption rates were determined with the reaction system which contained no hydrogen peroxide initially.
[d]After the enzymic exhaustion of the dissolved oxygen only hydrogen peroxide served as an electron acceptor.

by catalase the ordinary cytochrome oxidase reaction was resumed, the rate of the oxygen uptake becoming the same as that which had been registered before the peroxide addition.

The ratio of the hydrogen peroxide consumption rate to that of oxygen was 0.18-0.22. After exhaustion of the dissolved oxygen the rate of peroxide consumption increased 1.6-1.9 fold. This increase is comparable to that obtained previously under the same assay conditions but monitored spectrophotometrically (0.39 versus 0.27 μM H_2O_2/s in the absence and presence of oxygen, respectively). A decrease in the rate of oxygen uptake observed after addition of hydrogen peroxide suggests that a portion of the electron flux which otherwise is flowing exclusively to molecular oxygen is directed to the peroxide. The total electron flux pouring to both oxygen and hydrogen peroxide in Expt. 1 of TABLE I was 5.42 μM equivalents/s, higher than 5.2 μM equivalents/s obtained in the absence of the peroxide. Therefore, it is possible that the oxidase-catalyzed oxidation of ferrocytochrome *c* is accelerated on addition of hydrogen peroxide.

Effect of the Oxygen Concentration on the Cytochrome Oxidase-Catalyzed Oxidation of Ferrocytochrome c in the Presence of Hydrogen Peroxide-The oxidation of ferrocytochrome *c* catalyzed by cytochrome oxidase obeys an apparent first order kinetics. On the other hand, the peroxidatic oxidation under the anaerobic condition deviates from the first order kinetics, accompanying a gradual increase in the apparent rate constant (2). If hydrogen peroxide is added to the reaction mixture during aerobic turnover of cytochrome oxidase, the oxidation profile of ferrocytochrome *c* also changes in response to the onset of the peroxidation oxidation even in the presence of oxygen. Fig. 5 illustrates the effect of oxygen concentration

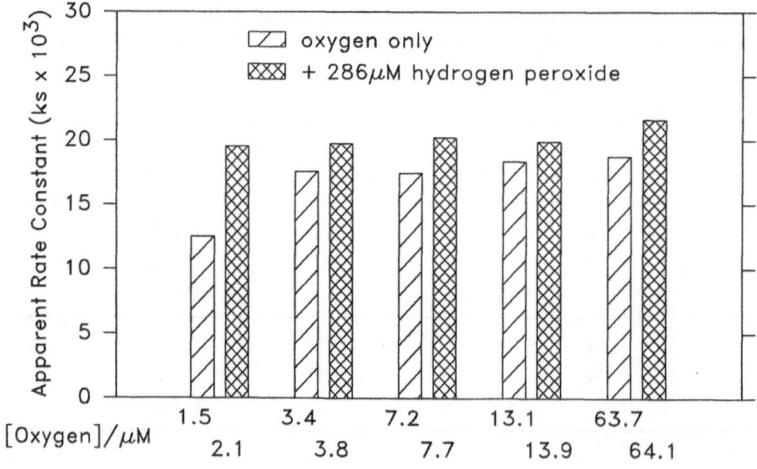

Fig. 5. Effect of the oxygen concentration on the cytochrome oxidase-catalyzed oxidation of ferrocytochrome *c* in the presence of hydrogen peroxide. The reaction mixture contained 8 μM ferrocytochrome *c*, 4.7 nM cytochrome oxidase with or without 287 μM hydrogen peroxide in 0.075 M potassium phosphate buffer, pH 6.0-1 mM EDTA. Prior to the addition of cytochrome *c* and the oxidase, the solution was bubbled with nitrogen gas containing a specified concentration of oxygen for 15 min at 25°C, and the reaction was initiated by addition of the oxidase (25).

on the apparent rate constant for the oxidation of ferrocytochrome c in the presence and absence of hydrogen peroxide. In the absence of hydrogen peroxide the rate constant remained unchanged at oxygen concentrations at and above 3.4 μM but decreased by 30% at 1.5 μM. On the other hand, the rate constant obtained in the presence of hydrogen peroxide was almost unchanged in the oxygen concentration range examined. The rate constant in the presence of hydrogen peroxide always surpassed that in its absence; that is, the relative ratio of the rate constant at and above 3.4 μM oxygen was 1.14 $\pm$ 0.03 on average but it increased to 1.56 at 1.5 μM oxygen. These results clearly indicate that even in the presence of oxygen the peroxidatic oxidation of ferrocytochrome c proceeds with hydrogen peroxide as electron acceptor and that the extent of contribution of the peroxidatic oxidation becomes large as the oxygen concentration is decreased.

Is Oxygen or Hydrogen Peroxide a Modulator of Cytochrome Oxidase?-An apparent K_m value of cytochrome oxidase towards hydrogen peroxide determined under the strictly anaerobic condition is 0.18 mM (2). This is much higher than that against molecular oxygen which is in a range of 10^{-6} to 10^{-8} M depending on measurement conditions (26). Such a large difference in K_m seems to hardly explain the utilization of hydrogen peroxide by cytochrome oxidase in the presence of a nearly equal concentration of oxygen. This difficulty may be overcome by postulating a certain kind of modulating effect of oxygen on cytochrome oxidase as to increase its affinity towards the peroxide. Or alternatively, hydrogen peroxide may lower the affinity of the enzyme towards oxygen. Additionally, it is also possible that latent cytochrome oxidase molecules are activated by hydrogen peroxide with itself serving as an electron acceptor because under ordinary cytochrome oxidase reaction not all of the redox centers in the oxidase may be functioning (see ref. 27 for details). This might account for the increased total electron flux when both hydrogen peroxide and oxygen act on cytochrome oxidase simultaneously. These speculations, however, need further studies for elaboration.

Apart from the mechanism underlying the dual function of cytochrome oxidase as both oxidase and peroxidase, it is worthwhile to speculate on the possible role of cytochrome oxidase as a scavenger of hydrogen peroxide produced in the cells of the aerobe. As far as the peroxide generated in mitochondria is concerned, it is conceivable that cytochrome oxidase located in the inner mitochondrial membrane utilizes the peroxide in situ taking advantage of its location before hydrogen peroxide escapes out of the membrane. If this reaction generates a chemiosmotic pressure of proton across the mitochondrial inner membrane as indicated by the model experiments using cytochrome oxidase incorporated into phospholipid vesicles (21, 22), cytochrome oxidase can be regarded as the most elaborate enzyme evolved to counter and tame hazardous oxygen reduction intermediates. Although it is not yet established whether cytochrome oxidase reacts with hydrogen peroxide generated in situ or not, spectral examinations of hemoglobin-free perfused rat livers recently have elicited novel spectral species that are formed under a continuous generation of hydrogen peroxide and may be ascribed to the "peroxy" and "ferryl" forms of cytochrome oxidase (28). Thus, understanding of the physiological significance of cytochrome oxidase in the peroxide metabolism is expected to see a great leap in a near future.

ACKNOWLEDGMENTS

This work was supported in part by Research Grant of Fujiwara Foundation of Kyoto University, Grants-in-Aid for Science Research on Priority Area of Bioenergetics, and grant 6387102 from the Ministry of Education, Science and Culture of Japan.

REFERENCES

1. Keilin, D. (1966) *The History of Cell Respiration and Cytochrome*, Cambridge University Press, Cambridge.
2. Orii, Y. (1982) *J. Biol. Chem.* **257**, 9246-9248.
3. Bickar, D., Bonaventura, J., and Bonaventura, C. (1982) *Biochemistry* **21**, 2661-2666.
4. Yamazaki, I. (1974) in *Molecular Mechanisms of Oxygen Activation* (Hayaishi, O., ed) pp. 535-558, Academic Press, New York.
5. Orii, Y., and Okunuki, K. (1963) *J. Biochem.* **54**, 207-213.
6. Lemberg, R., and Mansley, G. E. (1966) *Biochim. Biophys. Acta* **118**, 19-35.
7. Lemberg, R., and Stanbury, J. (1967) *Biochim. Biophys. Acta* **143**, 37-51.
8. Chance, B., Waring, A., and Powers, L. (1979) in *Cytochrome Oxidase* (King, T. E., Orii, Y., Chance, B., and Okunuki, K., eds) pp. 353-360, Elsevier/North-Holland, Amsterdam.
9. Chance, B., O'Connor, P., and Yang, E. (1982) in *Oxidases and Related Redox Systems* (King, T. E., Mason, H. S., and Morrison, M., eds) pp. 1019-1035, Pergamon Press, Oxford.
10. Orii, Y. (1982) in *Oxygenases and Oxygen Metabolism* (Nozaki, M., Yamamoto, S., Ishimura, Y., Coon, M. J., Ernster, L., and Estabrook, R. W., eds) pp. 137-149, Academic Press, New York.
11. Kumar, C., Naqui, A., and Chance, B. (1984) *J. Biol. Chem.* **259**, 11668-11671.
12. Wrigglesworth, J. M. (1984) *Biochem. J.* **217**, 715-719.
13. Gorren, A. C. F., Dekker, H., and Wever, R. (1985) *Biochim. Biophys. Acta* **809**, 90-96.
14. Gorren, A. C. F., Dekker, H., and Wever, R. (1986) *Biochim. Biophys. Acta* **852**, 81-92.
15. Witt, S. N., and Chan, S. I. (1987) *J. Biol. Chem.* **262**, 1446-1448.
16. Vygodina, T., and Konstantinov, A. A. (1987) *FEBS Lett.* **219**, 387-392.
17. Gorren, A. C. F., Dekker, H., Vlegels, L., and Wever, R. (1988) *Biochim. Biophys. Acta* **932**, 277-286.
18. Vygodina, T., and Konstantinov, A. A. (1989) *Biochim. Biophys. Acta* **973**, 390-398.
19. Orii, Y. (1988) *Chem. Scripta* **28A**, 63-69.
20. Orii, Y. (1988) *Ann. NY Acad. Sci.* **550**, 105-117.
21. Miki, T., and Orii, Y. (1986) *J. Biol. Chem.* **261**, 3915-3918.
22. Miki, T., and Orii, Y. (1987) *J. Biochem.* **100**, 735-745.
23. Orii, Y., Manabe, M., and Yoneda, M. (1977) *J. Biochem.* **81**, 505-517.
24. Allen, A. O., Hochanadel, C. J., Ghormley, J. A., and Davies, T. W. (1952) *J. Phys. Chem.* **56**, 575-586.
25. Orii, Y. manuscript in preparation.
26. Nicholls, P., and Chance, B. (1974) in *Molecular Mechanisms of Oxygen Activation* (Hayaishi, O., ed) pp. 479-534, Academic Press, New York.
27. Orii, Y., Matsumura, Y., and Okunuki, K. (1973) in *Oxidases and Related Redox Systems* (King, T. E., Mason, H. S., and Morrison, M., eds) pp. 666-672, University Park Press, Baltimore.
28. Orii, Y., Sakai, Y., and Ozawa, K. (1989) *Biochem. Biophys. Res. Commun.* in press.

DO HYDROGEN IONS SERVE

AS BIOENERGETIC MESSENGERS IN YEAST?

Arnošt Kotyk

Department of Membrane Transport
Institute of Physiology, Czechoslovak Academy of Sciences
142 20 Prague, Czechoslovakia

HISTORICAL BACKGROUND

It was in the late fifties[1] and early sixties[2] that the idea dawned upon the scientific community that cations, in particular sodium ions, can form a concentration-plus-potential gradient across cell membranes which can be used for driving the (secondary) transport of amino acids and sugars in the intestinal mucosa[3] and renal tubules[4] as well as in ascites cells.[5] The "sodium pump",[6,7] subsequently shown to be identical with the Na,K-adenosinetriphosphatase[6,7] was the magic wand of many animal physiologists for years to come, helping them to explain the conversions of energy taking place in animal cell membranes.

At the Prague Symposium on Membrane Transport and Metabolism[8] in 1960 Peter Mitchell put forth for the first time the idea that was to become the foundation for his chemiosmotic hypothesis, later theory, now a fact, that hydrogen ions can mediate energy conversions (in particular synthesis of ATP) in inner mitochondrial,[9] bacterial,[10] and thylakoid[11] membranes. Mitchell's followers and epigones extended this brilliant idea to the field of transport where it caught on and was then applied somewhat indiscriminately to every transport newly described, especially in microorganisms, even to cases where other mechanisms were involved, such as the bacterial phosphotransferase[12] or the binding-protein ATP-driven transports.[13]

To those in the field who were broad-minded enough to look back at pre-Mitchellian results it was clear that the involvement of hydrogen ions in many transports was strictly analogous to the involvement of sodium ions in almost as many transports in other cells. When this analogy was realized, suddenly the world of transport became divided into a half that used Na^+ and another half that used H^+ (ref.14). How often must we experience, both in science and in everyday life, oscillations of opinion that carry with them outbreaks of jealousy, animosity and loss of credibility!

Now, with more information available and owing to the general awareness that natural processes are beautifully diverse we accept the view that, indeed, H^+ (or H_3O^+) as well as Na^+ ions can serve as "driving" ions for uphill transports of both organic molecules and ions and that they do so quite often in one and the same cell (Table 1).

Table 1. Examples of Ion-Driven Transports
in the Biosphere

Type of cell or organelle	Substance transported

H^+-driven

Escherichia coli	D-Xylose
	L-Arabinose
	D-Galactose
	L-Fucose
	L-Rhamnose
	Lactose
	D-Glucose 6-phosphate
	Amino acids
	(by some 10 systems)
Saccharomyces cerevisiae	Trehalose
	Maltose
	Amino acids
	(by about 12 systems)
	Proline
	Sulfate
	Phosphate
	(high-affinity)
	Hypoxanthine, guanine
Chlorella vulgaris	Monosaccharides
	Nitrate
Mitochondria	Phosphate
	Glutamate
	Protein precursors
Vesicles in nerve endings	Acetylcholine
Chromaffin granules	Catecholamines

Na^+-driven

Escherichia coli	Glutamate
	Proline
	Melibiose
Halobacterium halobium	Amino acids
Klebsiella pneumoniae	Citrate
Salmonella typhimurium	Pantothenate
	Thiomethylgalactoside
Saccharomyces cerevisiae	Phosphate
	(low-affinity)
Liver cells	Thiamin
Intestinal mucosa	Ascorbate
	Monosaccharides
	Amino acids
Renal tubules	Monosaccharides
	Amino acids
Tumor cells	Amino acids
Nucleate erythrocytes	Amino acids

To complete the circle and make the scene fully symmetrical, a number of microorganisms are now known in which not only the transport can be driven by a Na^+ electrochemical potential gradient but also the synthesis of ATP[15] and flagellar motion.[16] This symmetry has a far-reaching corollary. Even if protons can be viewed as quantum particles (their tunneling through energy barriers can be envisaged even in biological systems) and even if they can move in aqueous solutions (via a charge relay) faster than predicted by their true diffusion it is obvious that in the energy-transducing systems where electrochemical potential gradients are generated, where secondary active transport is performed, where ATP is synthesized, where prokaryotic cells move their flagella, the hydrogen ion simply behaves like a cation, in a fashion comparable to that of a sodium ion. It is indeed here that arguments for the involvement of H_3O^+ rather than H^+ find fertile soil.[17]

KINETIC AND THERMODYNAMIC ASPECTS OF ION-DRIVEN TRANSPORTS

It is now clearly established that both Na^+- and H^+-driven transports are catalyzed by membrane-spanning integral proteins[18,19] and it is then obvious that the rates of transport can be described by the formalism of enzyme kinetics.[20,21] The accumulation ratios of an ion-driven solute can be derived in a straightforward way either by a kinetic approach or (for the limiting case of tightly coupled transport without any internal slips) by a thermodynamic approach.[22]

The rate expressions for solute transport are rather bulky and differ, depending on the sequence of the individual steps in the membrane cycle, thus:

(1) carrier + ion + driven solute $\xleftrightarrow{\text{translocation}}$ - driven solute - ion

(2) carrier + driven solute + ion $\xrightarrow{\text{translocation}}$ - ion - driven solute

(3) carrier + ion + driven solute $\xleftarrow{\text{translocation}}$ - ion - driven solute

(4) carrier + driven solute + ion $\xleftrightarrow{\text{translocation}}$ - driven solute - ion

Quite generally, the half-saturation constant and the maximum rate include concentrations of the driving ion at both sides of the membrane, as well as the membrane potential. To show the complexity of the expressions, that for the initial rate for system (3) defined above is presented:

$$J_{s_0} = c_{c_T} \frac{A \zeta^{1/2} c'_{ion}}{B \zeta^{1/2} + c'_{ion}(C \zeta^{-1/2} + D \zeta^{1/2} + E)} \cdot$$

$$\cdot \frac{c'_{solute}}{c'_{solute} + \dfrac{F + G\zeta^{1/2} + c'_{ion}(H + I\zeta^{-1/2} + J\zeta^{1/2}) + c''_{ion} K\zeta^{-1/2}}{B\zeta^{1/2} + c'_{ion}(C\zeta^{-1/2} + D\zeta^{1/2} + E)}}$$

where c_{CT} is the total carrier concentration, and $A, B, C, \ldots K$ are kinetic constants, composed of the individual rate constants, e.g. $C + S \rightleftarrows CS$, $CS + \text{ion} \rightleftarrows CSI$, $CSI' \rightleftarrows CSI''$, etc., and $\zeta^{1/2} = \exp(-nF\Delta\psi/RT)$. Throughout, single-primed concentrations refer to the cis, or starting, side of the membrane, the double-primed ones to the trans, or target, side of the membrane.

The diagnostic value of the various expressions is rather limited since manipulation of the various rate constants may yield qualitatively identical results with different intrinsic mechanisms[24] but, still, several "rules" can be defined.

1. Raising the external driving ion concentration will always increase J_{max} (except in the carrier + ion + solute sequence in local equilibrium) while raising the intracellular ion concentration will decrease it (except in the carrier + solute + ion sequence in local equilibrium).

2. Raising the membrane potential will always increase J_{max} and almost always decrease K_T (except in a steady-state system with the sequence carrier-solute-ion - ion - solute).

3. Effects of c'_{ion} as well as c''_{ion} on the K_T value are seen to be either positive or negative in any of the sequences, except carrier + ion + solute and carrier + solute + ion in local equilibrium where it is always negative.

The accumulation ratio of solute S has a more straightforward significance. It is in fact identical with the equilibrium constant of the reaction and, by Haldane's reasoning[22],[25], with a combination of the half-saturation constants K_T and of the maximum rates of transport J_{max}, such that

$$K_{eq} = c''_s/c'_s = \overset{\leftarrow}{K_T}\overset{\rightarrow}{J_{max}} / \overset{\rightarrow}{K_T}\overset{\leftarrow}{J_{max}}$$

At the same time, if the determination of ion concentrations and of the membrane potential are reliable, one can easily predict the maximum accumulation ratio of an ion-driven solute, thus

$$c''_s/c'_s = (c'_{ion}/c''_{ion})^n \exp(-nF\Delta\psi/RT)$$

where n is the number of cations taking part in each transport cycle (it is probably unity for most cases), F is the Faraday constant (96.5 kC mol^{-1}), $\Delta\psi$ is the membrane potential in V, R is the gas constant (8.314 J $mol^{-1}K^{-1}$) and T is the temperature in K. Here we arrive at several disconcerting differences between Na^+ and H^+ ions.

1. While a Na^+-free medium can be easily prepared a solution with less than 10^{-7} M H^+ ions is already detrimental to most cells.

2. While Na^+ ions play no role in enzyme (and transport) catalysis, it is a general property of all enzyme systems that they show an optimum pH for their function - this is a purely catalytic phenomenon but it can easily imitate effects that are due to _differences_ in pH and hence are of thermodynamic nature.

3. In all cell suspensions or tissue preparations, there is a tendency for intra-, _as well as_ extracellular buffering (by extrusion of CO_2, organic acids and direct H^+ shifts) which may obscure some important hydrogen ion movements. No such extracellular buffering exists with respect to Na^+.

SECONDARY (H^+-DRIVEN) TRANSPORTS IN YEAST

The above peculiarities, compounded by the high buffering capacity generated by yeast cells in the external medium and by the laboriousness and slowness of measuring the membrane potential,[25] have contributed to the hesitation in accepting the role of H^+ ions in yeast systems. Moreover, even the thermodynamical predictions of the maximum accumulation ratios were not fulfilled in experiments (Fig. 1).[26] Likewise, at very low suspension densities, accumulation ratios in the vicinity of 500:1 were observed, far in excess of the values based on measured Δ pH and membrane potential (Fig. 2).[27]

In addition to this discrepancy it was shown by a direct pH mapping of yeast cells[28],[29] that the pH profile across the yeast plasma membrane is gra-

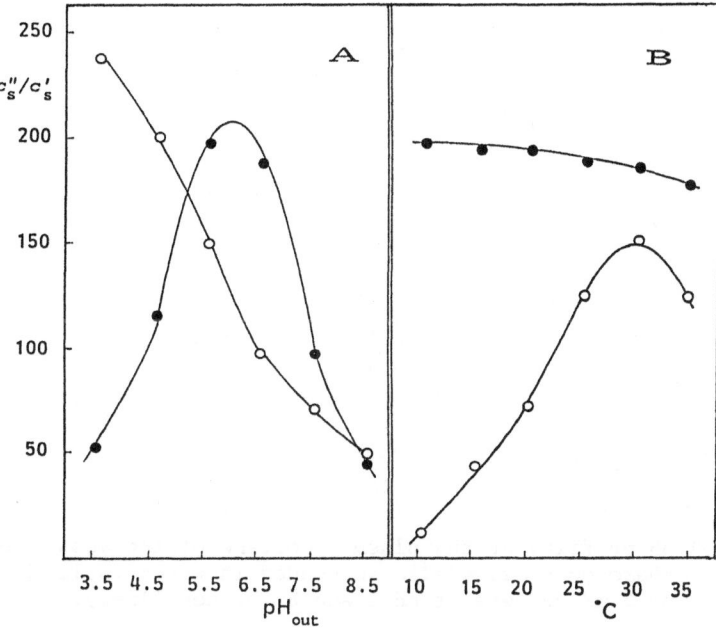

Fig. 1. pH dependence (A) and temperature dependence (B) of the accumulation ratio of 6-deoxy-D-glucose (initial $c_s' = 10\ \mu M$) in *Rhodotorula gracilis*. A at 30°C, B at pH 5.5. o Theoretical accumulation ratio based on Δψ and ΔpH estimated in parallel; ● observed values.

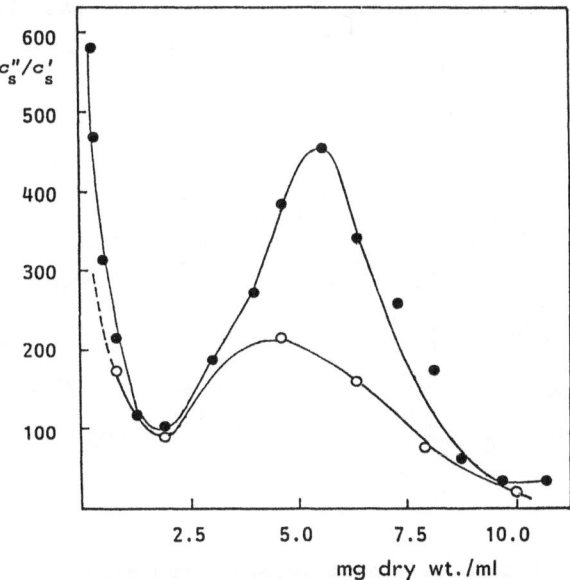

Fig. 2. Theoretical (o) and observed (●) accumulation ratios of 6-deoxy-D-glucose in the yeast *Lodderomyces elongisporus* (initial $c_s' = 10\ \mu M$), in dependence on the suspension density.

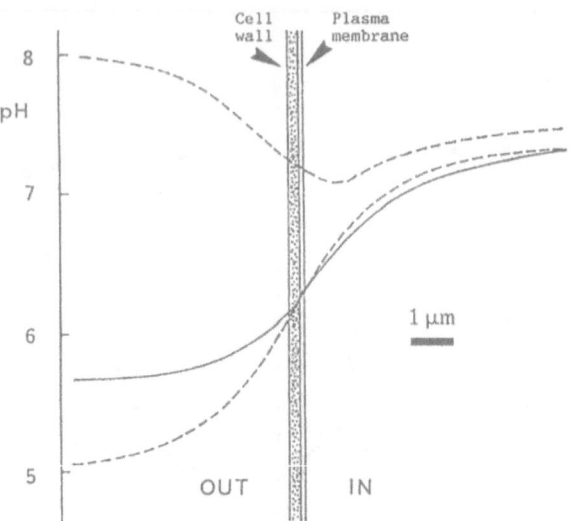

Fig. 3. pH profiles at the plasma membrane of the yeast *Sac-
        charomyces cerevisiae* suspended in water (solid line)
        of 5 mM buffers of pH 5 and 8 (broken lines).

dual and that the pH corresponding to the membrane thickness is negligible
(perhaps 0.01 pH unit) and thus does not contribute to the actual driving energy of the proton gradient (Fig. 3).

To save the H^+-symport theory it was necessary to invoke the view
that primary pH generators (the plasma membrane ATPase and the CO_2-gen-
erating and -hydrating system)[30,31] are in close vicinity of the secondary proton gradient dissipators,[32] a view that was later advanced in

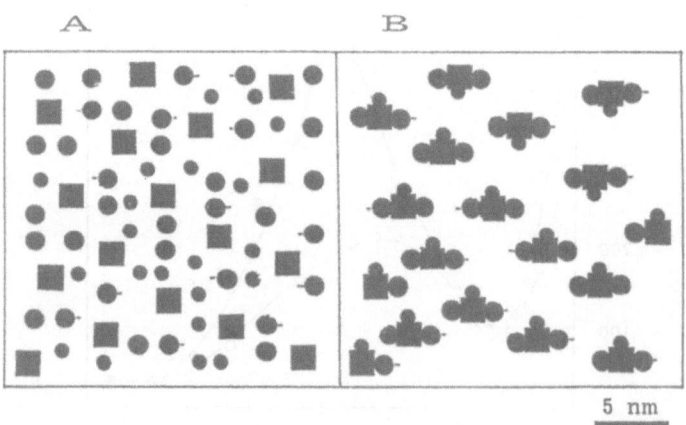

Fig. 4. Distribution of plasma membrane proteins of *Saccha-
        romyces cerevisiae* in a random way (A) and in clus-
        ters (B). Squares depict the primary ΔpH generator,
        the various circles the different transport-associ-
        ated proteins presumed to sense the local pH values
        in the membrane.

the mosaic-membrane hypothesis of Westerhoff et al.[33] This clustering (Fig. 4) received support not only from kinetic and recently fluorescence[34] data but it is in line even with freeze-etching studies of the yeast plasma membrane.[35]

If this is the way the primary and secondary proton movers are situated it follows that the membrane as a whole is not in equilibrium with respect to H+ ions and only in a pseudo-equilibrium with respect to the driven solute. After all, only in a dead cell would one expect a full equilibrium across the plasma membrane to prevail (and even that only in the absence of major degradative processes).

In spite of the overwhelming evidence (particularly by analogy with bacteria) in favor of proton symports in yeast the arguments adduced in their support (Fig. 5) have been based on indirect evidence. All the transient alkalifications and membrane potential changes could be qualitatively explained by assuming that the principal generator of ΔpH, the plasma membrane H+-ATPase, competes for ATP with transport systems which, instead of by the electrochemical potential gradient of H+, are driven by ATP hydrolysis. In such a case a partially competitive inhibition would

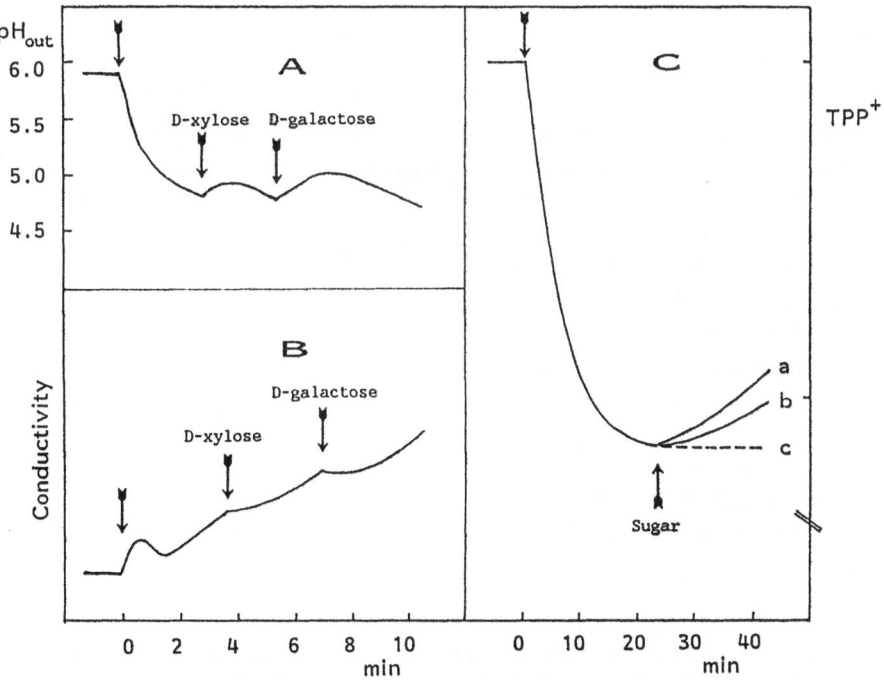

Fig. 5. Evidence for involvement of H+ ions in sugar transport in yeast. A - Extracellular pH trace in *Rhodotorula gracilis*; B - conductivity trace in *R. gracilis*; C - extracellular concentration of the tetraphenylphosphonium cation (TPP+), indicator of membrane electrical potential, in a suspension of the yeast *Metschnikowia reukaufii*; a - addition of 20 mM D-xylose, b - addition of 20 mM 3-O-methyl-D-glucose, c - no addition. Cells were always added at time zero. A based on ref.[36], B based on ref.[37].

ensue, just as is observed when two different transport systems utilize the same source of energy.[32]

USE OF HEAVY WATER IN YEAST TRANSPORT STUDIES

It was only quite recently that experiments, in which heavy water replaced ordinary water, addressed the involvement of protons in transport directly.[31] There it became clear that both the protonmotive activity of the plasma membrane ATPase and the proton-driven transports of solutes are affected by heavy water (Table 2) and hence that protons (or hydronium ions) are under physiological conditions directly involved in the transmembrane movements.

Table 2. Effect of D_2O on membrane-associated processes in yeast
R = ratio of rate in D_2O to rate in H_2O

Type of process	R
H^+ extrusion by plasma membrane ATPase	0.11 ± 0.04
ATP hydrolysis by plasma membrane ATPase	0.90 ± 0.10
Transport of 0.1 mM 6-deoxy-D-glucose in *Saccharomyces cerevisiae*[a]	1.00 ± 0.05
Transport of 0.1 mM 6-deoxy-D-glucose in *Lodderomyces elongisporus*	0.41 ± 0.15
Transport of 10 μM L-tryptophan in *Saccharomyces cerevisiae*	0.29 ± 0.13
"Spontaneous" acidification due to CO_2 generation and hydration[b]	0.08 ± 0.03

[a] Facilitated diffusion.
[b] Apparently an effect on carbonate dehydratase.

The lack of effect of D_2O on the facilitated diffusion of monosaccharides in baker's yeast and on the ATP-hydrolyzing activity demonstrated, that in the short-term experiments carried out here D_2O did not affect the catalytic protein conformation.

In conclusion, it appears that, after the years of hesitation concerning the true source of energy for nutrient transports in yeast, it is really the protons (or hydronium ions) which are directly associated with not only the primary transporter (the plasma membrane H^+-ATPase) but also with the secondary transports of solutes operating in the plasma membrane.

REFERENCES

1. E. Riklis and J. H. Quastel, Effects of cations on sugar absorption by
    isolated surviving guinea pig intestine, *Can. J. Biochem. Physiol.*
    36:347 (1958).
2. R. K. Crane, Hypothesis for the mechanism of intestinal active trans-
    port of sugars, *Fed. Proc.* **21**:892 (1962).
3. R. K. Crane, Na⁺-dependent transport in the intestine and other animal
    tissues, *Fed. Proc.* **24**:1000 (1965).
4. A. Kleinzeller and A. Kotyk, Cations and transport of galactose in kid-
    ney-cortex slices, *Biochim. Biophys. Acta* **54**:367 (1961).
5. J. A. Schafer and E. Heinz, The effect of reversal of Na⁺ and K⁺ elec-
    trochemical potential gradients on the active transport of amino a-
    cids in Ehrlich ascites tumor cells, *Biochim. Biophys. Acta* **249**:15
    (1971).
6. J. C. Skou, The influence of some cations on an adenosine triphosphat-
    ase from peripheral nerves, *Biochim. Biophys. Acta* **23**:394 (1957).
7. R. L. Post, C. R. Merritt, C. R. Kinsolving, and C. D. Albright, Mem-
    brane adenosine triphosphatase as a participant in the active trans-
    port of sodium and potassium in the human erythrocyte, *J. Biol. Chem.*
    235:1796 (1960).
8. A. Kleinzeller and A. Kotyk (eds), "Membrane Transport and Metabolism",
    Publ. House Czechosl. Acad. Sci., Prague (1961).
9. P. Mitchell, Coupling of phosphorylation to electron and hydrogen trans-
    fer by a chemiosmotic type of mechanism, *Nature (London)* **191**:144
    (1961).
10. F. M. Harold, Conservation and transformation of energy by bacterial
    membranes, *Bacter. Rev.* **36**:172 (1972).
11. R. A. Dilley, Coupling of ion and electron transport in chloroplasts,
    Curr. Topics Bioenerget. **4**:237 (1971).
12. W. Kundig and S. Roseman, Sugar transport. Isolation of a phosphotrans-
    ferase system from *E. coli*, *J. Biol. Chem.* **246**:1393 (1971).
13. E. Prossnitz, A. Gee, and G. Ferro-Luzzi Ames, Reconstitution of the
    histidine periplasmic transport system in membrane vesicles. *J. Biol.
    Chem.* **264**:5006 (1989).
14. V. P. Skulachev, "Membrane Bioenergetics", Springer, Berlin (1988).
15. P. A. Dibrov, R. L. Lazarova, V. P. Skulachev, and M. L. Verkhovskaya,
    The sodium cycle. Na⁺-dependent oxidative phosphorylation in *Vibrio
    alginolyticus*, *Biochim. Biophys. Acta* **850**:458 (1986).
16. N. Hirota, Y. Imae, Na⁺-driven flagellar motors of an alkaliphilic Ba-
    cillus strain YN-1, *J. Biol. Chem.* **258**:10577 (1983).
17. P. D. Boyer, Bioenergetic coupling to protonmotive force: Should we be
    considering hydronium ion coordination and not group protonation?
    Trends Biochem. Sci. **13**:5 (1989).
18. G. Semenza, M. Kessler, M. Hosang, J. Weber, and U. Schmidt, Biochem-
    istry of the Na⁺,D-glucose cotransporter of the small-intestinal
    brush-border membrane, *Biochim. Biophys. Acta* **779**:343 (1984).
19. H. R. Kaback, Proton electrochemical gradients and active transport :
    The saga of *lac* permease, *Ann. N. Y. Acad. Sci.* **456**:291 (1986).
20. W. Wilbrandt, Secretion and transport of non-electrolytes, *Symp. Soc.
    Exptl. Biol.* **8**:136 (1954).
21. W. D. Stein, An algorithm for writing down flux equations for carrier
    kinetics, and its application to co-transport, *J. Theor. Biol.* **62** :
    467 (1976).
22. A. Kotyk, Coupling of secondary active transport with $\Delta\tilde{\mu}_{H^+}$, *J. Bioener-
    get. Biomembr.* **15**:307 (1983).

23. A. Kotyk, Basic kinetics of membrane transport, *in* "Structure and Pro-
     perties of Cell Membranes", Gh. Benga, ed., CRC Press, Boca Raton
     (1985).

24. D. Sanders, U.-P. Hansen, D. Gradmann, and C. L. Slayman, Generalized
     kinetic analysis of ion-driven cotransport systems: A unified inter-
     pretation of selective ionic effects on Michaelis parameters, *J. Mem-
     brane Biol.* **77**:123 (1984).

25. M. Höfer, H. Huh, and A. Künemund, Membrane potential and cation per-
     meability. A study with a nystatin-resistant mutant of *Rhodotorula
     gracilis* (*Rhodosporidium toruloides*), *Biochem. Biophys. Acta* **735**:
     211 (1983).

26. A. Kotyk and J. Horák, Effects of pH and of temperature on saturable
     transport processes, *in* "Water and Ions in Biological Systems", A.
     Pullman, V. Vasilescu, and L. Packer, eds, Plenum Press, New York
     (1985).

27. A. Kotyk and D. Michaljaničová, Suspension density and accumulation
     ratio of sugars and amino acids in yeasts, *Folia Microbiol.* **32**:459
     (1987).

28. J. Slavík, Intracellular pH topography: Determination by a fluorescent
     probe, *FEBS Lett.* **156**:227 (1983).

29. J. Slavík and A. Kotyk, Intracellular pH distribution and transmemb-
     rane pH profile of yeast cells, *Biochim. Biophys. Acta* **766**:679
     (1984).

30. K. Sigler and M. Opekarová, CO_2-dependent K^+ efflux in yeast utiliz-
     ing endogenous substrates, *Cell. Mol. Biol.* **31**:195 (1985).

31. A. Kotyk, Heavy water and membrane transport in yeast, *in* "Structure,
     Function and Biogenesis of Energy Transfer Systems", Elsevier, Am-
     sterdam (1989).

32. A. Kotyk, Critique of coupled vs. noncoupled transport of nonelectro-
     lytes, *in* "Fifth Winter School on Biophysics of Membrane Trans-
     port"(J. Kuczera, J. Gabrielska, and S. Przestalski, eds), Agri-
     cultural Academy, Wrocław (1979).

33. H. V. Westerhoff, B. Andrea-Melandri, G. Venturoli, G. F. Azzone, and
     D. B. Kell, A minimal hypothesis for membrane-linked free-energy
     transduction. The role of independent, small coupling units, *Bio-
     chem. Biophys. Acta* **768**:257 (1984).

34. J. Slavík, P. Pauček, M. Souček, and A. Kotyk, Application of fluor-
     escent sugar analogues in the study of monosaccharide transport in
     yeast, *FEBS Lett.* in press (1989).

35. H. Moor and K. Mühlethaler, Fine structure in frozen-etched yeast
     cells, *J. Cell Biol.* **17**:609 (1963).

36. A. Kotyk and R. Metlička, Conductometry as a tool for studying ion
     transport in suspensions, *Studia Biophys.* **110**:205 (1986).

37. B. Aldermann and M. Höfer, The active transport of monosaccharides by
     the yeast *Metschnikowia reukaufii*: Evidence for an electrochemical
     gradient of H^+ across the cell membrane, *Exptl. Mycol.* **5**:121 (1981)

MOLECULAR ORGANIZATION AND REGULATION OF THE PROTONMOTIVE SYSTEM OF

MAMMALIAN ATP SYNTHASE

S. Papa, F. Guerrieri and F. Zanotti

Institute of Medical Biochemistry and Chemistry and Centre
for the Study of Mitochondria and Energy Metabolism, C.N.R.
University of Bari, Bari, Italy

INTRODUCTION

In bacteria the H^+ translocating membrane sector (F_o) of the F_o, F_1 H^+-ATP synthase is composed of three subunits (a, b, and c of the E.Coli enzyme (1). In mitochondria more than three proteins appear to constitute the F_o and stalk sectors (1), which are involved in H^+ conduction and/or coupling of H^+ translocation to the catalytic process in the F_1 moiety (Table 1) (2,3). Two of these are the products of the mitochondrial genoma:the ATPase 6, homologous to subunit a of E.Coli and likely to be involved as this (4-7) in H^+ conduction, and A6L in mammals (aapl in yeast), whose function is as yet unknown (1). Other 5 proteins have been identified in the mammalian enzyme which are encoded by nuclear genes (8,9). These include subunit c which is directly involved in H^+ conduction (10,11), OSCP and F_6 which contribute to connection of F_1 with F_o (1).

This paper is intended to summarize characteristics of proteins involved in H^+ translocation in bovine ATP synthase as deduced from our observations on the effect of selective removal and reconstitution of subunits in the complex, chemical modification of specific aminoacid residues and proteolytic cleavage of polypeptide segments.

RESULTS AND DISCUSSION

Fig.1 shows the SDS-PAGE of F_o purified by extraction with a zwitterionic detergent (CHAPS) from F_1-depleted bovine-heart submitochondrial particles (12). The preparation exhibits the characteristic components of F_o and the stalk listed in Table 1. It contained also two proteins of apparent Mr 46000 and 31000 which do not belong to F_o (see 12,13). The polypeptide bands of interest were electroeluted in glycerol and sequenced or used for raising polyclonal antibodies in rabbits (12,14). The band of apparent Mr 27000 was found from the aminoacid sequence (15) to correspond to the nuclear encoded protein characterized by Walker et al. (8) and considered by them to be analogous to the b subunit of E.Coli. We denominate this protein PVP from the first three N-terminal residues. The band of apparent Mr 25000 consisted of a closely spaced doublet, both components reacted, like

Papa et al.

Table 1. Subunits of the F_O and stalk moieties of bovine-heart ATP synthase

Genes	Calculated Mr (Kdaltons)	N-terminus	Denomination	Function
Nuclear gene	24.67	PVP (1 Cys)	PVP subunit	Binding of F_1 to F_O; H^+ conduction in F_O; involved in oligomycin and DCCD sensitivity.
More nuclear pseudogenes	20.97	FAK (1 Cys)	OSCP	Functional connection of F_1 to F_O
Mitochondrial gene	24.82	f-SFI	ATPase 6	Proton conduction ? Oligomycin sensitivity.
Nuclear gene	18.60	Ac-AGR (1 Cys)	d	Unknown
Nuclear gene	7.96	NKE	F_6	Binding of F_1 to F_O
Mitochondrial gene	7.96	f-MPQ	A6L	Unknown
Two Nuclear genes encoding different pre-sequences	8.00	DID (1 Cys)	c	Proton conduction
Nuclear gene	9.58	GSE	IF_1	Inhibits catalysis and H^+ translocation

the PVP protein and subunit c with the fluorescent thiol reagent N-(7 dimethylamino-4-methyl-coumarinyl)-maleimide (DACM). The lower band was identified by immunoblot with specific antisera as OSCP (14), the upper may correspond to the nuclear-encoded subunit d of ref.(8). The band of Mr 23000 was not labelled by thiol reagents and may represent the product of the ATPase 6 gene which has no codon for cysteine (16).

Removal of the ATPase inhibitor protein (17) from the F_1F_O complex in submitochondrial particles by chromatography on Sephadex results in twofold stimulation of the hydrolytic activity as well as of oligomycin-sensitive passive H^+ conduction (Table 2) (18,19). Both activities were brought back to control values by adding the purified inhibitor protein to the depleted particles. The inhibitor protein exerted the same effect when added to F_1 depleted particles supplemented with purified F_1.

Treatment of submitochondrial particles with silicotungstate, which results in the removal of F_6(20), caused 2.5 fold enhancement of the rate of oligomycin-sensitive proton conduction which was suppressed by the addition of purified F_6 (21) (see Table 2).

Fig.2 shows a titration of the inhibitory effect exerted by addition of the purified inhibitor protein on the ATPase activity and proton conductivity of depleted particles. Treatment of the isolated inhibitor protein with ethoxyformic anhydride (EFA) under conditions resulting in modification of one out of the five histidine residues of this protein (19) suppressed

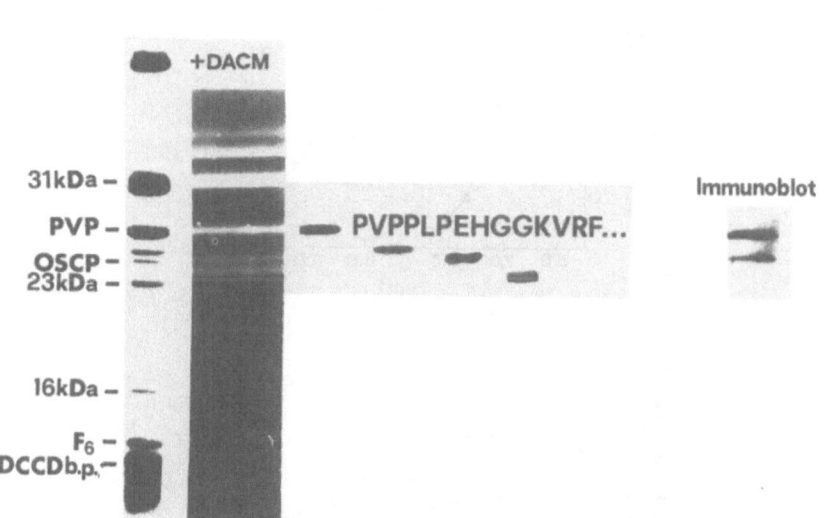

Fig.1. SDS-PAGE of F_o and isolated F_o polypeptides. Immunoblot with antisera against PVP protein and OSCP.
F_o was purified by extraction with CHAPS from F_1-depleted bovine-
heart submitochondrial particles (USMP) or from USMP treated with
the fluorescent thiol reagent DACM (22) as reported in (12). SDS-
PAGE, immunoblot analysis and isolation of individual F_o proteins
were carried out as reported in (14). Aminoacid sequence analysis
was carried out as reported in (15 and 22).

Table 2. Effect of F_1 inhibitor protein (IF_1) and of F_6 on ATPase activity
and oligomycin sensitive H^+ conduction in submitochondrial particles with various degrees of resolution of H^+-ATPase complex.

	ATPase activity(μmol ATP hydrolyzed$\cdot$min$^{-1}\cdot$ mg protein^{-1})		Anaerobic H^+release $1/t_{\frac{1}{2}}$ (s^{-1})		
	control	$+IF_1$	control	$+IF_1$	$+F_6$
ESMP	1.25	0.92	1.00	0.75	-
Sephadex-ESMP	2.88	0.90	2.00	0.91	-
Urea-ESMP	0.08	0.08	2.74	2.80	-
Urea-ESMP + F_1	1.00	0.50	1.67	1.11	-
Silicotungstate-ESMP	-	-	2.50	-	1.5

Preparations of ESMP, Sephadex treated ESMP, Urea ESMP, F_1 and IF_1 were carried out as described in (19). F_1 was prepared as described in (12), F_6 was
prepared as described in (21). Silicotungstate treated ESMP were prepared
by incubating ESMP (10 mg/ml) 10 min at 4°C in Sacc. 0.15 M, Tris sulphate
0.02 M (pH 8.0), silicotungstate 1.5% (pH 5.5). Then, after dilution with
10 volumes of Sucrose 0.25 M, the particles suspension was centrifuged 30
min at 105.000xg and washed with 0.25 M Sucrose + 5 mM DTT. ATPase activity
and oligomycin sensitive H^+ conductivity were determined as described in
ref.s 18, 19. Incubation of particles with IF_1 (4 μg/mg prot.) and F_6 (4
μg/mg prot.) were carried out at 25°C as described in ref.s 18, 19.

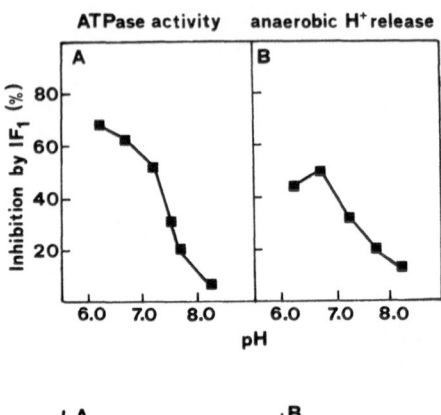

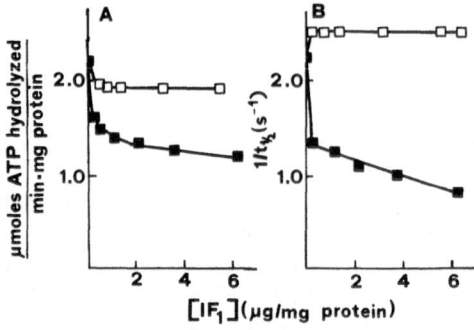

[IF$_1$] (µg/mg protein)

Fig.2. Titration of inhibition of purified inhibitor protein on ATPase
activity (Panels A) and on passive proton conductivity (Panels B)
of IF$_1$-depleted particles. Effects of diethylpyrocarbonate (EFA)
treatment of IF$_1$ and of pH. For preparations of Sephadex-EDTA
particles, purification of IF$_1$, its treatment with EFA, determination of ATPase activity, measurement of oligomycin sensitive proton conduction and pH dependence see ref.19. Symbols: ■—■ IF$_1$;
□—□ IF$_1$ modified by treatment with 0.5 mM EFA.

its inhibitory activity. Critical importance of a histidine residue is also
indicated by the pH dependence of the inhibitory activity which increased
with acidification of the reaction mixture with a pK$_a$ around neutrality typical of histidine-imidazol.

The above observations show that there are subunits, like the inhibitor protein and F$_6$, at the junction between F$_1$ and F$_0$ which prevent free
escape of H$^+$ from the M mouth of the transmembrane H$^+$channel and possibly
control their transfer to catalytic and/or allosteric sites in F$_1$.

Further evidence for such a role of junction proteins (components of
the stalk or coupling sector of the ATP synthase) is provided by the effects
exerted by certain aminoacid modifiers on proton conduction in submitochondrial particles. These are EFA, specific for histidine residues, and diamide
and Cd^{++}, which oxidize vicinal dithiols inducing formation of disulphide
bridges. Treatment of ESMP with each one these reagents (Fig.3) induces a
dramatic acceleration of passive H$^+$ conduction in the particles. Differently
from the promotion of H$^+$ conduction induced by FCCP, whose effect is oligomycin insensitive (the uncoupler promotes H$^+$ conduction in the phospholipid
bylayer) the enhanced H$^+$conductivity induced by the aminoacid reagents is

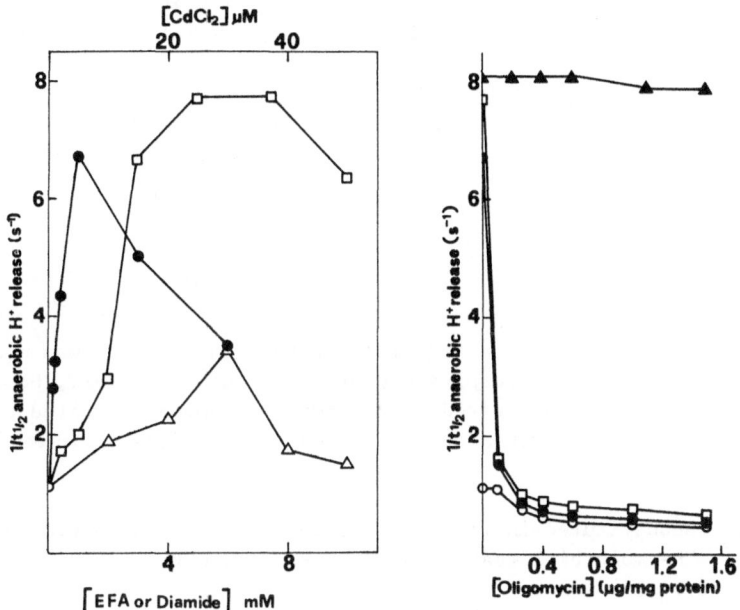

Fig.3. Titration of the effect of chemical modification of ESMP by aminoacid reagents on oligomycin sensitive H$^+$. ESMP were preincubated
with EFA (●—●) at the concentrations reported in the figure (left
panel) as described in ref.19 before oxygen induced proton translo-
cations. ESMP were preincubated 2 min with diamide (□—□) or Cd^{++}
(△—△) at the concentrations reported in the figure. The oligomycin
titration (right panel) was carried out on control (o—o) or ESMP
treated with 1 mM EFA (●—●); 5 mM diamide (□—□) or in presence
of 5 μM FCCP (▲—▲).

fully suppressed by oligomycin, this proving that it is, in fact, H$^+$ escape
from the F$_o$ channel to be promoted by the reagents. It can be noted that
the stimulatory effect of EFA, diamide and Cd^{2+} is followed, after maximal
stimulation at a critical concentration, by inhibition at higher concentrations. Evidently the thiol reagents modify two classes of residues, differently exposed to the reaction medium. Modification of superficial thiols
results in promotion of H$^+$ conduction. Modification, at higher concentrations, of thiols located more deeply in the membrane results in inhibition
of a step in H$^+$ conduction which obscures the stimulatory effect exerted on
more superficial steps of the process.

Evidence for these differential effects of the aminoacid modifiers is
further documented by comparing their effect on different types of submitochondrial particles from which various components of the ATP synthase complex had been selectively removed (Table 3). The stimulatory effect of EFA
was lost in particles deprived of the inhibitor protein, this confirming
that the stimulatory effect exerted on H$^+$ conduction was specifically due
to modification of this protein. Interestingly enough the stimulatory effect exerted by diamide, which, differently from that exerted by EFA, was
retained after removal from the particles of F$_1$, was lost after further
treatment of these particles with ammonia which reportedly removes OSCP (24).

Table 3. Effect of various aminoacid reagents on oligomycin sensitive H^+
conduction in submitochondrial particles and F_O liposomes

Reagents	H^+ release $1/t_{\frac{1}{2}}$ (sec^{-1})				
	ESMP sec^{-1}	Sephadex sec^{-1}	USMP sec^{-1}	ASMP sec^{-1}	F_O Liposomes sec^{-1}
None	1.0	2.0	2.0	0.62	2.50
Diamide,5 mM	7.7	–	6.6	0.66	1.88
EFA, 1 mM	6.7	2.0			
DACM, 200 μM	0.48				0.70

For F_O liposomes preparations see ref.(12). Proton conductivity was measured as anaerobic release of respiratory proton gradient in ESMP, Sepha-
dex-ESMP (Sephadex) and urea-treated ESMP (USMP). In ammonia treated USMP
(ASMP) and in F_O-liposomes proton conduction was measured as valinomycin
+ K^+ induced H^+ release.

In isolated F_O reconstituted in liposomes, diamide caused inhibition of proton conduction. The monothiol reagent DACM caused inhibition of H^+ conduction
both in ESMP and F_O-liposomes.

It is conceivable that the stimulatory effect caused by diamide possibly
results from oxidation and disulphide-bridging of thiol groups in neighbor
superficial subunits. The inhibitory effect exerted by high concentrations
of diamide in ESMP, as well as in F_O-liposomes may, be due to modification of
those thiols in membrane integral proteins, whose modification by monofunc-
tional thiol reagents (see Table 3., ref.25, 26) results in inhibition of H^+
conduction.

Thiol groups in F_O components are very limited in number, there is only
1 Cys residue each in the PVP protein, protein d of ref.(8) and in c subunit
(see Fig.1). According to Sanadi (27), critical dithiol is present in Factor B, an additional putative component of F_O.

Information on the constitutens of mitochondrial F_O and their function
have been obtained by selective enzymatic digestion of membrane proteins
and reconstitution with the native isolated components (12,15,22). Treatment
of submitochondrial particles with trypsin results, only after removal of F_1,
in digestion of the PVP protein and OSCP (Fig.4). Immunoblots with anti-PVP
serum of F_O extracted from trypsin digested USMP, show that the PVP protein
can be digested to an immunoreactive fragment some kDa smaller. The fragment retains the N-terminal region as well as the only Cys residue in this
protein (reactive with ^{14}C-NEM) at position 197 (16) 17 residues away from
the carboxy-terminal Met (Fig.5).Thus trypsin cleaves off selectively the
carboxy-terminal region of PVP down to Lys 202 or Lys 206. It can be noted
from the loss of ^{14}C-NEM binding that also the 31 kDa protein is digested
by trypsin, subunit c, on the other hand, retains ^{14}C-NEM binding.

The experiment of Fig.6 shows that progressive digestion of PVP by
trypsin is linearly associated to depression of proton conduction (15). It
should be noted that, for complete digestion of PVP, there still remains a
substantial proton conductivity, which was, however, insensitive to oligomycin and DCCD (see below and Fig.s 7 and 8).

When F_O extracted from trypsin digested particles was incorporated in
liposomes it exhibited a markedly depressed, oligomycin-insensitive, H^+ con-

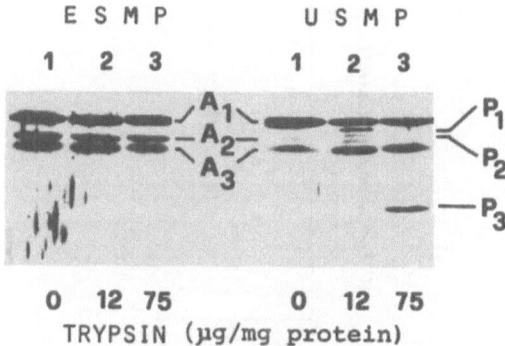

Fig.4. Immunoblot analysis of trypsin digestion of PVP protein and OSCP in
ESMP and USMP. Immunodecoration with antiserum against 25-27 kDa
protein fractions of F_O (see ref.14) was carried out as described
in ref.14. ESMP or USMP were prepared and treated with trypsin (at
the concentrations reported in figure) as reported in ref.14. A_1
identifies the PVP protein; A_2 OSCP and A_3 does not correspond to
any F_O protein (see ref.14).P_1,P_2 and P_3 are the products of tryptic digestion of PVP protein.

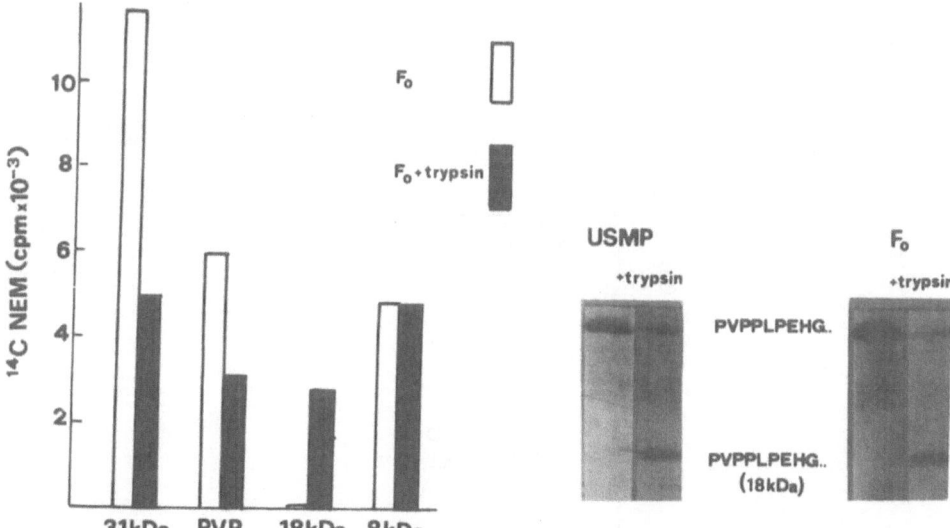

Fig.5. Immunoblot analysis of trypsin digestion of PVP protein in USMP and
F_O,isolated from USMP,before and after trypsin digestion.Analysis of
binding of ^{14}C-NEM to F_O proteins.USMP and F_O were prepared as reported in (12).Trypsin(50 µg/mg protein)digestion was carried out for 20
min as described in ref.14 and immunoblot analysis of USMP(20 µg) or
F_O(20 µg)was carried out with specific anti-PVP protein serum as described in (12). 18 kDa protein indicates the specific product of
trypsin digestion. Left panel reports the binding of ^{14}C-NEM to F_O
proteins. F_O was isolated from ^{14}C-NEM treated control and trypsin
treated USMP. Then 50 µg protein of F_O were subjected to SDS-PAGE and
the polypeptide isolated from the gel as reported in ref.(26) and
used for determination of radioactivity. On the abscissa the apparent
molecular masses of isolated proteins are reported.

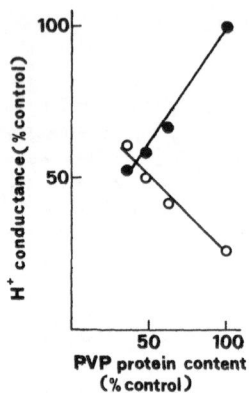

Fig.6. Relationship between H^+ conduction in USMP and the content of PVP
protein. Effect of oligomycin. For USMP preparations, immunoblot
analysis, trypsin treatment, measurement of oligomycin sensitive
H^+ conduction and evaluation of PVP protein content see ref.14,15.
Symbols:(●—●) control; (o—o) + 1 μg/mg particles protein oligomycin.

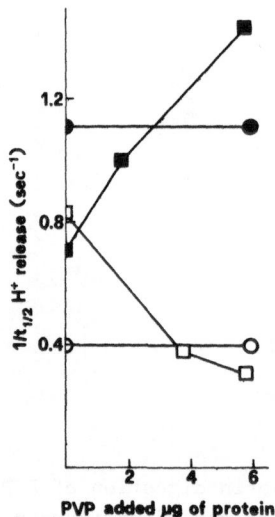

Fig.7, Reconstitution of H^+ conduction and oligomycin sensitivity of F_O
extracted from trypsin USMP by addition of isolated PVP protein.
For reconstitution conditions see ref.s (12,15). Symbols: (●—●)
F_O prepared from control USMP; (o—o)F_O prepared from control USMP
+ oligomycin (2 μg/mg F_O).(■—■)F_O isolated from trypsin treated
USMP (50 μg/mg particle protein);(□—□)F_O isolated from trypsin treated USMP + oligomycin (2 μg/mg F_O).

Table 4. Restoration of H^+ conduction and oligomycin sensitivity by
addition of purified F_O proteins to USMP treated with trypsin.

	$1/t\frac{1}{2}$ H^+ release(s^{-1})	
	Control	+Oligomycin (2 µg/mg protein)
USMP	1.82	0.25
Tryp.-USMP	0.91	0.80
Tryp.-USMP + PVP	2.00	0.57
Tryp.-USMP + 18 kDa protein	1.00	0.90
Tryp.USMP + Mr 31 kDa protein	1.00	0.91
Tryp.-USMP + PVP DACM	2.00	0.59

Tryp.-USMP were preincubated for 15 min at 25°C with purified PVP, 31 or
18 kDa proteins (6 µg/mg particle protein) before measurement of passive
proton conduction. PVP-DACM protein purified from DACM-treated USMP (for
details see ref.22).

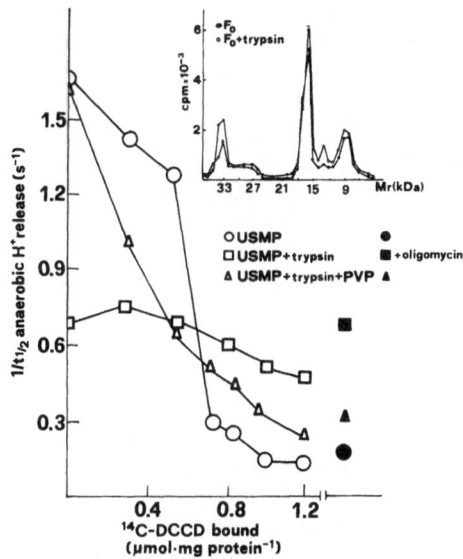

Fig.8. Effect of trypsin digestion of USMP on sensitivity of H^+ conduction
to DCCD (open symbols) and oligomycin (closed symbols). Effect of
reconstitution with isolated PVP protein (triangles). Insert refers
to ^{14}C-DCCD binding to F_O polypeptides extracted from control (•)
and trypsin treated USMP (o). For USMP preparation, isolation of F_O,
trypsin treatment of USMP, isolation of PVP protein and analysis
of ^{14}C-DCCD binding see ref.(12). Symbols: o—o control (• + 2 µg
oligomycin/mg particles protein); □—□ trypsin (50 µg/mg protein)
treated USMP (■ + 2 µg oligomycin/mg particle protein); △—△ trypsin
treated USMP + 0.2 µg PVP protein/mg particle protein (▲ + oligomycin
2 µg/mg particle protein).

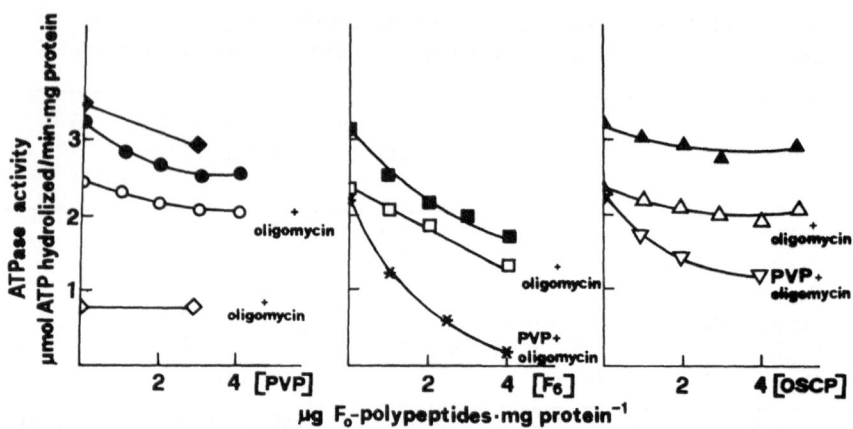

Fig.9. ATPase activity in trypsinized USMP supplemented with purified F_1.
For trypsin (50 µg/mg USMP protein) digestion and other conditions
see ref.(12,14,21). After trypsin digestion isolated PVP protein(●)
or F_6 (■), or OSCP (▲) were added before F_1 (12,14). ◆—◆,
control USMP supplemented with PVP protein. Where indicated oligomycin (2 µg/mg particle protein) was added (symbols: ◇—◇ control
USMP reconstituted with F_1 in the presence of PVP; o—o trypsinized
USMP reconstituted with F_1 in the presence of PVP; □—□ trypsinized
USMP reconstituted with F_1 in the presence of F_6; x—x trypsinized
USMP reconstituted with F_1 in the presence of PVP (3 µg/mg particle
protein) and F_6 at the concentrations reported in the figure; △—△
trypsinized USMP reconstituted with F_1 in the presence of OSCP; ▽—▽
trypsinized USMP reconstituted with F_1 in the presence of PVP (3 µg/
mg particle protein) and OSCP at the concentrations reported in the
figure).

ductivity as compared to that of F_o extracted from untreated particles(Fig.7).
Proton conduction in liposomes reconstituted with digested F_o was progressively restored to the control values of undigested F_o by the addition of increasing amounts of isolated PVP protein, which also restored oligomycin
sensitivity (15) (Fig.7). These effects of the intact PVP protein were
highly specific in that the truncated protein of apparent Mr 18000 or the
protein of Mr 31000 were totally ineffective (Table 4). Furthermore it can
be noted that modification of Cys-197 by DACM in the isolated PVP did not
affect its capacity to restore oligomycin sensitive H^+ conduction (Table 4).
This is consistent with previous observations showing that the inhibition
by NEM of proton conduction in F_o-liposome was directly correlated with the
binding of ^{14}C-NEM to subunit c (and/or subunit d) but not to the binding
to the PVP protein (band of apparent Mr 27000 in ref.26).
     Trypsin digestion of F_1 depleted particles did not affect the total
binding of ^{14}C-DCCD to F_o (12) or its specific binding to the Mr 8000 protein (DCCD-binding protein (10,11) or to a band of Mr 16000 which may be a
dimer of the Mr 8000 protein (see ref.11,12) (Fig.8). Trypsin digestion re-

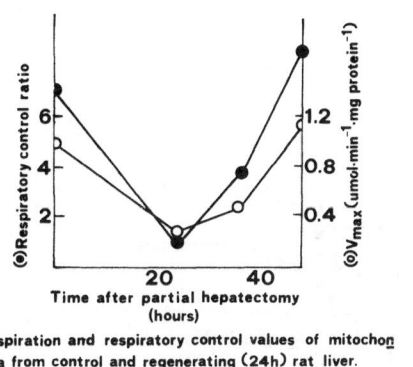

Respiration and respiratory control values of mitochon
dria from control and regenerating (24h) rat liver.

Mitochondria	State IV	State III (+ADP)
	ng O·min^{-1}·mg protein^{-1}	
Control	14.5	65.2
Control+oligomycin	10.9	19.5
Regenerating	69.7	69.7
Regenerating+oligomycin	20.8	20.8

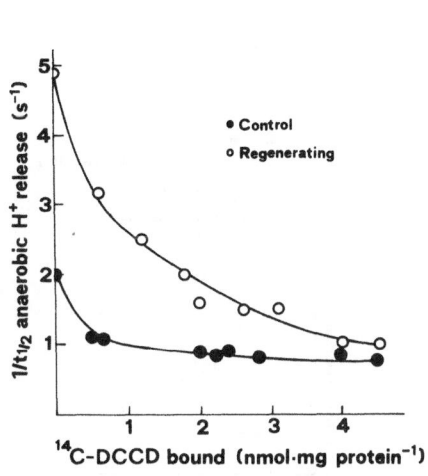

Fig.10. ATPase activity, respiratory control and H$^+$ conduction in regenerating rat-liver mitochondria (from ref.32). For experimental conditions and liver regeneration see ref.(32).

sulted, however, at equal levels of ^{14}C-DCCD binding to F$_o$, in a loss of
its inhibitory action on H$^+$ conduction. Inhibition could then be restored
together with oligomycin sensitivity, when the intact purified PVP protein
was added back to the digested particles.

Fig.9 illustrates the effect of addition of isolated PVP protein, F$_6$
and OSCP on the functional binding of soluble F$_1$ to depleted particles
(USMP) digested by trypsin. Functionally correct binding of F$_1$ to F$_o$ in the
particles is evaluated by reconstitution of oligomycin sensitivity of the
ATPase activity of F$_1$. Trypsin digestion of USMP almost completely abolished
the oligomycin-sensitivity of the ATPase activity of F$_1$ added to the particles. In the control, F$_1$ reconstituted with USMP was 80% inhibited by oligomycin. The addition of PVP protein, OSCP or F$_6$ which per se caused some inhibition of the ATPase activity of reconstituted F$_1$ (F$_6$ was the most effective in this respect), added individually did not promote oligomycin sensitivity. However, combined addition of PVP protein and OSCP and even better
of PVP protein and F$_6$ were effective in restoring oligomycin sensitivity.

Thus it can be concluded that OSCP, F$_6$ and the carboxy-terminal region
of the PVP protein are all required for the functionally correct binding of
F$_1$ to F$_o$ and coupling of the hydrolytic process to transmembrane H$^+$ conduction. In addition, the carboxy-terminal segment of the PVP protein, which
extends at least in part out of the M surface of the membrane and enters in
contact with F$_1$ is essential for a correct functional organization of transmembrane proton channel in F$_o$. In fact DCCD binding to the 8 and 16 kDa components of F$_o$ results in inhibition of H$^+$ conduction only in the presence

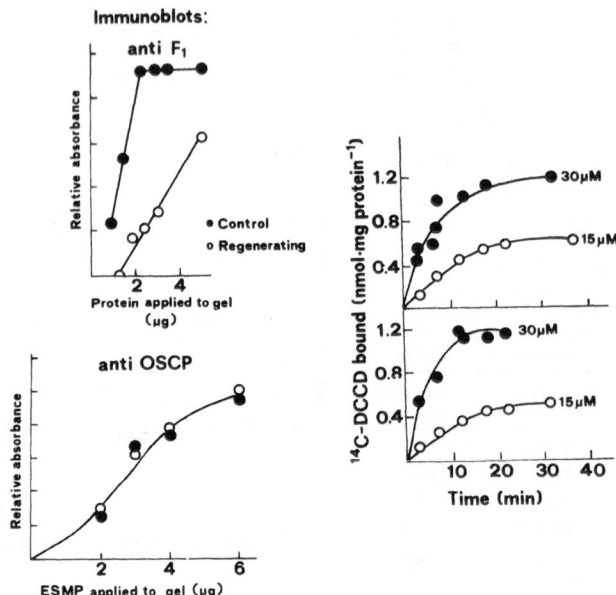

Fig.11. Immunoblot titration of the content of F_1 and OSCP and analysis
     of ^{14}C-DCCD binding in submitochondrial particles obtained from
     normal and regenerating rat-liver. For experimental conditions
     and rat-liver regeneration see ref.(32). Immunoblot was run as
     described in ref.(32) by antisera against bovine F_1 and OSCP.

of the PVP protein. The PVP protein may thus play a central role in organizing the transmembrane H^+ channel in F_o and in controlled H^+ conduction
from the H^+ channel to catalytic and/or allosteric sites in F_1.

A final aspect that will be briefly dealt with here concerns the factors controlling the assembly of the F_oF_1 complex in the membrane. F_1 and
F_o will each constitute per se a powerful energy dissipating system, the
first catalysing rapid hydrolysis of ATP, the second futile dissipation of
ΔH^+ in the coupling membrane. It was initially felt that, possibly to avoid
energy dissipation, biogenesis of F_1 and F_o and their membrane assembly occurred in a concerted process (29). More recently strains of E. Coli
have been described which are capable of expressing independently the F_1 or
the F_o moiety (30).

Our group has looked at this problem by following the levels and the
activity of mitochondrial F_1 and F_o during liver regeneration after extensive hepatectomy (31,32). This is a particularly interesting system since
it allows to follow "in vivo" biogenesis of enzymes in a tissue which, after
an initial phase during which the residual cells get prepared to divide and
express antigenic factors typical of foetal tissues (phase of retro-diffe-
rentiation) (33), starts to rapidly proliferate reproducing in about 48
hours the original mass of the organ when proliferation stops.

The principal observations of these studies are summarized in Fig.10
and 11. In the first 24 hours after hepatectomy there is a dramatic decrease of the ATPase activity (estimated from determinations of Vmax in submitochondrial particles) which is accompanied by a loss of respiratory control in mitochondria isolated from the residual liver. The loss of respira-

tory control is due to enhancement of state 4 respiration which approaches state 3 rate in the control. This enhancement of state 4 respiration is completely abolished by oligomycin indicating it to result from enhanced proton conductivity in F_0. Measurements in submitochondrial particles showed that the oligomycin (32) and DCCD sensitive passive proton conductivity in F_0 was enhanced by more than two fold in the regenerating liver, 24 hr after hepatectomy, this enhancement being suppressed by DCCD (Fig.10).

Immunoblot titration with an antibody against the β subunit of F_1 (Fig.11), and the lower yield of purified F_1(32), showed that the decrease of the ATPase activity and the enhanced proton conductivity in F_0 is in fact due to considerable decrease in the content of F_1 in the regenerating liver.

On the contrary immunoblot titrations showed that there was no decrease in the regenerating liver in the content of OSCP as compared to the control liver nor in the content of subunit c as indicated from the extent of ^{14}C-binding to submitochondrial particles (Fig.11).

These observations thus provide evidence that in mitochondria, like in bacteria, there are conditions in which the F_0 moiety can be fully expressed and assembled in the membrane in a functional state in the absence of F_1. The decreased content of F_1, which may result from a defective expression and/or assembly in the membrane, occurs during the retro-differentation stage of the regenerating liver and is then restored to normal levels during the replicative phase of the hepatocytes. The normal content of OSCP and subunit c suggests that these components are not responsable for the decreased level of F_1 associated to the membrane. Further exploitation of this interesting system may, indeed, help to characterize the factors and the events involved in the expression of the F_1 and F_0 moiety and in their correct assembly in the membrane.

ACKNOWLEDGEMENTS
This work was supported by grant n.88.00794.44 of Consiglio Nazionale delle Ricerche, Italy.

REFERENCES

1. Senior, A.E. (1988) Physiological Reviews 68, 177-232
2. Papa, S. (1989) in "Organelles of Eukaryotic Cells:Molecular Structure and Interactions" (Tager,J.M., Guerrieri,F., Azzi,A. and Papa,S. eds.) Plenum Press, N.Y. (in press).
3. Papa, S., Guerrieri, F., Zanotti,F., Houstek, J., Capozza, G. and Ronchi, S., (1988) in "Molecular Basis of Biomembrane Transport" (Palmieri, F. and Quagliariello, E.,eds.) Elsevier Science Publisher BV, Amsterdam, pp.249-259.
4. Von Meyenburg, K., Jørgensen, B.B., Michelsen, O., Sørensen, L. and McCarty, J.E.C. (1986) EMBO Journal 4, 2357-2362.
5. Lightowlers, R.N., Howitt, S.M., Hetch, L. and Cox, G.B. (1988) Biochim. Biophys.Acta 933, 241-248.
6. Eya, S., Noumit, , Maede, M. and Futai,M. (1988) The Journal of Biol. Chem.263, 10056-10062.
7. Paule, C.R. and Fillingame, R.H. (1989),Arch.of Biochem. and Biophys. 274, 270-284.

8. Walker, J.E., Runswick, M.J. and Poulter, L. (1987) J.Mol.Biol.197, 89-100.
9. Walker, J.E., Gey, N.J., Powell, S.J., Kostina, M. and Dyer, M.R.(1987) Biochemistry 26, 8613-8619.
10. Hoppe, J. and Sebald, W. (1984) Biochim.Biophys.Acta 768, 1-22.
11. Kopecky, J., Guerrieri, F., and Papa, S. (1983) Eur.J.Biochem.131,17-24.
12. Guerrieri, F., Capozza, G., Houstek, J., Zanotti, F., Colaianni, G., Jirillo, E. and Papa, S. (1989) FENS Lett.250, 60-66.
13. Houstek, J., Svoboda, P., Kopecky, J., Kuzela, S. and Drahota,Z.(1981) Biochim.Biophys.Acta 634, 331-339.
14. Houstek, J., Kopecky, J., Zanotti, F., Guerrieri, F., Jirillo,E., Capoz za, G. and Papa, S. (1988) Eur.J.Biochem.173, 1-8.
15. Zanotti, F., Guerrieri, F., Capozza, G., Houstek, J., Ronchi, S. and Papa, S. (1988) FEBS Lett.237, 9-14.
16. Fearnley, I.M. and Walker, J.E. (1986) EMBO Journal 5, 2003-2008.
17. Pullman, M.E. and Monroy, G.C., (1963), J.Biol.Chem.238, 3762-3769.
18. Guerrieri, F., Scarfò, R., Zanotti, F., Che, Y.W. and Papa, S. (1987) FEBS Lett.213, 67-72.
19. Guerrieri, F., Zanotti, F., Che, Y.W., Scarfò, R. and Papa, S. (1987) Biochim.Biophys.Acta 892, 284-293.
20. Racker, E., Horstman, L.L., Kling, D. and Fesseden-Raden, J.M. (1969) J.Biol.Chem.244, 6668-6674.
21. Kanner, B.I., Serrano, M., Kadrach, M.A. and Racker, E. (1976) Biochem. Biophys.Res.Comm.69, 1050-1056.
22. Papa, S., Guerrieri, F., Zanotti, F., Houstek, J., Capozza, G., and Ronchi, S. (1989) FEBS Lett.249, 62-66.
23. Zanotti, F., Guerrieri, F., Scarfò, R., Berden, J. and Papa, S. (1985), Biochem.Biophys.Res.Comm.132, 985-990.
24. Mac Lennan, D.H. and Tzagaloff, A. (1968) Biochemistry, 7, 1603-1610
25. Guerrieri, F. and Papa, S. (1982) Eur.J.Biochem.128, 9-13.
26. Zanotti, F., Guerrieri, F., Che, Y.W., Scarfò, R. and Papa, S. (1987) Eur.J.Biochel.164, 517-523.
27. Sanadi, D.R. (1982) Biochim.Biophys.Acta 683, 39-56
28. Pansini, A., Guerrieri, F. and Papa, S. (1978) Eur.J.Biochem.92, 544-551.
29. Perlin, D.S., Cox, D.N. and Senior, A.E. (1983), J.Biol.Chem.258, 9713-9800.
30. Ario, J.P., Klionsky, J. and Simoni, R.D. (1985) J.Biol.Chem.260, 11207-11215.
31. Buckle, M., Guerrieri, F. and Papa, S. (1985), FEBS Lett.188,345-351.
32. Buckle, M., Guerrieri, F., Pazienza, A. and Papa, S. (1986) Eur.J.Bio-chem. 155, 439-445.
33. Uriel, J. (1979) Advances Cancer Res.29, 127-174.

SITUATION OF ARCHAEBACTERIAL ATPASE AMONG ION-TRANSLOCATING ATPASES[1]

Yasuo Mukohata and Kunio Ihara

Department of Biology, Faculty of Science
Nagoya University
Nagoya 464-01 Japan

SUMMARY

Among the ATPases so far isolated from archaebacteria, only the halo-bacterial ATPase has been proven to be the (catalytic) part of ATP synthase. This ATPase/synthase is H^+-translocating and has unique chracteristics as enzyme/protein.

An antibody raised against this halobacterial ATPase cross-reacted to the ATPases from other archaebacteria, thermoacidophile and methanogen. From various additional points of view, all the archaebacterial ATPases can be regarded as a family, and categolized into "A-type" ATPase (Archae-ATP-synthase)[2].

Immunoblotting also showed that the A-type ATPase is close to the V-type one, so-called anion-sensitive H^+-ATPase e.g., of vacuolar membranes (specifically sensitive to nitrate). This finding would support the concept of endosymbiotic evolution with a primitive archaebacterium as the host. In contrast, the A-type ATPase, esp. of halobacteria, was found to be very much less related to the F-type ATPase (i.e., F_oF_1 ATP synthase) or to the P-type (e.g., Na,K-ATPase and Ca^{2+}-ATPase) than the V-type one. This finding together with unique characteristics of the A-type ATPase, such as azide insensitivity, made the halobacterial ATP synthase the first exception from the common understanding that the F-type ATPase is present ubiquitously in all aerobic organisms to synthesize ATP.

The amino acid sequences deduced from the DNA sequences of the cloned genes for the ATPases of halobacteria and other archaebacteria, fully confirmed the immunoblotting data described above.

On the basis of these observations an attempt is made to discuss the evolution of H^+-ATPase.

1) This work was supported in part by Grant-in-Aid for Scientific Research in Priority Areas of "Bioenergetics (Energetics of Extremophiles)" to Y.M. from the Ministry of Education, Science and Culture of Japan.
2) In order to describe and discuss this new group of ATPase/synthase of archaebacteria, we proposed the name "archae"-ATP-synthase (1) or the "A-type ATPase" as well as F-type (azide sensitive), V-type (nitrate sensitive) and P-type (vanadate-sensitive) ATPases categorized earlier (2). Here the differences among different types of ATPases are emphasized to discuss the distance among them, although all these ATPases have diverged from an ancestry ATPase and thus they should carry the common parts to some extents.

INTRODUCTION

Extremely halophilic archaebacteria, such as <u>Halobacterium</u> <u>salinarium</u> (<u>halobium</u>) live on amino acids by respiration (3). They also synthesize ATP by proton motive force produced by light-energized bacteriorhodopsin (4) and/or halorhodopsin (5). Although molecular features of bacteriorhodopsin and halorhodopsin have been revealed to great extents, the key enzyme, ATP synthase, involved in the ATP synthesis has not been unveiled. When the sophisticated ATP synthesis with proteoliposomes of bacteriorhodopsin and F_oF_1-ATPase proved the chemiosmotic theory (6), F_oF_1-ATPase was readily speculated in the ATP synthesis in halobacteria. In order to prove or disprove this speculation, two lines of investigation were initiated; characterization of ATP synthesis <u>in</u> <u>situ</u> and isolation of ATPase which, on the analogy to F_1-ATPase in F_oF_1 ATP synthase, would be the head piece of the synthase. These investigation gave us unexpected results, e.g., azide insensitivity, acidic pH optima, high salt dependency, on the nature of the halobacterial ATP synthase (7,8) as much as of the ATPase (9).

The unique enzymes were further studied by three different means and found; First, enzymatically the ATPase was identified as the (catalytic) part of the ATP synthase (1); Second, immunochemically this halobacterial ATPase was closely related to ATPases from other archaebacteria, and surprisingly, also to so-called anion-sensitive H^+-ATPase from plant vacuoles, but very little to F_1-ATPases of various origin (10); Third, molecular-biologically the amino acid sequence deduced from the gene for the ATPase confirmed the above immunochemical results by its homology to other ATPases (11). In this article, we summarize the feature of the halobacterial ATP synthase which is the only ATP synthase so far known among all ATP synthesis systems in archaebacteria, and will attempt to discuss the situation of this ATP synthase in evolution of H^+-ATPases.

MATERIALS AND METHODS

Cells of <u>Halobacterium</u> <u>salinarium</u> (<u>halobium</u>) R_1mR (12) was grown as previously described (5), harvested and sonicated to prepare cell envelope vesicles as described (7).

For the ATP synthase assay, the vesicles were stuffed by sonication with the reaction medium (1 M NaCl, 80 mM $MgCl_2$, 3 mM ADP, 12 mM Pi and 10 mM PIPES pH 6.8 and, if needed, with an inhibitor, $NaNO_3$, azide, NEM or NBD-Cl). The stuffed vesicles were washed and then examined for ATP synthesis by the base-acid transition (7,8) at $\Delta pH = 2.8$ and $30^\circ C$ in the substrate-free stuffing medium (final 3 mg protein/ml). ATP was determined by a luciferin-luciferase method as described in (13).

For the ATPase assay, the vesicles were suspended in the reaction medium (final 1 mg protein/ml) of the optimal conditions (1 M Na_2SO_4, 10 mM $MnSO_4$, 40 mM MES pH 5.8, 4 mM ATP and, if needed, with an inhibitor, $NaNO_3$, azide, NEM or NBD-Cl) with additional $C_{12}E_9$ (nonaethyleneglycol dodecylether;0.025 %/mg protein) to solubilize vesicles and develop ATPase activity (1). The reaction mixtures were incubated for 30 min at $38^\circ C$ and liberated Pi was determined (14).

The ATPase was isolated from the vesicles as described in (9). The two subunits of the ATPase were separated on SDS-PAGE (electrophoresis on polyacrylamide gel in the presence of sodium dodecylsulfate) and extracted from sliced gels electrophoretically (15). The isolated ATPase and each subunit (in 1 % sodium dodecylsulfate) were mixed separately into a complete or an incomplete adjuvant (about 1 mg protein/ml) and individual mixtures were injected to three male rabbits three times each with two week intervals. Blood was withdrawn from the rabbits three weeks after the final injection and the sera were collected. The polyclonal anti-ATPase, anti-α-subunit and anti-β-subunit antibodies were prepared by fractional precipitation with $(NH_4)_2SO_4$ and by passing ion-exchange columns (DE-52; Pharmacia).

Chloroplast F_1-ATPase (CF_1) was isolated from spinach chloroplasts (16). A membrane fraction rich in tonoplasts was prepared from fresh red beet Beta vulgaris (17). The sarcoplasmic reticulum from rabbit skeletal muscle was a gift of Dr. Taibo Yamamoto, Osaka University, isolated ATPase of Sulfolobus acidocaldarius was a gift from Dr. Masasuke Yoshida, Tokyo Institute of Technology. Membrane preparations of Methanobacterium thermo-autotrophicum and Methanococcus vannielli were gifts from Dr Fuchs, University of Ulm and Dr. Y. Koga, Industrial Medical College, respectively.

Western blotting was carried out by a routine method (18). One set of proteins on the gel (1 mm thick) after SDS-PAGE was stained by Coomassie Brilliant Blue. The other set was transfered on CLEARBLOT-P film (Atto Co.), reacted with the anti-ATPase antibody, and then developed by alka-linephosphatase-labeled goat-anti-rabbit-antibody (Bio Rad) with Nitro Blue-tetrazorium and disodium 5-bromo-4-chloro-3-indolylphosphate.

Molecular cloning of the ATPase of halobacteria was carried out as follows. The ATPase and then its α- and β-subunit were obtained as above. The subunits were separately digested either trypsin or lysylendo-peptidase and peptide fragments were fractioned by FPLC (Pharmacia, PEP-RPC). Each fragment was subjected to a gas phase amino acid sequencer (Applied Biosystems, 477A). On the basis of the obtained partial amino acid sequences, two set of DNA probes of 64 mix of 17mer (for α-subunit) and of 32 mix of 17mer (for β-subunit) were synthesized (Applied Biosystems, 381A). Construction of a genomic BamHI library of H. halobium into pUC18, tranformation and screening were carried out by an ordinary method (19). Recombinant plasmid DNA was sequenced by the dideoxy chain termination method.

ATP and ADP were purchased from Yamasa Shoyu Co., protein standards from Sigma. The adjuvants and other chemicals of highest grade available were obtained from Nakarai Chemicals except for NaCl in culture media from Japan Tobacco Industry. Tap water was used in culture media.

RESULTS AND DISCUSSION

In Table 1 the characteristics of the ATP synthase in situ and ATPase of Halobacterium salinarium (halobium) are summarized (7-9). The ATP syn-thase is characterized on the cell envelope vesicles. The synthase is most active at pH 6.8 and in a supporting salt solution of 1 M NaCl, although the cytosol in which the enzyme is immersed is almost saturated KCl. The enzyme requires Mg^{2+} at around 100 mM $MgCl_2$ where F_oF_1 ATP synthase is completely inhibited. The isolated ATPase is most active at pH 5.8 and in 1.5 M Na_2SO_4 or Na_2SO_3, in which the enzyme is prevented from dissociation (9) and would be able to maintain its active architecture of subunits. The ATPase activity is 3-times higher in Mn^{2+} than in Mg^{2+}. Under these very unusual assay conditions for the ATPase, other ATP hydrolysing enzymes (e.g., 5'-nucleotidase, apyrase) cannot remain active. The cell envelope vesicles are also almost inactive (Fig. 1), unless $C_{12}E_9$ (less than 0.05 %) was added to make the vesicles broken and the ATPase released from the inner face of the vesicle membrane (9). The subunit sizes of the ATPase estimated by SDS-PAGE are 86 (denoted to be α) and 64 kDa (β)(9). Similar figures (87 and 60 kDa) have been reported for the ATPase from Halobacterium saccharovorum (20,21). The holo-ATPase is about 320 kDa by gel filtration (9). Both synthase and hydrolase are not inhibited by

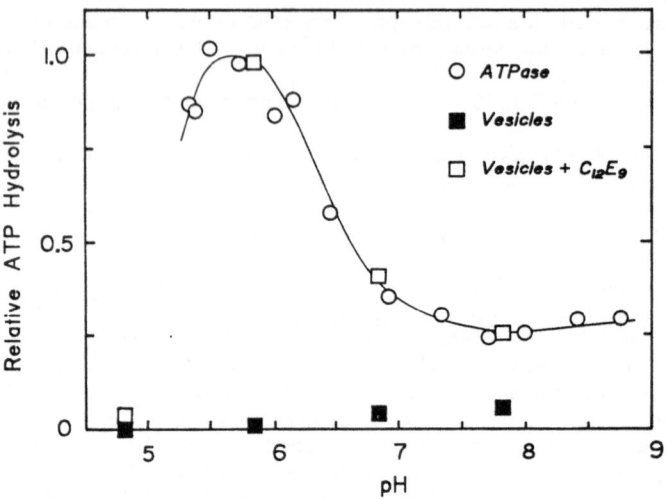

Fig. 1. The pH dependence of the isolated ATPase
of <u>Halobacterium</u> <u>salinarium</u> (<u>halobium</u>) and of the
cell envelope vesicles with or without nonionic
detergent $C_{12}E_9$.

Table 1. Some specific features of the ATP synthase <u>in</u> <u>situ</u> and the
isolated ATPase of <u>Halobacterium</u> <u>salinarium</u> (<u>halobium</u>) (1,7,8,9)

	ATP synthase	ATPase
Molecular size (kDa)		α 64 (86)[a]
		β 52 (64)
pH optimum	6.8	5.8
	H^+-translocating	
Supporting salt	NaCl (1 M)	Na_2SO_4 (1.5 M)
Divalent cation	Mg (100 mM)	Mn (10 mM)
Inhibitors		
Azide	insensitive	insensitive
Vanadate	insensitive	insensitive
Nitrate	insensitive	I_{50} = 3 mM
Inhibiting		
Chemical Modifier	DCCD,NEM,NBD-Cl	DCCD,NEM,NBD-Cl

a) The smaller values were obtained from the amino acid sequences
deduced from genomic DNA analysis (11), while the larger values
were originally estimated (9) by SDS-PAGE.

either azide, a specific inhibitor of (the F-type) F_oF_1 ATP synthase and F_1-ATPase (22) or vanadate, a specific inhibitor of (the P-type) E_1E_2 ATPase (23) that forms the phospho-enzyme intermediate. Both enzymes are inhibited by NEM, NBD-Cl and DCCD (8,9). The DCCD-binding polypeptides were isolated and found to be 78 kDa and 12 kDa by SDS-PAGE (24). The degree of DCCD-labelling on the 78 kDa polypeptide paralleled to inhibition of ATP synthesis (24). Therefore, one or both of these DCCD-binding peptides would be the membrane part(s) of the ATP synthase.

To determine whether the ATPase is part of the ATP synthase as in the case of F_1-ATPase to F_oF_1 ATP synthase, the cell envelope vesicles were first stuffed with NBD-Cl or NEM, incubated then divided into two portions. One portion of the vesicles were stuffed with substrates by sonication and assayed for ATP synthase after pH-jump (7). The other was solubilized with $C_{12}E_9$ and assayed for ATPase. As shown in Fig. 2 the relative remaining activities of both ATP synthase and ATPase are coincident even with a partial stimulation in the presence of ADP, a competitor to the modifier (7,8). Including other data with NEM a correlation factor of 0.988 can be obtained (1). This indicates that the modifier inactivated the one enzyme, the ATP synthase in situ, from which a subunit cluster of 320 kDa was released by $C_{12}E_9$ as (inactivated) ATPase. Therefore, we now identify the ATPase released from halobacterial plasma membrane as the part (most probably the catalytic head piece) of the ATP synthase. As listed in Table 1, the ATPase/synthase is quite unique from the established knowledge on F_oF_1 ATP synthase (the F-type ATPase).

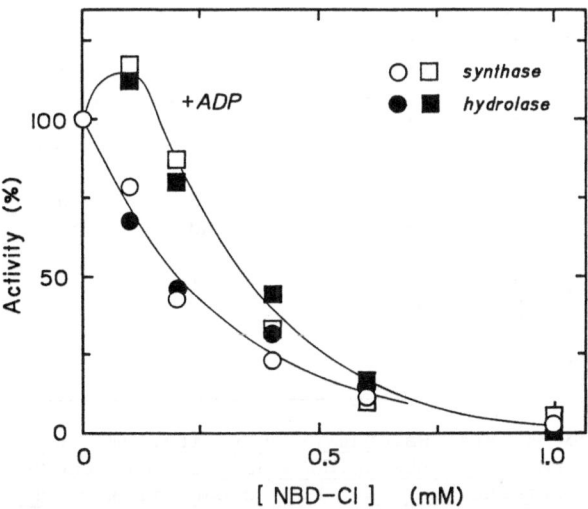

Fig. 2. **Remaining activities of ATP synthesis and hydrolysis after modification of the cell envelope vesicles of Halobacterium salinarium (halobium) with NBD-Cl in the presence or absence of 3 mM ADP.**

The polyclonal antibody raised against halobacterial ATPase cross-reacted (10) not only to its antigen ATPase but also to the ATPase of another archaebacteria <u>Sulfolobus</u> <u>acidocaldarius</u> (65 and 51 kDa (25,26)) and, strikingly, to the ATPase of red beet tonoplast (67 and 57 kDa (27)). In contrast, the antibody cross-reacted little to CF_1 (major two are 59 and 56 kDa (28)) and EF_1 (55 and 50 (29,30)) or Ca^{2+}-ATPase (110.5 kDa (31)). Since each ATPase was applied on SDS-PAGE in a similar range of mass (several μg enzyme/well, even larger amounts for enzymes of weaker reaction; except for the antigen ATPase that was used one half μg/well), the staining would reflect the immunochemical affinity. Furthermore, although there should be several common sites among ATPases for immuno-chemical reaction such as the nucleotide binding site, the blotting was carefully carried out (ran SDS-PAGE on the same gel; transfered on the same film; cross-reacted and stained simultaneously) and repeated, and found that long before F_1 and Ca^{2+}-ATPase were stained (i.e. the color was

Table 2. Immunochemical cross-reactivity among four types of ATPases.(10,40)

ATPase/ subunit		Antibody against			
		Hal ATPase α	β	Sul ATPase β	CF_1 (holo)
A Hal	α	+++	–	–	–
	β	–	+++	++	–
Sul	α	++	–	+	–
	β	–	++	+++	–
Met	α	++	–	–	–
	β	–	++	++	–
V Vac	α	++	–	–	–
	β	–	++	++	–
F CF_1	α	–	–	nd	+++
	β	–	–	nd	+++
TF_1	α	nd	nd	++	+
	β	nd	nd	+	+
P Cal		–	–	nd	–

Hal and Sul are <u>Halobacterium</u> <u>salinarium</u> (<u>halobium</u>) and <u>Sulfolobus</u> <u>acidocaldarius</u>. Met denotes <u>Methanosarcina</u> <u>barkeri</u>, <u>Methanobacterium</u> <u>thermoautotrophicum</u> and <u>Methanococcus</u> <u>vannielli</u>, which showed identical results. Vac denotes vacuolar ATPases from <u>Beta</u> <u>vulgaris</u>. CF_1 and TF_1 are F_1 ATPase from chloroplasts of <u>Spinacia</u> <u>oleracea</u> and Thermophilic bacterium PS3, respectively. Cal is Ca^{2+}-ATPase from sarcoplasmic reticulum of rabbit skeletal muscle. nd, not determined.

developed) due to these common sites, ATPases of sulfolobus and vacuolar
membrane as well as that of halobacterial plasma membrane were stained
clearly. The additional and more detailed data are summarized in Table 2.
The antibodies raised separately against α - and β-subunit of halobacterial ATPase were tested for their cross-reaction.It became clear that
among A- and V-type ATPases their (largest) α-subunits are closely
related to each other and their (second large) β-subunits as well.

On the basis of these results, three conclusions can be drawn;

(1) All the archaebacterial ATPases can be grouped in a family, we
would say, the A-type ATPase, named after archae-ATP synthase (1). These
ATPases from other archaebacteria also carry common inhibitor sensitivity;
insensitive to azide and vanadate and sensitive to nitrate (25,32). They
have subunits in which the largest two have very similar molecular sizes
(See Table 3). Since the antigen, halobacterial ATPase is, as shown above,
the (catalytic) head piece of the ATP synthase, those immunochemically
reacted ATPases from other archaebacteria would also be parts of their ATP
synthases. From another point of view, the common presence of the A-type
ATPase as the key enzyme of energy transduction implies that halobacteria,
methanogens and thermoacidophiles can be grouped as archaebacteria.

(2) The A-type ATPase is related closely to the V-type ATPase. The
V-type ATPase has been identified on vacuole membranes (33,34), chromaffin
granules (35) and other low-density membranes (36). The V-type ATPases are
known to be specifically inhibited by nitrate (I_{50} = 10 mM (37)). The
halobacterial ATPase, but not synthase, is also very much sensitive to
nitrate (Fig. 3). These intracellular organelles where A/V type ATPases
locate have single membrane structure. This will be discussed in relation
to biological evolution below.

(3) The A-type ATPase is very remotely related to the F-type ATPase.
In separate experiments, repeated attempts to identify the F-type ATPase
in archaebacteria, by Western blotting with various antibodies against the
F-type ATPase, by Southern blotting with the gene segments for the F-type

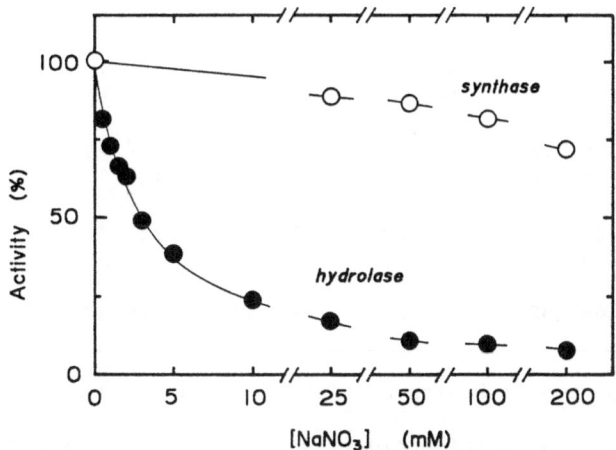

Fig. 3. Effects of $NaNO_3$ on the activities of ATP
synthesis of the cell envelope vesicles of Halobacterium salinarium (halobium) and of ATP hydrolysis of $C_{12}E_9$-solubilized vesicle ATPase.

ATPase (38) and others have not been successful. All these data strongly suggest that in archaebacteria (at least halobacteria) ATP is synthesized by this A-type ATPase/synthase instead of the F-type one which is absent. This gives the first exception from the widely believed expression in biochemistry text books i.e., F_oF_1 ATPase is the sole and ubiquitous ATP synthase coupled to respiratory and/or photosynthetic electron transport.

It should be added here that as shown in Table 2 the antibody against ATPase of <u>Sulfolobus acidocaldarius</u> exceptionally cross-reacted to F-type ATPase (39,40). This suggests that among the A-type ATPases sulfolobus ATPase is a little closer to the F-type one. The antibody raised against the β -subunit of sulfolobus ATPase cross-reacted to the α -subunit of TF_1 ATPase stronger than to the β -subunit (40) (Table 2). This may also suggest a liitle difference of sulfolobus ATPase from other A-type ATPases (This will be discussed further elsewhere (11)).

The DNA sequences of the gene for the α- and β -subunits of the halobacterial ATPase, and the amino acid sequences deduced from them (11) supported the above immunochemical observations, and the numerical data on homology analyses were obtained. Table 3 shows percentage homologies in amino acid sequences of the two large subunits of three types of ATPases. The close relationships among the A-type ATPases, and those between the

Table 3. Amino scid sequence homologies among the major two subunits of the ATPases of archaebacteria (A-type), mold vacuole (V-type) and <u>E. coli</u> (F-type).

		Hal α	Hal β	Sul α	Sul β	Met α	Met β	Vac α	Vac β	EF_1 α	EF_1 β
Hal	α		23	49	27	63	26	50	28	27	25
	β	23		27	55	23	65	24	54	23	24
Sul	α	49	27		25	50	25	48	27	23	26
	β	27	55	25		25	57	23	53	23	28
Met	α	63	23	50	25		28	53	27	27	25
	β	26	65	25	57	28		26	56	26	25
Vac	α	50	24	48	23	53	26		25	24	25
	β	28	54	27	53	27	56	25		26	23
EF_1	α	27	23	23	23	27	26	24	26		24
	β	25	24	26	28	25	25	25	23	24	
Mw		**64**	**52**	**66**	**51**	**64**	**50**	**67**	**57**	**55**	**50**

The major two subunits of ATPases are denoted as α and β . Hal, Sul and Met are <u>Halobacterium</u> <u>salinarium</u> (<u>halobium</u>) (11), <u>Sulfolobus</u> <u>acidocaldarius</u> (41,42) and <u>Methanosarcina</u> <u>barkeri</u> (43),respectively. Vac and EF_1 are <u>Neurospora</u> <u>crassa</u> (44,45) and <u>E. coli</u> (29,30), respectively. Homologies were computed by a software programmed by Genetyx and the figures were rounded. On the bottom line the molecular masses are also shown in kDa.

A-type ATPases and the V-type one are shown quite clearly. The homology between the F-type ATPase and any one of those A- or V-type ones is quite low. Also, among the A- and the V-type ATPases, homologies between the individuals of the α-subunits are high and those of the β-subunits as well. These are consistent with those observed immunochemically.

It should be noted that the molecular masses calculated from the amino acid sequences for the α- and β-subunit of halobacterial ATPase are 64 and 52 kDa, respectively. These values are much smaller than the Mr values estimated from SDS-PAGE (86 and 64 kDa, respectively). The differences are beyond the margins expected in the SDS-PAGE method. This is mostly due to abundance of acidic amino acids (D+E $\sim$ 20 %). The Mr values became smaller on SDS-PAGE after these subunits were amidized (46). Therefore, the DCCD-binding polypeptide of 78 kDa (24) may be the α-subunit in which a portion of acidic amino acids was modified with DCCD. This fits to the data of parallel inactivation of ATP synthesis with DCCD labelling (24), if the α-subunit is catalytic as suggested by <u>Methanosarcina</u> ATPase (32). In the V-type ATPases the α-subunit has been assigned for the catalytic subunit (47-51).

Now we found that the V-type ATPase in eukaryotes originates most likely from the A-type one in primitive archaebacteria. As discussed above, the A-type and the F-type ATPases are remotely related with each other and the A-type ATPase is still used in archaebacteria as ATP synthase. Such a distribution of ATPase/synthase in the present biosphere would be illustrated in Fig. 4.

A primigenial H^+-ATPase may be formed by repeated gene duplication from a primitive H^+-translocating PPase. This ancestry H^+-ATPase would have evolved into a proto-F-type ATPase in ancestry photosynthetic microorganisms and also into a proto-A-type in ancestry thermoacidophiles. When oxygen concentration in the primeval atmosphere increased, an F-type ATP-

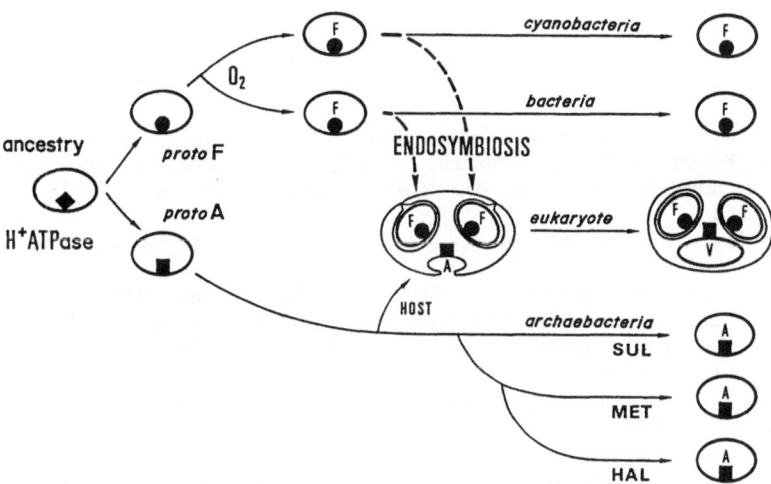

Fig. 4. Schematic pathways of ATPases in evolution which explain the newly confirmed relationships among ATPases in the present biosphere and the orientation of each ATPase on the membrane. (For details see text)

ase became to present in ancestry microorganisms which could live on oxidative phosphorylation. After these photosynthetic and respiratory microorganisms have been incorporated into a host archaebacteria (like thermoacidophiles) by endosymbiosis, those F-type ATPases remained in chloroplasts and mitochondria, respectively, as in the present eukaryotic cells.
These F-type ATPase thus extrudes into stroma and matrix, respectively,
which were the cytosol of the symbionts. These organelles have double
membrane structure because the outer membrane originated from the plasma
membrane of the host upon endocytosis. Meanwhile, the host cell membrane,
where the A-type ATPase located, invaginated and finally formed vacuoles.
This process would be important for the cells especially when organisms
formed multicellular structures and became hard to communicate to the
exterior milieu. The vacuole was the nearest exterior where the cell could
excrete (and later cells store the substances for metabolism and regulation). Vacuole (and organelle membranes which carry the V-type ATPases)
has a single membrane structure originated from the host cell plasma
membrane and the A-type ATPase extrudes into the cytosol of the host cell.
       In eukaryotes, the F-type ATPases synthesize ATP, while the V-type
ones hydrolyse ATP as a proton pump. This "right ATPase in the right place"
would have happened when "efficiency" became important in biological
evolution. The "efficiency" here is the H^+/ATP ratio. The threshold proton
motive force for the F-type ATP synthase is -180 mV (52), while that for
the halobacterial A-type ATP synthase is -100 mV (7). This indicates the
H^+/ATP ratio is higher in the A-type. In other words, the A-type ATPase is
suitable for a H^+ pump (pumps more protons per ATP hydrolysed) and the
F-type for an ATP synthase (causes less proton debt per ATP synthesized).
Therefore, after the respiratory and/or photosynthetic redox systems
became to supply sufficient size of pmf of over -180 mV, the F-type ATP
synthase would have been selected as the ATP synthase. However, since
archaebacteria do not carry the F-type ATPase, they should synthesize ATP
by their rather old-fashioned A-type ATPase.

REFERENCES

1. Mukohata, Y. and Yoshida, M. (1987) J. Biochem. **102**, 797 - 802
2. Pedersen, P.L. and Carafoli, E. (1987) Trends Biochem. Sci. **12**, 146 - 150
3. Larsen, H.(1984) in "Bergey's Mannual of Systematic Bacteriology" 9th edition (Krieg, N.R. ed.) pp 261 - 267, Williams & Wilkins, Baltimore
4. Danon, A. and Stoeckenius, W. (1974) Proc. Natl. Acad. Sci. USA **71**, 1234 - 1238
5. Matsuno-Yagi, A. and Mukohata, Y.(1977) Biochem. Biophys. Res. Commun. **78**, 237 - 243
6. Racker, F. and Stoeckenius, W. (1974) J. Biol. Chem. **249**, 662 - 663
7. Mukohata, Y., Isoyama, M. and Fuke, A. (1986) J. Biochem. **99**, 1 - 8
8. Mukohata, Y. and Yoshida, M. (1987) J. Biochem. **101**, 311 - 318
9. Nanba, T. and Mukohata, Y. (1987) J. Biochem. **102**, 591 - 598
10. Mukohata, Y., Ihara, K., Yoshida, M., Konishi, J., Sugiyama, Y. and Yoshida, M. (1987) Arch. Biochem. Biophys. **259**, 650 - 653
11. Ihara, K. and Mukohata, Y. submitted
12. Mukohata, Y., Matsuno-Yagi and Kaji, Y. (1980) in "Saline Environment" (Morishita, M. & Masui, M. eds.) pp31 - 37, Business Center Acad. Soc. Japan, Tokyo
13. Chapman, J.D., Webb, R.G. and Borsa, J. (1971) J. Cell Biol. **49**, 229 - 233
14. Taussky, H.H. and Schorr, E. (1953) J. Biol. Chem. **202**, 675 - 685

15. Hunkapiller, M.W., Lujan, E., Ostrander, F. and Hood, L.E. (1983) Methods Enzymol. **91**, 227 - 236
16. Lien, S. and Racker, E. (1971) Methods Enzymol. **23**, 547 - 561
17. Bennett, A.B., O'Neill, S.D. and Spanswick, R.M. (1984) Plant Physiol. **74**, 538 - 544
18. Towbin, H., Staehelin, T. and Gordon, J. (1979) Proc. Natl. Acad. Sci. USA **76**, 4350 - 4354
19. Maniatis, R.T., Fritsh, E.F. and Sambrook, J. (1982) Molecular Cloning -a laboratory mannual, Cold Spring Harbor Lab., New York
20. Kristjansson, H., Sadler, M.H. and Hochstein, L.I. (1986) FEMS microbiol. Rev. **39**, 151 - 157
21. Hochstein, L.I., Kristjansson, H. and Altekar, W. (1987) Biochem. Biophys. Res. Commun. **147**, 295 - 300
22. Roisin,M.P. and Kepes, A.(1973) Biochim. Biophys. Acta, **305**, 249 - 259
23. Cantley, L.C.Jr., Josephson, L., Warner, R., Yanagisawa, M., Lechene, C. and Guidotti, G. (1977) J. Biol. Chem. **252**, 7421 - 7423
24. Mukohata, Y., Isoyama, M., Fuke, A., Sugiyama, Y. Ihara, K., Yoshida, M. and Nanba, T. (1987) in "Perspectives of Biological Energy Transduction" (Mukohata, Y., Morales, M.F. & Fleischer, S. eds.) pp 331 - 338, Academic Press, Tokyo
25. Konishi, J., Wakagi, T., Oshima, T. and Yoshida, M. (1987) J. Biochem. **102**, 1379 - 1387
26. Lübben, M. and Schäfer, G. (1987) Eur. J. Biochem. **164**, 533 - 540
27. Manolson, M.F., Rea, P.A. and Poole, R.J. (1985) J. Biol. Chem. **260**, 12273 - 12279
28. Nelson, N., Deters, D.W., Nelson, H. and Racker, E. (1973) J. Biol. Chem. **248**, 2049 - 2055
29. Walker, J.E., Saraste, M. and Gay, N.J. (1984) Biochem. Biophys. Acta **768**, 164 - 200
30. Walker, J.E., Gay, N.J., Saraste, M. and Eberle, A.N. (1984) Biochem. J. **224**, 799 - 815
31. Brandl, C.J., Green, N.M., Korczak, B. and MacLennan, D.H. (1986) Cell **44**, 597 - 607
32. Inatomi, K. (1986) J. Bacteriol. **167**, 837 - 841
33. Uchida, E., Ohsumi, Y. and Anraku, Y. (1985) J. Biol. Chem. **260**, 1090 - 1095
34. Marin, B. (1983) Plant Physiol., **73**, 973 - 978
35. Moriyama, Y. and Nelson, N. (1987) J. Biol. Chem. **262**, 9175 - 9180
36. Young, P-H.G., Qiao, J-Z. and (1988) Proc. Natl. Acad. Sci. USA, **85**, 9590 - 9594
37. Bowman, B.J. and Bowman, E.J. (1986) J. Memb. Biol. **94**, 83 - 97
38. Dharmavaram, R., and Konisky, J. (1987) J. Bacteriol. **169**, 3921 - 3925
39. Lübben, M., Lünsdorf, H. and Schäfer, G. (1987) Eur. J. Biochem. **167**, 211 - 219
40. Konishi, J., Oshima, T., Wakagi, T., Uchida, E., Ohsumi, Y., Anraku, Y., Matsumoto, T., Wakabayashi, T., Mukohata, Y., Ihara, K., Inatomi, K., Kato, K., Ohta, T., Allison, W.S. and Yoshida, M. (1989) J. Biochem. in press
41. Denda, K., Konishi, J., Oshima, T., Date, T. and Yoshida, M. (1988) J. Biol. Chem. **263**, 6012 - 6015
42. Denda, K., Konishi, J., Oshima, T., Date, T. and Yoshida, M. (1988) J. Biol. Chem. **263**, 17251 - 17254
43. Inatomi, K., Eya, S., Maeda, M. and Futai, M. (1989) J. Biol. Chem. **264**, 10954 - 10959
44. Bowman, E.J., Tenney, K. and Bowman, B.J. (1988) J. Biol. Chem. **263**, 13994 - 14001
45. Bowman, B.J., Allen, R., Wechser, M.A. and Bowman, E.J.(1988) J. Biol. Chem. **263**, 14002 - 14007

46. Okutani, S.Y., Ihara, K. and Mukohata, Y. submitted
47. Arai, H., Berne, M., Terres, G., Terres, H., Puopolo, K. and Forgac, M. (1987) Biochem. **26**, 6632 - 6638
48. Uchida, E., Ohsumi, Y. and Anraku, Y.(1988)J. Biol. Chem. **263**, 45 - 51
49. Percy, J.M. and Apps, D.K. (1981) Biochem. J. **239**, 77 - 81
50. Moriyama, Y. and Nelson, N. (1987) J. Biol. Chem. **262**, 14723 - 14729
51. Mandala, S. and Taiz, L. (1987) J. Biol. Chem. **262**, 15780 - 15789
52. Hirata, H., Ohno, K., Sone, N., Kagawa, Y. and Hamamoto, T. (1986) J. Biol. Chem. **261**, 9839 - 9843

STRUCTURE AND CHEMICAL MODIFICATION OF

PIG GASTRIC $(H^{+}+K^{+})$-ATPase

Masatomo Maeda, Shigehiko Tamura, and Masamitsu Futai

Department of Organic Chemistry and Biochemistry
Institute of Scientific and Industrial Research
Osaka University, Ibaraki, Osaka 567, Japan

SUMMARY

This paper summarizes our recent studies on pig gastric $(H^{+} + K^{+})$-ATPase. The amino acid sequence (from cDNA) of the pig enzyme was closely homologous to that of the rat enzyme. Functionally important amino acid residues for catalysis and ion translocation are pointed out. Chemical modification with pyridoxal 5'-phosphate suggested that Lys-497 may be located in the catalytic site or in its vicinity. DCCD inhibited the enzyme, possibly by cross-linking amino acid residues in the hydrophobic region of the enzyme.

INTRODUCTION

Gastric $(H^{+}+K^{+})$-ATPase is localized in plasma membranes of the secretory surface of parietal cells, and is responsible for acid secretion [1]. This enzyme, classified as P-type ATPase (E_1E_2-type ATPase), forms a phospho-enzyme intermediate during catalysis, and catalyzes electroneutral exchange of intracellular H^{+} and extra-cellular K^{+} coupled with hydrolysis of ATP [2]. A large quantity of membrane vesicles enriched with this ATPase can easily be prepared from pig gastric mucosa, and the kinetic properties of the enzyme have been studied extensively [3]. The effects of various protein chemical reagents on this enzyme have also been studied, although their sites of modification were not determined [2, 4, 5]. Thus the pig enzyme is ideal for use in studies on the role(s) of gastric $(H^{+}+K^{+})$-ATPase in acid secretion by combined physiological and molecular biological approaches.

This article summarizes our recent studies on the primary structure and chemical modification of the gastric $(H^{+}+K^{+})$-ATPase, and discusses the topological organization and functional residues of the enzyme based on the amino acid sequence deduced from complementary DNA (cDNA) [6]. We found that pyridoxal 5'-phosphate (PLP) modified a specific Lys residue (Lys-497) in the nucleotide binding site of the enzyme [7, 8]. The enzyme was inhibited by dicyclohexylcarbodiimide (DCCD), like H^{+}-ATPase (P-type ATPase) of Neurospora crassa plasma membranes [9] and F_oF_1 ATPase of oxidative phosphorylation [10]. From these findings, a model of the catalytic site of the enzyme was proposed [8].

1 MGKAENYELY QVELGPGPSG DMAAKMSKKK AGRGGGKRKE KLENMKKEME INDHQLSVAE   60
  LEQKYQTSAT KGLSASLAAE LLLRDGPNAL RPPRGTPEYV KFARQLAGGL QCLMWVAAAI 120
  CLIAFAIQAS EGDLTTDDNL YLALALIAVV VVTGCFGYYQ EFKSTNIIAS FKNLVPQQAT 180
  VIRDGDKFQI NADQLVVGDL VEMKGGDRVP ADIRILQAQG RKVDNSSLTG ESEPQTRSPE 240
  CTHESPLETR NIAFFSTMCL EGTAQGLVVN TGDRTIIGRI ASLASGVENE KTPIAIEIEH 300
  FVDIIAGLAI LFGATFFIVA MCIGYTFLRA MVFFMAIVVA YVPEGLLATV TVCLSLTAKR 360
  LASKNCVVKN LEAVETLGST SVICSDKTGT LTQNRMTVSH LWFDNHIHSA DTTEDQSGQT 420
  FDQSSETWRA LCRVLTLCNR AAFKSGQDAV PVPKRIVIGD ASETALLKFS ELTLGNAMGY 480
  RERFPKVCEI PFNSTNKFQL SIHTLEDPRD PRHVLVMKGA PERVLERCSS ILIKGQELPL 540
  DEQWREAFQT AYLSLGGLGE RVLGFCQLYL SEKDYPPGYA FDVEAMNFPT SGLSFAGLVS 600
  MIDPPRATVP DAVLKCRTAG IRVIMVTGDH PITAKAIAAS VGIISEGSET VEDIAARLRV 660
  PVDQVNRKDA RACVINGMQL KDMDPSELVE ALRTHPEMVF ARTSPQQKLV IVESCQRLGA 720
  IVAVTGDGVN DSPALKKADI GVAMGIAGSD AAKNAADMIL LDDNFASIVT GVEQGRLIFD 780
  NLKKSIAYTL TKNIPELTPY LIYITVSVPL PLGCITILFI ELCTDIFPSV SLAYEKAESD 840
  IMHLRPRNPK RDRLVNEPLA AYSYFQIGAI QSFAGFTDYF TAMAQEGWFP LLCVGLRPQW 900
  ENHHLQDLQD SYGQEWTFGQ RLYQQYTCYT VFFISIEMCQ IADVLIRKTR RLSAFQQGFF 960
  RNRILVIAIV FQVCIGCFLC YCPGMPNIFN FMPIRFQWWL VPMPFGLLIF VYDEIRKLGV 1020
  RCCPGSWWDQ ELYY*   1034

Fig. 1. Amino acid sequence of pig gastric (H⁺+K⁺)-ATPase deduced from the nucleotide sequence.

The consensus sequences of the phosphorylation site of cAMP-dependent protein kinase are boxed. The lysine-rich sequence and that around the phosphorylation site are underlined. Wavy-underlines indicate hypothetical membrane spanning regions determined from the hydropathy profile. D-386 (phosphorylation site), K-497 (PLP binding site), K-518 (FITC binding site), K-708 [corresponding to the adenosine triphosphopyridoxal binding site in Ca^{2+}-ATPase (22)] and K-736 [corresponding to the FSBA binding site in the α subunit of $(Na^+ + K^+)$-ATPase (21)].

EXPERIMENTAL PROCEDURES

The construction of a DNA library and screening of cDNA for (H^++K^+)-ATPase were described previously (6). Membrane vesicles enriched in (H^++K^+)-ATPase were obtained from gastric mucosa of freshly slaughtered hogs (7). (H^++K^+)-ATPase activity was assayed at 37°C in 20 mM PIPES-triethanolamine (pH 6.8) (7). Vesicles were incubated with various concentrations of PLP (followed by $NaBH_4$ treatment) at 25°C, DCCD, or 1-ethyl-3-(3-dimethylaminopropyl)carbodiimide (EDC) at 37°C, and were then used for ATPase assay or determination of modified protein or amino acid residues, as described in detail previously (6-8).

RESULTS AND DISCUSSION

Amino Acid Sequence of Pig Gastric (H^++K^+)-ATPase

Pig gastric cDNA encoding (H^++K^+)-ATPase was cloned, and the amino acid sequence of the enzyme was deduced from the nucleotide sequence (6). The enzyme consists of 1034 amino acid residues (114,285 dalton) including the initiation methionine (Fig. 1). Alignment of the amino acid sequence of pig (6) and rat (11) (H^++K^+)-ATPases indicated about 98 % homology (identical residues). We recently cloned genomic DNA coding for human gastric (H^++K^+)-ATPase (12); and found that the human enzyme also showed about 98 % homology with the pig enzyme. About 63 % of the amino acid residues of the pig (Na^++K^+)-ATPase (13) were identical with those of pig (H^++K^+)-ATPase (6), indicating that these two enzymes are closely related. On the other hand, the pig (H^++K^+)-ATPase was less homologous with sarcoplasmic Ca^{2+}-ATPase (about 30 % homology) (14), although comparison of the two enzymes from the same origin has not yet been possible. Consensus sequences for sites phosphorylated by cAMP-dependent protein kinase, (K-R-X-X-S) and (R-R-X-S) (15), were found at residues 359 to 363 (K-R-L-A-S) and residues 950 to 953 (R-R-L-S), respectively, in the pig enzyme (Fig. 1). These sites may be phosphorylated during the trans-membrane regulation of the enzyme with histamine, gastrin or acetylcholine (16). Three Asn residues (Asn-225, 493 and 730) similar to known N-glycosylation sites (N-X-S/T) were also found. Consistent with this finding, the enzyme separated by polyacrylamide gel electrophoresis could be stained with carbohydrate reagent (unpublished observation). The amino acid sequences conserved in P-type ATPases may be essential for the common catalytic steps. Ten amino acid residues (I-C-S-D-K-T-G-T-L-T) (Fig. 1) around the phosphorylation site are highly conserved in (H^++K^+)-, (Na^++K^+)-, Ca^{2+}-, K^+-, and H^+-ATPases (17), although bacterial K^+-ATPases have slightly different sequences [L-L-L-D-K-T-G-T-L-T in Escherichia coli (18) and I-M-L-D-K-T-L-T in Streptococcus faecalis (19)]. The Asp (D) residues underlined above are phosphorylated during catalytic cycles, and correspond to Asp-386 of pig gastric (H^++K^+)-ATPase. A similar sequence is present in the β subunit of F_oF_1 ATPases

including that of E. coli (I–T–S–T–K–T–G–S–I–T), although the Asp
(D) residue phosphorylated in P–type ATPases corresponds to Thr
(T) of the β subunit. Mutagenesis studies suggested that the
sequence is also essential for normal catalysis by the F_oF_1 ATPase
(17).

Similarly, we located the fluorescein isothiocyanate (FITC)
binding site (20) at Lys–518. The (p–fluorosulfonyl)benzoyl adenosine (FSBA) binding site in the α subunit of (Na^++K^+)–ATPase (21)
and the adenosine triphosphopyridoxal binding site of Ca^{2+}–ATPase
(22) correspond to Lys–736 and Lys–708 residues, respectively, of
the pig gastric enzyme (Fig. 1, 2). These residues are located in
the central hydrophilic region of the enzyme and may form the
catalytic site of the enzyme.

The (H^++K^+)–ATPase may traverse the membrane seven times
judging from its hydropathy profile (Fig. 2). The Asp or Glu
residues (Asp–303, Asp–878 and Glu–344) in the membrane spanning
region may form pathways for translocation of H^+ and K^+ ions.
As shown in Fig. 1, like the α subunit of (Na^++K^+)–ATPase, the
(H^++K^+)–ATPase has a lysine rich sequence in the amino terminal
region, $K_{1-2}X_{0-2}K_{1-3}X(GXGGG)K/R_{1-3}X(K/R)_{1-2}X_{1-4}Z_2XK_2E$, where Z is
an Asp, Glu or Asn residue (6). It is noteworthy that a cluster of
glycine residues (G–X–G–G–G) is found only in (H^++K^+)–ATPases (6,
11). The lysine–rich sequence is not found in other related cation–
transporting ATPases. As discussed previously (6), this lysine–rich
sequence may have a regulatory role(s) in binding and occlusion of
the monovalent cation.

We could locate functionally important amino acid residues
which may be interesting for future application of site–directed
mutagenesis. Chemical modification experiments of the ATPase became
more fruitful as the amino acid sequence and a membrane–spanning
model of the enzyme (Fig. 2) became available.

Lys–497 of Pig Gastric (H^++K^+)–ATPase Is Modified with PLP

On incubation of gastric membrane vesicles with PLP,
(H^++K^+)–ATPase was the only protein modified: about one mole of
this reagent was incorporated into one mole of the ATPase (7). The
PLP–modified enzyme did not show K^+–dependent ATP hydrolysis or
formation of a phospho–enzyme intermediate, and the gastric vesicles
containing the modified enzyme had no ability to catalyze H^+
uptake. ATP protected the enzyme from the modification with PLP,
but poor substrates, such as GTP, CTP and UTP, had no significant
protective effect. ADP also had a protective effect against
modification, but phosphate and AMP did not. The reactive ATP
analogues, adenosine tri– and di– phosphopyridoxal were also
inhibiory, showing essentially similar kinetics (unpublished
observation). These results suggest that PLP modifies a specific Lys
residue(s) located in the catalytic site or a region in its vicinity.

These results prompted us to identify the reactive Lys
residue(s). A peptide labeled with radioactive PLP could be
released from gastric vesicles by chymotrypsin treatment. Only two
labeled peptides N–S–T–N–K–F (corresponding to residues 493–498 of
the ATPase) and S–T–N–K–F (residues 494–498) were isolated quantitatively, and Lys–497 was concluded to be the binding site of PLP
(8). This Lys residue is conserved in (H^++K^+)–ATPase of pig (6)
and rat (11), the α subunit of (Na^++K^+)–ATPase from various
sources (13, 23–27) and Ca^{2+}–ATPases of rabbit sarcoplasmic

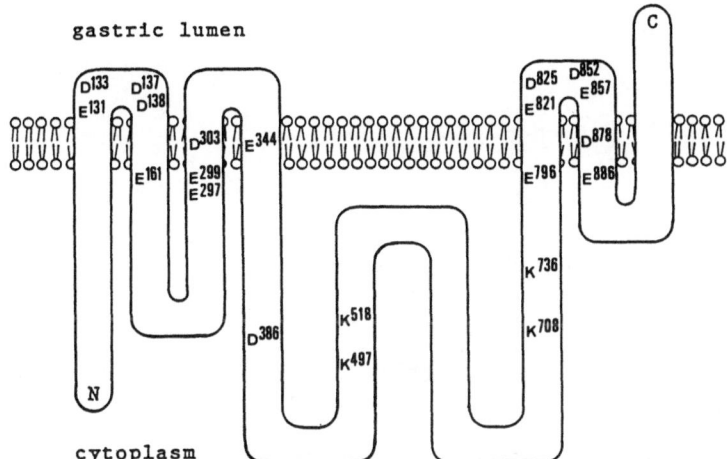

Fig. 2. Hypothetical transmembrane orientation of pig gastric (H⁺+K⁺)-ATPase and the residues potentially important for cation transport.

The possible orientation of membrane spanning regions is based on the hydropathy profile of the enzyme. Acidic residues in and near the transmembrane segments are indicated. The phosphorylation site and important residues identified in chemical modification experiments (see legends to Fig. 1) are also indicated.

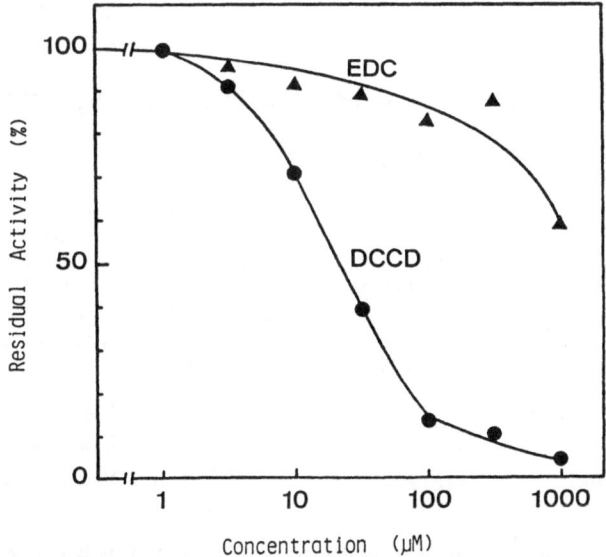

Fig. 3. Specific inhibition of the K⁺-dependent ATPase activity by a hydrophobic carbodiimide.

The reaction mixture for ATPase assay (with gastric vesicles and without ATP) was preincubated with various concentrations of DCCD or EDC at 37°C for 10 min. Then ATPase activity was measured in the presence of ATP (7).

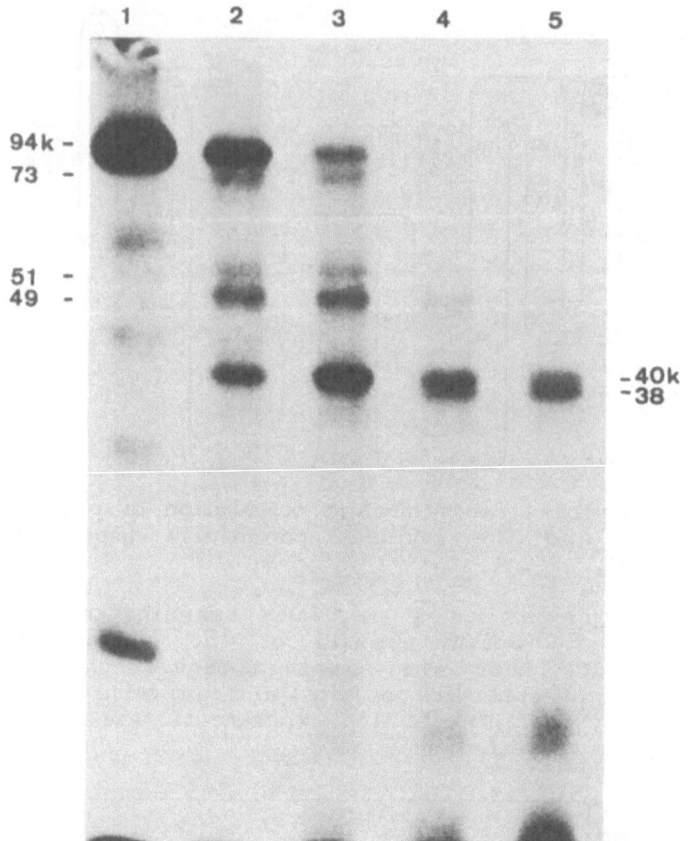

Fig. 4. Specific modification of $(H^+ + K^+)$–ATPase by $(^{14}C)DCCD$.

Gastric vesicles (2 mg/ml) were incubated in 5 mM PIPES (pH 6.8) and 0.25 M sucrose containing 100 μM (^{14}C)DCCD (57 mCi/m mol) at $37^{\circ}C$ for 15 min. Then they were precipitated (100,000 x g, 30 min) and resuspended in the same buffer but without DCCD. Vesicles (100 μg/20 μl) were treated with various amounts of chymotrypsin (0, 1, 2, 5, 10 μg from lanes 1 to 5) on ice for 20 min. After digestion, aliquots (20 μg of vesicle protein) were analyzed by SDS–polyacrylamide gel-electrophoresis followed by autoradiography.

F_oF_1-ATPase $\underline{c}$ subunit (E. coli) ^{56}V M G L V $\underline{D}$ A I P M I A V

                                                       I   II II   II I   II II I

$(H^+ + K^+)$-ATPase (pig stomach) ^{298}I E H F V D I I A G I A I

                                                    II I   II I   II   I II

H^+-ATPase (N. crassa plasma membrane) ^{124}I Q F V M $\underline{E}$ G A A V L A A

Fig. 5. Comparison of the sequence around residue Asp^{303} of pig gastric $(H^+ + K^+)$-ATPase with those around the DCCD binding sites of the E. coli F_oF_1-ATPase $\underline{c}$ subunit and N. crassa plasma membrane H^+-ATPase.

Identical residues are indicated by II, and conservative ones by I.

reticulum (14) and rat brain (28) and human teratoma (29) plasma membranes. Judging from the kinetic studies discussed above, this Lys residue may be located in the catalytic site or in its vicinity. The organization of amino acid residues including Lys-497 in the catalytic site was proposed (8).

Modification of Pig (H^++K^+)-ATPase with DCCD

As discussed above, the Asp or Glu residues in the membrane spanning region may participate in translocation of cations (Fig. 2). The hydrophobic carbodiimide DCCD is known to inhibit H^+ translocation through the intrinsic membrane sector F_o of H^+-ATPase of oxidative phosphorylation (10). DCCD also inhibits the P-type H^+-ATPase of N. crassa (9). Thus it was of interest to study the effect of DCCD on (H^++K^+)-ATPase. As shown in Fig. 3, DCCD inhibited this enzyme almost completely, whereas the hydrophilic carbodiimide EDC was only slightly inhibitory. These results suggested that DCCD reacted specifically with an amino acid residue(s) in the hydrophobic region of the enzyme. To determine the location of the DCCD reactive residue(s), we labeled the ATPase with radioactive DCCD. As shown in Fig. 4, (H^++K^+)-ATPase was preferentially labeled with DCCD in gastric vesicles, and limited proteolysis indicated that DCCD bound covalently to a specific region of the enzyme. Cosistent with this finding, residues 298-310 are similar to the region around the DCCD-reactive Asp and Glu residues, respectively, of the c subunit of E. coli F_oF_1 (10) and N. crassa H^+-ATPase (9) (Fig. 5). However, incorporation of about 0.1 mole of radioactive DCCD per mole (H^++K^+)-ATPase was sufficient to inhibit the ATPase activity completely. Thus there was no apparent correlation between the amount of DCCD incorporated and the inhibition of the enzyme activity. As no intermolecular cross linking of enzyme was detected, these results suggest that DCCD inhibited ATPase activity by forming intramolecular cross-links between amino acid residues, possibly in membrane segments. The protein did not become radioactive after intramolecular cross-linking by DCCD (30), presumably because DCCD played only a catalytic role in the cross-linking reaction between carboxyl and amino moieties of the protein. It will be of interest to determine the amino acid residues cross-linked by DCCD, and such studies should be important in determining the organization of different trans-membrane domains of the enzyme.

The cDNA of the enzyme obtained in this study may be useful for detailed studies using molecular biological procedures. Site-directed mutagenesis of Lys-497 and other residues identified by chemical modifications will be of interest. Further studies should be focused especially on the differences between this enzyme and $(Na^+ + K^+)$-ATPase.

ACKNOWLEDGEMENTS

We thank H. Maki for technical assistance in DCCD-modification experiments. This research was supported in part by grants from the Ministry of Education, Science and Culture of Japan, and the Foundation for Promotion of Pharmaceutical Science (to M. M.).

REFERENCES

1. Faller, L., Jackson, R., Molinowska, D., Mukidjam, E.,
    Rabon, E., Saccomani, G., Sachs, G., and Smolka, A. (1982)
    Ann. N.Y. Acad. Sci. **402**, 146–163.
2. Sachs, G., Chang, H. H., Rabon, E., Schackman, R., Lewin,
    M., and Saccomani, G. (1976) J. Biol. Chem. **251**, 7690–7698.
3. Wallmark, B., Stewart, H. B., Rabon, E., Saccomani, G., and
    Sachs, G. (1980) J. Biol. Chem. **255**, 5313–5319.
4. Ray, T. K., and Forte, J. G. (1976) Biochim. Biophys. Acta
    443, 451–467.
5. Saccomani, G., Barcellona, M. L., and Sachs, G. (1981) J.
    Biol. Chem. **256**, 12405–12410.
6. Maeda, M., Ishizaki, J., and Futai, M. (1988) Biochem.
    Biophys. Res. Commun. **157**, 203–209.
7. Maeda, M., Tagaya, M., and Futai, M. (1988) J. Biol. Chem.
    263, 3652–3656.
8. Tamura, S., Tagaya, M., Maeda, M., and Futai, M. (1989) J.
    Biol. Chem. **264**, 8580–8584.
9. Sussman, M. R., Strickler, J. E., Hager K. M., and Slayman,
    C. W. (1983) J. Biol. Chem. **262**, 4569–4573.
10. Futai, M., and Kanazawa, H. (1983) Microbiol. Rev. **47**,
    285–312.
11. Shull, G. E., and Lingrel, J. B. (1986) J. Biol. Chem. **261**,
    16788–16791.
12. Maeda, M., Ohshiman, K., Tamura, S., and Futai, M.,
    submitted.
13. Ovchinnikov, Y. A., Modyanov, N. N., Broude, N. E.,
    Petrukhin, K. E., Grishin, A. V., Arzamazova, N. M.,
    Aldanova, N. A., Monastyrskaya, G. S., and Sverdlov, E. D.
    (1986) FEBS Lett. **201**, 237–245.
14. MacLennan, D. H., Brandl, C. J., Korczak, B., and Green, N.
    M. (1985) Nature **316**, 696–700.
15. Krebs, E. G., and Beavo, J. A. (1979) Ann. Rev. Biochem. **48**,
    923–959.
16. Ueda, S., Oiki, S., and Okada, Y. (1986) Biomed. Res. **2**,
    105–108.
17. Noumi, T., Maeda, M., and Futai, M. (1988) J. Biol. Chem.
    263, 8765–8770.
18. Hesse, J. E., Wieczorek, L., Altendorf, K., Reicin, A. S.,
    Roruv, E., and Epstein, W. (1984) Proc. Natl. Acad. Sci.
    U.S.A. **81**, 4746–4750.
19. Solioz, M., Mathews, S., and Fürst, P. (1987) J. Biol. Chem.
    262, 7358–7362.
20. Farley, R. A., and Faller, L. D. (1985) J. Biol. Chem. **260**,
    3899–3901.
21. Ohta, T., Nagano, K., and Yoshida, M. (1986) Proc. Natl.
    Acad. Sci. U.S.A. **83**, 2071–2075.
22. Yamamoto, H., Tagaya, M., Fukui, T., and Kawakita, M.
    (1988) J. Biochem. (Tokyo) **103**, 452–457.
23. Shull, G. E., Shwartz, A., and Lingrel, J. B. (1985) Nature
    316, 691–695.
24. Shull, G. E., Greeb, J., and Lingrel, J. B. (1986)
    Biochemistry **25**, 8125–8132.
25. Kawakami, K., Ohta, T., Nojima, H., and Nagano, K. (1986) J.
    Biochem. (Tokyo) **100**, 389–397.
26. Takeyasu, K., Tamkun, M. M., Renand, K. J., and Fambrough,
    D. M. (1988) J. Biol. Chem. **263**, 4347–4354.

27. Kawakami, K., Noguchi, S., Noda, K., Takahashi, H., Ohta, T., Kawamura, M., Nojima, H., Nagano, K., Hirose, T., Inayama, S., Hayashida, H., Miyata, T., and Numa, S. (1985) Nature **316**, 733-736.
28. Shull, G. E., and Greeb, J. (1988) J. Biol. Chem. **263**, 8646-8657.
29. Verma, A. K., Filoteo, A. G., Stanford, D. R., Wieben, E. D., Penniston, J. T., Strehler, E. F., Fisher, R., Heim, R., Vogel, G., Mathews, S., Strehler-Page, M.-A., James, P., Vorherr, T., Krebs, J., and Carafoli, E. (1988) J. Biol. Chem. **263**, 14152-14159.
30. Taylor, S. S., and Buechler, J. A. (1989) Biochemistry **28**, 2065-2070.

INHIBITION OF THE UNCOUPLER-INDUCED MITOCHONDRIAL ATP-HYDROLYSIS BY THE COOPERATIVE WORK OF THE ATPase INHIBITOR, 9K PROTEIN AND 15K PROTEIN *

Tadao Hashimoto, Haruo Mimura, Yukuo Yoshida
Naoki Ichikawa and Kunio Tagawa

Department of Physiological Chemistry, Medical School
Osaka University, Kita-ku, Osaka 530, Japan

SUMMARY

Yeast mitochondrial F1Fo-ATPase has three regulatory proteins, ATPase inhibitor, 9K protein and 15K protein. Binding of ATPase inhibitor to the enzyme is facilitated and stabilized by cooperative work of 9K protein and 15K protein [Hashimoto, T. et al.(1986) J. Biochem. 99, 251-256]. In the present study, we constructed mutant yeasts lacking in ATPase inhibitor, 9K protein or 15K protein. ATP-synthesizing activities of mitochondria from these mutant yeasts were similar to that of wild-type mitochondria. An uncoupler, CCCP, induced ATP hydrolysis in ATPase inhibitor-deficient mitochondria but not in normal mitochondria. Mitochondria from 9K protein-deficient and 15K protein-deficient cells also exhibited the uncoupler-induced ATP hydrolysis, although their rates were smaller than that of the inhibitor-deficient mitochondria. Submitochondrial particles from the 15K protein-deficient cells exhibited low ATPase activity. However, its ATPase activity gradually increased during the incubation in diluted buffer, while that of wild-type cells remained low. These observations strongly suggest that the ATPase inhibitor acts only to inactivate the ATP-hydrolyzing activity of F1Fo-ATPase when the membrane potential disappeared, and that 15K protein and 9K protein act to reinforce this action of the inhibitor protein.

* This work was supported by Grant-in-Aid for Scientific Research (No.01580161) and in part by Grant-in-Aid for Scientific Research on Priority Areas of "Bioenergetics" to K.T. from Ministry of Education, Science and Culture of Japan.

The abbreviations used are: A.A, antimycin A; CCCP, carbonyl cyanide m-chlorophenylhydrazone; F1-ATPase, soluble mitochondrial ATPase; F1Fo-ATPase, mitochondrial ATP synthase.

Bioenergetics, **Edited by C. H. Kim and T. Ozawa, Plenum Press, New York, 1990**

INTRODUCTION

An intrinsic ATPase inhibitor found in mitochondria is considered to be involved in the regulation of oxidative phosphorylation(1,2). The inhibitor protein has been isolated from various sources including beef heart(3), rat liver(4) and yeast mitochondria(5-7), and the gene has been cloned in beef heart mitochondria(8). The inhibitor binds to purified F_1-ATPase and F1Fo-ATPase in a 1:1 molar ratio in the presence of ATP and Mg^{2+} forming a complex inactivated completely. Regulatory mechanism of the enzyme activity by the inhibitor has been examined by the studies with submitochondrial particles(4,9-11). The initial rate of ATP synthesis is influenced by the amount of ATPase inhibitor bound to the particles. The inhibitor-rich particles exhibit lag of several minutes to attain the maximum rate, while the inhibitor-depleted particles do not(4). Therefore, it seems likely that release from inhibition by the ATPase inhibitor precedes ATP synthesis in mitochondria. Early work by Van de Stadt et al. showed that submitochondrial particles became active to hydrolyze ATP when incubated with a respiratory substrate, suggesting that release of the inhibitor protein from the binding site of the enzyme occurs in an energy-dependent fashion(2). Actually, the release of the inhibitor from the enzyme by the energization of membranes has been observed(4). However, there remain some questions on the function of the inhibitor such that it takes several minutes for the release of the inhibitor upon onset of energization(4) and that only a small fraction of the inhibitor is released under the experimental conditions(9).

We have found two protein factors, 9K protein and 15K protein, in yeast mitochondria, which facilitate and stabilize the binding of ATPase inhibitor to F1Fo-ATPase(12,13). The factors act so efficiently on the F1Fo-ATPase complex that they can also play important roles in regulation of oxidative phosphorylation, by controlling binding to and release from ATP synthase of the inhibitor protein. Since these two factors, as well as the inhibitor, are water-soluble proteins but have affinity for mitochondrial inner membrane, it is very difficult to prepare submitochondrial particles free of the factors, which are very desired for the investigation. For this purpose, we cloned genes for these regulatory proteins and constructed mutant yeasts lacking in each of these protein factors. In this paper, we describe ATPase activity in mitochondria from these mutant cells.

EXPERIMENTAL PROCEDURES

Construction of Mutant Cells----Mutant yeast cells lacking in the ATPase inhibitor, 9K protein or 15K protein were constructed by transformation of S. cerevisiae strain DKD-5D (a, trp1, leu2, his3) with the respective cloned genes which were disrupted in vitro by the insertion of marker genes. Mutations were confirmed by Southern and Western blotting analyses.

Preparation of Mitochondria-----Yeast cells were cultivated in a medium containing 1% yeast extract, 1% polypeptone and 2% lactate, pH 4.0. Cells were harvested at mid log phase and mitochondria were prepared according to the method of Daum et al.(15).

Assay of ATP Synthesis and Hydrolysis---- ATP synthesis and hydrolysis were assayed at 25°C by measuring changes in concentrations of adenine nucleotides in the reaction mixture. Mitochondria were introduced to a medium containing 1 mM ADP or ATP, 5 mM potassium phosphate, pH 6.5, 5 mM succinate, 0.3 M mannitol and 0.1% bovine serum albumin in a final volume of 2 ml. For the determination of adenine nucleotides, samples of 50 ul were pipeted out at a time interval and subjected into 950 ul of 0.3 M sodium phosphate, pH 3.0, precooled at 0°C to quench the reaction. Nucleotides were determined by HPLC as described previously(16). Since the samples of mitochondria contained adenylate kinase, net phosphorylation or hydrolysis was calculated from the value of ΔATP-ΔAMP to compensate extra ATP formed by the adenylate kinase.

Protein Assay---- Protein concentration was determined by the method of Lowry et al.(17) with bovine serum albumin as the standard.

RESULTS

Activation of F1Fo-ATPase in Submitochondrial Particles ----Previously we reported that F1Fo-ATPase-inhibitor complex could be stabilized by both the stabilizing factors, 9K protein and 15K protein, and that omission of either of them caused the activation of the enzyme(13). To examine the stabilizing action of these factors, submitochondrial particles were prepared by sonication. The sonication induced ATPase activity to some extent, probably releasing ATPase inhibitor, 9K protein or 15K protein from the particles. As shown in Fig. 1, particles prepared from cells lacking in the inhibitor exhibited ATPase almost fully activated. Particles from wild-type cells exhibited low ATP-hydrolyzing activity which was not activated throughout the incubation. The similar latency of the enzyme was observed in particles from 9K protein-deficient cells, while the enzyme of 15K protein-deficient particles gradually turned to active and reached to the maximum level after 4 hours. The lack of these protein factors in mitochondria did not affect the content of F1Fo-ATPase.

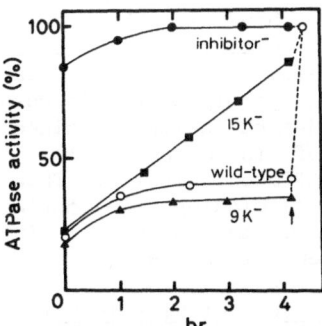

Fig. 1. Activation of ATPase in submitochondrial particles from various kinds of mutant yeasts.
Submitochondrial particles(1.0 mg) were incubated at 25°C in 100 mM Tris-acetate buffer, pH 7.2 in a final volume of 2 ml. At the time indicated, sample of 50 ul was used for assay of ATPase activity. At the end of the incubation (arrow), each sample of 0.1 ml was mixed with an equal volume of 1.0 M Tris-acetate, pH 8.0 and incubated for 40 min to activate the ATPase completely.

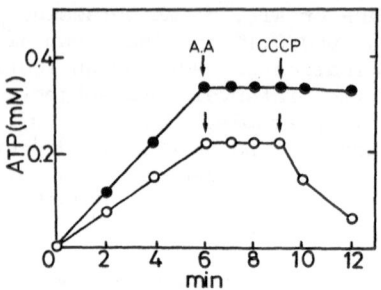

Fig. 2. ATP synthesis and hydrolysis of normal and ATPase
inhibitor-deficient mitochondria.
Mitochondria(0.6 mg) were incubated at 25°C in a medium
containing 0.3 M mannitol, 5 mM potassium phosphate, pH 6.5,
5 mM succinate, 0.1% bovine serum albumin and 1 mM ADP in a
final volume of 2 ml. At the time indicated, adenine
nucleotides in medium were measured by HPLC as described
in text. Antimycin A(2 nmol) and CCCP(2 nmol) were added
as indicated.   Control mitochondria;   ATPase inhibitor-
deficient mitochondria.

   ATP Synthesis and Hydrolysis in Mitochondria of Wild-type and ATPase
Inhibitor-deficient Yeast Cells----- Since yeast cells lacking in ATPase
inhibitor were able to grow on a non-fermentable carbon source, such as
lactate or glycerol, it is clear that ATPase inhibitor is not involved in
mitochondrial oxidative phosphorylation per se. Actually as shown in Fig.
2, mitochondria from the mutant cells were able to phosphorylate ADP at a
similar rate to that of mitochondria from wild-type cells. Antimycin A
completely blocked the phosphorylation and slightly induced ATP-hydrolyzing
activity of the mutant mitochondria. The rate of hydrolysis was less than
0.01 umol/min/mg protein indicating that the block of respiration did not
induce activation of ATPase. In contrast, an uncoupler, CCCP, induced
mitochondrial activity to hydrolyze ATP in mutant but not in wild-type.
Since mitochondria of wild-type cells contained the same amount of F1Fo-
ATPase with mitochondria of the mutant cells, the lack of activation is
ascribed to the action of ATPase inhibitor. Thus, it is likely that ATPase
inhibitor released from the binding site during ATP synthesis turns to
bind to the original site of the enzyme to inhibit ATP hydrolysis when
mitochondrial membrane potential is lost.

   Uncoupler-induced ATPase Activity in Mitochondria from Various Kinds
of Mutant Yeasts-----As shown in Fig. 3, ATPase inhibitor-deficient
mitochondria actively hydrolyzed ATP when membrane potential was lost by
the addition of CCCP. The specific activity was calculated to be 0.7
umol/min/mg protein, which is nearly full activity of the membrane-bound
F1Fo-ATPase. Mitochondria from 9K protein-deficient and 15K protein-
deficient cells also showed ATP-hydrolyzing activities, although·their
rates were smaller than that of mitochondria from inhibitor-deficient

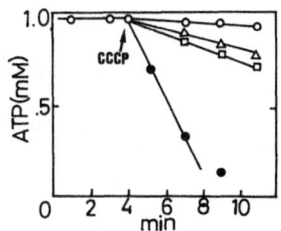

Fig. 3. Uncoupler-induced ATP hydrolysis.
Mitochondria(0.6 mg) were incubated in the medium as described
for Fig. 2, except for 1 mM ATP instead of ADP. At the time
indicated, adenine nucleotides were measured.
Symbols of ○, △, □, ● indicate mitochondria from wild-type,
15K protein-deficient, 9K protein-deficient and ATPase
inhibitor-deficient cells, respectively.

cells. Since each of ATPase inhibitor, 9K protein and 15K protein exists
in yeast mitochondria in an equimolar ratio to F1-ATPase, probably forming
a regulatory substructure of F1Fo-ATPase(18), it seems likely that defect
in either 9K protein or 15K protein weaken the prompt binding of the ATPase
inhibitor to the enzyme responding to the de-energization of mitochondrial
membranes.

DISCUSSION

     In the present work we found that ATP-synthesizing activity of
mitochondria of mutant yeast lacking in ATPase inhibitor was similar to
that of mitochondria from wild-type cells. Thus, loss of the inhibitor
protein seems to cause no impairment in oxidative phosphorylation machin-
ery. This was also the case with 9K protein-deficient and 15K protein-
deficient mitochondria. These observations are compatible with those that
the ATPase inhibitor is released from F1Fo-ATPase depending on energization
of the membranes prior to phosphorylation(2,4,9-11). The uncoupler-induced
ATP-hydrolysis, however, was observed in mitochondria only from mutant
cells. As shown in Fig. 3, ATP-hydrolyzing activity was induced in ATPase
inhibitor-deficient mitochondria immediately after when membrane potential
was depleted by the addition of CCCP. From its specific activity it was
estimated that the enzyme was fully activated to hydrolyze ATP, while in
normal mitochondria it remained inactive under the conditions. It is,
therefore, inferred that the inhibitor protein in normal mitochondria is
liberated from the enzyme during ATP synthesis and rebinds quickly to the
binding site of the enzyme upon de-energization of membranes, promptly
enough to inhibit virtual ATP hydrolysis. The similar uncoupler-induced
ATPase activity was observed in mutant mitochondria lacking in either 9K
protein or 15K protein despite the presence of the inhibitor protein.
Thus, consistently with our earlier observations, it may be concluded that
the stabilizing factors can facilitate the binding of inhibitor protein to
the enzyme and without the aid of these factors the binding hardly takes
place properly to leave the enzyme incompletely inactivated. None of
these factors and inhibitor protein is required for performance of oxidative phosphorylation but they act in concert not to hydrolyze ATP futilely
during cessation of its synthesis.

REFERENCES

1. Asami, K., Juntti, K., and Ernster, L.(1970)
   Biochim. Biophys. Acta 205, 307-311
2. Van de Stadt, R.J., de Boer, B.C., and Van Dam, K.(1973)
   Biochim. Biophys. Acta 292, 338-349
3. Pullman, M.E., and Monroy, G.C.(1963)
   J. Biol. Chem. 238, 3762-3769
4. Schwerzmann, K., and Pedersen, P.L.(1981)
   Biochemistry 20, 6305-6311
5. Ebner, M., and Maier, K.L.(1977)
   J. Biol. Chem. 252, 671-676
6. Satre, M., de Jerphanion, M.M., Heut, J., and Vignais, P.V.(1975)
   Biochim. Biophys. Acta 387,241-255
7. Hashimoto, T., Negawa, Y., and Tagawa, K.(1981)
   J. Biochem. 90, 1151-1157
8. Walker, J.E., Gay, N.J., Powell, S.J., Kostina, M., and Deyer, M.R.
   (1987) Biochemistry 26, 8613-8619
9. Klein, G., and Vignais, P.V.(1983)
   J. Bioenerg. Biomembr. 15, 347-362
10. Powers, J., Crofts, R.L., and Harris, D.A.(1983)
    Biochim. Biophys. Acta 724,128-141
11. Lippe, G., Sorgato, M.C., and Harris, D.A.(1988)
    Biochim. Biophys. Acta 933, 1-11
12. Hashimoto, T., Yoshida, Y., and Tagawa, K.(1983)
    J. Biochem. 94, 715-720
13. Hashimoto, T., Yoshida, Y., and Tagawa, K.(1984)
    J. Biochem. 95, 131-136
15. Daum, G., Bohni, P.C., and Schatz, G.(1982)
    J. Biol. Chem. 257, 13128-13133
16. Kamiike, W., Watanabe, F., Hashimoto, T., and Tagawa, K.(1982)
    J. Biochem. 91, 1349-1359
17. Lowry, O.H., Rosebrough, N.J., Farr, A.L., and Randall, R.J.
    (1951) J. Biol. Chem. 193, 265-275
18. Okada, Y., Hashimoto, T., Yoshida, Y., and Tagawa, K.(1986)
    J. Biochem. 99, 251-256

RECONSTITUTION OF H+ATPase INTO PLANAR PHOSPHOLIPID BILAYERS AND ITS

KINETIC ANALYSIS

Hajime Hirata and Eiro Muneyuki*

Department of Biochemistry
Jichi Medical School
Tochigi, Japan 329-04

SUMMARY

The proton-translocating ATPase of the thermophilic bacterium PS3 (TFoF1) was incorporated into planar phospholipid bilayers, and its electrogenicity and kinetic characteristics were examined. A short-circuit current of up to 1 nA/cm² was generated upon the addition of ATP. The generation of the electric current was progressively suppressed by inhibitors of TF1. From the reversal potential, the electrogenicity of TFoF1 was indicated to be some 180 mV. The relationship between the electric current induced by ATP and the concentration of ATP revealed that the magnitude of the electric current followed simple Michaelis-Menten type kinetics and the Km was found to be 0.14 mM under the conditions studied. Furthermore, there was no apparent dependence of the Km on externally applied membrane potential. These results suggest that the voltage dependence resides in some steps that defines the apparent Vmax rather than Km in the reaction cycle.

INTRODUCTION

Proton-translocating ATPase (H⁺-ATPase) is a complex enzyme consisting of two major portions designated as F1 and Fo. This enzyme converts the electrochemical potential of protons to the chemical potential of ATP by combining the ATP hydrolytic reaction and proton translocation across the membrane, or vice versa. Although the enzymatic reaction catalyzed by F1-ATPase has been studied extensively by several groups (1-4), the most important part of the mechanism, that is how the flow of protons through the Fo portion is linked to the ATP hydrolytic reaction in the F1 portion, is still unknown. In order to investigate the mechanism of this large and complicated enzyme, both ATPase activity and proton-pumping activity must be analyzed under strictly controlled conditions. In this respect, a planar bilayer system would be the most suitable, since both sides of the membrane are accesible and controllable.

In this report we describe a method to incorporate the TFoF1 molecules into planar phospholipid bilayers and demonstrate the generation of electric current in response to ATP addition. Furthermore, utiliz-

----

*Present address : Faculty of Technology, Saitama University, Urawa,
        Saitama 338

ing this system, kinetic analysis of proton translocation under the influence of membrane potential was carried out, and the mechanism of coupling of the ATP hydrolytic reaction to the proton translocation is discussed.

EXPERIMENTAL PROCEDURES

Preparations---TFoF1 was purified from membranes of the thermophilic bacterium PS3 and reconstituted with soybean phospholipids (asolectin) at the protein:lipid ratio of 1:20 or 1:40 (w/w) by the detergent dialysis method as described previously (5). Purified TF1 was supplemented at the concentration of 0.5 mg/ml during the dialysis.

Formation of Planar Phospholipid Bilayers Containing TFoF1---All procedures described below were carried out at room temperature. TFoF1 was reconstituted into the preformed bilayer by the modified fusion method as previously described (6,7).

Electrical Measurement---All procedures and measurements were carried out at room temperature in a laboratory made shield box. The electric current was measured with a laboratory-made current-voltage (I-V) converter which contained a Teledyne 1035 operational amplifier with a 10 Gohm feedback resistor. The I-V converter was connected to the trans chamber via an Ag-AgCl electrode through a salt bridge. The electrical potential of the trans chamber was held at virtual ground, and a command voltage was fed to the cis chamber via a similar electrode. From the I-V converter, the out-put signal was fed to a buffer amplifier and then distributed to a monitoring oscilloscope or a strip chart recorder.

For estimation of the membrane capacitance, a square wave voltage of 1 mV at 80 Hz was applied to the cis chamber, and the capacitive current across the planar bilayer after conversion to voltage by the I-V converter was monitored. Electric currents generated by the TFoF1 reconstituted in the planar bilayer were measured with the I-V converter and recorded after being filtered at 0.1 or 0.3 Hz on an RC low pass filter.

RESULTS AND DISCUSSION

Reconstitution of TFoF1 into Planar Bilayer from Proteoliposomes--- Solvent-free planar bilayers were formed by the method of Montal and Mueller (8) at an aperture (diameter, 0.2 mm) in a Teflon film (25 μm thick) separating two Teflon chambers (internal volume of each chamber is about 1.5 ml). The aqueous solution contained 20 mM Tris-SO_4, pH 8.0, and 2 mM $MgSO_4$. The proteoliposomes were added to one side of the bilayer (designated as the cis side) followed by the addition of $CaCl_2$ (about 30 mM, final concentration). As shown in Fig. 1, the adsorption of the proteoliposomes to the planar bilayer occurred in less than 10 min, which was monitored by measuring the electric capacitance of the bilayer. When the electric capacitance decreased to approximately 60% of the original value (0.7 μ farad/cm^2), the water level of the other side of the bilayer (trans side) was lowered to "peel off" the monolayer from the bilayer. Large noises observed between 20 min and 22 min were due to the manipulation. A slight increase in the electric capacitance, which reflects the enlargement of the bilayer area due to the hydrostatic pressure difference between two chambers, was followed by an abrupt shut down in the electric capacitance indicating that the water level in the trans chamber was completely below the aperture. Then, the water level was re-elevated to reform the bilayer. After this manipulation, the electric capacitance of the bilayer returned to the original value, indicating that the proteoliposomes

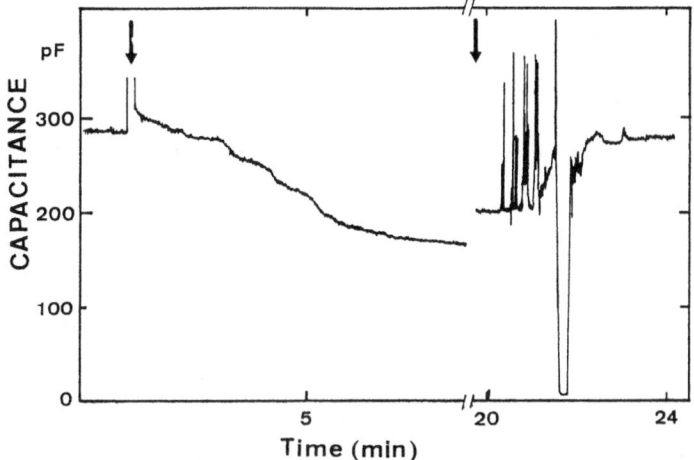

Fig. 1. **Change of membrane capacitance during adsorption of proteoliposomes onto a preformed planar phospholipid bilayer followed by the manipulation.** The planar bilayer was formed from asolectin-hexan solution as previously described (6). Proteoliposomes (30 μ l) were then added into the cis chamber followed by the addition of 30 μ l of 1 M CaCl$_2$ at the first arrow indicated. After completion of the adsortion (the second arrow), the water level of the trans side was slowly lowered and re-elevated (see text). Membrane capacitance was monitored as described under "EXPERIMENTAL PROCEDURES".

attached to the bilayer were completely fused to the planar bilayer. Figure 2 speculatively outlines the overall procedure.

Generation of an Electric Current by TFoF1 in the Planar Bilayer---A typical recording of the electric current generated by adding ATP is shown in Fig. 3. Addition of ATP to the cis chamber (2 mM) resulted in rapid development of an electric current that reached a steady state level of about 30 fA. The sign of the electric current corresponded to a flow of positive charges from the cis chamber to the trans chamber. The steady state current ceased when the substrate was removed by perfusion. The same level of current was generated by a second addition of ATP, this was abolished by adding the F1-ATPase inhibitor, NaN$_3$ (3 mM final concentration). This effect was reversed on removal of the inhibitor by extensive perfusion, and almost the same level of current was observed when ATP was added again. Addition of ATP to the trans chamber failed to induce a current (data not shown). The amplitude of the electric current varied with different independently formed planar bilayers (20-400 fA). These results indicate that the orientation of TFoF1 in the planar bilayer is asymmetric and the catalytic moiety (TF1 portion) faces only to the cis side of the bilayer. Furthermore, although data not shown, these electrical responses were not affected by the ionic species or ionic strength (e.g. 0.1 M KCl, 0.1 M NaCl, etc.) in the medium, suggesting that the positively charged species translocating across the bilayer are protons.

If TFoF1 molecules (Mr 550 kDa) in the proteoliposomes are transferred perfectly to the planar bilayer at the same lipid to protein ratio in the proteoliposomes in the aperture, approximate number of TFoF1 is ex-

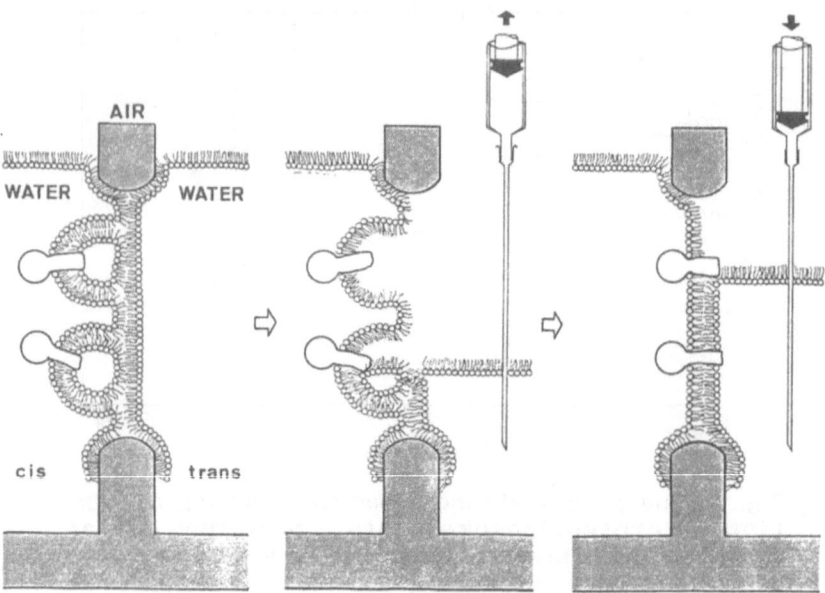

Fig. 2. **A speculative outline of incorporation of TFoF1 into a planar bilayer by the modified fusion method.** After TFoF1 containing proteoliposomes were adsorbed onto a preformed planar bilayer (left panel), the water level of the trans side was lowered to "peel off" the monolayer from the bilayer, resulting in the disruption of proteoliposomes at the aperture (middle panel). Then, the water level of the trans side was re-elevated to form the bilayer containing TFoF1 (right panel).

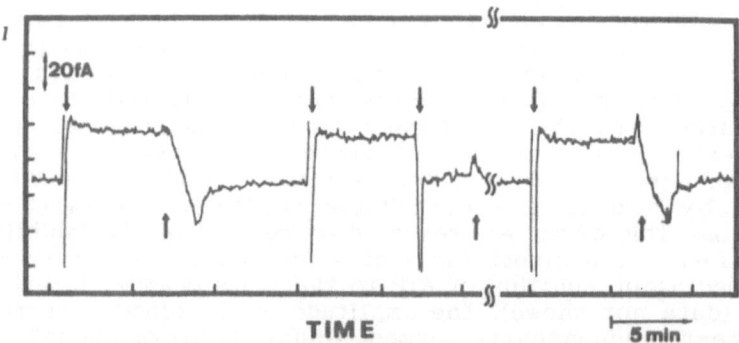

Fig. 3. **Generation of steady state electric current by TFoF1 in a planar bilayer.** ATP was added at a final concentration of 1.8 mM at the time indicated by the first arrow and removed by perfusion as indicated by the second arrow. ATP and NaN_3 were added at the times indicated by the third and fourth arrows, respectively. ATP and NaN_3 were removed by perfusion during the interval indicated by the broken line, and then ATP was added again.

pected to be in the order of 10^6 molecules/aperture based on the following considerations: (i)the area of the bilayer at the aperture is 3×10^{-4} cm²; (ii)the protein-to-lipid ratio in the planar bilayer is 1:25,000 (mol/mol); and (iii)the area occupied by one phospholipid molecule is 60 Å². However, the electric current of 50 fA corresponds to a flow of some 10^6 positive charges/s which leads to the number of TFoF1 molecules in the planar bilayer being in the order of 10^3-10^4, based on the assumption that the turnover number of 10 μ mol of ATP hydrolyzed/min/mg protein and 3 protons transported/ATP under the conditions of this study. Thus, the number of active TFoF1 reconstituted into the planar bilayer would be less than 1% of that expected from the calculations described above.

Effect of Applied Voltage on ATP-induced Current---The effect of applied voltages on the generation of electric current by TFoF1 in the planar bilayer is shown in Fig. 4. The steady state current slightly increased when the cis chamber (TF1 side) was held at higher voltages than the trans chamber, while it decreased almost linearly when the holding potential was reversed. At -180 mV (negative in the cis chamber), the current was almost entirely suppressed. The results indicate that the electrogenicity of TFoF1 is some 180 mV under the condition studied.

Assuming that the standard free energy change of ATP hydrolysis is -30.5 kJ/mol and (ADP)(Pi)/(ATP) is 10^{-7}M under the present conditions, the free energy change of ATP hydrolysis (ΔG_{ATP})is calculated to be some -720 mV. At an equilibrium, the following equation is established.

$$n\Delta \mu H^+ + \Delta G_{ATP} = 0$$

Here, n is the number of protons transported/mole of ATP hydrolyzed. When the $\Delta \mu H^+$ of 180 mV, as estimated above, is introduced into this equation, then,

$$n = 720 \text{ mV}/180 \text{ mV} = 4.0$$

Since this calculation is based on many presumptive factors, the n value of 4.0 does not necessarily have a strict meaning. More precise measurements under rigorously controlled conditions are obviously required to estimate the real H⁺/ATP ratio.

Relationship between the Electric Current and ATP Concentration--- Using the system described above, we studied the dependence of the steady state current on the ATP concentration. Successive addition of small amounts of ATP resulted in the stepwise increase of steady state current. As shown in Fig. 5a, the current vs. ATP concentration relation revealed saturation kinetics, and the Lineweaver-Burk plot (Fig. 5b) was linear in the ATP concentration range between 60 and 1800 μ M. From these results, the Km value was found to be 140 μ M. The steady state electric current should reflect the real velocity of proton translocation, and hence its kinetics should be that of the coupling ATPase reaction. The value of 140 μ M found in the present study was close to the values for the trisite (9), multisite (10), Km₂ (4), and so on. This value also corresponds well to the value obtained with TFoF1 proteoliposomes (11). On the other hand, when the ATPase activity of TFoF1 used in the present study was measured, at least two Km values, 140 μ M and 1.2 mM, were demonstrated by nonlinear root mean square fitting (Fig. 5c). The higher value corresponds well to the Km₃ previously reported by others (4). However, the results of the above-described electrical measurements indicate that the ATPase activity with the higher Km (1.2 mM) may not participate in the proton translocation.

Recently, we have shown that both the ATP hydrolysis reaction and proton translocation exhibited negative cooperativity over the ATP concentration range of 1-2000 μ M (12), with two Km values of about 200 and 10 μ M. When the scheme of the alternate binding change model of Boyer's group (10) is considered, these results apparently indicate that not only

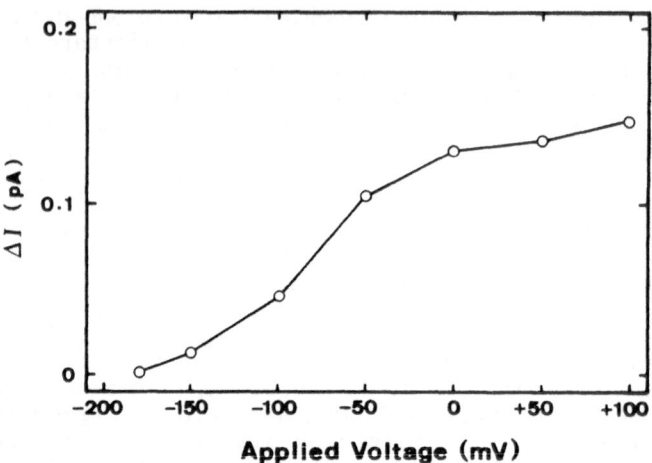

Fig. 4. **Effect of applied voltage on the ATP-induced electric current.** The measurement of electric current induced by ATP addition was the same as described in Fig. 3 except that voltage was applied from an external source as indicated. The signs of voltages indicate the voltages in the cis chamber (TF1 side). The membrane resistance was 300 Gohm. The steady state current (Δ I) after the ATP addition was plotted against the applied voltage.

the trisite cycle, which was originally proposed as the coupling step between proton translocation and the ATPase reaction by the "binding change", but also at least the bisite cycle should be coupled to proton translocation. In the present study, such negative cooperativity was not observed. However, this does not necessarilly mean that only the step with a Km of 140 μ M is the coupling step, since the lowest substrate concentration of ATP, which could induce a distinguishable electric current, was some 60 μ M. In order to see the negative cooperativity, the substrate concentration should be lowered to at least 1 μ M, which is not practical at present and obviously required a further elaboration of the experimental system.

Voltage Dependence of Km---The voltage dependence of the Km value was examined under voltage-clamped conditions. If the binding of ATP per se causes simultaneous proton translocation, the dissociation constant of ATP should be largely affected by the membrane potential, resulting in a change of the Km value. However, as shown in Table I, the Km values were almost independent of membrane potential applied in the range between +34.8 and -34.8 mV.

As shown above, TFoF1 is regarded as a voltage source of 180 mV, and the magnitude of the electric current, that is the rate of the coupled ATPase reaction, is almost linearly controlled by an externally applied membrane potential. Thus, when a membrane potential of +30 mV and -30 mV is applied, the electric current ideally becomes 7/6 and 5/6 of that at 0 mV, respectively. On the other hand, it was not known whether the membrane potential affects the kinetic parameters, Km and/or Vmax. The present results revealed that the Km was not much affected by membrane potential. In fact, if the membrane potential solely affects the Km value while the Vmax remains unchanged, the denominator of the Michaelis-Menten equation should become 6/7 or 6/5 at +/-30 mV.

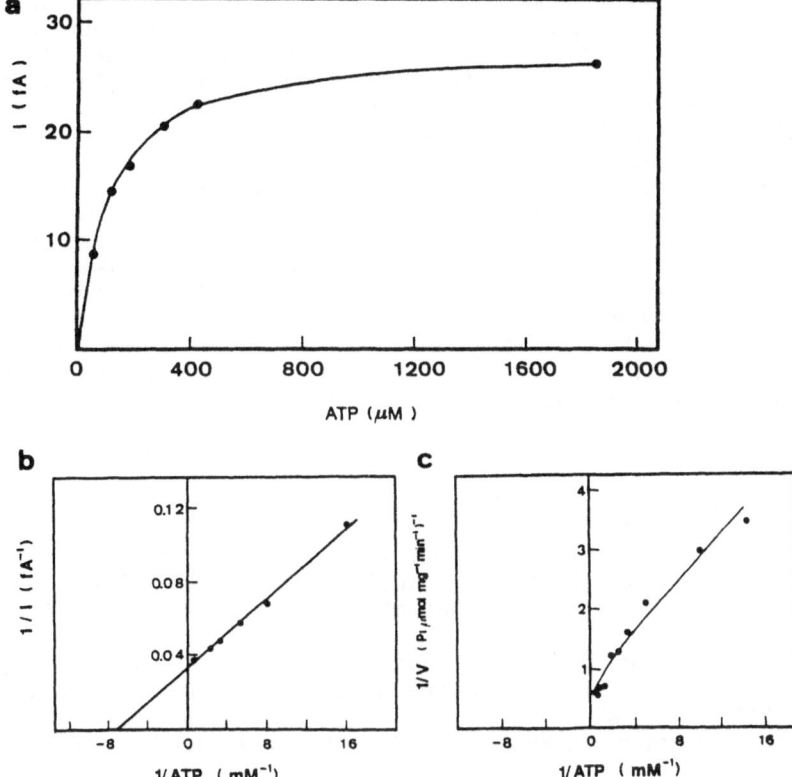

Fig. 5. **Kinetics of proton translocation and ATP hydrolysis reactions.** a, steady state current vs. ATP concentration; b, Lineweaver-Burk plot of the results in a; c, Lineweaver-Burk plot of ATP hydrolysis by TFoF1.

$$V = Vmax*S/(Km + S)$$

If values of S and Km at 0 mV are assumed to be 2000 and 140 μ M, respectively, the Km values would become some 600 μ M at -30 mV and even a negative value at +30 mV to account for the linear current-voltage relationship. Clearly, this was not the case, and thus we conclude that the major voltage dependence resides in the step which determines the apparent Vmax rather than Km.

TABLE I
Michaelis constants for proton translocation at different values of membrane potential

Vm[a]	Km[b]	No. of experiments[c]
mV	μ M	
+34.8	130	5
0	140	8
-30.0	150	2
-34.8	120	9

[a] Applied membrane voltage.
[b] Km values are averages for independent experiments shown in [c].

Although the detailed mechanism of the proton translocation is still unknown, we can exclude at least the possibility that the coupling of the proton translocation to ATPase reaction occurs only at the ATP binding step to the enzyme. Actually, if the coupling resided strictly at the step of ATP binding, the Vmax would not have any voltage dependence and the Kd, subsequently the Km, for ATP would exhibit large voltage dependence. However, it was not the case. On the other hand, it has been reported that the Km values for ADP and Pi are $\Delta \tilde{\mu} H^+$ dependent under the condition of oxidative phosphorylation (13). Thus, together with our previous results (12), we would like to tentatively propose that, under the steady state conditions, the proton translocation is coupled at some latter step in the ATPase reaction. However, we cannot exclude the possibility that the binding of ATP to the enzyme involves two steps and the second step is coupled with proton translocation or the proton translocation is coupled at more than one step of ATPase reaction. Obviously, the elucidation of the coupling mechanism still remains as a very challenging problem.

REFERENCES

1. Boyer, P.D., Cross, R.L., and Momsen, W. (1973) Proc. Natl. Acad. Sci. **70**, 2873-2839
2. Grubmeyer, C., and Penefsky, H.S. (1981) J. Biol. Chem. **256**, 3718-3727
3. Takabe, T., and Hammes, G.G. (1981) Biochemistry **20**, 6859-6864
4. Wong, S.Y., Matsuno-Yagi, A., and Hatefi, Y. (1984) Biochemistry **23**, 5004-5009
5. Kagawa, Y., and Sone, N. (1979) Methods Enzymol. **55**, 364-372
6. Hirata, H., Ohno, K., Sone, N., Kagawa, Y., and Hamamoto, T. (1986) J. Biol. Chem. **261**, 9839-9843
7. Muneyuki, E., Kagawa, Y., and Hirata, H. (1989) J. Biol. Chem. **264**, 6092-6096
8. Montal, M., and Mueller, P. (1972) Proc. Natl. Acad. Sci. USA **69**, 3561-3566
9. Cross, R.L., Grubmeyer, C., and Penefsky, H.S. (1982) J. Biol. Chem. **257**, 12101-12105
10. Gresser, M.J., Myers, J.A., and Boyer, P.D. (1982) J. Biol. Chem. **257**, 12030-12038
11. Rögner, M., and Gräber, P. (1986) Eur. J. Biochem. **159**, 255-261
12. Muneyuki, E., and Hirata, H. (1988) FEBS Lett. **234**, 455-458
13. Hatefi, Y., Yagi, T., Phelps, D.C., Wong, S-Y., Vik, S.B., and Galante, Y.M. (1982) Proc. Natl. Acad. Sci. USA **79**, 1756-1760

A MITOCHONDRIAL CARRIER FAMILY FOR SOLUTE TRANSPORT

Martin Klingenberg

Institute for Physical Biochemistry
University of Munich
Goethestrasse 33
8000 Munich 2, FRG

INTRODUCTION

It is generally accepted that mitochondria have been incorporated into eukaryotic cells from prokaryote progenitors. The similarity to prokaryotes is reflected clearly in the structure of the mitochondrial DNA. The most important compounds which are carried by mitochondria in the eukaryote cell are the systems of oxidative phosphorylation, comprising the respiratory chain and the ATP synthesis. Although the function of these components are quite similar, these complexes have acquired a considerable amount of additional peptides of still largely unknown function. The membrane carriers for anionic metabolites are entirely new components with clear functions having evolved on the transition from the procaryotes to the mitochondria.

Whereas in bacteria cells the metabolism is autonomous and the metabolite traffic across the membrane is confined to the uptake of substrate from the environment, the intracellular localization of mitochondria within the eukayotic cell induces an important metabolic traffic between mitochondria and cytosol. Most obvious is of course the ATP supply from oxidative phosphorylation, but also the oxidative substrates and amino acids branch in their metabolic network across the mitochondrial membrane. In addition, the thermodynamic conditions of metabolism in the two different progenitors of the eukaryotic cells have been quite divergent, particularly with respect to the phosphorylation potential (1) and the redox potential (2). The symbiosis in the eukaryotic cell made it necessary to coordinate these metabolic differences which are retained in the eukaryotic intra– and extra–mitochondrial compartments. All these requirements are met by the carriers inserted into the mitochondrial membrane. Since they have emerged with the development of the eukaryotic cell, they are the most characteristic mitochondrial components. In line with this origin is the localization of the carrier genes in the nucleus.

MITOCHONDRIAL CARRIERS

The most obvious carriers in the mitochondria are those linked to oxidative phosphorylation, in particular the ATP supply to the cytosol. Other carriers translocate intermediates of the tricarboxylic acid cycle and amino acids. There are about 12 different anion carriers presently defined in the mitochondrial membrane. So far, the primary structures of the carriers for ADP and ATP (AAC) (3) and for phosphate (PIC) have been elucidated (4). Both these carriers are elementary components for all mitochondria. It was therefore a great surprise that a very peculiar component of mitochondria, occurring only in a very highly specialized organ, the brown adipose tissue, has also a structure similar to the AAC and PIC (5). The uncoupling protein (UCP) is an H^+/OH^- translocator short–circuiting H^+ generated by the respiratory chain for producing heat. With this function UCP is also a translocator and thus may belong to the mitochondrial carrier group. However different

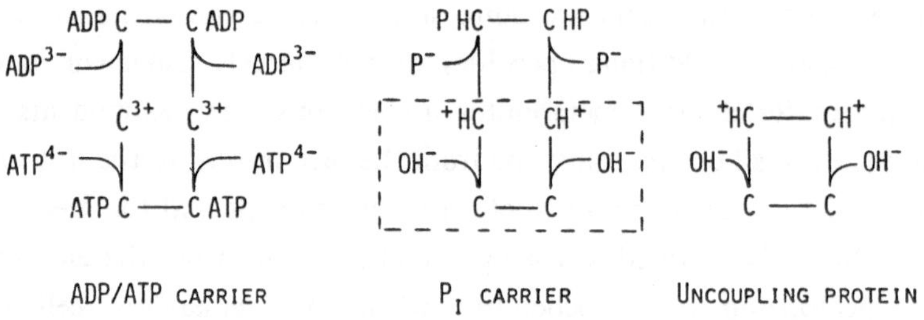

Fig. 1. The functional and putative evolutionary relationship between the mitochondrial carriers. UCP may have emerged from H^+–substrate co–transporter or equivalent OH^-–substrate exchange by deletion of the substrate binding site and by allowing translocation of the positively charged carrier form.

from the AAC, UCP prevents ATP production. Another contrarian feature between both carriers is that UCP translocates about the smallest ion (H^+/OH^-), whereas AAC catalyses the transport of about the largest substances actively moved through biomembranes (Fig. 1).

In evolutionary terms UCP must have evolved very late as it occurs only in mammalians and therefore may be the very late offspring of a fundamental mitochondrial group of proteins. We have proposed as a result of the primary structure similarity to AAC and PIC that UCP can be like an H^+–substrate cotransporter from which the substrate binding necessary for transport has been eliminated (6). Thus the protein is left only with an H^+/OH^- transport capability.

The structure of these mitochondrial carriers has a similar length of around 300 residues. As judged from the primary structure these proteins, although intrinsic transmembrane proteins, are relatively hydrophilic as judged from the primary structure. Yet, a large hydrophobic surface can be deduced from the large detergent micelle which is formed after solubilization of all these proteins (7, 8). Thus the overall structure of these proteins also seems to have been similar. At least for the the AAC and UCP it is known that both proteins form homodimers which can bind about 160 to 180 molecules of the detergent Triton X–100 (7, 8). This indicates that the major portion of the surface is hydrophobic. Also both molecules have half–site reactivity because they bind only one molecule inhibitor per homodimer. This common overall structure of both proteins is reflected also in the behaviour towards absorptive material during chromatography. Thus, both proteins do not absorbe hydroxylapatite in contrast to the majority of other mitochondrial proteins and can therefore be purified very easily (9, 10). The phosphate carrier cannot be purified in an undenatured state because it cannot be protected by these inhibitors towards the detergent influence.

STRUCTURE

By analysing the primary structure of these carriers, i.e. by diagon plots, a surprising similarity is found between three about equally long segments, each about 100 residue long (4, 5, 11, 12). This repeat structure is found in all three carriers. Notably there is not only a similarity within each carrier between three segments, but also a similarity of the segments between the various carriers. In fact, the inter–carrier similarity of the segments is more striking than the intra–carrier similarity. Thus the primary structure of the three carriers can be arranged by introducing deletions or insertions such that 9 repeats become aligned to a maximum of similarities. In this arrangement certain residues are conserved strikingly throughout the 9 domains. There are acidic residues in the relative position 30, of proline in position 28, and of glycine in position 72. In 8 out of 9 domains are conserved the basic residues in position 32, aromatic residues in position 91. In 7 out of 9 domains are conserved aromatic residues in position 50 and 70, and basic residues in position 34, 62, 71 and 96, and acidic residues in position 64. There are additional less frequent highly significant

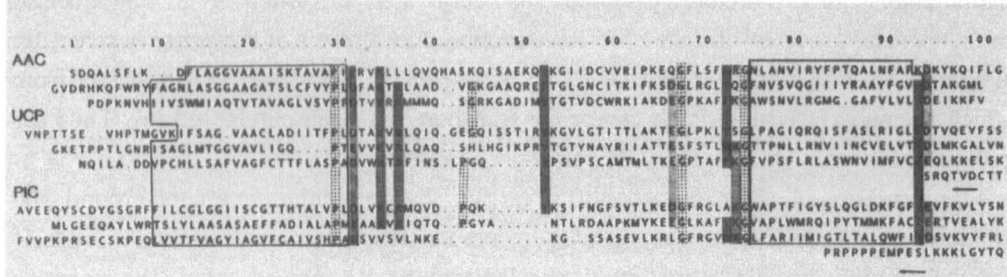

Fig. 2. Alignment of the amino acid sequences of the three mitochondrial carriers AAC, PIC and UCP according to the triplication of about 100—residue—long domains. The alignment searches for maximal similarity and maximal conservation of residues between the 9 domains of the three carriers by allowing for insertions and deletions.

conservations of charged aromatic or α—structure braking residues, such as glycine and proline (Fig. 2).

In general, the primary structure of membrane protein is analysed for transmembrane folding by elucidating hydrophobic α helices of about 20 residues length in these carriers. Simple hydrophobicity plots, however, do not allow a unique assignment of these stretches because of the wide distribution of hydrophilic residues. More successful was the use of a computer programme searching for amphipathic α and β stretches. In this way we could assign 6 amphipathic α and 1 β stretch (5). With the realization that the carriers are constructed out of 3 somewhat similar domains, a more compelling procedure for assigning the α—helical stretches became available. From the alignment of the triplicate structure we can conclude that the transmembrane α structures are conserved in a similar position within each domain. By the alignment of the 9 domains of the 3 different carriers a more significant assignment of the α—helical stretches is possible than merely by the hydropathic plots. Thus there appear to exist two transmembrane α structures in each domain out of which the first is somewhat more amphipathic or hydrophilic than the second. The two helical stretches are separated by a central hydrophilic region in each domain. Here most of the charged residues are concentrated. Further charged residues are found within the stretches which connect the three domains and in the N— and C—terminal.

It can be concluded that these carriers are basically constructed of three somewhat similar building blocks of each about 100 residue which span twice the membrane by two α—helical stretches. Each building block contains a central domain which probably is connected to the translocation channel. The 3 building blocks can be visualized to be arranged around a two— or threefold axis within each subunit (13). Two subunits again are arranged around the twofold symmetric axis transferring the membrane. It seems that these carriers have evolved by gene triplication from a common ancestor gene, coding for about 100 residues by gene triplication (Figs. 3, 4).

A Mitochondrial Carrier Family

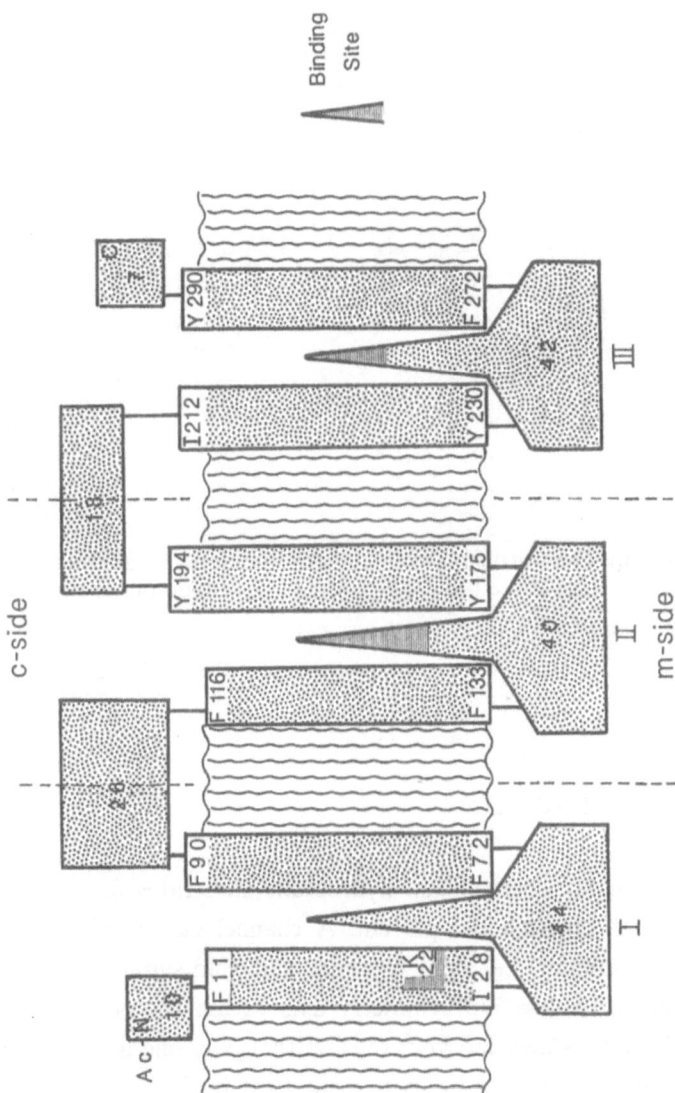

Fig. 3. Proposed secondary structure and folding arrangement of the AAC from bovine heart. Numbers refer to the length of the section or to the sequence location of the transmembrane α helices. The triplicate structure is indicated by the numbering of the three repeat domains of Fig. 2 with I, II and III.

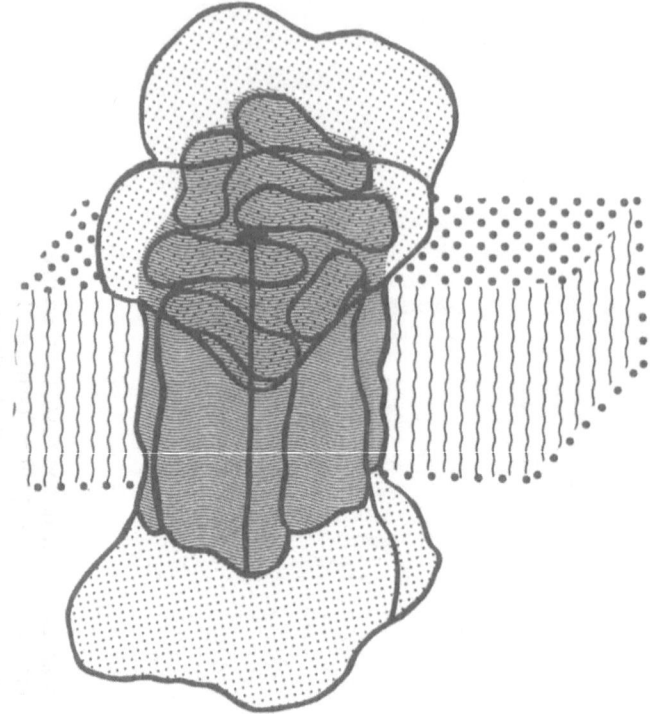

Fig. 4. A common model for the mitochondrial carrier. The dimer is arranged around a twofold axis. The construction of three building blocks of the monomer is indicated.

THE TRANSLOCATION CHANNEL

The transport of the highly charged and hydrophilic substrates for these carriers requires a hydrophilic path through the protein. A channel can be visualized to be formed either within each subunit, i.e. along a pseudo threefold symmetry axis formed by the 3 domains or along the twofold symmetry axis within the homodimer. Some evidence for the second choice is deduced from the fact that only one ligand can bind to the homodimer of the carrier. This can be most easily explained by assuming that the binding occurs at the twofold axis and thus inhibits the binding of the second ligand to the opposing subunit. The structure must not provide only a binding center, but also the gates around this binding center directed to both faces of the membrane.

Photoaffinity labeling studies using 2–azido ATP have shown that part of the binding center in the AAC is in the region around 162 to 165, i.e. in the central hydrophilic section of the second repeat domain. The assignment was more unique in yeast AAC where the equivalent region between 172 and 200 is labeled.

An important contribution to assigning the folding but also candidates for the translocation path came from probing the AAC with a lysine reagent. With this approach lysine groups were defined which are accessible both from the c–side and m–side in different states of the carrier gates (16). These groups are located in the central hydrophilic section of the second repeat domain. Other positive charges conspicuously localized within the second α–helical stretch of each domain are due to arginine. By site–directed mutagenesis at the yeast carrier it was found that replacement of the arginine in the first domain partially inactivates the AAC (unpublished results).

ISOFORMS OF THE CARRIERS

It was early noted for the AAC that antibodies raised against the bovine heart AAC did not crossreact with the AAC from liver, and only weakly from kidney. On the other hand the bovine heart AAC antisera did crossreact with the AAC from rat heart or other species (17, 18). This indicated that there is a larger organ than species specificity of the AAC. In more recent years it was shown that the human AAC can occur in three different isoforms (19, 20, 21). One of these differs from the other two by 18 %. Expression of different isoforms in different organs seems to be highly probable.

It was then more surprising that the AAC in yeast also occurs in two isoforms which, however, are quite differently expressed (22). In wild type only the AAC2 is expressed and the AAC1 gene appears to be silent. Only after delition of the AAC2 gene the AAC1 is weakly expressed. Also the AAC1 is less active than the AAC2. Thus in normal wild type yeast ADP/ATP transport completely depends on the AAC2. The role of the AAC1 is still unknown since AAC1 is also not expressed in other vegetative forms of the yeast such as in anaerobic or sporulating yeast cells. It can be expected that also for the other carriers isoform gene families exist. It will be highly interesting to see to what extent functional differences in these isoforms are geared to the peculiar intracellular demands on the ATP transport. For example the regulation of the selection by the AAC of ADP versus ATP by the membrane potential might be dependent on certain structural elements which are modified by subtle changes in the carrier structure.

CONCLUSIONS

Mitochondria are equipped with a specific and characteristic family of carriers which must have emerged concomitantly with the symbiosis of the mitochondria in the eukaryotic cell. These are uniquely mitochondrial components which cannot be expected to have any predecessors in prokaryos. In contrast, other components of oxidative phosphorylation in mitochondria have prokaryotic predecessors. In line with this arguing, there is no similarity in the structure of the mitochondrial carriers with the bacterial carriers for sugar or amino acids etc.(4). Because of the central rule of these carriers in eukaryotic metabolism there might have evolved structural subtleties adapted also to the peculiar demands of various issues.

The mitochondrial carriers have a comparatively simple structure as compared to other solute carrier in eukayotic bacterial membranes. In addition, an internal structural triplication may suggest that they have emerged from a unique ancestral gene which coded only for about 100 amino acids. Thus the mitochondrial carrier may provide a minimum structure required for solute transport through biomembranes. For these reasons they are favourable candidates for elucidating structure–functional relationships.

ACKNOWLEDGEMENT

This work was supported by a grant of the Deutsche Forschungsgemeinschaft Kl 134/24.

REFERENCES

1.   Klingenberg, M. (1976) in: The Enyzmes of Biological Enzymes: Membrane Transport, Martonosi A.N. (ed.), Plenum Publ. Corp., New York

2.   Bücher, T., Klingenberg, M. (1958) Angew. Chemie 70, 552–570

3.   Aquila, H., Misra, D., Eulitz, M., Klingenberg, M. (1982) Hoppe–Seylers Z. Physiol. Chemie 363, 345–349

4.   Aquila, H., Link, T.A., Klingenberg, M. (1987) FEBS Lett 212, 1–9

5.   Aquila, H., Link, T.A., Klingenberg, M. (1985) EMBO J. 4, 2369–2376

6.   Klingenberg, M. (1988) in: Molecular Basis of Biomembrane Transport, Palmieri, F. and Quagliariello (eds.), Elsevier, Amsterdam

7.   Hackenberg, H., Klingenberg, M. (1980) Biochemistry 19, 548–555

8.   Lin, C.S., Hackenberg, H., Klingenberg, M. (1980) FEBS Lett. 113, 304–306

9.   Riccio, P., Aquila, H., Klingenberg, M. (1975) FEBS Lett. 56, 133–138

10. Lin, C.S., Klingenberg, M. (1980) FEBS Lett. 113, 299–303

11. Saraste, M., Walker, J.E. (1982) FEBS Lett 144, 250–254

12. Runswick, M.J., Powell, S.J., Nyren, P., Walker, J.E. (1987) EMBO J. 6, 1367–1373

13. Klingenberg, M. (1989) Arch. Biochem. Biophys. 270, 1–14

14. Dalbon, P., Brandolin, G., Boulay, F., Hoppe, J., Vignais, P.V.(1988) Biochemistry 27, 5141–5149

15. Mayinger, P., Winkler, E., Klingenberg, M. (1989) FEBS Lett. 244, 421–426

16. Bogner, W., Aquila, H., Klingenberg, M. (1986) Eur. J. Biochem. 161, 611–620

17. Eiermann, W., Aquila, H., Klingenberg, M. (1977) FEBS Lett. 74, 209–214

18. Schultheiss, H.P., Klingenberg, M. (1984) Eur. J. Biochem. 143, 599–605

19. Battini, R., Ferrari, S., Kaczmarek, L., Calabretta, B., Chen, S., Baserga, R. (1987) J. Biol. Chem. 262, 4355–4359

20. Neckelmann, N., Li, K., Wade, R.P., Schuster, R., Wallace, D.C. (1987) Proc. Natl. Acad. Sci. USA 84, 7580–7584

21. Houldsworth, J., Attardi, G. (1988) Proc. Natl. Acad. Sci. USA 84, 377–381

22. Lawson, J.E., Douglas, M.G. (1988) J. Biol. Chem. 263, 14812–14818

THE CALCIUM PUMP OF THE PLASMA MEMBRANE: STRUCTURE/ FUNCTION RELATIONSHIPS

E. Carafoli, P. James, E. Zvaritch*, and E.E. Strehler

Laboratory of Biochemistry, Swiss Federal Institute of Technology (ETH), 8092 Zurich, Switzerland, and

*Permanent Address: Shemyakin Institute of Bioorganic Chemistry, USSR Academy of Sciences, 117 871 GSP Moscow, USSR

SUMMARY

The plasma membrane Ca pump is a P-type ATPase which shares with the other pumps of this type the essential mechanistic properties. At variance with the other pumps of this group, it is activated by calmodulin. It is a single polypeptide chain of molecular mass about 134 kDa, which has now been cloned in both rat and human tissues. It contains ten putative transmembrane domains, and protrudes with most of its mass into the intracellular ambient. The calmodulin binding domain is located next to the C-terminus. Several isoforms of the pump have now been described, differing particularly in the (regulatory) C-terminal portion. In the absence of calmodulin the pump can still be activated by alternative treatments, e.g., the exposure to acidic phospholipids (or polyunsaturated fatty acids), a phosphorylation mediated by the cAMP dependent protein kinase, or a controlled proteolytic treatment, e.g., with trypsin. The latter protease degrades the pump in sequence to fragments of molecular mass 90, 85, 81, and 76 kDa. The 90 to 85 to 81 kDa transition gradually impairs the response to calmodulin and then its binding to the pump. The transition from 81 to 76 kDa abolishes also the response to acidic phospholipids. Trypsin cuts the pump in and around the calmodulin binding domain in producing the fragments of 90 to 81 kDa, between putative transmembrane domains two and three in producing the fragment of 76 kDa.

INTRODUCTION

The plasma membrane Ca-ATPase (see 1, 2 for general reviews) is one of the P-type ion pumps (3, 4). Its properties may be summarized as follows: Its molecular mass is about 134 kDa, it forms an aspartyl phosphate during the reaction cycle and as a result is inhibited by vanadate. A simplified reaction scheme of the pump is shown in Figure 1. The ATPase is a proton exchanger, but it is not established whether the Ca/ATP ratio is 2:1 or 1:1. It exists in two states of calcium affinity, low and high, the latter resulting from the interaction with calmodulin: this property has been exploited for purifying the enzyme on calmodulin affinity columns (5) (Figure 2).

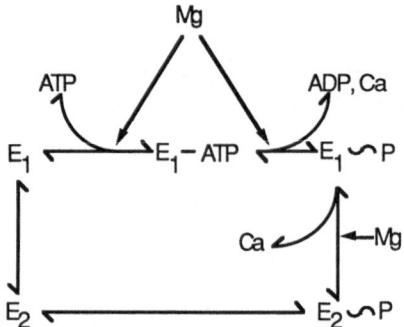

Figure 1. The reaction cycle of the plasma membrane Ca^{2+} pump. The pump is visualized to exist in two different conformational states E_1 and E_2, the Ca^{2+} translocating step probably occurring during the $E_1 \sim P \rightarrow E_2 \sim P$ transition. Mg influences both the formation and the degradation of the phosphorylated intermediate.

RESULTS AND DISCUSSION

To isolate the enzyme, plasma membranes from erythrocytes (5) and also from a variety of other cells (e.g., heart, 6) were treated with EDTA to remove the calmodulin bound to the enzyme, solubilized with Triton X-100 and applied to a calmodulin column. Proteins not bound to calmodulin were removed by washing the column with a Ca-containing buffer, followed by an EDTA elution that removed the ATPase. The purified protein had an estimated molecular mass on Na-dodenyl-sulphate (SDS) gels of approximately 138 kDa and was found to be functional, i.e., it retained calmodulin sensitivity, and could be reconstituted in liposomes as a Ca-transporting system (7). The purified enzyme was also shown to be activated by a number of treatments alternative to calmodulin, among them the exposure to acidic phospholipids (or polyunsaturated fatty acids), controlled proteolytic treatments and a cAMP-dependent phosphorylation.

The activation by proteolysis has been used to map functional domains in the pump (8, 9). Figure 3 shows that chymotrypsin, after a transient and

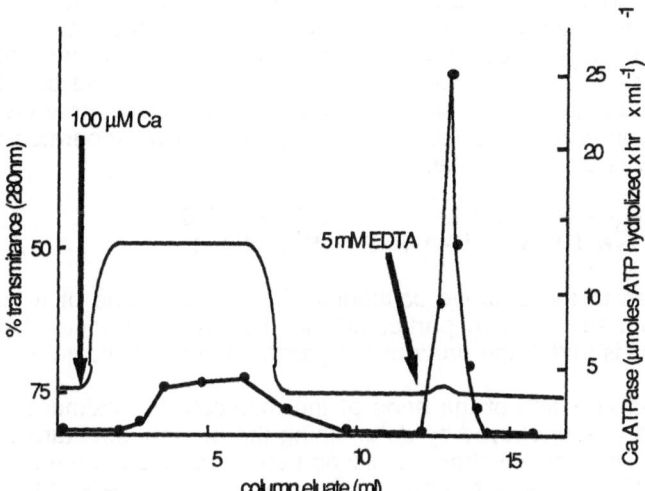

Figure 2. The purification of the Ca^{2+} pump from the erythrocyte membrane using a calmodulin column (modified from ref. 5).

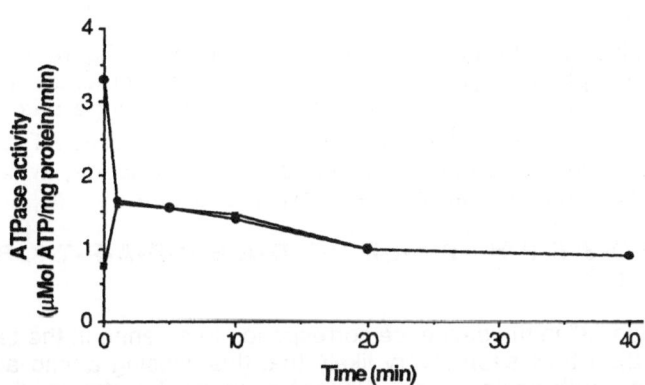

Figure 3. Chymotrypsin proteolysis of the purified human erythrocyte Ca pump. The experimental conditions are described in (9). Chymotrypsin (1:10 weight ratio to the ATPase) was applied for the times indicated.

incomplete stimulation, caused the enzyme to become calmodulin-insensitive after about 1 min (9). During this time the enzyme became degraded into many fragments ranging in molecular mass from about 120 to 15 kDa. In gel overlay experiments calmodulin bound only to a fragment of about 12 kDa. Since the higher molecular mass enzyme fragments failed to bind calmodulin, chymotrypsin evidently removed the calmodulin-binding domain from one of the ends of the ATPase molecule and not from its middle. CNBr cleavage of the purified pump labeled with a radioactive, cleavable, photoactivatable, bifunctional cross-linker bound to calmodulin produced a labeled fragment which was separated by HPLC and sequenced. The structure obtained:

E-L-R-R-G-Q-I-L-W-F-L-G-R-N-R-I-Q-
T-K-I-K-V-V-N-A-F-S-S-S-L-H-E-F

shows homologies to the putative calmodulin binding domains of a number of calmodulin-modulated proteins, particularly in the predominance of positively charged amino acids and in the propensity to form an amphiphilic helix.

The question at which of the ends of the molecule the calmodulin-binding domain is located was answered by determining the primary structure of the Ca-ATPase (10, 11). This has now been achieved both in rat and human tissues. The sequence of the Ca-ATPase determined in human teratoma cells (11) had 1220 amino acid residues, corresponding to a molecular weight of ~134 kDa, and is shown in Figure 4.

The sequence shows that the calmodulin-binding domain is located next to the C-terminus. The aspartic acid which forms the phosphoenzyme during the reaction cycle (aspartic acid 475) and the lysine which binds the ATP antagonist fluorescein isothiocyanate (FITC) (lysine 601) are shown by a circle. Ten putative transmembrane helices, identified with the usual Kyte-Doolittle algorithm, are shown boxed.

On the C-terminal side of the calmodulin-binding domain Figure 4 shows a serine that undergoes phosphorylation by the cAMP-dependent kinase (12, 13). The identification of this serine was achieved by labeling a large batch of the purified erythrocyte ATPase with γ^{32}P-ATP and splitting the labelled ATPase with CNBr (14). HPLC analysis identified a single labeled peptide, which was subjected to sequence analysis yielding the following structure:

T-H-P-E-F-R-I-E-D-S-E-P-H-I-P-L-I-D-D-T-D-A-E-D-D-A-P-T-K-R-N-S-
[X]-P-P-P-S-P-D-K-N

The gap labeled "X" in the sequence corresponds to a serine in the teratoma sequence (Figure 2): it thus seems very likely that this missing amino acid is a serine also in the erythrocyte pump. Phosphorylation by the cyclic AMP-dependent kinase increases the affinity of the pump for Ca (Figure 5). The activation is less complete than with calmodulin and is somehow competed for by the latter: possibly when calmodulin is bound to its site a steric hindrance prevents the kinase from interacting with its substrate site efficiently.

A comparison of the peptides obtained by proteolytic fragmentation of the erythrocyte Ca-ATPase with the deduced sequence of the teratoma enzyme indicates the existence of different isoforms in the two tissues. Two isoforms have

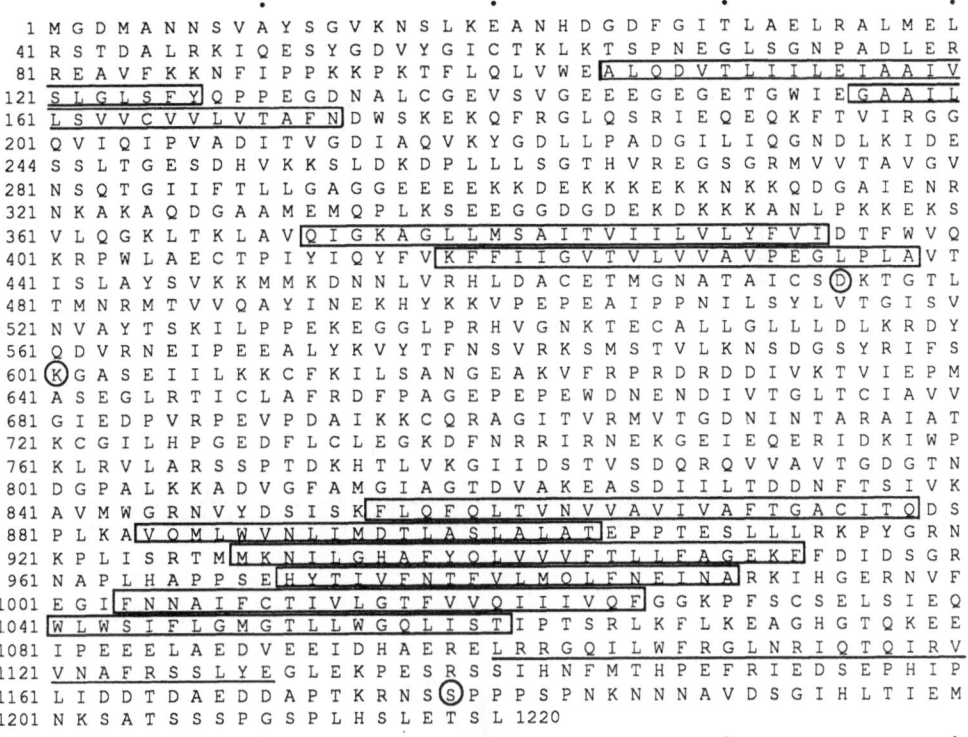

```
   1 M G D M A N N S V A Y S G V K N S L K E A N H D G D F G I T L A E L R A L M E L
  41 R S T D A L R K I Q E S Y G D V Y G I C T K L K T S P N E G L S G N P A D L E R
  81 R E A V F K K N F I P P K K P K T F L Q L V W E A L Q D V T L I I L E I A A I V
 121 S L G L S F Y Q P P E G D N A L C G E V S V G E E E G E G E T G W I E G A A I L
 161 L S V V C V V L V T A F N D W S K E K Q F R G L Q S R I E Q E Q K F T V I R G G
 201 Q V I Q I P V A D I T V G D I A Q V K Y G D L L P A D G I L I Q G N D L K I D E
 244 S S L T G E S D H V K K S L D K D P L L L S G T H V R E G S G R M V V T A V G V
 281 N S Q T G I I F T L L G A G G E E E E K K D E K K K E K K N K K Q D G A I E N R
 321 N K A K A Q D G A A M E M Q P L K S E E G G D G D E K D K K K A N L P K K E K S
 361 V L Q G K L T K L A V Q I G K A G L L M S A I T V I I L V L Y F V I D T F W V Q
 401 K R P W L A E C T P I Y I Q Y F V K F F I I G V T V L V V A V P E G L P L A V T
 441 I S L A Y S V K K M M K D N N L V R H L D A C E T M G N A T A I C S D K T G T L
 481 T M N R M T V V Q A Y I N E K H Y K K V P E P E A I P P N I L S Y L V T G I S V
 521 N V A Y T S K I L P P E K E G G L P R H V G N K T E C A L L G L L L D L K R D Y
 561 Q D V R N E I P E E A L Y K V Y T F N S V R K S M S T V L K N S D G S Y R I F S
 601 K G A S E I I L K K C F K I L S A N G E A K V F R P R D R D D I V K T V I E P M
 641 A S E G L R T I C L A F R D F P A G E P E P E W D N E N D I V T G L T C I A V V
 681 G I E D P V R P E V P D A I K K C Q R A G I T V R M V T G D N I N T A R A I A T
 721 K C G I L H P G E D F L C L E G K D F N R R I R N E K G E I E Q E R I D K I W P
 761 K L R V L A R S S P T D K H T L V K G I I D S T V S D Q R Q V V A V T G D G T N
 801 D G P A L K K A D V G F A M G I A G T D V A K E A S D I I L T D D N F T S I V K
 841 A V M W G R N V Y D S I S K F L Q F Q L T V N V V A I V A F T G A C I T Q D S
 881 P L K A V Q M L W V N L I M D T L A S L A L A T E P P T E S L L L R K P Y G R N
 921 K P L I S R T M M K N I L G H A F Y Q L V V V F T L L F A G E K F F D I D S G R
 961 N A P L H A P P S E H Y T I V F N T F V L M Q L F N E I N A R K I H G E R N V F
1001 E G I F N N A I F C T I V L G T F V V Q I I I V Q F G G K P F S C S E L S I E Q
1041 W L W S I F L G M G T L L W G Q L I S T I P T S R L K F L K E A G H G T Q K E E
1081 I P E E E L A E D V E E I D H A E R E L R R G Q I L W F R G L N R I Q T Q I R V
1121 V N A F R S S L Y E G L E K P E S R S S I H N F M T H P E F R I E D S E P H I P
1161 L I D D T D A E D D A P T K R N S S P P P S P N K N N N A V D S G I H L T I E M
1201 N K S A T S S S P G S P L H S L E T S L 1220
```

Figure 4. Primary structure of the plasma membrane (human teratoma) Ca pump. The domain assignments are described by Verma et al (11). The pump contains 1220 amino acids, corresponding to a molecular mass of 134,363. The aspartic acid which forms the phosphoenzyme, the lysine which binds FITC, and the serine which is phosphorylated by the cAMP-dependent kinase are circled. The calmodulin binding domain (residues 1100-1130) is underlined. The 10 putative transmembrane domains are boxed, two stretches (residues 22-33 and 296-321) show homologies to the EF-type Ca-binding loops of proteins like parvalbumin (11).

also been deduced from cDNA in rat brain (10). Isoforms exist also in other plasma membranes: by screening a fetal human skeletal muscle cDNA library with a probe from the teratoma sequence clones were detained whose sequences differed from that of the teratoma cDNA in a region that follows the codon for amino acid 1117, in the calmodulin-binding domain (15). The difference consists of the insertion of 87 bp, (5 clones) or 114 bp (1 clone) after the codon for this amino acid. The calmodulin-binding domain of these two isoforms would thus be altered, probably affecting calmodulin reactivity. The two human skeletal muscle isoforms arise from alternative splicing of RNA involving a C-terminal exon, showing that more than one splicing mode occurs: one of the two isoforms characterized in rat brain (10) has a longer insert in this region (154 bp) corresponding to the full exon. More recently, a second isoform, in addition to the one typical of human teratoma cells, has been found to be expressed in human red cells (16), and the sequence of yet another isoform has been deduced from cDNA in rat brain (17).

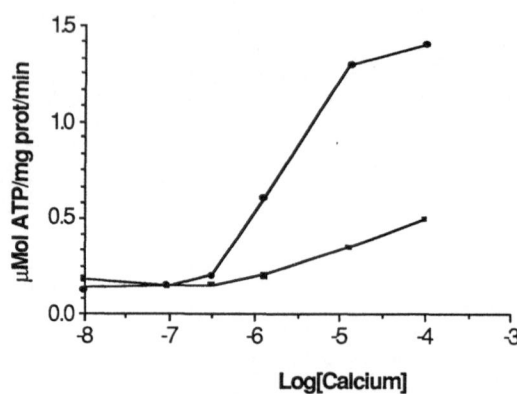

Figure 5. Activation of the purified plasma membrane calcium pump by cAMP-dependent phosphorylation. The experiment describes the activation of the pump from erythrocytes (9).

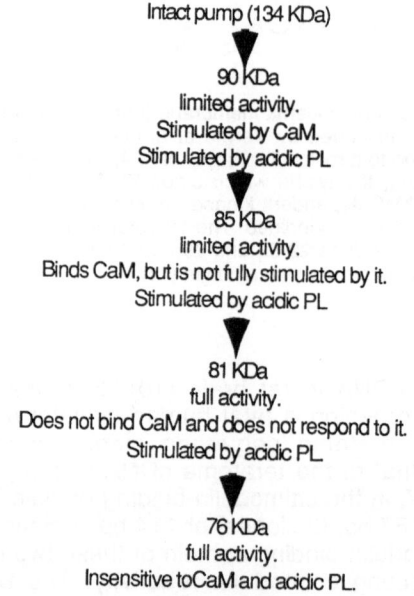

Figure 6. A scheme describing the properties of the main fragments of the pump produced by trypsin (8, 18-20). PL = acid phospholipids

In activating the purified ATPase, trypsin progressively reduces its mass to fragments of 90, 85, 81, 76 kDa (8, 18, 19, 20). The first fragment is rapidly transient, the others are relatively more stable, and can be made to accumulate by performing the proteolysis under special conditions (18, 19). Preparations enriched in these fragments display a modified response to calmodulin and acidic phospholipids, as summarized in Figure 6. Recent work in which the C- and N-termini of these fragments have been determined (21) have now permitted to locate them within the primary structure of the enzyme molecule (Figure 7). In producing all these fragments trypsin cleaves the enzyme on the cytoplasmic side, and progressively removes portions from its C-terminus. In producing the 90 kDa fragment the portion removed is C-terminal to the calmodulin binding domain, explaining why the fragment has normal binding and response to calmodulin. The C-terminal fragment of the 85 kDa fragment is within the calmodulin binding domain, in line with the finding that this fragment still binds calmodulin, but reacts incompletely to it. The C -termini of the 81 and 76 kDa fragments coincide, and are N-terminal to the calmodulin binding domain: this explains why these fragments don't bind calmodulin, and shows that their different response to acidic phospholipids is evidently related to their N-terminus. The N-termini of the 90, 85, and 81 kDa fragments are located between transmembrane helices two and three, that of the 76 kDa fragment 44 residues downstream: thus, this highly charged segment of 44 amino acids is involved in the response to acidic phospholipids. Alignment studies of the sequence of the erythrocyte Ca pump with those of other P-type ion pumps shows that this domain is only present in the former, rationalizing the special sensitivity of the plasma membrane Ca^{2+}

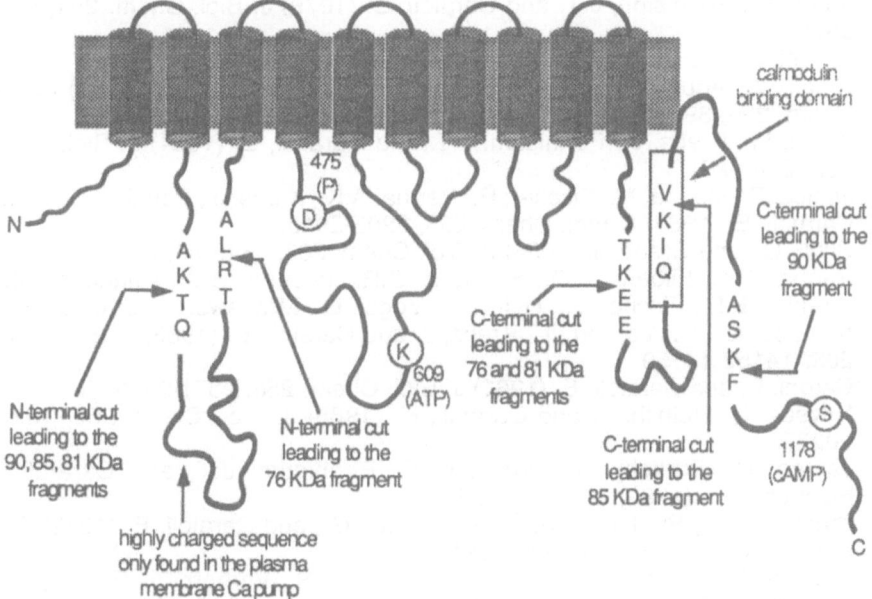

Figure 7. N- and C-termini of the main fragments of the pump produced by trypsin. The positions of the phosphoenzyme-forming aspartic acid, of the FITC-binding lysine, of the calmodulin binding domain, and of the serine which is phosphorylated by the cAMP-dependent kinase are indicated.

pump to acidic phospholipids. One important corollary of these proteolytic studies is whether the product N-terminal to all these fragments, i.e., the sequence from the N-terminal of the pump to the cleavage points (s) between transmembrane helices two and three (which is clearly visible as a separate entity in SDS polyacrylamide gels) also becomes separated under nondenaturing conditions. While there is no doubt that the segments of the pump located downstream of the C-terminal cuts produced by trypsin in or around the calmodulin binding domain become separated (see the trypsin studies on the pump in vivo, 22), the N-terminal portions of the pump molecule probably remain associated with the larger fragments. This is indicated by experiments aimed at producing the fragments in the isolated state. It is also in line with the conservation of some of the N-terminal portion of the structure among different P-type ion pumps, a finding suggesting that they may be important to function. If the 90-85, 81, and 76 kDa fragments do not separate from the N-terminal portion of the pump under non-denaturing conditions, the different phospholipid reactivity of the two latter fragments evidently reflects the structural derangement of the highly charged sequence of 44 amino acids, rather than its removal from the pump molecule.

REFERENCES

1. Schatzmann, H.J. (1982) In: "Membrane Transport of Calcium," E. Carafoli, ed., Acad. Press, London.
2. Carafoli, E, Pyshiological Rev. (1990) in press.
3. Pedersen, P.L. and Carafoli, E. (1987) Trends in Biochem. Sci. 12, 146-160
4. Pedersen, P.L. and Carafoli, E. (1987) Trends Biochem. Sci. 12, 186-189.
5. Niggli, V., Penniston, J.T. and Carafoli, E. (1979) J. Biol. Chem. 254, 9955-9998.
6. Caroni, P. and Carafoli, E. (1981) J. Biol. Chem. 256, 3263-3270.
7. Niggli, V., Adunyah, E.S., Penniston, J.T. and Carafoli, E. (1981) J. Biol. Chem. 256, 395-401.
8. Zurini, M., Krebs, J., Penniston, J.T. and Carafoli, E. (1984) J. Biol. Chem. 259, 618-627.
9. James, P., Maeda, M., Fischer, R., Verma, A.K., Krebs, J., Penniston, J.T. and Carafoli, E. (1980) J. Biol. Chem. 263, 2905-2910.
10. Shull, G., and Greeb, J. (1988) J. Biol. Chem. 263, 8646-8657.
11. Verma, A.K., Filoteo, A.G., Stanford, D.R., Wieben, E.D., Penniston, J.T., Strehler, E.E., Fischer, R., Heim, R., Vogel, G., Mathews, S., Strehler-Page, M.A., James, P., Vorherr, T., Krebs, J. and Carafoli, E. (1988) J. Biol. Chem. 263, 14152-14159.
12. Caroni, P. and Carafoli, E. (1981) J. Biol. Chem. 256, 9371-9373.
13. Neyses, L., Reinlib, L. and Carafoli, E. (1985) J. Biol. Chem. 260, 10283-10287.
14. James, P., Pruschy, M., Vorherr, T., Penniston, J.T. and Carafoli, E. Biochemistry. (1989) 24, 4253-4258.
15. Strehler, E.E., Strehler-Page, M.-A., Vogel, G., and Carafoli, E. (1989) Proc. Natl. Acad. Sci. USA. 86, 6908-6912.
16. Strehler, E.E. , James, P., Fischer, R., Heim, R., Vorherr, T., Filoteo, A.G., Penniston, J.T. and Carafoli, E. (1990) J. Biol. Chem. 265, 2835-2842.
17. Greeb, J., and Shull, G., (1989) J. Biol. Chem. 18569-18576.
18. Benaim, G., Zurini, M., and Carafoli, E. (1984) J. Biol. Chem. 259, 8471-8477.
19. Benaim, G., Clark, A., and Carafoli, E. (1986) Cell Calcium. 4, 175-186.
20. Enyedi .A., Flura, M., Sarkadi, B., Gardos, G., and Carafoli, E. (1987) J. Biol. Chem. 262, 6425-6430.
21. Zvaritch, E., James, P., Vorherr, T., Falchetto, R., Modyanov, N., and Carafoli. E. Biochemistry. (submitted).

Part I. 3. Other Related Topics

THE 9-kDa POLYPEPTIDE WITH IRON-SULFUR CENTERS A/B IN SPINACH PHOTOSYSTEM I WITH SPECIAL REFERENCE TO ITS STRUCTURE AND TOPOGRAPHIC CONSIDERATION IN THYLAKOID MEMBRANE

Hirozo Oh-oka, Yasuhiro Takahashi and
Hiroshi Matsubara

Department of Biology, Faculty of Science
Osaka University, Toyonaka, Osaka 560, Japan

SUMMARY

The 9-kDa polypeptide in the photosystem I (PS I) complex is iron-sulfur protein carrying centers A and B, whose sequence has the typical distribution of cysteine residues found in bacterial-type ferredoxins, Cys-X-X-Cys-X-X-Cys-X-X-X-Cys-Pro, in the two distinct regions. The 9-kDa polypeptide has been successfully isolated with iron-sulfur clusters under anaerobic conditions. It contains 8.5 atoms of non-heme iron and 8.0 atoms of inorganic sulfide per mol, and shows an absorption spectrum similar to those of bacterial-type ferredoxins. Topological studies of the 9-kDa polypeptide has been conducted by examining the results of alkaline and chaotropic ion treatments, tryptic digestion, and cross-linking of thylakoid membranes supplemented with immunoblotting techniques. It appears that the 9-, 14- and 19-kDa polypeptides in the PS I complex are peripheral proteins situated in close to each other on the stromal side of the membranes. The 9-kDa polypeptide with centers A and B is stable within a specific environment, in which the polypeptide is embedded under the two other subunits, the 14- and 19-kDa polypeptides.

INTRODUCTION

The photosystem I (PS I) complex is a supramolecular complex which has a transbilayer organization in thylakoid membranes (1). It is composed of at least 7 subunits in higher plants, namely, two core subunits (59- and 63-kDa polypeptides) and 5 small subunits (9-, 10-, 14-, 16- and 19-kDa polypeptides) (2,3). This complex accomplishes the efficient conversion of quantum energy to the ultimate chemical product, NADPH, through the transfer of electrons from reduced plastocyanin to ferredoxin. In a series of electron transfer reactions, 6 intrinsic redox centers (P700, A_0, A_1, X, A and B) are involved (4), and vigorous efforts have been focused on the identification of the respective chemical species and subunits, particularly in regard to the iron-sulfur centers X, A and B (3,5). The topology of these subunits in the PS I complex has been also explored by proteolytic digestion, chemical labeling, and cross-linking studies (6,7). However, the information obtained was insufficient to permit the elucidation of the interrelationships among subunits in the PS I complex and, thereby, an understanding of the mechanism of electron transfer in the PS I complex.

Bioenergetics, Edited by C. H. Kim and
T. Ozawa, Plenum Press, New York, 1990

We have studied, for the past few years, on characterizing some properties of the 9-kDa polypeptide (8,9), which is now considered to carry centers A and B. This iron-sulfur protein is soluble in saline and is very sensitive to the oxygen when dissociated from the PS I complex. It must exist in a stable and bound (or buried) form *in situ* if it is to function as an electron carrier, and it probably forms a stable complex with other protein subunits embedded in the membranes. Therefore, we explored the topography of the 9-kDa polypeptide, including an analysis of interactions with other subunits in the PS I complex, by using three different approaches, namely, alkaline and chaotropic ion treatments, tryptic digestion and cross-linking of thylakoid membranes. In this paper, we demonstrate some structural and spectrophotometric properties of the 9-kDa polypeptide, and present a topological model in which the 9-, 14- and 19-kDa polypeptides in the PS I complex of spinach are arranged on the basis of the results of the various treatments of thylakoid membranes mentioned above.

MATERIALS AND METHODS

Preparation of Intact Thylakoid Membranes

Intact chloroplasts were obtained from spinach leaves essentially by the method of Mills and Joy (10). After suspension of chloeoplasts in 20 mM Tricine-KOH (ph 7.9) to lyse the chloroplast envelope, intact thylakoid membranes were collected by centrifugation at 12,000 xg for 5 min. Membranes were washed with 20 mM Tricine-KOH (pH 7.9) containing 10 mM NaCl and 100 mM sorbitol, and they were then suspended in the same buffer. For the cross-linking experiments, 25 mM K-phosphate (pH 7.4) was used as a buffer solution instead of 20 mM Tricine-KOH (pH 7.9).

Various Treatments of Thylakoid Membranes

Alkaline and Chaotropic Ion Treatments. Thylakoid membranes were first washed with distilled water. Treatment with 0.1 N NaOH was carried out according to the procedure described by Piccioni et et al.(11). Chaotropic ion treatment was peformed as follows. Thylakoid membranes were suspended in 0.1 M Tris-HCl (pH 8.0) at a concentration of chlorophyll (Chl) of 400 μg/ml, and an equal volume of a solution of 2 or 4 M Na-thiocyanate (NaSCN) in 0.1 M Tris-HCl (pH 8.0) was added to this suspension. After gentle stirring for 30 min at 4°C, the mixture was diluted with an equal volume of distilled water and centrifuged at 12,000 xg for 10 min. The suspernatant was dialyzed against distilled water and lyophilized. The pellet was washed with distilled water.

Tryptic Digestion. Intact thylakoid membranes (1 mg Chl/ml) were incubated with trypsin (100 μg/mg Chl/ml) in 20 mM Tricine-KOH (pH 7.9) buffer containing 10 mM NaCl and 100 mM sorbitol, at 4°C for 10 and 60 min. The reaction was terminated by the addition of phenylmethylsulfonyl fluoride (PMSF) to a final concentration of 5 mM, and the reaction mixture was left on ice for 60 min. After a 5-fold dilution with the same Tricine-KOH buffer, membranes were collected by centrifugation at 12,000 xg for 10 min, and solubilized in the sample buffer for electrophoresis. Disrupted membranes were prepared by treating intact thylakoid membranes (1 mg Chl/ml) with 0.2% (final concentration) Triton X-100 for 30 min at 4°C, and tryptic digestion of these membranes was carried out in the same way as that of intact membranes. However, the sample buffer for electrophoresis was added directly to the reaction mixture after termination of the digestion.

Chemical Cross-linking Experiments. The stock solutions of cross-linking reagents of 400 mM ethyleneglycolbis(succinimidylsuccinate) (EGS), 4% tolylene-2,4-diisocyanate (TDIC), and 4% hexamethylenediisocyanate (HMDIC) were freshly prepared by dissolving the compounds in dried N,N-dimethylformamide. Intact thylakoid membranes (0.6 mg Chl/ml) were suspended in 25 mM K-phosphate (pH 7.4) buffer containing 10 mM NaCl, 2mM $MgCl_2$ and 100 mM sorbitol, and they were then treated with 4 mM EGS, 0.04% TDIC or 0.04% HMDIC for 5 min at 24°C. The reaction was quenched by addition of a solution of glycine to a final concentration of 0.2 M. After a 5-fold dilution of the reaction mixture with the same buffer, membranes were collected by centrifugation at 12,000 xg for 10 min. All the above procedures were carried out at 4°C unless otherwise stated.

Preparation of Antibodies against the 9-, 14- and 19-kDa Polypeptide

The 9- and 19-kDa polypeptide were purified by preparative SDS-PAGE from the n-butanol extract of the PS I complex. Each purified polypeptide was used as an antigen to raise polyclonal antibodies in New Zealand white rabbits. A mixture of antibodies against the 14- and 19-kDa polypeptides was also obtained after injection of the whole PS I complex into a rabbit. Purification of antibodies by an affinity method was achieved according to the protocol described by Kelly et al.(12).

RESULTS AND DISCUSSION

Structural Studies of the 9-kDa Polypeptide

The complete amino acid sequence of the 9-kDa polypeptide was determined by using a combination of gas-phase sequencer and conventional procedures (8). The total number of residues was 80 and the molecular weight was calculated to be 8,894. The correctness of the polypeptide sequence is supported by a comparison with those deduced from the chloroplast DNA sequences of liverwort *frxA* and tobacco *psaC* genes (Fig. 1) (13, 14). They are very homologous to each other: 7 amino acid replacements between the spinach and liverwort sequences and only one between the spinach and tobacco sequences. Therefore, the 9-kDa polypeptide isolated from spinach PS I complex should be encoded by a corresponding gene in spinach chloroplasts. The distribution of cysteine residues typically seen in the bacterial-type ferredoxins, Cys-X-X-Cys-X-X-Cys-X-X-X-Cys-Pro, was present in the two regions at 10-21 and 47-58 residues of spinach polypeptide, indicating that this polypeptide is an iron-sulfur protein capable of carrying 2 [4Fe-4S] clusters with a redox function in the PS I complex. Since one of the three iron-sulfur centers in PS I complex, center X, is associated with the core subunits (15), the 9-kDa polypeptide probably forms center(s) A and/or B.

The amino acid sequence of the 9-kDa polypeptide was compared with those of two other bacterial-type ferredoxins (16, 17), as shown in Fig. 1. This leads to an insertion of 8 amino acid residues in the middle region of the chloroplast polypeptide and an extension of the C-terminal region as compared with bacterial-type ferredoxins. As far as the sequence of the 9-kDa polypeptide is concerned, it is fairly hydrophilic as a whole. In fact, it was extractable from the membranes and became

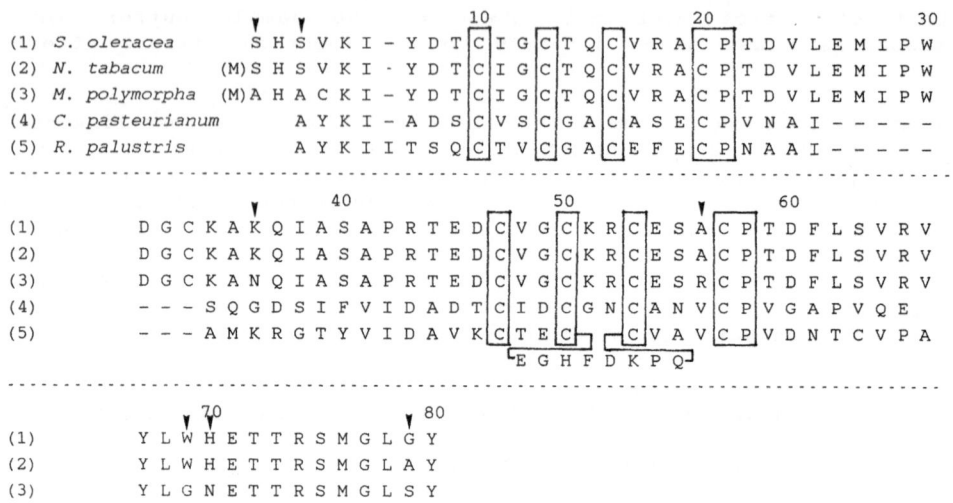

Fig. 1. Comparison of the sequences of the 9-kDa polypeptides ((1)-(3)) and those of bacterial-type ferredoxins ((4),(5)). Several gaps are inserted to make the alignment homologous. The typical cysteine and proline residues are framed. Arrowheads ▼ show amino acid replacements among the 9-kDa polypeptides. (1), *Spinacia oleracea* (8); (2), *Nicotiana tabacum* (14); (3), *Marchantia polymorpha* (13); (4), *Clostridium pasteurianum* (16); (5), *Rhodopseudomonas palustris* (17).

soluble in a saline solution (see below). However, it is noteworthy that there are hydrophobic amino acid residues in those insertion and extension regions, namely, -Met-Ile-Pro-Trp- and -Tyr-Leu-Trp-X-X-X-X-X-X-Met-Gly-Leu-Gly-Tyr, respectively, and these regions may interact with thylakoid membranes or other subunits of PS I complex.

Absorption Spectrum of the 9-kDa Polypeptide with Iron-Sulfur cluster(s)

The 9-kDa polypeptide was isolated with iron-sulfur cluster(s) from thylakoid membranes of spinach essentially according to the method of Wynn and Malkin (18), except that all the manipulations were performed in anaerobic systems. This led a reproducible result and high yield of the preparation. Figure 2 shows the absorption spectrum of the 9-kDa polypeptide, which is similar to those of bacterial-type ferredoxins. But the absorbance in the visible region changed rapidly when it was exposed to the air (data not shown). The ratio of the absorbance at 390 nm to that at 280 nm was 0.6, and the molar extinction coefficient at 390 nm was calculated to be 32,000 $M^{-1}cm^{-1}$ from estimation of its concentration by amino acid analysis. The analysis for contents of non-heme iron and inorganic sulfide in this protein gave 8.5 and 8.0 atoms per mol, respectively. We, therefore, consider that this iron-sulfur protein has two [4Fe-4S] clusters.

Topography of Subunits around the 9-kDa Polypeptide

All of the present results involved analysis by immunoblotting (19) after SDS-PAGE performed in the buffer system of Schägger et al.(20). The 33-kDa protein and the 19-kDa polypeptide had relative mobilities corresponding to molecular masses of 30 kDa and 21 kDa, respectively, in

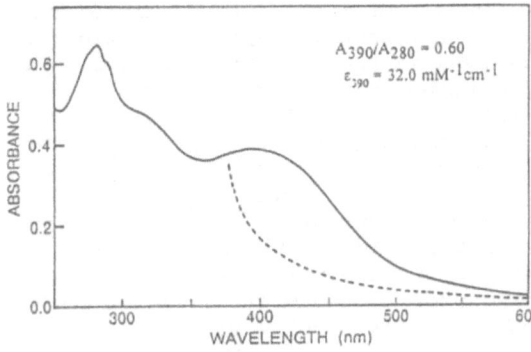

Fig. 2. Absorption spectra of the 9-kDa polypeptide with iron-sulfur clusters. Reduction of the polypeptide was carried out by the addition of a small amount of solid sodium dithionite (----).

this buffer system. Our 14- and 19-kDa polypeptides correspond to the subunits III and II named by Münch et al.(21), respectively, on the basis of the close similarities in their amino acid compositions and identical amino-terminal sequences (8).

Figure 3 shows the effects of alkaline and chaotropic ion (SCN⁻) treatments of thylakoid membranes on the extraction of the 9-, 14- and 19-kDa polypeptides. Although the three polypeptides were not extracted in the soluble fraction after treatment with 1 M NaSCN, they were recovered in the soluble fraction after treatment with 0.1 N NaOH or 2 M NaSCN. These polypeptides, therefore, appeared to be peripheral polypeptides (11, 22), as anticipated from the amino acid sequences of the three polypeptides which were hydrophilic as a whole and included no hydrophobic region typical of membrane-spanning proteins (8, 21, 23).

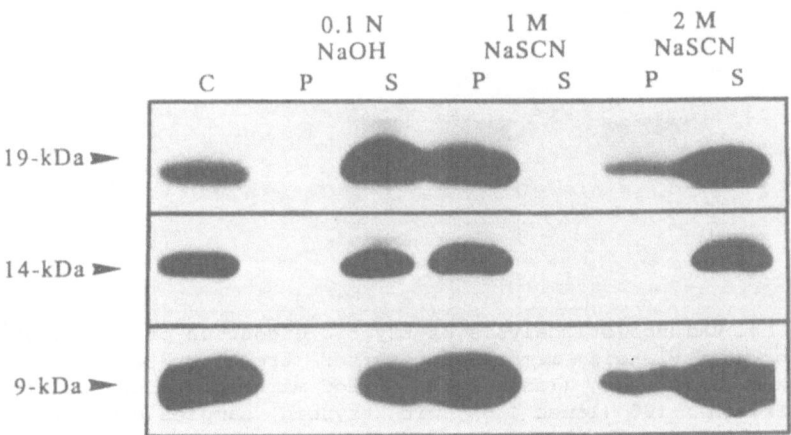

Fig. 3. Immunoblot analysis of the effects of alkaline and chaotropic iontreatments of thylakoid membranes on the extraction of the 9-, 14- and 19-kDa polypeptides. C, P and S refer to control, pellet and supernatant, respectively. Samples equivalent to 10 μg Chl were applied to each lane.

To probe further the location of these 9-, 14- and 19-kDa polypeptides in thylakoid membranes, we carried out tryptic digestion of the membranes. The 33-kDa protein involved in the evolution of oxygen in photosystem II particles was used as a marker protein that resides on the luminal side of thylakoid membranes (24). Since the 33-kDa protein in the untreated intact thylakoid membranes was not digested with trypsin after 60 min at 4 °C (Fig. 4, A, lane 3), the integrity of the membranes was held in our experimental conditions. After tryptic digestion of untreated intact thylakoids, the 19-kDa was partially degraded with removal by cleavage of a fragment of about 2 kDa. By contrast, the 14-kDa polypeptide was degraded almost completely and a fragment of about 10 kDa could be seen in trace amounts. These profiles were the same as those obtained by tryptic digestion of thylakoids pre-treated with 0.2% Triton X-100. The pre-treatment was sufficient to destroy the integrity of the membranes to the extent that the 33-kDa protein was completely digested (Fig. 4, A, lanes 4-6). Therefore, it is reasonable to consider that both the 14- and 19-kDa polypeptides exist on the stromal side of thylakoid membranes and do not traverse them. However, the 9-kDa polypeptide, being now considered to be an iron-sulfur protein associated with centers

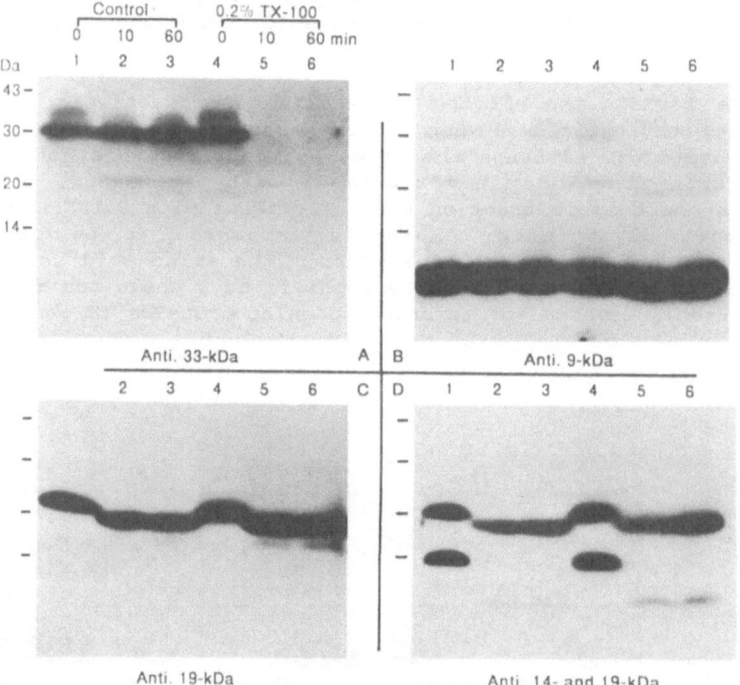

Fig. 4. Immunoblot analysis of tryptic digestion of intact or disrupted thylakoid membranes. After treatment of intact thylakoid membranes (lanes 1-3), or of membranes disrupted by 0.2% Triton X-100 (lanes 4-6), with trypsin, samples equivalent to 10 μg Chl were analyzed with antibody against the 33-kDa protein (panel A), antibody against the 9-kDa polypeptide (panel B), antibody against the 19-kDa polypeptide (panel C) and a mixture of antibodies against the 14- and 19-kDa polypeptides (panel D), respectively. Duration of incubation with trypsin was 0 min (lanes 1 and 4), 10 min (lanes 2 and 5) and 60 min (lanes 3 and 6), respectively.

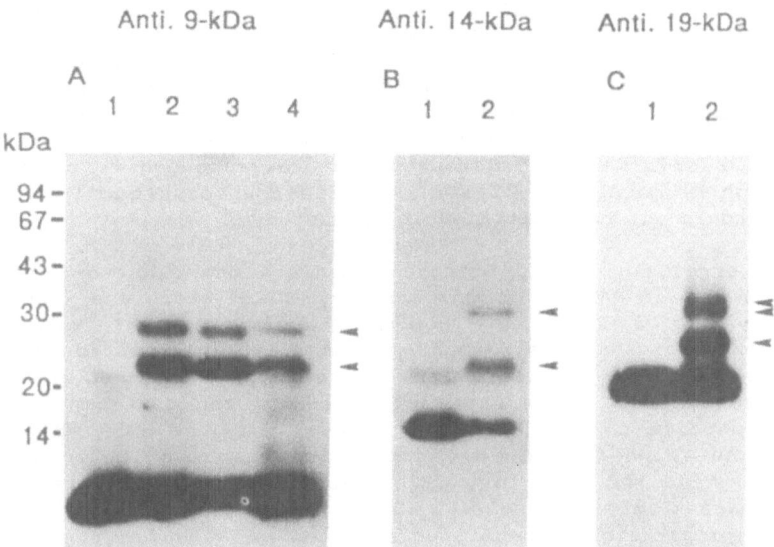

Fig. 5. Immunoblot analysis of treatment of thylakoid membranes with cross-linking reagents. Samples equivalent to 10 μg Chl were analyzed with antibody against the 9-kDa polypeptide (panel A), antibody against the 14-kDa polypeptide (panel B) and antibody against the 19-kDa polypeptide (panel C), respectively. Lane 1, untreated; lanes 2, 3 and 4, treated with EGS, TDIC and HMDIC, respectively. Arrowheads ◄ show cross-linked products.

A and B, was inaccessible to tryptic digestion whether or not the thylakoids were pre-treated with 0.2% Triton X-100. The relatively non specific protease, proteinase K, also did not digested the 9-kDa polypeptide (data not shown).

Cross-linking experiments were conducted to identify subunits located close to the 9-kDa polypeptide in the PS I complex. In order to determine precisely the interactions between subunits *in situ*, thylakoid membranes themselves were directly treated with three different bifunctional cross-linking reagents rather than preparations of isolated PS I complex. EGS is a hydrophilic reagent, while TDIC and HMDIC are hydrophobic reagents. In spite of the difference in their properties (e.g. hydrophobicity and distance between two functional groups), the treatments with these three reagents gave similar cross-linking patterns, as revealed by immunoblot analysis with the antibody specific for the 9-kDa polypeptide (Fig. 5, A). Because the yield of cross-linked products was relatively low, it was necessary to purify the antibody to decrease non-specific reactions, as mentioned in "MATERIALS AND METHODS". Two cross-linked species were detected at positions that corresponded to apparent molecular masses of 22 and 28 kDa, respectively. These two species were revealed to be products of cross-linking between the 9- and 14-kDa, and the 9- and 19-kDa polypeptides, respectively, as identified by using antibodies specific for the 14- and 19-kDa polypeptides (see

Fig. 5, B and C). In this case, the antibody raised against the 14-kDa polypeptide was also purified from the mixture of antibodies against the 14- and 19-kDa polypeptides in order to specify the antigenicity of cross-linked products. An additional species with an apparent molecular mass of 32 kDa was found to be a product of cross-linking between the 14- and 19-kDa polypeptides. With the antibody against the 19-kDa polypeptide, a spot was also detected that corresponded to a molecular mass of slightly more than 32 kDa, but the counterpart of the 19-kDa polypeptide could not be identified in this study.

Since ferredoxin is considered to mediate the electron flow from center A and/or B to NADP$^+$ through ferredoxin-NADP$^+$ reductase (FNR), some subunits in the PS I complex could be expected to interact with ferredoxin. Zanetti and Merati found that a 20 kDa polypeptide in the PS I complex was cross-linked with ferredoxin by a water-soluble carbodiimide (25). We reexamined their report, and confirmed that the 19-kDa polypeptide in our present study was the one which they identified. As shown in Fig. 6, ferredoxin was cross-linked with the 19-kDa polypeptide and FNR, to yield cross-linked species with apparent molecular masses of 33 and 48 kDa, respectively. Although this cross-linked preparation of membranes retained an adequate capacity for

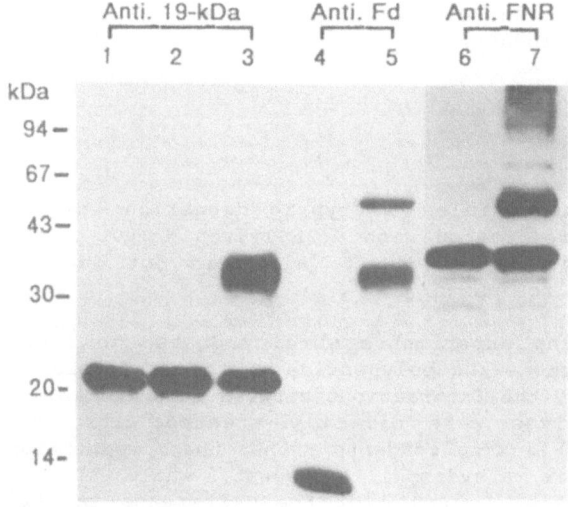

Fig. 6. Immunoblot analysis of cross-linking experiment with ferredoxin and thylakoid membranes. Samples equivalent to 10 µg Chl were analyzed with antibody against 19-kDa polypeptide (lanes 1-3), antibody against ferredoxin (lanes 4 and 5) and antibody against FNR (lanes 6 and 7). Lanes 1, 4 and 6, untreated; lane 2, thylakoids treated with EDC alone; lane 3, 5 and 7, thylakoids treated with EDC and ferredoxin. Experimental conditions were the same as those by Merati and Zanetti (25).

the photoreduction of cytochrome *c* but not NADP$^+$ (25), it is unknown whether the 19-kDa plypeptide is directly involved in transporting electrons from the 9-kDa polypeptide, with centers A and B, to ferredoxin as an electron carrier, or whether the 19-kDa polypeptide merely serves to bring the 9-kDa polypeptide and ferredoxin near to each other so that electrons can be transferred efficiently from the former to the latter.

Figure 7 shows a model of the structural arrangement of subunits in the PS I complex, based on the results described in the present paper. The three subunits, the 9-, 14- and 19-kDa polypeptides, are situated near each other on the surface of the stromal side of thylakoid membranes, although our 14-kDa polypeptide, which corresponds to subunit III, has been suggested by others to be a protein located on the luminal side (20). As mentioned above, the 9-kDa polypeptide became very sensitiveto the oxygen after it was dissociated from the PS I complex, when it gave different EPR signals from those given *in situ* (9, 18). The 9-kDa polypeptide appears, therefore, to be stabilized in the intact PS I complex by being embedded in a particular environment that protect the iron-sulfur centers from destruction, as indicated from the inaccessibility of the 9-kDa polypeptide to protease. Alternatively, the polypeptide may stabilize its own centers by interacting with other redox centers or subunits, such as the 14- and 19-kDa polypeptide, although we have no further data to explain how the subunits in the PS I complex might interact with each other. The 10- and 16-kDa polypeptides are also present in the PS I complex, but their chemical properties are, as yet, unknown. However, they are probably hydrophobic, because they were not extracted by treatment with *n*-butanol of the PS I complex (8).

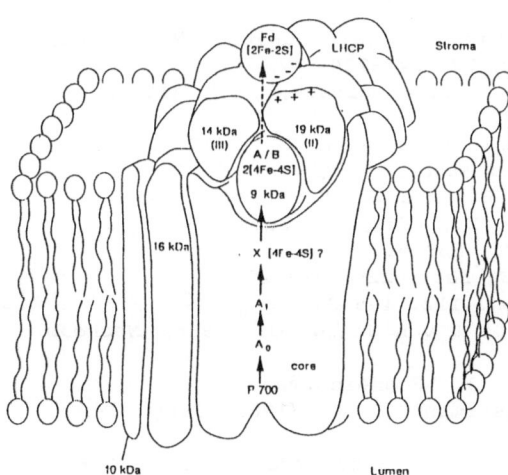

Fig. 7. A model of the arrangement of subunits around the 9-kDa polypeptide in the PS I complex.

ACKNOWLEDGEMENTS

We wish to thank Dr. K. Saeki for his useful advice and discussion on anaerobic systems to prepare the 9-kDa polypeptide with iron-sulfur clusters. This work was supported in part by Grant-in-Aid for Encouragement of Young Scientists to H.O. and for Scientific Research on the Priority Area of "Molecular Mechanism of Photoreceptor (No. 62621506)" to H.M. from the Ministry of Education, Science and Culture of Japan. H.O. is also grateful for a JSPS Fellowship for Japanese Junior Scientists.

REFERENCES

1. Malkin, R. (1982) *Annu. Rev. Plant Physiol.* **33**, 455-479
2. Sakurai, H. & San Pietro, A. (1985) *J. Biochem.* **98**, 69-76
3. Malkin, R. (1986) *Photosynth. Res.* **10**, 197-200
4. Haehnel, W. (1984) *Annu. Rev. Plant Physiol.* **35**, 659-693
5. Høj, P. B. & Møller, B. L. (1986) *J. Biol. Chem.* **261**, 14292-14300
6. Andersson, B., Anderson, J. M. & Ryrie, I. J. (1982) *Eur. J. Biochem.* **123**, 465-472
7. Ortiz, W., Lam, E., Chollar, S., Munt, D. & Malkin, R. (1985) *Plant Physiol.* **77**, 389-397
8. Oh-oka, H., Takahashi, Y., Kuriyama, K., Saeki, K. & Matsubara, H. (1988) *J. Biochem.* **103**, 962-968
9. Oh-oka, H., Takahashi, Y., Matsubara, H. & Itoh, S. (1988) *FEBS Lett.* **234**, 291-294
10. Mills, W. R. & Joy, K. W. (1980) *Planta* **148**, 75-83
11. Piccioni. R., Bellemare, G. & Chua, N.-H. (1982) in *Methods in Chloroplast molecular Biology* (Edelman, M. et al., eds.) pp.985-1014, Elsevier Biomedical Press, Amsterdam - New York - Oxford
12. Kelly, J. L., Greenleaf, A. L. & Lehman, I. R. (1986) *J. Biol. Chem.* **261**, 10348-10351
13. Ohyama, K., Fukuzawa, H., Kohchi, T., Shirai, H., Sano, T., Sano, S., Umesono, K., Shiki, Y., Takeuchi, M., Chang, Z., Aota, S., Inokuchi, H. & Ozeki, H. (1986) *Nature* **322**, 572-574
14. Hayashida, N., Matsubayashi, T., Shinozaki, K., Sugiura, M., Inoue, K. & Hiyama, T. (1987) *Curr. Genet.* **12**, 247-250
15. Parrett, K. G., Mehari, T., Warren, P. G. & Golbeck, J. H. (1989) *Biochim. Biophys. Acta* **973**, 324-332
16. Matsubara, H., Hase, T., Wakabayashi, S. & Wada, K. (1980) in *The Evolution of Protein Structure and Function* (Sigman, D. S. & Brazier, M. A. B., eds.) pp.245-266, Academic Press, New York
17. Minami, Y., Wakabayashi, S., Yamada, F., Wada, K., Zumft, W. G. & Matsubara, H. (1984) *J. Biochem.* **96**, 585-592
18. Wynn, R. M. & Malkin, R. (1988) *FEBS Lett.* **229**, 293-297
19. Burnette, W. N. (1981) *Anal. Biochem.* **112**, 195-203
20. Schägger, H., Link, T., Engel, W. D. & von Jagow, G. (1986) *Methods Enzymol.* **69**, 129-141
21. Münch, S., Ljungberg, U., Steppuhn, J., Schneiderbauer, A., Nechushtai, R., Beyreuther, K. & Herrmann, R. G. (1988) *Curr. Genet.* **14**, 511-518
22. Hatefi, Y. & Hanstein, W. G. (1969) *Proc. Natl. Acad. Sci. U.S.* **62**, 1129-1136
23. Lagoutte, B. (1988) *FEBS Lett.* **232**, 275-280
24. Murata, N. & Miyao, M. (1985) *TIBS* **10**, 122-124
25. Zanetti, G. and Merati, G. (1987) *Eur. J. Biochem.* **169**, 143-146

A PHOTO-SIGNAL TRANSDUCING PHOTORECEPTOR (STENTORIN)

IN STENTOR COERULEUS: A BRIEF REVIEW

Pill-Soon Song
Department of Chemistry and
Institute for Cellular & Molecular Photobiology
University of Nebraska
Lincoln, NE 68588

SUMMARY

Stentorin acts as the photosensor molecule in the ciliate *Stentor coeruleus*. This unicellular protozoan is most sensitive to red light (610 nm). *Stentor* also senses the direction of light propagation, as evidenced by their light-avoiding and negative phototactic swimming behaviors. This aneural photosensory phenomenon is triggered by the photoreceptor stentorin. The possible involvement of a light-induced proton release from the photoreceptor as the primary mechanism of light-signal processing has been discussed. The primary sensory signal, in the form of proton release, triggers subsequent transduction steps that include calcium ion influx from the extracellular medium. It is proposed that the calcium ion influx causes the *Stentor* cell to reverse its ciliary beating and subsequently steer away from the light trap.

INTRODUCTION

Living organisms possess extremely sensitive sensors by which they probe their various environmental signals such as chemicals (chemotaxis), heat (thermotaxis), gravitational force (gravitropism), magnetic field (magnetotaxis) and radiation (phototaxis, photophobic response, etc). In this lecture, we discuss one example of the biological radiation sensor, namely, stentorin in an aneural unicellular protozoan ciliate *Stentor coeruleus*. The organisms perceive specific wavelength light as the stimulus signal, which is then processed through a series of transduction reactions leading to the output, i.e. response, as shown in Scheme I:

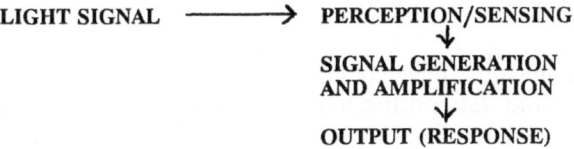

Scheme I. Photosensory transduction in light-sensitive organisms.

Stentor coeruleus is a unicellular protozoan ciliate. This organism exhibits an extremely sensitive response to visible wavelength light, particularly red light of 610-620 nm (for review, see Song (ref.1). In the dark, the cell swims forward by beating rows of cilia in a clockwise direction, which propels the cell to rotate in the same direction. When the cell encounters visible light, it suddenly stops as the ciliary beating reverses its beating direction, and turns away from the light source. This type of photoresponse is called a "photophobic response," and it occurs whenever the cell encounters a sudden increase in light intensity, e.g. from darkness to light and from dim light to bright light. In addition, *Stentor coeruleus* is capable of sensing the direction of light propagation. Thus, the cell swims parallel to the direction of light propagation. This type of photoresponse is called "negative phototaxis." The two types of responses ensure that the ciliate can efficiently escape from the photodynamically harmful radiation (2).

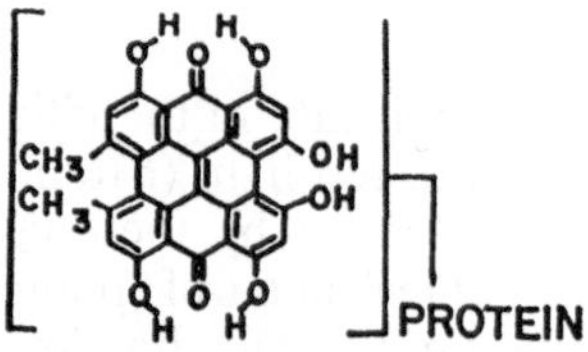

(S-OH)

Scheme II.   The possible structure of the chromophore in stentorin. The site of linkage to apoprotein is unknown.

The photosensing processes of *Stentor coeruleus* can be described within the general framework of the photosensory transduction chain shown in Scheme I. For example, in the photophobic response of *Stentor coeruleus*, light signal is perceived by stentorin localized in the pigment granule. A signal transduction then leads to the reversal of the direction of ciliary beating that is entailed in the stop-turning motion of the cell away from the source of light. Here, we briefly describe the photosensory transduction which utilizes stentorin as the photosensor pigment (Scheme II).

STENTORIN

Stentorin Molecules

The stentorin molecules have been isolated and partially characterized.[1] However, information and data on the molecular weight, the chromophore identity and its stoichiometry, the binding site and chromophore-apoprotein linkage, the amino acid sequence and the photochemical reactivity are still incomplete. The absorption spectrum of stentorin suggests that the chromophore of the stentorin photoreceptor is similar to hypericin (3). In the native stentorin molecule, the chromophore is presumably linked to the apoprotein via an acid-labile bond. Scheme II shows a plausible structure of the chromophore in stentorin. One of the most striking features of the stentorin chromophore is its poly-hydroxyl group.

Available evidence indicates that there are two different forms of stentorin, one small molecular weight species "stentorin I" and one large molecular mass assembly "stentorin II". The latter appears to be a functionally active species, whereas the former is either a non-functional species or an antenna pigment (4). Here, we explore the mechanism of photosensing in *Stentor coeruleus* by the sensor pigment stentorin.

Signal Generation

No evidence for a photochemical cycle similar to the rhodopsin photocycle has been reported with stentorin. On the basis of indirect evidence, we proposed that the photophobic response in *Stentor coeruleus* is triggered by a signal transducing mechanism which is initiated by a light-induced proton release/dissociation from the photoreceptors in the pigment granules to the cytoplasm (1,3,5). More recently, picosecond fluorescence kinetics of native and denatured stentorins have shown that only the native photoreceptor assembly releases protons efficiently during its excited state lifetime, with an apparent excited state pK^* of ca. 2.5 for the hydroxyl groups of the stentorin chromophore. Scheme III outlines the process of proton dissociation/release from the excited stentorin ($S\text{-}OH^*$).

$$S\text{-}OH + h\nu \text{---}> \qquad S\text{-}OH^* \qquad \text{(light absorption)}$$

$$S\text{-}OH^* \text{----------}> \qquad S\text{-}OH + hv_f \quad \text{(fluorescence emission, } k_f\text{)}$$

$$S\text{-}OH^* \text{---------}> \qquad S\text{-}OH + heat \text{ (heat dissipation, } k_h\text{)}$$

$$S\text{-}OH^* \text{--------}> \qquad S\text{-}O^{-\bullet} + H^+ \quad \text{(proton dissociation, } k_d\text{)}$$

$$S\text{-}O^{-\bullet} + H^+ \text{-->} \qquad S\text{-}OH^* \qquad \text{(proton association, } k_a\text{)}$$

$$S\text{-}O^{-\bullet} \text{----------}> \qquad S\text{-}O^- + hv_f \quad \text{(anion fluorescence, } k_{af}\text{)}$$

$$S\text{-}O^{-\bullet} \text{----------}> \qquad S\text{-}O^- + heat \quad \text{(heat dissipation, } k_{hd}\text{)}$$

$$S\text{-}O^- + H^+ \text{--->} \qquad S\text{-}OH \qquad \text{(reassociation, } k_{ra}\text{)}$$

$$S\text{-}OH \text{----------}> \qquad S\text{-}O^- + H^+ \quad \text{(dissociation, } k_{da}\text{)}$$

Scheme III. Proton dissociation/release from the excited state stentorin (S-OH)

A recent picosecond time-resolved fluorescence study provides a direct evidence for the proposed release of protons from the photoreceptor assembly (4). We suggest that a transient proton release induced by the light stimuli serves as an initial signal in the photosensory transduction chain for the photophobic response in *Stentor coeruleus*.

Signal Amplification

An initial signal resulting from light perception in *Stentor coeruleus* can be a transient proton release from the excited stentorin in the pigment granules to the cytoplasm. A depolarizing effect of the intracellular and/intraciliary pH drop may then trigger the opening of Ca^{2+} channels, as outlined in Scheme IV.

A light-induced action potential of the ciliate has been recorded, consistent with the calcium ion influx from the extracellular medium to the cytoplasm when the cell is subjected to the light stimuli. The action potential cannot be induced by light in the presence of protonophores and can be completely quenched by calcium blocking agents, in qualitative agreement with Schemes III and IV (3).

LIGHT SIGNAL--->   SIGNAL PERCEPTION--->   SIGNAL GENERATION
               (stentorin)   (protons)

---> SIGNAL AMPLIFICATION & TRANSMISSION
    (opening of Ca^{2+} channels & influx)

---> MECHANOTRANSDUCTION
    (ciliary reversal)

Scheme IV.   A tentative scheme for a step-up photophobic response in *Stentor coeruleus*

In addition to the inhibitory effects of specific calcium blocking agents on the photophobic responses of the ciliate, calimycin stimulates the photophobic sensitivity of the ciliate, apparently due to its enhancement of calcium permeability across the cellular membrane (1). Other drugs that facilitate calcium permeability and thus enhance the photosensitivity of the *Stentor coeruleus* cell include caffeine and alpha-phosphatidic acid.

In Scheme IV, the link between the Ca^{2+} influx and the dynein ATPase activity in the ciliary contractile axoneme has not been established. The contractile mechanism of km fibers and myonemes is operative in the cell body contraction of *Stentor coeruleus* upon mechanical and electrical stimulation, and is based on Ca^{2+} influx as the driving force for the contractile phenomenon. However, the role of Ca^{2+} ions in the contractile axonemes of cilia and ciliary basal bodies remains to be elucidated (for review, see ref. 3).

ACKNOWLEDGEMENTS

The author's work described in this chapter has been supported by the NIH grant NS-15426.

REFERENCES

1. Song, P.S. (1983) *Annu. Rev. Biophys. Bioengin.*, **12**, 35-68.
2. Yang, K.C., Prusti, R.K., Walker, E.B., Song, P.S., Watanabe, M. and Furuya, M. (1986) *Photochem. Photobiol.*, **43**, 305-310.
3. Song, P.S. (1981) *Biochim. Biophys. Acta*, **639**, 1-29.
4. (a) Kim, I.H., Florell, S., Lee, K.W., and Song, P.S. (1989) *Biochim. Biophys. Acta*, submitted.
   (b) Song, P.S., Kim, I.H., Tamai, N., Yamazaki, T. and Yamazaki, I. (1989) *Biochim. Biophys. Acta*, submitted .
5. Song, P.S. "The Biology of Photoreception," D. Cosens and D. Vince-Prue, Eds., Cambridge Univ. Press, Cambridge, (1983) 503-520.

THEORY AND APPLICATIONS OF THE KINETICS OF SUBSTRATE REACTION

DURING IRREVERSIBLE MODIFICATION OF ENZYME ACTIVITY

Chen-Lu Tsou

National Laboratory of Biomacromolecules
Institute of Biophysics Academia Sinica.
Beijing, China

SUMMARY

Kinetics of enzyme inhibition has been studied systematically and from a unified schemeit has been shown that the
concept of substrate competition applied toboth reversible
and irrreversible inhibitions. The effect of substrate on
the inhibition rate constants can be used as kinetic
criteria for irreversible inhibitions. Equations are
presented for the substrate reaction in the presence of the
modifier and it is shown that the apparent rate constant,
A, for irreversible modification reactions can be determined
in one experiment by following the formation of products in
presence of the inhibitors. Experimental examples are given
to show that constants thus obtained are generally in
agreement with those given by conventional methods. The
above approach for modification kinetics has been applied
to enzyme inactivation during denaturation, enzyme activation, slow reversible inhibition and enzyme reactions
involving two substrates.

INTRODUCTION

Enzyme inhibition has always been an important field of study
owing not only to its usefulness in providing valuable information
on fundamental aspects of enzymatic catalysis and metabolic pathways but also to its implications in pharmacology and toxicology.
Compared to reversible inhibition[1,2], the kinetics for irreversible inhibition has received relatively little attention. Recent
developments have however shown that irreversible modification of
enzyme activity is important in giving definitive information
of the nature of functional groups essential to enzymatic catalysis that is not possible to obtain with reversible inhibitors
[3]. In this respect, the so-called affinity probes[4,5] are
particularly useful not only in mechanistic studies[6], but also
as prospective enzyme targeted drugs[7].

A systematic study of the kinetics of irreversible modification of enzyme activity was presented some years ago[8,9]. Based

on a unified scheme, the concept of substrate-inhibitor competition is now shown to be applicable to irreversible as well as
reversible inhibitions. From the equations derived for the substrate reaction in the presence of the modifier, the apparent rate
constant for the irreversible modification of enzyme activity can
often be obtained in a single experiment[10] and the substrate
competition types ascertained from kinetic criteria based on the
substrate effect on the rate constants for the irreversible
inhibition reactions. This kinetic approach has also been applied
to inactivation during enzyme denaturation[11], enzyme activation
[12], inhibition studies of enzyme reactions involving two substrates[13] and slow binding reversible inhibitors[14].

THEORETICAL

As theoretical treatment has been reviewed recently[15],
only a brief summary will be given here for the convenience of the
readers. The following scheme applies to both reversible and
irreversible modification reactions of the enzyme, E, by the
modifier, Y, including activation and inhibition in presence of
the substrate, S:

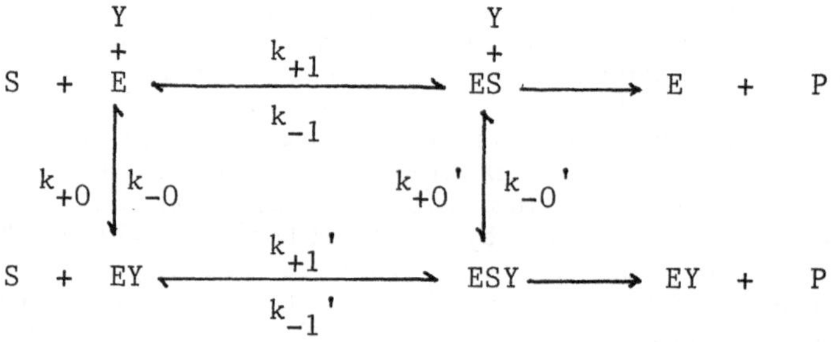

$$
\begin{array}{ccccccc}
 & & Y & & & Y & \\
 & & + & & & + & \\
S & + & E & \underset{k_{-1}}{\overset{k_{+1}}{\rightleftharpoons}} & ES \longrightarrow & E & + & P \\
 & & & & & & \\
k_{+0} \downarrow k_{-0} & & & k_{+0}' \downarrow k_{-0}' & & & (1) \\
 & & & & & & \\
S & + & EY & \underset{k_{-1}'}{\overset{k_{+1}'}{\rightleftharpoons}} & ESY \longrightarrow & EY & + & P
\end{array}
$$

It can be shown that the product, P, formed when t approaches
infinity, $[P]_\infty = v/A[Y]$ where A is the apparent rate constant for
the binding of the inhibitor with the enzyme and v the rate of the
substrate reaction in absence of the inhibitor. The asymptote of a
plot of [P] against t gives the value of $[P]_\infty$ as shown in
Fig. 1. As [Y] is known and v can be separately determined in
experiments in the absence of the inhibitor, the apparent rate
constant, A, is easily obtained. Alternatively as it can be
shown that:

$$\log([P]_\infty - [P]) = \log[P]_\infty - 0.43A[Y]t \qquad (2)$$

and a plot of $\log([P]_\infty - [P])$, as indicated in the shaded area in
Fig. 1, versus t gives a straight line with a slope of $-0.43A[Y]$.
Furthermore, by suitable plotting of the data at different
substrate concentrations, the substrate competition types as well
as the microscopic inhibition rate constants, k_{+0} or k_{+0}', can then
be determined.

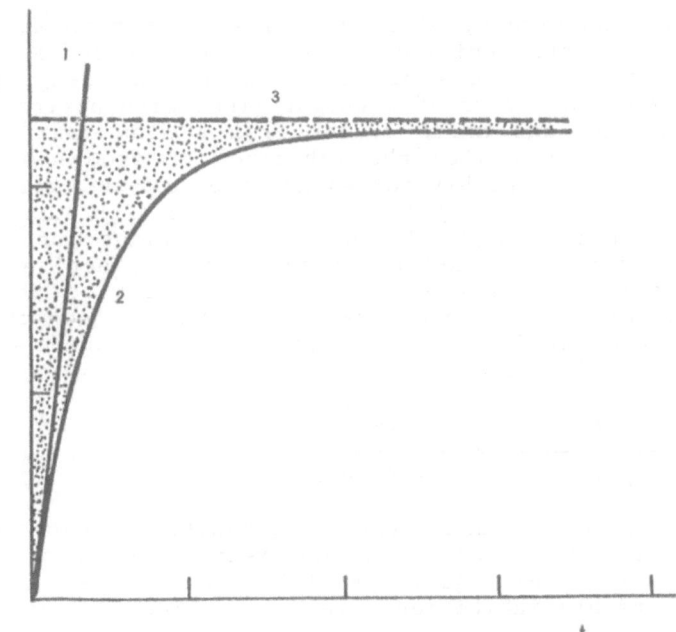

Fig. 1. Reaction of the substrate in the presence of an irreversible inhibitor. 1, Control without inhibitor, 2, and 3, the progress curve and the asymptote. The shaded area indicates their difference ([P] and $[P]_\infty$).

It is known that some modifiers bind reversibly with the enzyme to form complex EY before the irreversible modification step leading to the inactivated enzyme, EY'.

$$[E] + [Y] \rightleftharpoons [EY] \longrightarrow [EY'] \qquad (3)$$

Unlike non-complexing inhibitions[15], the apparent rate constant for complexing inhibitions is related to [S] and [Y] as follows:

$$\frac{1}{A} = \frac{1 + [S]/K_M + [Y]/K_o}{k_{+3}/K_o} \qquad (4)$$

A plot of $1/A$ against [Y] gives a straight line with a positive slope. Both the binding constant, K_o and the inactivation rate constant k_{+3}, can then be obtained.

As reaction scheme 1 applies to both reversible and irreversible inhibitions, it is clear that the concept of substrate competition is applicable to both types of inhibition. For irreversible inhibitors, substrate binding decreases, does not affect, or increases binding rate between the inhibitor and the enzyme for inhibitions of the competitive, non-competitive and uncompetitive types respectively, whereas for reversible inhibitors, the apparent association constant is affected.

For some inhibitors, plots of [P] against time at different
concentrations of the inhibitor give a series of curves which do
not approach finite values of [P]   with   the lengthening   of
reaction time but a series of straight lines with positive slopes.
Such a situation   can arise either from a slow   but reversible
binding of the enzyme with the inhibitor or from some residual
activity of the irreversibly formed modified enzyme, EY. These two
possibilities can be easily differentiated as with increasing [Y]
the slopes of the straight lines should decrease for reversible
inhibitions and remain constant if EY is partly active. From the
equation for product formation when the reaction between the
inhibitor and the enzyme is reversible it can be shown that when t
becomes sufficiently large, the progress curve becomes a straight
line as shown in Fig. 2.

$$[P] = \frac{vBt}{(A[Y] + B)} + \frac{vA[Y]}{(A[Y] + B)} \qquad (5)$$

The apparent forward, A and reverse, B, inhibition rate constants
can be obtained from its slope and x-axis intercept. Alternatively, a similar semilogarithym plot of the shaded area(Fig. 2)
as for irreversible inhibitions yields a series of straight lines
and the rate constants can then be obtained therefrom[14].

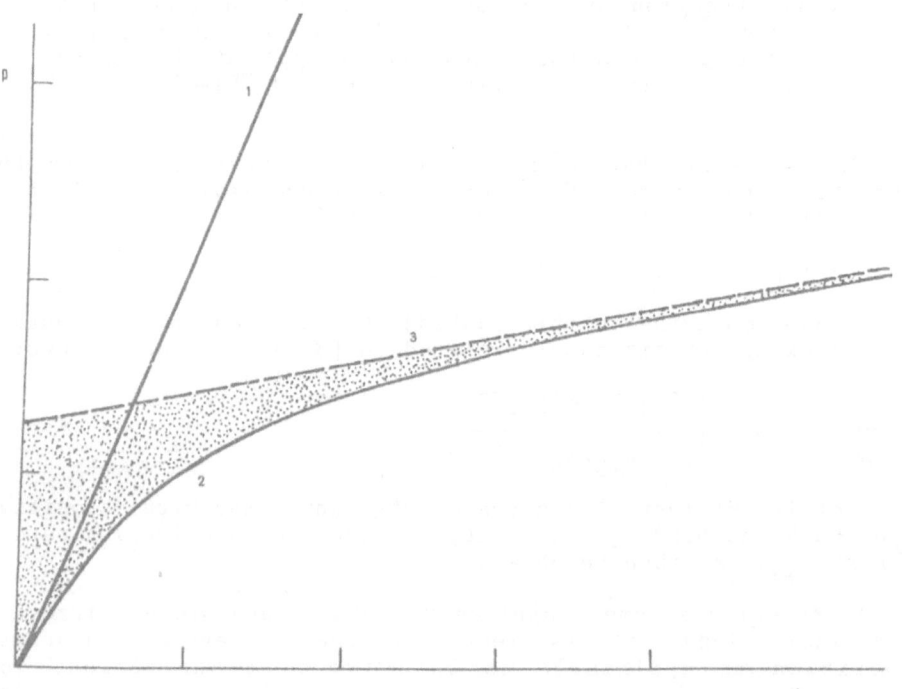

Fig. 2. Reaction of the substrate in the presence of a
reversible inhibitor. The control is indicated by and the
shaded area is the difference between the asymptote 3 and
the progress curve 2.

APPLICATIONS

1. Determination of irreversible inhibition rate constants

The approach of the study of irreversible inhibition kinetics by following the course of the substrate reaction in the presence of the modifier have been applied for the determination of inhibition rate constants in this[10,16] and other laboratories [17-21]. In some cases, the inhibitors complex specifically to enzymes at the active sites prior to the irreversible modification step and these inhibitors appear to have great prospects as therapeutic agents. From the kinetic equations derived for the substrate reaction during the action of this type of inhibitors, both the microscopic rate constant and the binding constant can be obtained from the apparent rate constant obtained at different concentrations of [Y]. Comparison of the rate constants obtained by the present method with those recorded in the literature is summarized in Table 1. The rate constants obtained are in satisfactory agreement with those recorded in the literature if available.

Table 1   Binding and rate constants for the irreversible inactivation of certain enzymes.

Enzyme	modifier	binding constant, M^{-1}	inhibition constant, min^{-1}
chymotrypsin	PMSF	36,000	0.32
	TPCK	5,290	0.22
acetylcholine esterase	DFP	3,900	2.1
	paraoxon	12,000	18
creatine kinase	urea, 4 M		780
trypsin	STI	1,350,000	0.095

Abbreviations used:DFP, diisopropyl fluorophosphate; PARA-OXON, diethyl 4-nitrophenyl phosphate; PMSF, phenylmethane sulfonylfluoride; STI, soybean trypsin inhibitor; TPCK, L-1-[(p-toluenesulfonyl)-amino]-2-phenylethyl chloromethyl ketone.

2. Inactivation kinetics during denaturation

Although the importance of conformational integrity for the activity of an enzyme is generally recognized and the kinetics of unfolding of protein molecules extensively studied, very few attempts have been made to compare the activity with the conformational changes during unfolding of enzyme molecules brought about by denaturating agents. This is at least partly because of the lack of a suitable method to measure quantitatively fast inactivation reactions and the present method appears to be ideally suitable. The inactivation of creatine kinase in guanidineHCl and urea are fast reactions; the course can be followed by the substrate reaction in presence of the denaturant with a stopped

flow apparatus[11,22] and the inactivation rate constant determined. The rate constant for urea inactivation is included in
Table 1. Comparisons of the rate constants for inactivation and
unfolding clearly show that the inactivation rates of the enzyme
are several orders of magnitude greater than the rates of conformational changes as followed by UV absorbance, fluorescence and CD
changes, and by exposure of buried SH groups[23]. It appears that
the active site is located in a limited region of the molecule
more sensitive to guanidineHCl or urea than the enzyme as a whole.
A slight disturbance to the correct spatial geometry of the
functional groups responsible for the catalytic mechanism of the
enzyme destroys its activity before any gross conformational
changes can be detected by the methods employed[23.24].

3. Activation kinetics

It is evident that scheme 1 applies to both inhibition and
activation and the substrate reaction during enzyme activation can
be treated in a similar way. The reactivation of acetylcholine
esterase inhibited by organophosphorous compounds[12] and the
reactivation of guanidineHCl inactivated enzymes have been treated
by the above approach[25] giving useful kinetic constants apart
from shedding some light on the reactivation mechanism. However,
the reactivation kinetics are usually much more complicated than
that shown in scheme 1 and in each case the kinetics should be
treated by its own specific scheme.

4. Slow binding reversible inhibitors

Naturally occurring proteinase inhibitors are ubiquitous in
animals, plants and microorganisms playing an important role in
the regulation of many physiological processes[26]. Earlier kinetic studies of the inhibition reaction have been made mostly by
the conventional method of measuring the activity remaining of
aliquots of the enzyme-inhibitor incubation mixtur at definite
time intervals yielding only the rate constant for the binding of
the inhibitor with the enzyme[27]. The inhibition of trypsin by
pancreatic(PTI), ovomucoid(OTI) and soybean(STI) inhibitors have
now been studied by the above method[28] of following the kinetics
of the substrate reaction in presence of the inhibitor and the
results obtained are consistent with the following conclusions:
a)the enzyme binds with PTI irreversibly to form an inactive
complex, b)the binding of OTI to trypsin is reversible and c)an
intermediate is formed before the stable inactive complex with STI
and both steps are reversible. The respective microscopic rate
constants are determined by suitable plots of the apparent rate
constant A, under different substrate and inhibitor concentrations
and part of the data obtained are included in Table 1.

5. Reactions involving two substrates

Irreversible inhibitions of enzyme reactions involving two
substrates have to be frequently dealt with and the corresponding
kinetic equations for such reactions have been derived[16].
Moreover, it has been shown that irreversible inhibition kinetics
can be used to distinguish between different mechanisms for
substrate binding sequences.

Creatine kinase is a dimeric enzyme composed of identical subunits with8 Cys residues two of which are capable of reacting with a number of modification reagents and believed to be essential for the activity ofthe enzyme. The kinetics of inactivation by iodoacetamide and DTNB, followed by the substrate reaction as proposed before for the evaluation of the apparent rate constant shows the monophasic nature of the pseudo first order inactivation reaction with both inhibitors in accord with that shown in Fig. 1. However, the apparent rate constants for the irreversible inhibition of enzymes catalyzing reactions with two substrates are, necessarily, complex functions of the respective microscopic constants. Nevertheless, by measurements under different substrate and inhibitor concentrations and suitable treatment of the data, the microscopic rate constants for the reaction of the inhibitor with the free enzyme and the various enzyme-substrate complexes can all be obtained as shown in Table 2. The rate constants for the inactivation and the modification reactions were also determined by the conventional method and results obtained are in reasonable agreement. Because of space limitations, for detailed kinetic treatment the readers are referred to earlier publications[13,15,28].

Substrate protection against inactivation of enzymes has always been discussed in a qualitative sense. The microscopic rate constants for the free enzyme, the enzyme-creatine, enzyme-ATP and the ternary complexes listed in Table 2 now provide a quantitative measure of substrate protection.

Table 2. Inactivation rate constants of creatine
kinase by iodoacetamide and DTNB.

Form of enzyme reacting	Second order rate constants $M^{-1}s^{-1}$	
	iodoacetamide	DTNB, x 10^{-4}
E	7.6	1.1 12.0[a] 4.4[b]
E-Mg^{2+}-ATP	14.4	1.7
E-Creatine	2.2	1.1
E-Mg^{2+}-ATP-Creatine	1.5	<0.05

[a]By the conventional method of taking aliquots of the reaction mixture at time intervals and assay for enzyme activity,
[b]Rate constant for the modification reaction measured directly at 412 nm.

It can be seen that iodoacetamide is non-competitive to ATP but weakly competitive with respect to creatine in that this substrate gives about 3 fold protection to its inhibition whereas DTNB is non-competitive to both substrates. The formation of the

ternary complex gives only slightly further protection against
iodoacetamide but almost complete protection against DTNB. It
should also be pointed out here that by the conventional method,
it would be difficult to determine the rate constant for the reaction of the inhibitor with the ternary complex.

5. Discussion

Conventional method for the determination of rate constants
for the irreversible modification of enzyme activity is to take
aliquots from an enzyme-modifier incubation mixture at definite
time intervals and assay for the activity. This method is not only
laborious but also too slow to be applied to fast reactions with a
half life of, say, a few seconds. In the present approach, the
apparent rate constant can be obtained in one single experiment
and with a stopped flow apparatus first order rate constant in the
order of 10 s^{-1} can be easily obtained as shown for the inactivation of creatine kinase by urea or quanidineHCl[11,22]. The
simplicity of the present method can be a great asset in the
comparison of the inhibitory power of a large number of compounds
on a key enzyme as could be the case in the screening of possible
chemotherapeutic agents. In this respect, the presence of substrate during determination of the rate constant for the modification reaction would be most desirable as during the in vivo
action of these chemotherapeutic agents, the presence of substrate
is inevitable. From the equations derived, it is now possible to
determine the the microscopic rate constants for the binding of
the modifier with the free enzyme and with the various enzyme
substrate complexes.

The original treatment was for both reversible and irreversible inhibitions[8,9] and it is now shown experimentally that for
kinetics of inhibitions by slow reversible inhibitors the present
approach is equally applicable. Earlier studies on the kinetics of
trypsin inhibition by its specific inhibitors were made mostly by
the conventional methoh of taking aliquots from the incubation
mixture at different time intervals and assay for activity
remaining[27]. Because of the gradual dissociation of the enzyme-
inhibitor complex upon dilution especially in the presence of the
substrate, even if precautions are taken to avoid unnecessary
dilution and to take as far as possible the initial velocity
during the measurements of the activity remaining, the reactivation induced by the substrate cannot be completely avoided by
the conventional method[14].

The effect of the substrate on the inactivation reactions
has been described as substrate protection without a quantitative
criteria which can be easily determined experimentally. This can
now be quantitatively defined by comparing the microscopic
inactivation rate constants of the enzyme-substrate complexes with
that of the free enzyme as discussed in a previous section based
on results given in Table 2.

Many physiological processes involve irreversible changes of
enzyme activity and the present approach can be easily adopted to
problems where kinetics of rapid irreversible changes in enzyme

activity during these processes are to be studied, for instance, the kinetics of zymogen activation. It is also known that many important metabolic processes are regulated by enzymes that exist in active and inactive forms interconvertible through covalent modification of the enzyme molecules. The phosphorylation and dephosphorylation of protein kinase is a well known example. The kinetics of these reactions are ideally amenable to analysis by the present method.

REFERENCES

1. Laidler, K. J. and Bunting, P. S.(1973) "The Chemical Kinetics of Enzyme Action", 2nd ed., pp. 175-180, Clarendon Press, Oxford.
2. Cornish-Bowden, A. (1979) "Fundamentals of Enzymes Kinetics", pp. 73-74, Butterworths, London.,
3. Wold, F. (1977) Methods Enzymol. 46, 3-14.
4. Baker, B. R. (1967) "Design of Active-Site-Directed Irrever-sible Enzyme Inhibitors", John Wiley and Sons, New York.
5. Plapp, B. V. (1982) Methods Enzymol. 87, 469-499.
6. Walsh, C. T. (1983) Trends Biochem. Sci. 8, 254-257.
7. Heby, O. (1985) Adv. Enz. Regulation, 24, 103-124.
8. Tsou, C.-L. (1965) Acta Biochim. Biophys. Sin. 5, 398-408.
9. Tsou, C.-L. (1965) Acta Biochim. Biophys. Sin. 5, 409-417.
10. Tian, W. X. and Tsou, C.-L. (1982) Biochemistry 21, 1028-1032.
11. Yao, Q.-Z., Zhou, H.-M., Hou, L.-X. and Tsou, C.-L. (1982) Sci. Sin. (Eng. Ed.) 25B, 1296-1302. ,
12. Liu, W., Zhao, K. Y. and Tsou, C.-L. (1985) Eur. J. Biochem. 151, 525-529.
13. Wang, Z. X., Preiss B. and Tsou, C.-L. (1988) Biochemistry, 27, 5095-5100.
14. Zhou, J. M., Liu. C. and Tsou, C.-L. (1989) Biochemistry, 28, 1070-1076.
15. Tsou, C.-L. (1988) Adv. Enzymol. 61, 381-436.
16. Liu, W. and Tsou, C.-L. (1986) Biochim. Biophys. Acta, 870, 185-190.
17. Leytus, S. P., Toledo, D. L. and Mangel, W. F. (1984) Biochim. Biophys. Acta. 788, 74-86.
18. Harper, J. W., Hemmi, K. and Powers, J. C.(1985) Biochemistry, 24, 1831-1843.
19. Silverberg, M., Longo, J. and Kaplan, A. P. (1986) J. Biol. Chem. 261. 14965-14968.
20. Wijnands, R. A., Muller, F. and Visser, A. J. W. G. (1987) Eur. J. Biochem. 163, 535-544.
21. Crawford, C., Madson, R. W., Wikstrom, P. and Shaw, E. (1988) Biochem. J. 253, 751-758,
22. Yao, Q. Z., Tian, M. and Tsou, C.-L. (1984) Biochemistry, 23, 2740-2744 .
23. Tsou, C.-L. (1986) Trends Biochem. Sci., 11, 427-429.
24. Tsou. C.-L. (1989) Chin. Sci. Bull. 34, 793-799.
25. Liu, W. and Tsou, C.-L. (1987) Biochim. Biophys. Acta, 916, 465-473.
26. Laskowski, M., Jr. and Kato, I. (1980) Ann. Rev. Biochem. 49, 593-626.
27. J. Engel, U. Quast, H. Heumann, G. Krause and E. Steffin *in* "Proteinase Inhibitors". (H. Fritz, H. Tshesche, L. J. Greene

and E. Truscheit, eds.) pp. 412–419, Springer-Verlag, Berlin, 1974.

28. Wang, Z. X. and Tsou, C.-L. (1987) J. Theoret. Biol. 127, 253–270.   11111

HIGH PRESSURE KINETIC STUDIES ON CERTAIN HEMOPROTEINS

Claude Balny

Institut National de la Santé et de la Recherche Médicale
INSERM U 128 - CNRS - B.P. 5051, 34033 Montpellier Cedex - France

SUMMARY

Pressure, as a perturbing agent, is one of the most powerful tools to investigate the thermodynamic parameters of biochemical (and chemical) reactions. By using the transition state theory, the thermodynamic quantity $\Delta V^{\ddagger}$ was determined under different experimental conditions for some reactions involving various heme proteins. Results are discussed in terms of changes in protein conformation induced by changes in solvent, temperature or pressure which affected the reactions rates. These were determined using the high-pressure low-temperatures stopped-flow method.

INTRODUCTION

Since the pioneering work of Ogunmola *et al* (1), a number of papers has been published concerning pressure effect on heme proteins. Works were devoted to spin equilibria (1,2), conformational changes (1), redox reactions (3), denaturation (4) and binding of small ligands to hemoglobin, myoglobin, peroxidases and other cytochromes (5). Many techniques, including optical spectroscopies, NMR, EPR, etc ... were used. Recently, dynamics of the spin transition of substrate-bound ferric cytochrome P-450 versus temperature, pressure and medium were reported. For this, the pressure jumps method was applied and a thermodynamic description of conformational changes provided some insight into the internal properties of this cytochrome in terms of substates (6, 7).

The purpose of this communication is to summarize our recent findings concerning the effects of pressure on the binding of CO to some ferrous heme proteins, on the reaction of substrates with HRP, cytochrome *c* peroxidase and hydroxylamine oxidoreductase, on dithionite reduction of cytochromes and on intermolecular electron transfer. Our goal is to discuss the thermodynamic values obtained. Whereas these entities in themselves are rather uninformative, their variation on the perturbation of the system under study can lead to useful information. On the other hand, although the parameter $\Delta V^{\ddagger}$ presently lacks a full theoretical basis (8), it is hoped that the collection of accurate thermodynamic data for several proteins could ultimately lead to an understanding of this parameter.

The abbreviations used are : HRP, horseradish peroxidase; HAO, hydroxylamine oxidoreductase; CcP, cytochrome *c* peroxidase; EtOOH, ethyl peroxide; EGOH, ethylene glycol; DMSO, dimethyl sulfoxide.

EXPERIMENTAL PROCEDURES

To record kinetics we have adapted the stopped-flow method to both high pressure (up to 2 kbar) and low temperatures (-30° C). The apparatus incorporates some features of previous stopped-flow devices built for high pressure work (9) but with modifications making it suitable for absorption (10) and fluorescence (11) spectroscopy of fast reactions at subzero temperatures. The general design consits of a driving mechanism, two vertical drive syringes, a mixing chamber, an observation chamber and a waste syringe. To avoid thermal artifacts, temperature and pressure homogeneities are maintained by housing the whole apparatus in a high pressure thermostated bomb. The dead time is < 5 ms, data nearly independent of the pressure. The volume of the cell is 30 μl ; 120 μl of reaction mixture is used at each injection.

The apparatus is monted either on an Aminco DW2 spectrophotometer or on a Union Giken Model RA415/RA401 fast response spectrophotometer. This second apparatus allows one to record rapid scan measurements (maximum recording speed : 95 nm/ms). In fluorescence mode, the device is adapted to a special spectrofluorimeter designed in the laboratory.

The enzymes used were either prepared according to published procedures or purchased from Sigma. For most of the experiments, Tris/HCl was chosen as buffer since the concentration of hydrogen ions in the system is almost independent of pressure (12). The actual pH in hydro organic media was estimated according to procedures published elsewhere (13). The preparation of anaerobic reaction mixtures and reduction of the enzymes were achieved as previously described (14).

Exploitation of data

The thermodynamic parameters concerning a kinetic constant were interpreted in terms of the transition state theory (15) assuming that between two states A and B, there is a labile complex $A^{\ddagger}$ named transition state :

$$A \quad \underset{}{\overset{k_+}{\rightleftharpoons}} \quad A^{\ddagger} \quad \overset{k_-}{\leftharpoondown} \quad B \qquad (1)$$

The subscripts + and - refer to forward and reverse reaction, respectively. By varying temperature and pressure one can obtain estimates of the thermodynamic quantities relating to the activated complexes, namely : $\Delta H^{\ddagger}$, $\Delta S^{\ddagger}$ and $\Delta V^{\ddagger}$ (variation of enthalpy, entropy and volume, respectively). A complete description of the thermodynamic considerations has been published elsewhere (16). Concerning the apparent activation volumes ($\Delta V^{\ddagger}$), they were calculated from the pressure-dependence of rate constant k according to :

$$\delta \ln k / \delta P = - \Delta V^{\ddagger} / RT \qquad (2)$$

where P is the pressure ; T the absolute temperature and R the gas constant ($82 \; ml \cdot atm.K^{-1} \cdot mol^{-1}$ with 1 atm = 101.3 kPa)

Kinetics remarks

The general "induced-fit" theory implicitly means that the binding of a ligand / or substrate to the protein is a two-step process : formation of a collision complex (rapid attainment of equilibrium) followed by an isomerization step :

$$E + S \xrightarrow{K_1} ES \underset{k_{-2}}{\overset{k_2}{\rightleftharpoons}} E*S \qquad (3)$$

S being either the substrate or the ligand.

Spectroscopic measurements of the formation of E*S provide the first-order kinetic constant:

$k_{obs} = k_{-2} + k_2 [S] / K_1 + [S]$, where K_1 is the dissociation constant.

When [S] is increased, a plateau for k_{obs} must be reached.

Origin of activation volume in proteins

In opposition to simple chemical reactions, it is very difficult to give a precise physical description of $\Delta V^{\ddagger}$. It should be considered as a thermodynamic parameter associated with the reacting species and the solvent environment. The sources of volume changes during enzyme reactions and protein subunit association or dissociation events have been discussed (17). The measured $\Delta V^{\ddagger}$ can be considered as the sum of several components : chemical event, configuration term, intramolecular interaction and solvation.

RESULTS AND DISCUSSION

Binding of carbon monoxide to heme proteins

Several studies on the effects of pressure on the binding of oxygen and/or CO to heme proteins have been published (18, 19). Present work has been undertaken to study the properties of the elementary rate constants describing the binding of CO to reduced horseradish peroxidase under various conditions of temperature, pressure and solvent (20). The kinedics at 423 nm were studied under these conditions, i.e. in water, in 40 % EGOH (in a temperature range 20 to -10° C) and at pressure up to 1200 bar. Whatever the conditions, the linearity of k_{obs} as a function of CO concentration (up to 0.5 mM) means that K_1 and k_2 (see Experimental Procedures section) remain high. Only $k_+ = k_2 \cdot K_1$, the slope of k_{obs} as a function of [CO], has been obtained. Values and activation volume for k_+ are summarized in Table 1.

Similar determinations have been obtained for binding of CO to ferrous P460 (14). The heme-like chromophore P460 is part of a site of hydroxylamine oxidoreductase. This enzyme (HAO) of the ammonia-oxidizing bacterium *Nitrosomonas* (21) catalyzes the oxidation :

$$NH_2OH + H_2O \rightarrow HNO_2 + 4 e^- + 4 H^+ .$$

P460 binds substrate, extracts electrons and then passes them to the many *c* hemes of the enzyme.

Ferrous P460 (dithionite reduction) of HAO binds CO with an accompanying shift in the absorption maximum from 463 nm to 447 nm. k_+ has been obtained and data are summarized in Table 1.

Table 1 .   Values and activation volume for k_+ under different experimental conditions
for the reaction of CO with ferrous HRP and ferrous P-460 respectively.
Values are given ± 10 %. For both HRP and P460, the buffer used was 0.1 M
Tris pH 7.5 (from ref. 16 and 20).

heme protein	Medium	Temperature °C	k_+ $M^{-1} s^{-1}$	$\Delta V^{\ddagger}$ $ml \cdot mol^{-1}$
HRP	water	20	6200	- 23.5
		4.5	1000	-27
	40 % EGOH	20	2870	-7
		-10	200	-15
P460	water	20	2740	-36.5
		3.5	1100	-32
	40 % EGOH	20	7500	-23.5
		3.5	3400	-14

For both enzymes, HRP and HAO respectively, the solvent effects were rather small for the association constant. This independence of solvent would occur if the interaction between CO and ferrous heme involved a hydrogen bond between an amino acid group and the carbonyl oxygen of CO. The latter bond is not strongly solvent dependent. For both enzymes also, the values of $\Delta V^{\ddagger}$ were negative, with the same order of magnitude and $- \Delta V^{\ddagger}$ decreased in the presence of 40 % EGOH. These deserve further comments. The $\Delta V^{\ddagger}$ observed is the sum of several components : contribution of the covalent bond formation ($\cong -10 \; ml \cdot mol^{-1}$) and the contribution of solvation. The variation observed in $\Delta V^{\ddagger}$ under different conditions suggest that solvent reorganization is an important factor responsible for the modulating effect observed in presence of ethylene glycol.

In the temperature range exploited, the present measurements of the kinetics for both proteins were accurate enough to show that the behaviour of the enzymes fit the Maxwell equations remarkably well : $(\delta\Delta H^{\ddagger} / \delta P)_T = \Delta V^{\ddagger} - T(\delta\Delta V^{\ddagger} / \delta T)_P$ and $- (\delta\Delta S^{\ddagger} / \delta P)_T = (\delta\Delta V^{\ddagger} / \delta T)_P$. This merely shows that the results are internally consistent (22).

Moreover, data obtained with P460 suggested that the CO binding site which has little access to solvent is buried in the protein and is not susceptible to change in protein conformation.

Pressure effect on substrate binding

Our aim was to obtain the thermodynamic parameters of a single step on the HRP, CcP and HAO reaction pathways, respectively, via the formation of the complexes between these enzymes and their substrates.

1 . Reaction of peroxides with CcP and HRP. Some years ago, the effects of pressure at room temperature on the reactions of HRP with hydrogen cyanide and hydrogen peroxide were reported (23). This was the first study in this field involving pressure dependence upon the formation of intermediate species for the heme class of enzymes.

To extend these studies we had undertaken the investigation of the reaction of CcP with hydroperoxide which follows the general scheme (10):

$$E + S \rightleftharpoons ES$$

ES is highly stable at room temperature in the absence of reductants or ferrocytochrome. The reaction occurred in two phases. The fast phase was assigned to the reaction of native active (pulsed) CcP with peroxides whereas the slow phase was due to the presence of an inactive (aged, resting) enzyme (24). For the first phase, data are reported in Table 2.

Table 2 . Values and activation volume for k_+, K_1 and k_2 ($k_+ = K_1 \cdot k_2$) under different experimental conditions for the reaction of ethyl hydroperoxide with CcP and HRP respectively. Buffer for CcP : 0.1 M Bistris, pH 5.8; buffer for HRP : 0.1M Tris/HCl, pH 8.7. Values are given ± 10 %. (from ref. 10 and 14).

Protein	temperature °C	k_+ x 10^{-6} M^{-1} s^{-1}	K_1, k_2	$\Delta V^{\ddagger}$ ml · mol^{-1}
CcP (water)	15 2	5 3		9.5 9.2
HRP (water)	19.5 1	1.2 1.1		3 14.5
HRP (60 % DMSO)	- 23	0.0015	$K_1 = 3.8 \; 10^3 \; M^{-1}$ $k_2 = 0.65 \; s^{-1}$	30 11 19

The $\Delta V^{\ddagger}$ obtained are positive and the values are larger for the slow phase (slow conformational modification) than for the rapid process.

Concerning the binding of EtOOH with HRP, in water at 2° C, the behaviour of the reaction was similar to the preceding results obtained for the fast phase of the CcP reaction : positive $\Delta V^{\ddagger}$ were observed (16). The exception concerned the temperature effect which modulated $\Delta V^{\ddagger}$ (see Table 2). This observation had been exploited and we had demonstrated, as with the alkaline form, the neutral form of the HRP binds peroxide substrates in two steps. It was the combined use of organic solvents (60 % DMSO) and low temperatures (down to - 23° C) which revealed saturation kinetics. According to scheme (3), in water and organic solvents at temperatures above -10° C, K_1 was too small and k_2 too large to be measured. The thermodynamic parameter $\Delta V^{\ddagger}$ (together with $\Delta H^{\ddagger}$ and $\Delta S^{\ddagger}$) for K_1 and k_2 was obtained (see Table 2). Data had been interpreted as follows : the transition ES $\rightarrow$ compound I described by k_2 is very sensitive to the medium conditions whereas K_1, which describes the formation of the collision complex ES, is very much less sensitive. The large $\Delta V^{\ddagger}$ value for k_2 (similar to the $\Delta V^{\ddagger}$ of the slow phase of the CcP reaction : 12 - 14 ml · mol^{-1}) suggests large structural rearrangements which may involve the hydration shell at the protein-solvent interface.

Table 3 . Activation volume for k_1 and k_2 for the reduction reaction of hydroxylamine with HAO. Values are given ± 10%. Buf:0.1 Tris/HCl, pH 7.5.(from ref.14)

Medium	Temperature °C	$\Delta V^{\ddagger} k_1$ ml · mol^{-1}	$\Delta V^{\ddagger} k_2$ ml · mol^{-1}
water	20.7	-3.5	
	3.5	-2	
40 %	20.7	57	-30
EGOH	3.5	10	16
	-15	1.5	30

2 . Reduction of the c hemes of HAO. The kinetics of reduction of some of the 24 c hemes of the ferric HAO by its substrate hydroxylamine (NH_2OH), as seen by the change in absorbance at 552 nm, are biphasic with apparent first-order rate constants k_1 and k_2 (25). One explanation for the two phases of reduction of c hemes could invoke two sites of reaction of substrates with the enzyme. An alternative model could suggest that the enzyme can exist in at least two states differing in the rate of heme-heme electron transfer. Values of $\Delta V^{\ddagger}$ were determined for the two phases as a function of the temperature in water and in 40 % EGOH. If in aqueous solution, the pressure effect on k_1 was small (k_2 was not accessible in water) both k_1 and k_2 were pressure sensitive in 40 % EGOH (see Table 3). It has been demonstrated that a clear conformational change occurs at approximatively 0°C. This indicated that the physical orientation of the electron transfer was modified. The application of high pressure magnified the phenomena, only slightly detectable at atmospheric pressure. The biphasic nature of the Arrhenius curves was enhanced at high pressure. For k_1, for example, at 800 bar, whereas at subzero temperatures, $\Delta H^{\ddagger}$ remained 51 kJ · mol^{-1}, the activation energy was -18 kJ · mol^{-1} in the positive temperature range. It means that at this pressure, the velocity of the reaction was faster at 0° C than at room temperature (14). Thus, the interactions of the redox centers involved are particularly susceptible to a change in protein structure as well as the nature of the solvent. In the temperature range 20 - 0° C, the measurements of the kinetics of reduction show that the behaviour of the enzyme fits the Maxwell equation well. On the other hand, this equation was not followed over the entire temperature range 20.7 to -15° C supporting also the modification in the conformation of the protein as a function of temperature.

Pressure effect on dithionite reduction of heme proteins

Two proteins have been investigated : the mammalian cytochrome c and the ferric c - hemes of the tetraheme cytochrome c -554 from *Nitrosomonas europaea* (26). For cytochrome c, whatever the pH, the value of $\Delta V^{\ddagger}$ related to dithionite reduction increased as a function of decreasing temperature ($\Delta V^{\ddagger}$ = 5.5 and 19 ml · mol^{-1} at 20 and 4° C respectively, pH 9). The cause of the positive value of $\Delta V^{\ddagger}$ was interpreted as a pressure-induced displacement of the iron-bound ligand, temperature dependent.

For cytochrome c -554, the dithionite reduction is biphasic. Based on the heat of activation, the rate limiting step in the slow phase involves a change in the conformation of the protein (27). The kinetics of the slow phase (the fast phase was too fast to be recordable in our conditions) were examined as a function of pressure at several values of pH (5 to 9.5) and temperature (20° C and 4° C). The activation volume depends on the pH in a manner parallel to the response of spin state of the cytochrome to pH (see Table 4). By analogy with the results with mammalian cytochrome c , the present data suggest that cytochrome c -554 might have greater conformational flexibilty at alkali pH than at acidic pH. Moreover, at basic pH, the proteins appears to be specially susceptible to changing structure as the temperature is decreased (reflected by reversal of the sign of $\Delta V^{\ddagger}$).

Table 4 .   Values of $\Delta V^{\ddagger}$ for reduction of hemes of cytochrome c - 554 by dithionite as a function of temperature and pH. Values are given ± 10 %. LS : low-spin. Buffers : pH 5 : 0.1 M acetate ; pH 9.5 : 0.1 M Tris/HCl. (from ref. 26)

pH	spin state	$\Delta V^{\ddagger}$, 20° C (ml · mol^{-1})	$\Delta V^{\ddagger}$, 4° C (ml · mol^{-1})
5	50 % LS	28	22
9.5	75 % LS	36	- 14

Intermolecular electron transfer

As preliminary experiments to study the effect of pressure on the reduction of cytochrome c -554 catalyzed by HAO (the cytochrome c -554 being the first step in the transfer of electrons from HAO to a terminal oxidase and ammonia monooxygenase (28)), we have investigated the effects of elevated hysdrostatic pressure on the reduction of cytochrome c catalyzed by HAO. The rate of electron ransfer between HAO, catalytically reduced by NH_2OH, and cytochrome c -554 is strongly dependent on ionic strength (I), k_{obs} decreased when I increased. The dependence of $\Delta V^{\ddagger}$ on the ionic strength has been determined. It has been found $\Delta V^{\ddagger}$ = -24, -17 and -5.5 ml · mol^{-1} for ionic strength 20, 200 and 420 mM respectively. At high ionic strength, the pressure effect was smaller than at low ionic strength, a medium condition where the reduction is highly accelerated by the pressure which strengthen electrostatic interaction.

This observation is in agreement with several studies carried out on model systems. It is known that for ionic interaction, the carrying out of new charges is accompanied by a volume decrease. However, we must point out that large spectral changes have been observed if cytochrome oxidase is subjected to pressure, a phenomenon associated with the inhibition of electron transport within the oxidase (29).

CONCLUSIONS AND PERSPECTIVES

We have surveyed some aspects of the action of pressure on several heme proteins. Despite the results collected in this laboratory and elsewhere, the interpretation remains phenomenological and needs further accumulation of experiments and data. In the future, we would like to ascribe a physical significance to the volume changes observed. To date, for the reactions described above, the $\Delta V^{\ddagger}$ is considered as a thermodynamic parameter which closely follows Maxwell's equations. In opposition to simple chemical reactions (30), it is very difficult to give a precise physical description of $\Delta V^{\ddagger}$. This value is the reflect of the effects of pressure on both the chemical catalytic act and on the structural conformation changes of the protein. However, the potential of perturbations by pressure as a tool to understanding biochemical reactions mechanisms is supported by recent reports (for review , see 31). For many systems the clearest interpretations of $\Delta V^{\ddagger}$ are in terms of protein conformation, denaturation, spin state equilibrium and/or solvation.

By collecting pressure data as many quantities as possible, one might eventually be able to come to conclusions as to the general concepts concerning the dynamics of the processes under study. This is one of the few avenues available for approching the problem of protein dynamics in solution.

ACKNOWLEDGEMENTS

I thank Profs. P. Douzou, A.B. Hooper, T. Yonetani, Drs. T. Barman, C. Kim, P. Masson, F. Travers and Mr. J.L. Saldana for their help and stimulating and fruitful discussions. This work was in part supported by grants from the Direction des Recherches, Etudes et Techniques (DRET n° 89037) and La Fondation pour la Recherche Médicale Française.

REFERENCES

1 . Ogunmola, G.B., Zipp, A., Chen, F and Kauzmann, W (1977) *Proc. Natl. Acad. Sci. USA,* **74,** 1-4.

2 . Hui Bon Hoa, G. and Marden, M.C. (1982) *Eur. J. Biochem.* **124,** 311-315

3 . Heremans, K. (1982) *Ann. Rev. Biophys. Bioeng.* **11,** 1-21.

4 . Marden, M.C., Hui Bon Hoa, G. and Stetzkowski-Marden, F. (1986) *Biophys. J.* **49,** 619-627.

5 . Alden, R.G., Satterlee, J.D., Mintorovitch, J., Constantinidis, I., and Ondrias, M.R. (1989) *J. Biol. Chem.* **264,** 1933-1940.

6 . Marden, M.C. and Hui Bon Hoa, G. (1982) *Eur. J. Biochem.* **129,** 111-117.

7 . Beece, D., Eisensten, L., Frauenfelder, H., Good, D., Marden, M.C., Reinish, L., Reynolds, A.H., Sorensen, L.B. and Yue, K.T. (1980) *Biochemistry* **19,** 5147-5157.

8 . Morild, E. (1981) *Adv. Prot.. Chem.* **34,** 93-166.

9 . Heremans, K. Snauwaert, J. and Rijkenberg, J. (1980) *Rev. Sci. Instrum.* **51,** 806-808.

10 . Balny, C., Saldana, J.L. and Dahan, N. (1984) *Anal. Biochem.* **139,** 178-189.

11 . Balny, C., Saldana, J.L. and Dahan, N. (1987) *Anal. Biochem.* **163,** 309-315.

12 . Neuman, R.C., Kauzmann, W. and Zipp, A.(1973) *J. Phys. Chem.* **77,** 2687-2691.

13 . Douzou, P. (1977) *Cryobiochemistry, an Introduction* Academic Press, N.Y.

14 . Balny, C. and Hooper, A.B. (1988) *Eur. J. Biochem.* **176,** 273-279.

15 . Glasstone, S., Laidler, K.J. and Eyring, H. (1941) *The theory of rate processes .* McGraw-Hill, N.Y.

16 . Balny, C., Travers, F., Barman, T. and Douzou, P. (1987) *Eur. Biophys. J.* **14,** 375-383.

17 . Low, P.S. and Somero, G.N. (1975) *Proc. Natl. Acad. Sci. USA,* **72,** 3014-3018.

18 . Caldin, E.F. and Hasinoff, B.B. (1975) *J. Chem. Soc. Faraday Trans.* **3,** 515-527.

19 . Frauenfelder, H. and Wolynes, P.G. (1985) *Science .* **229,** 337-345.

20 . Balny, C. and Travers, F. (1989) *Biophys. Chem.* **33,** 237-244.

21 . Andersson, K.K. and Hooper, A.B. (1983) *FEBS Lett.* **164** , 236-240.
22 . Hamann, S.D. (1984) *Aust. J. Chem.* **37**, 867-869.
23 . Ralston, I.M., Dunford, H.B., Wauters, J. and Heremans, K. (1981) *Biophys. J.* **36**, 311-314.
24 . Balny, C., Anni, H. and Yonetani, T. (1987) *FEBS Lett.* **221** , 349-354.
25 . Hooper, A.B., Tran, V.M. and Balny, C. (1984) *Eur. J. Biochem.* **141**, 565-571.
26 . Balny, C. and Hooper, A.B. (submitted).
27 . DiSpirito, A.A., Balny, C. and Hooper, A.B. (1987) *Eur. J. Biochem.* **162**, 299-304.
28 . Yamanaka, T. and Shinra, M. (1974) *J. Biochem. (Tokyo)* **75**, 1265-1273.
29 . Kornblatt, J.A., Hui Bon Hoa, G. and Heremans, K. (1988) *Biochemistry* **27**, 5122-5128.
30 . Asano, T. and Le Noble, W.J. (1978) *Chem. Rev.* **78**, 407-489.
31 . Balny, C., Masson, P. and Travers, F. (1989) High Press. Res. (in press).

THE PURIFICATION AND PROPERTIES OF GLYCEROL-3-PHOSPHATE

DEHYDROGENASE IN THE MITOCHONDRIAL INNER MEMBRANE

Qi-shui Lin, Shan-ping Shi and Jian-hua Liu

Shanghai Institute of Biochemistry, Academia Sinica
320 Yue-yang Road
Shanghai 200031
China

SUMMARY

Glycerol-3-phosphate dehydrogenase (E.C.1.1.99.5) was solubilized from rabbit skeletal muscle mitochondria by Triton X-100 and purified through hydroxyapatite column chromatography, DEAE-Sepharose CL-6B column chromatography and sucrose density gradient ultracentrifugation. The preparation was electrophoretically pure. The apparent molecular weight of the enzyme polypeptide was 69,000. There were 1.7 mg Triton X-100 and 26 μg phospholipid per mg protein of the preparation. Each molecule contained one molecule of FAD. The absorption spectra were studied after the Triton X-100 had been replaced by octylglucoside. It was observed that the absorption spectra altered upon the addition of the substrate, indicating that the reduction of the enzyme induced conformational changes. If the sulfhydryl groups of the enzyme were modified by iodoacetamide, two negative absorption peaks appeared at 256 nm and 305 nm, concomitant with the loss of enzyme activity. Chemical modification of amino side chain and tryptophan residues was also studied.

INTRODUCTION

Mitochondria glycerol-3-phosphate dehydrogenase (E.C.1.1.99.5), located on the outer surface of the inner membrane[1], catalyzes the dehydrogenation of glycerol-3-phosphaste to yield dihydroxyacetone phosphate with concomitant reduction of coenzyme Q. It is generally accepted that the mitochondrial glycerol-3-phosphate dehydrogenase and the NAD^+ linked glycerol-3-phosphate dehydrogenase in the cytosol constitute an α-glycerolphosphate cycle which enables the cytosol NADH to be oxidized by mitochondria [2-4]. The content of glycerol-3-phosphate dehydrogenase in mitochondria is relatively low. Only a few studies

---

Abbreviations: FAD, flavin adenline dinucleotide; NAD^+, nicotinamide adenime dinucleotide; Tris, [(tris hydroxymethyl] -aminomethane); DCPIP, dichlorophenolindolphenol; EDTA, ethylenediaminetetraacetic acid; Pi, inorganic phosphate; SDS, sodium dodecyl sulfate; TNBS, 2, 4, 6-trinitrobenzene sufonic acid.

concerning the purification of mitochondrial membrane bound glycerol-3-phosphate dehydrogenase were reported and the results were so far not satisfactory[5-7]. The purification method described in this communication was simple in manipulation and reproducible as compared with purification methods reported in the literature, and the preparation obtained was electrophoretically pure. The results, part of which was reported earlier[8], are of significance to further study of membrane enzymes.

EXPERIMENTAL PROCEDURES

Isolation of mitochondria. Rabbit spinal and great muscles were ground and to each 100 g ground muscles were added 1,000 ml 0.05 M Tris-HCl buffer, pH 7.4, containing 0.1 M KCl, 5 mM $MgSO_4$, 1.5 mM $CaCl_2$ and 1 mM mercaptoethanol. The mixture was homogenized in a Waring Blender with two 30-s bursts at high speed. The homogenate was centrifuged at 2,400 x g for 20 min. The resulting supernatant was centrifuged at 12,000 x g for 40 min. The pellet was washed with the above-mentioned buffer and centrifuged at 1,000 x g for 20 min, the supernatant was again centrifuged at 12,000 x g for 40 min, and the resulting sediment consisting of isolated mitochondria was kept in liquid nitrogen.

Determination of enzyme activity. Enzyme activity was determined spectrophotometrically, using DCPIP as an acceptor, the reaction medium containing 40 mM phosphate buffer, pH 7.4, 40 mM glycerol-3-phosphate, 120 µM DCPIP, 2 mM KCN and the total volume being 3.0 ml. The reaction was started by adding the substrate and the absorbance change at 600 nm was recorded by a Hitachi 557 dual wavelength double beam spectrophotometer.

Determiniation of differential absorption spectra of the enzyme. The differential absorption spectra of the enzyme were determined on a Hitachi 557 dual wavelength double beam spectrophotometer. When the substrate was added to the enzyme solution, the enzyme was reduced. At the same time, the product dihydroacetone phosphate was formed. Since there was no electron acceptor present in the reaction medium, the reduced enzyme could not be reoxidized. The concentration of the product could not be higher than that of the enzyme. For practical measurements of both the control and the assay, two cuvettes with optical paths of 1 cm each were used, one containing the enzyme and the other the buffer. When the substrate was added to the enzyme solution in the assay cuvette, an appropriate amount of dihydroxyacetone phosphate was added to the enzyme solution in the control cuvette. Thus, the absorption caused by dihydroxyacetone phosphate was subtracted. All other reagents were added into the buffer-containing cuvette and equivalent concentrations of the enzyme and the substrate and the product were maintained in the two cuvettes.

RESULTS AND DISCUSSION

Enzyme purification. The isolation of mitochondria from rabbit muscle was a very important step of the purification process. The skeletal muscle, abounding with muscle protein, contained a relatively low amount of mitochondria. The modification we made for the isolation of mitochondria involved: the addition of a small amount of Ca^{2+} to the homogenizing medium to prevent nuclear rupture and EDTA to avoid the agglutination of mitochondria caused by Ca^{2+} and Mg^{2+}; raising the ratio of homogenate; and an additional low speed centrifugation after harvesting the mitochondria through differential centrifugation.

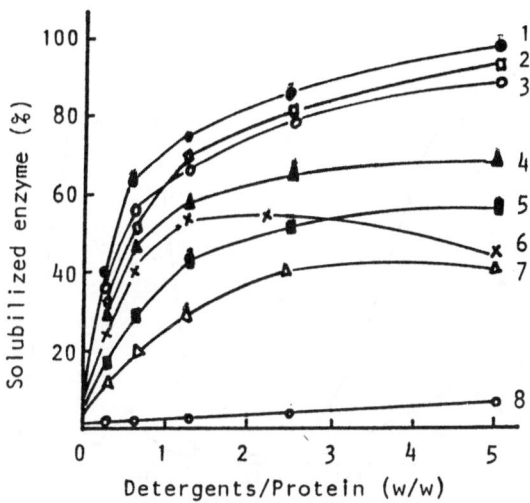

Fig. 1 **Solubilization of glycerol-3-phosphate dehydrogenase from rabbit muscle mitochondria by detergents.**
1. Triton X-100; 2. JFC; 3. O-P 10; 4. Brij 58; 5. Cholate; 6. Deoxycholate; 7. Lubrol WX; 8. Tween 20.

The solubilization effects of different detergents including Triton X-100, JFC, O-P 10, Cholate, Deoxycholate, Brij 58, Lubrol WX and Tween 20 were compared (Fig. 1). Tween 20 was almost unable to solubilize the enzyme and deoxycholate would inactivate the enzyme. All other detergents exhibit a certain solubilization effect and Triton X-100 was the best of all. A high salt concentration would enhance the solubilization effect of Triton X-100. Accordingly, 0.5 M NaC1 was included in the medium.

The three steps of purification including hydroxyapatite column chromatography, DEAE-Sepharaose CL-6B column chromatography and sucrose density gradient ultracentrifugation are summarized in Table 1.

Table 1. Purification of rabbit muscle mitochondria glycerol-3-phosphate dehydrogenase

	Activity (units)	Protein (mg)	Specific Activity (units/mg.)	Purification (fold)	Yield (%)
Mitochondria	113	1390	0.081	1.0	100
Triton X-100 extract	130	1180	0.11	1.3	115
Hydroxypatite	60	12.3	4.86	60	53
DEAE-Sepharose CL 6B	27	2.7	10.0	123	23.9
Ultracentrifugation	10.8	0.68	15.9	196	9.6

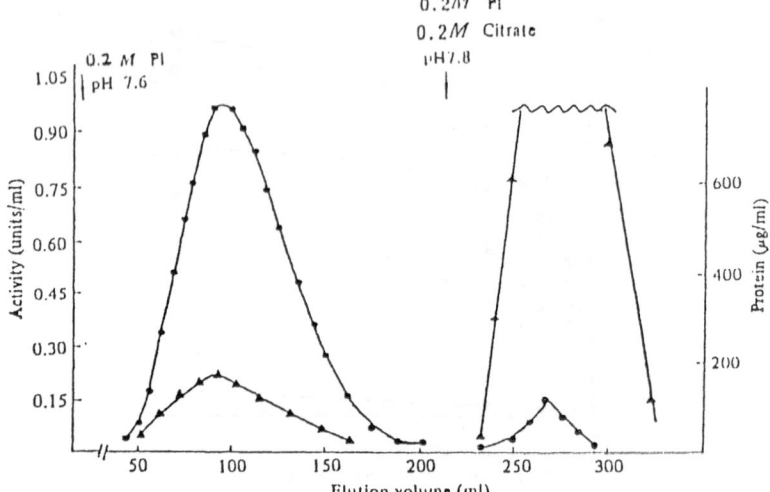

Fig. 2. Chromatography of glycerol-3-phosphate dehydrogenase from rabbit muscle mitochondria on hydroxyapatite. ● — ●, activity; ▲ — ▲, protein.

As shown in Table 1, the major purification step was hydroxyapatite column chromatography. the crux of a successful separation is that the equilibrium buffer must be at pH 7.6. Although the enzyme would not be absorbed at this pH, and under low concentrations of salt, most of the other proteins would still be absorbed on the column, thus leading to a highly efficient separation (Fig. 2). If the enzyme was allowed to be absorbed on the column and then eluted, a considerable amount of the enzyme would be inactivated.

The DEAE-Sepharose CL-6B column chromatography not only gave further purification but also had the advantage of getting rid of excess Triton X-100, thus facilitating the subsequent sucrose density gradient ultracentrifugation separation. 20 mM dithiothreitol was added to the gradient medium to prevent agglutination of the enzyme during centrifugation. This also indicated that at the low concentration of Triton X-100, aggregation was probably related to the formation of the intermolecular disulfide bond.

The gradient SDS-polyacrylamide slab gel electrophoresis of the purified enzyme gave a single protein band. On Ultrogel AcA-34 gel filtration, the purified enzyme appeared to have a symmetrical activity and protein peak. Thus, the purity of the preparation was satisfactory.

Properties. Some important properties are summarized in Table 2. It was clear from the SDS-polyacrylamide electrophoresis that the enzyme was composed of a single species of polypeptide. The molecular weight of the enzyme-Triton X-100 complex was 200,000 as calculated from the S value of the complex. The molecular weight of the enzyme polypeptide was 69,000, and there was 1.7 mg Triton X-100 per mg of protein in the complex, i.e. 117,000 per polypeptide chain which was equivalent to the size of a Triton X-100 micelle[8]. Taking into account the phospholipid in the complex, the molecular weight of the complex

Table 2. Properties of Purified Glycerol-3-Phosphate Dehydrogenase

Molecular Weight	69,000
FAD Content	1 mole/65,000 g protein
Triton X-100 Binding	1.7 mg/mg protein
Phospholipid Binding	0.025 mg/mg protein
Stokes' Radius	59 Å
Substrate Reducible Fraction	88%
Essential Sulfhydryl Group	1
Essential Amino Group	1
Essential Tryptophan Residue	1

should be 188,000, quite close to what was calculated from the S value. Thus it was clear that the enzyme in the complex existed as a monomer. Results from the determination of FAD showed that each 65,000 gram protein contained one mole FAD. This coincides with the fact that the molecular weight of glycerol-3-phosphate dehydrogenase was about 69,000. Garrib and McMurray reported the purification of glycerol-3-phosphate dehydrogenase from rat liver mitochondria and considered that the enzyme consists of four identical subunits of M_r 74,000 [8]. However, they determined molecular mass through gel filtration and HPLC but not through analytical ultracentrifugation. Moreover, they did not determine the Triton binding as what we have done, so the assumption that the enzyme is a tetramer is without solid grounding. It was reasonable to suggest that glycerol-3-phosphate dehydrogenase *in situ* also existed in monomeric form.

The sulfhydryl groups of glycerol-3-phosphate dehydrogenase were determined using DTNB. There are four SH groups per enzyme molecule. Among those two are located on the surface. One of the surface SH groups with less reactivity with DTNB is necessary for enzyme activity. Glycerol-3-phosphate facilitates the modification of the SH group, and hence potentiates the inactivation of the enzyme. The SH group modified enzyme has an increased K_m, 14.3 mM compared with 10.6 mM of the natural enzyme, and is unstable in comparison with the natural enzyme.

N-bromosuccinimide has been used as a modifier to determine the number of tryptophan residues per glycerol-3-phosphate dehydrogenase molecule as three; only one of them is exposed on the surface, which is essential for enzyme activity. The modification reagent 2, 4, 6-trinitrobenzene sulfonic acid TNBS has been used to detect 41 amino groups in the enzyme. One amino group showed the highest reactivity and was the first to be modified; when this was modified, all enzyme activity disappeared. The substrate protects this group and prevents TNBS inhibition. It is suggested that this amino group is located at the substrate binding site.

Absorption spectra. The absorption spectra of purified glycerol-3-phosphate dehydrogenase are shown in Fig. 3. In the ultraviolet region, there was an absoprtion peak at 282 nm and a negative peak at 254 nm. Moreover, the 414 nm absorption peak was

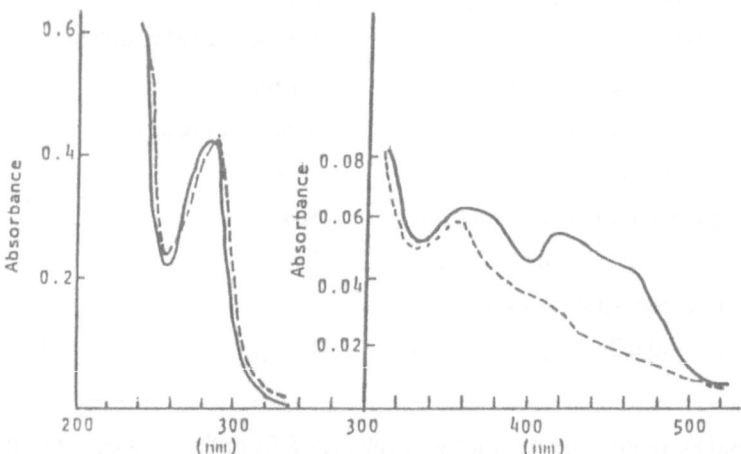

Fig. 3. Absorption spectra of rabbit muscle mitochondrial glycerol-3-phosphate dehydrogenase. The enzyme concentration was 12 µM. There was 1% Octylglucoside and 0.2 M NaCl present in the 5 mM MPS buffer (pH 7.4). The same buffer was used for the control.

— Purified enzyme; --- Enzyme reduced by substrate.

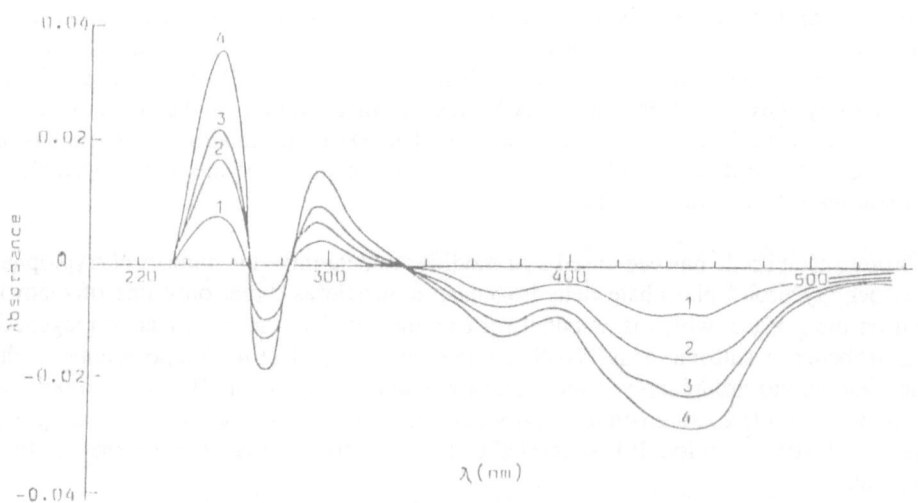

Fig. 4. Differential absorption spectra of rabbit muscle mitochondrial glycerol-3-phosphate dehydrogenase in the presence of different concentration of substrate. Substrate concentrations for the curves were: 1) $6 \times 10^{-5}M$ 2) $6 \times 10^{-4}M$ 3) $6 \times 10^{-3}M$ 4) $6 \times 10^{-2}M$.

related to non-haem iron, and the 360 nm absorption peak and a shoulder in the region of 460-465 nm were related to the prosthetic group FAD. Upon the addition of the substrate, the enzyme molecule was reduced and the absorption at 360, 414, and 460-465 nm greatly decreased. Based on calculations from the reduction spectra by dithionite, 88% of the purified enzyme could be reduced by the substrate. In general, the ultraviolet absorption spectra did not change substantially.

The differential absorption spectra of the enzyme in the presence of different concentrations of the substrate were shown in Fig. 4. In the ultraviolet region, there were two peaks at 296 and 254 nm and one negative peak at 276 nm. The height of those peaks was related to the concentration of the substrate. The negative peaks at 370 and 454 nm were related to the reduction of FAD. The absorption peak at 296 nm probably reflected the greater exposure of the tryptophanyl residue and the appearance of the negative peak at 276 nm was probably due to further hindrance of the tyrosyl residue. The absorption peak at 254 nm was probably related to the exposure of the phenylalnyl residue. Altogether, the appearance of these differential spectra indicated that in the vicinity of the enzyme-substrate binding center conformational changes had occurred.

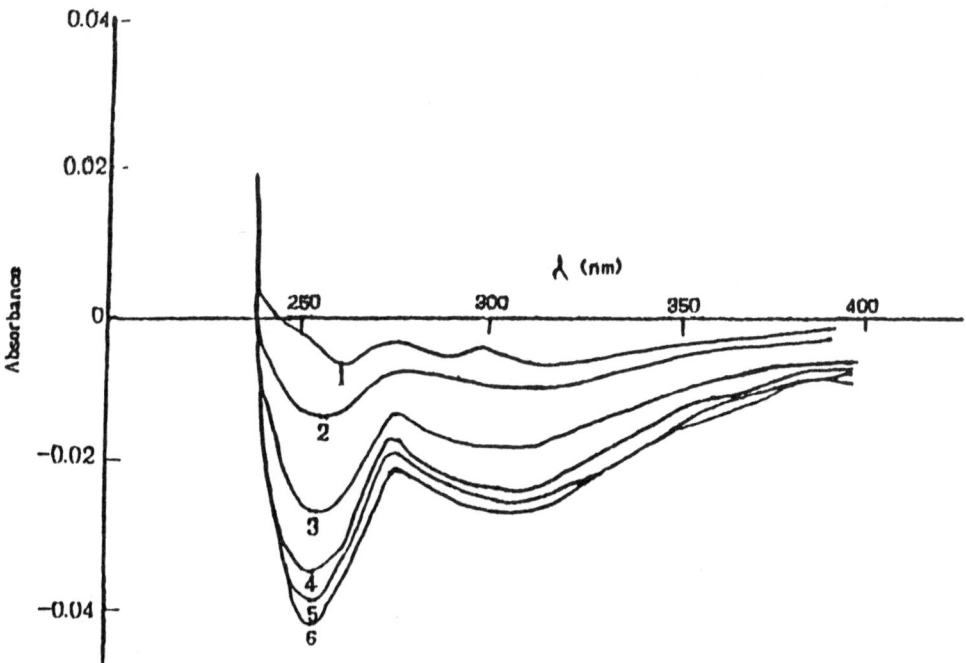

Fig. 5. Effect of iodoacetamide on the differential spectra in the ultraviolet region of rabbit muscle mitochondria glycerol-3-phosphate dehydrogenase. The reaction medium contained 0.05 M phosphate buffer, pH7.4, 1% octylglucoside, 0.4 M NaC1 and 0.12M iodoacetamide. The reaction was carried out at 37°C. Absorption spectra were scanned on a Hitachi 557 dual wavelength double beam spectrophotometer after adding Iodoacatamide at different time intervals. Identical amounts of enzyme and iodoacetamide were present in the control light path. Scanning times and corresponding residual activity were as follows:
1) 10 min, 85%; 2) 20 min, 70%; 3) 30 min, 60%;
4) 45 min, 55%; 5) 60 min, 50%; 6) 70 min, 40%.

   The influence of the modification of the sulfhydryl group on enzyme conformation was studied by observing the effect of iodoacetamide on ultraviolet differential spectra. The result is shown in Fig. 5. Two negative peaks at 256 and 305 nm appeared as the SH group was modified. If the reaction time was prolonged, the amount of modified SH group increased, the enzyme activity decreased and the peaks became higher. After 60 min. the peak reached their maxima and the residual activity was only 5%. It indicated that the modification of SH groups by iodoacetamide would induce delicate changes in the enzyme conformation, leading to its inactivation.

REFERENCES

1. Klingenberg, M., (1970) *Eur. J. Biochem.,* **13,** 247-252.
2. Hatefi, Y., and Stiggall, D.L., (1976) *The Enzymes, 3rd Ed., Vol. 13,* pp. 256-260. Academic Press, New York.
3. Salganicoff, L. and Fukami, M.H., (1972) *Arch. Biochem. Biophys.* **153.** 726-735.
4. Swierczynski, J. et al., (1976) *Biochim. Biophys. Acta.* **429,** 46-54.
5. Cole, E.S. et al., (1978), *J. Biol. Chem.,* **253,** 7952-7959.
6. Garrib, A. and McMurray, W.C., (1986) *J. Biol. Chem.,* **261,** 8042-8048.
7. Beleznai, Z. and Jancsik, V., (1987) *Biochem. Intl.,* **15,** 55-63.
8. Shi, S., Liu, J., and Lin, Q.S. (1986) *Sci. Sin. Ser. B, (Engl. Ed.),* **29,** 1027-1038.
9. Garrib, A., and McMurray, W.C. (1988) *J. Biol. Chem.* **263,** 19821-19826.

Part II. Molecular Biology

SUBUNIT 8 OF YEAST MITOCHONDRIAL ATP SYNTHASE: BIOCHEMICAL GENETICS AND MEMBRANE ASSEMBLY

Phillip Nagley, Rodney J. Devenish, Ruby H. P. Law,
Ronald J. Maxwell, Debra Nero and Anthony W. Linnane

Department of Biochemistry and Centre for Molecular Biology
and Medicine, Monash University
Clayton, Victoria 3168, Australia

SUMMARY

Subunit 8 is a small integral membrane protein of the proton-translocating F_0 sector of the mitochondrial ATP synthase complex. We here review our current understanding of the structure, expression and membrane integration of this protein, which is naturally encoded by the mitochondrial aap1 gene in bakers' yeast Saccharomyces cerevisiae. Genetic, biochemical and immunological analyses of yeast mutants deficient in subunit 8 production have begun to reveal the role of subunit 8 in the assembly and function of the mitochondrial ATPase complex. A recent major advance has been the recoding of the gene encoding subunit 8 to achieve its relocation to the nucleus such that nuclearly encoded subunit 8 can be demonstrated to functionally assemble into the mitochondrial ATPase complex. Further, the expression of subunit 8 in vitro, in the form of a chimaeric precursor bearing an N-terminal cleavable presequence, has permitted study of the import of the protein into isolated mitochondria and its assembly into the enzyme complex. The powerful combination of in vivo and in vitro approaches has now led to the systematic manipulation of subunit 8 using site-directed mutagenesis in order to gain further insight into its structure and function.

INTRODUCTION

In the mitochondrial proton-translocating ATP synthase (mtATPase), the membrane-embedded F_0 sector consists of three hydrophobic subunits. Two of these, subunits 6 and 9, contain transmembrane helices with charged amino acids which are thought to determine proton conductivity across the membrane. Homologues of these two subunits are found in all F_1F_0-ATPases, namely those of bacteria, mitochondria and chloroplasts [1,2]. The mtATPase of fungi and metazoa contain a third hydrophobic F_0 subunit known variously as subunit 8 (fungi) or A6L (metazoa). This subunit is characterized by its short length (about 50 to 70 residues), a central non-polar region representing a single transmembrane stem, and a polar C-terminus with a predominance of positively charged residues [2].

Most of the detailed information on the structure and function of subunit 8 has resulted from intensive investigations on bakers' yeast,

Saccharomyces cerevisiae. Since the initial identification of subunit 8 as an integral component of mtATPase and the product of the yeast mitochondrial aap1 gene [3,4], the application of a integrated array of genetic, biochemical and immunochemical techniques has indicated a critical role for subunit 8 in the assembly of the functional F_O sector [5]. Subsequent efforts in gene construction, gene expression and protein targeting have led to the functional transfer of the subunit 8 gene to the yeast nucleus [6] which has opened up a new chapter in the systematic manipulation of mitochondrial bioenergetic systems. It is the purpose of this article to review these developments and to indicate some new directions in the dissection of mitochondrial membrane assembly and (function.

STRUCTURE, EXPRESSION AND MEMBRANE INTEGRATION OF SUBUNIT 8

A class of mitochondrial mutants defective in mtATPase function was shown by Macreadie and colleagues [3] to be deficient in subunit 8, the responsible mutations being shown to map outside of the previously identified mitochondrial genes oli1 and oli2 encoding subunits 9 and 6, respectively. The newly identified aap1 gene in mtDNA that harbored these mutations was analyzed at the level of nucleotide sequencing [4] and was found to encode a polypeptide 48 amino acids in length. The predicted amino acid sequence was confirmed by direct analysis of purified subunit 8 protein [7]. Whilst known earlier as a "10 kilodalton protein" [3,8], subunit 8 in fact has an M_r of 5,800. The apparent gel mobility of 10 kDa for subunit 8 was erroneously assigned on the basis of the more rapid gel mobility of subunit 9, whose previously determined sequence [9] indicated an M_r of 7,800. In fact, subunit 9 has subsequently been shown to have anomalous gel electrophoretic properties and often migrates with an apparent size of 3.5 kDa, while subunit 8 migrates at a rate corresponding to its M_r [10].

In wildtype yeast mitochondria, subunit 8 is expressed in the form of dicistronic mRNA molecules which include the aap1 reading frame, 706 nucleotides downstream of which is found the oli2 reading frame encoding subunit 6 [11]. These long candidate messages (about 4.5 or 3.9 kb in size) [12] arise by post-transcriptional cleavage of a precursor mRNA at least 16 kb in length, whose promoter lies upstream of the oxi3 gene encoding cytochrome oxidase subunit I [13,14]. This processing yields composite aap1-oli2 mRNA molecules, with a 5'-untranslated region (about 900 or 300 nucleotides) preceding the aap1 reading frame. This mode of transcription, whereby the mature mRNA is generated by RNA processing, yielding long flanking untranslated sequences in mRNA, is typical of mitochondrial gene expression in S. cerevisiae [15].

There is very little information concerning the transcriptional or translational regulation of expression of the aap1 gene. Moreover, it is not at all certain whether there is any functional significance to the dicistronic yeast mitochondrial mRNA encoding both subunits 8 and 6, which has parallels in mtDNA of other organisms [11], notably the overlapping A6L and subunit 6 reading frames in mammalian mtDNA [16]. The biosynthesis of subunit 8 and 6 can be effectively disconnected under several circumstances. Thus, in oli2 mutants of yeast defective in subunit 6 biosynthesis, subunit 8 is produced and is also assembled into mtATPase [5,17]. In aap1 mutants which are unable to make subunit 8, subunit 6 is continues to be synthesized [18], although the latter subunit is not assembled into mtATPase [5,17]. Finally, in a reconstructed yeast strain in which subunit 8 is expressed solely from a nuclear gene and imported into mitochondria [19], subunit 6 is produced and is assembled with the imported subunit 8 into the F_0 sector to produce functional mtATPase [6].

Following its biosynthesis on mitochondrial ribosomes, subunit 8 is delivered into the inner mitochondrial membrane. The overall hydrophobic character of subunit 8 as judged by its amino acid sequence, is consistent with the ready solubility of this polypeptide in chloroform/methanol. In other words, subunit 8 is a proteolipid [7]. The precise physical relationship of subunit 8 to the lipid bilayer (Fig. 1A) of the inner membrane has been assessed on the basis of both theoretical considerations and biochemical experimentation.

Using the method of Von Heijne and colleagues [20] to identify hydrophobic stretches in the protein, Velours and colleagues [7] predicted residues 15-35 of subunit 8 to be located in the inner mitochondrial membrane. Various other predictive methods for structural motifs consistently indicated an α-helical structure for this segment of the protein and Velours and colleagues [7] reached the conclusion that subunit 8 very likely traverses the membrane once by this α-helical segment. The C-terminal portion of subunit 8 from residue 36 was predicted to be outside the membrane because of its content of 3 basic residues.

Concerning the orientation of subunit 8, Velours and Guerin [21] subsequently demonstrated that the C-terminal moiety of subunit 8 is located on the matrix side of the inner mitochondrial membrane. Subunit 8 was chemically modified using the polar, lipid-insoluble and non-penetrating reagent isethionylacetimidate, which specifically reacts under physiological conditions with primary amines. Sodium bromide-treated submitochondrial particles (inside-out inner membrane vesicles) were prepared, in which were exposed the hydrophilic protein domains of integral mitochondrial membrane proteins that normally face the aqueous environment of the mitochondrial matrix. It was found by Velours and Guerin [21] that the affinity reagent bound to the ξ amino group of the only lysine residue (position 47) of subunit 8; this residue contains the only free amino group in yeast subunit 8 since the N-terminal methionine residue of this polypeptide is blocked by N-formylation (cf. Fig. 1A).

In our laboratory we have also sought to predict precisely those residues making up the transmembrane stem. First, a Kyte-Doolittle [22] hydropathy analysis of the yeast subunit 8 sequence, using a 7 amino acid window (Fig. 2A), showed a distinct hydrophobic peak with its N-terminal boundary close to residue 15, but with a much less clearly demarcated C-terminal boundary. The analysis of Velours and colleagues [7] indicated a more polar C-terminal tail than is apparent in Fig. 2A. A recent version of the Goldman, Engelman and Steitz (GES) hydrophobicity algorithm [23] has also been utilized in our analysis because it calculates the hydrophobicity of each amino acid residue as if it were situated in an α-helical transmembrane stem. It is assumed that a stretch of 19-20 amino acids is sufficient to span the lipid bilayer. Further, this GES algorithm has been used to make predictions of the position of the transmembrane stems of subunits comprising the photosynthetic reaction centre of Rhodospirillum viridis that are in close agreement with determinations made using X-ray crystallography [23].

The pattern revealed by GES analysis for subunit 8 of S. cerevisiae (Fig. 2B) is not readily definitive of the transmembrane stem. This plot

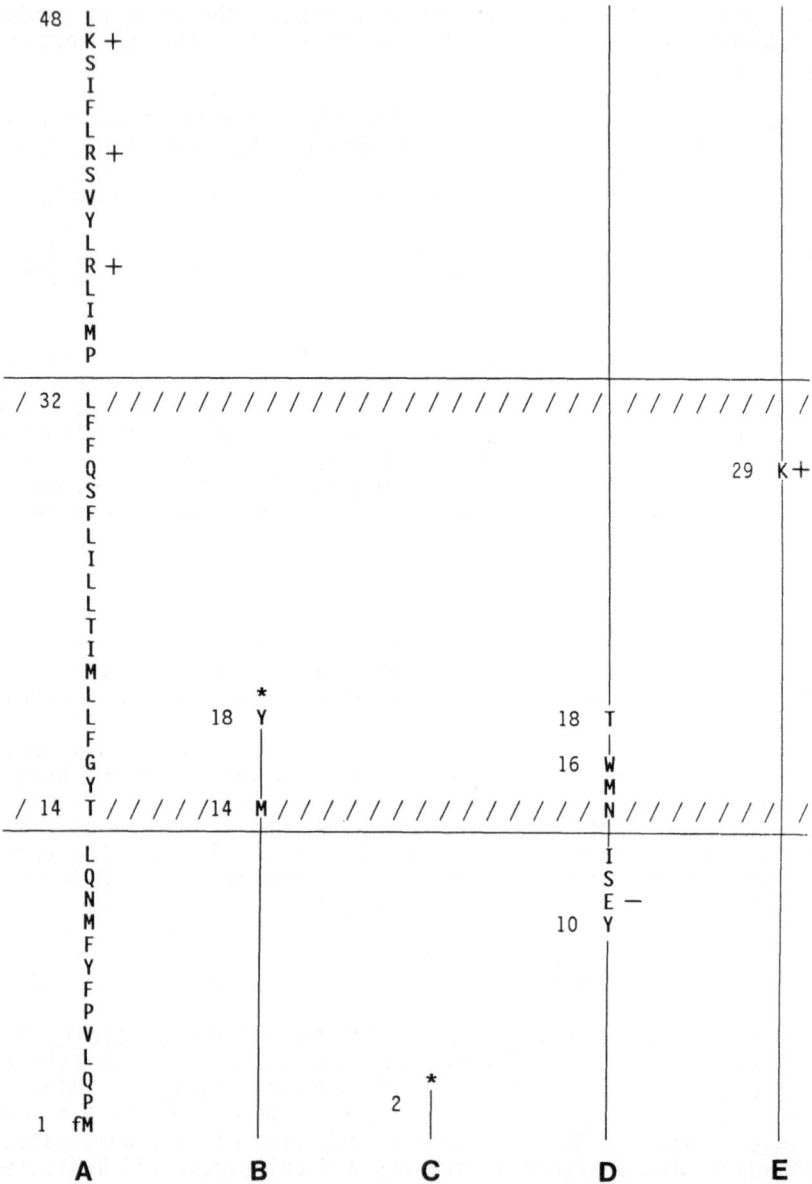

Fig. 1. Topology of mtATPase subunit 8 in S. cerevisiae: natural and synthetic mutants. The orientation of the wildtype subunit 8 and proposed boundaries with respect to subunit 8 of the lipid bilayer of the inner mitochondrial membrane (enclosed within hatched regions) are discussed within the text. The N-terminal residue is N-formyl methionine (fM); all other amino acids are indicated using the standard single letter code. The full sequence is shown only for the wildtype protein; for all others only those amino acids changed as a consequence of mutation or reversion events are shown. Other indications: + or -, charged amino acids; *, in-frame stop codons. A, wildtype; B, aap1 mit⁻ mutant M26-10; C, aap1 mit⁻ mutant M31; D, revertant M26-10R7; E, synthetic mutant Q29→K.

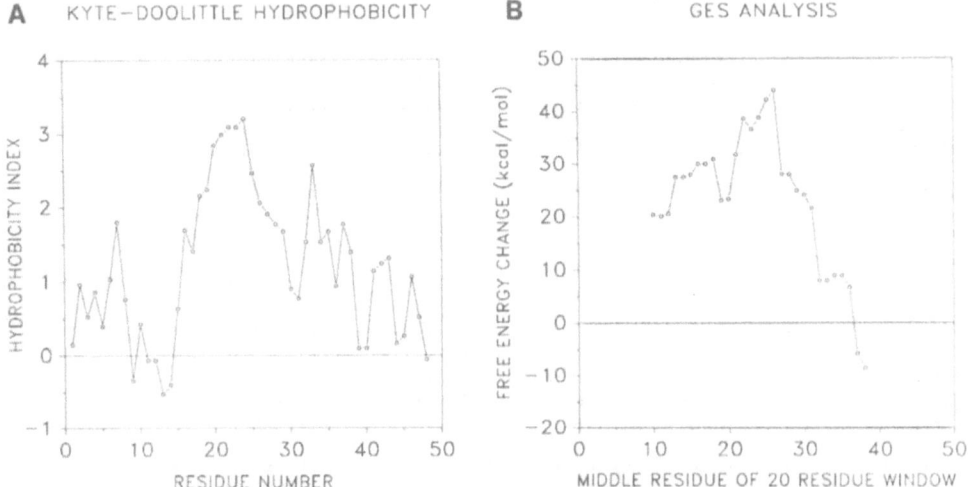

A KYTE—DOOLITTLE HYDROPHOBICITY   **B** GES ANALYSIS

Fig. 2. Hydropathy plots of mtATPase subunit 8 from S. cerevisiae. A, hydropathy plot constructed using the algorithm of Kyte and Doolittle [22] using a 7 amino acid window. B, hydropathy plot constructed using the GES algorithm described by Engelman, Steitz and Goldman [23] using a 20 amino acid window. The latter was obtained using the FOAM-PC program kindly provided by Dr. R. M. Macnab, Yale University.

shows multimodal hydrophobic peaks, implying that all but the C-terminal portion of the protein is of great hydrophobic character. The two most prominent peaks are centred on residues 22 and 26, with a further prominent hydrophobic zone towards the N-terminus. As the intention is to accurately position the transmembrane stem, clarification of the situation may be achieved by making both comparative sequence and hydropathy plots for subunit 8 of different organisms. In Fig. 3 is presented a sequence comparison of subunit 8 for five different fungi. The sequence of each organism has been compared to S. cerevisiae subunit 8 in Fig. 3A. All sequences have been aligned at the N-terminal methionine residue and no deletions/insertions have been made to accommodate the alignment. This comparison reveals conservation of positively charged C-terminal regions as such (Fig. 3B); note that the filamentous fungi Neurospora crassa and Aspergillus nidulans have positive charges further into the central region of the protein than does S. cerevisiae and its closer relatives Torulopsis glabrata and Schizosaccharomyces pombe. Extensive sequence divergence is also evident in the membrane-spanning hydrophobic portion of the protein as well as in the C-terminal tail regions of the protein. The sequence of bovine A6L is shown for comparison in Fig. 3A, and is seen to be almost totally diverged in sequence from that of yeast or fungal subunit 8. Nevertheless, the overall distribution of hydrophobic and charged residues is broadly conserved with respect to fungal subunit 8 proteins (Fig. 3B), although as suggested below from hydropathy plots, the transmembrane stem of A6L may lie closer to the N-terminus of this protein. Indeed Walker and colleagues [27] have pointed out that the 'MPQL' sequence homology at the N-terminus of mammalian A6L and fungal subunit 8 may not reflect the most appropriate alignment of these proteins to obtain the maximal homology overall, which after adjustment is still not extensive.

A

```
                    10        20        30        40        50        60
S. cerevisiae  MPQLVPFYFMNQLTYGFLLMITLLILFSQFFLPMILRLYVSRLFISKL
T. glabrata    ....I...........L..ITV.........................
S. pombe       .........I.I.SF...IFTV..YIS.VYV..RYNE.FI..SI..S.
A. nidulans    .......F.V..VVFA.IVLTV.IYA..KYI..RL..T.I..IY.N..
N. crassa      .........V.EI.FT.VIITLMVYIL.KYI..RFV..FL..T.....SDISKK

B. bovis       ....DTSTWLTMILSM..TLFIIFQ.KVSKHNFYHNPELTPTKMLKQNTPWETKWTKIYLPLLLPL
```

B

```
                    10        20        30        40        50        60
S. cerevisiae  ..............................+....+....+.
T. glabrata    ..............................+....+....+.
S. pombe       ............................+..-....+.....
A. nidulans    ...........................+....+..+....+.....+.
N. crassa      ...........-...............+....+..+....+....+..-..++

B. bovis       ....-...................+..++...+..-....+..+......-.+..+.........
```

Fig. 3. Sequence comparisons and charge distributions in mtATPase subunit 8 from different organisms. A, the complete amino acid sequence (standard single letter code) is given for subunit 8 of S. cerevisiae. For other species amino acids homologous to those for S. cerevisiae subunit 8 are denoted by a dot, and differing amino acids by their single letter code symbol. No gaps have been introduced into sequences to achieve alignment of homologous amino acids. B, position of charged amino acids in sequences of subunit 8 and A6L. Positively and negatively charged residues are indicated by + and -, respectively; all other residues are represented by a dot. Sources for amino acid sequences were as follows: S. cerevisiae [4], S. pombe (B. F. Lang personal communication), A. nidulans [24], N. crassa [25], B. bovis (A6L) [26]. The residue number is indicated at the top of each panel.

The comparative hydropathy plots (Fig. 4) using the GES algorithm serve to clarify the position of the C-terminal membrane boundary in the subunit 8 proteins. It is evident that the plot for S. cerevisiae subunit 8 is atypical in the sense that its hydrophobic region extends much further towards the C-terminus than do those of subunit 8 from the fission yeast S. pombe (Fig. 4A) and both the filamentous fungi, A. nidulans (Fig. 4B) and N. crassa (Fig. 4C). Subunit 8 protein from these organisms lack the most prominent peak of S. cerevisiae centred on residue 26. Note that the comparison between subunit 8 of S. cerevisiae and T. glabrata is not informative because the sequences are almost identical (Fig. 3A), and the hydropathy plots almost exactly superimpose (data not shown). It is also striking that the bovine A6L protein (Fig. 4D) follows a very similar pattern for its hydrophobic domain as the fungal subunit 8 (cf. [7]), except that its charged polar C-terminal region (Fig. 3B) is substantially more extended. Indeed, bovine A6L seems to have a truncated N-terminal membrane-exterior domain relative to fungal subunit 8 proteins; note the scale displacement of the A6L analysis at the top of Fig. 4D.

The interim conclusion we reach from the above considerations is to locate the transmembrane stem of S. cerevisiae subunit 8 so as to be centred on residue 23. With a 19 amino acid transmembrane stretch as suggested by Engelman and colleagues [23], this implies residues 14 to 32 of subunit to be buried in the bilayer (Fig. 1A). This assignment

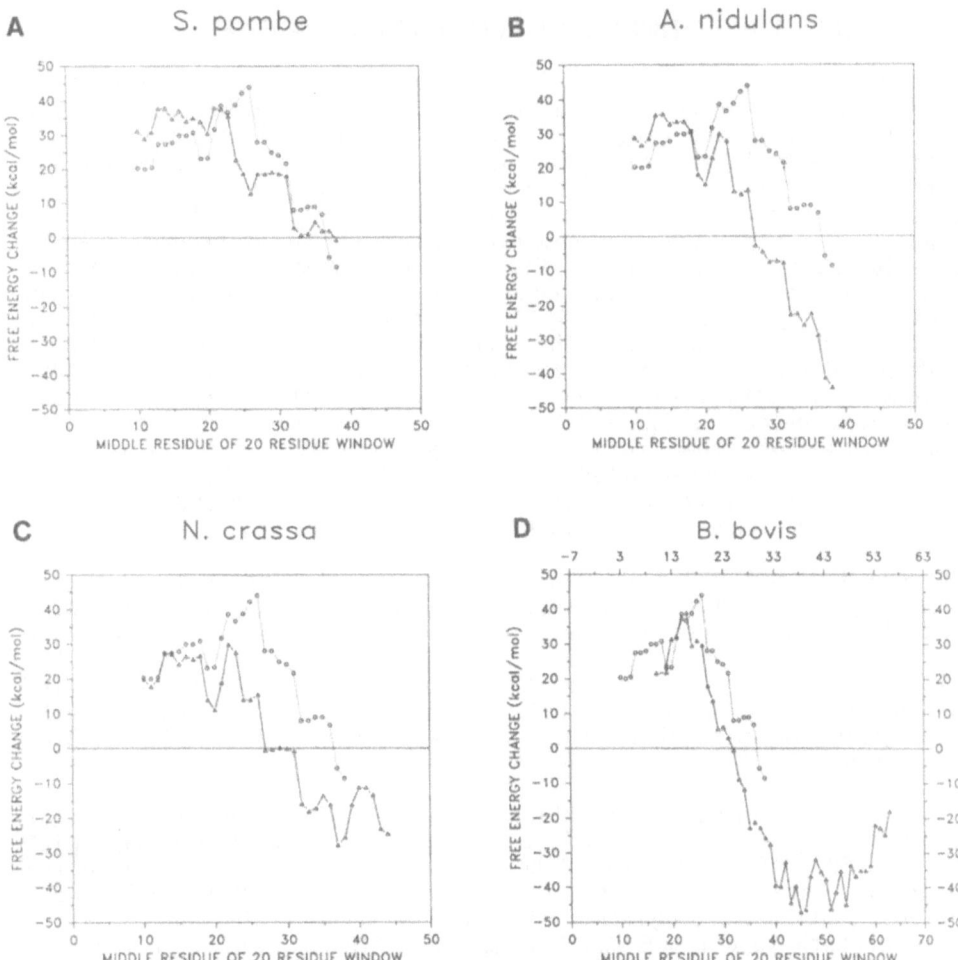

Fig. 4. Comparative hydropathy plots of mtATPase subunit 8 from different organisms. Hydropathy plots were constructed using the GES algorithm of Engelman, Steitz and Goldman [23], using a 20 amino acid window (cf. Fig. 2B). A, S. pombe; B, A. nidulans; C, N. crassa; D, B. bovis. For each organism the hydropathy plot for subunit 8 (triangles, solid line) is shown in direct comparison to that of S. cerevisiae (circles, broken line). The abscissa is identical for each comparison except in D, where an adjusted scale for B. bovis is used (top of panel).

balances two sorts of indicators. First is the proposition to shift the position of the C-terminal boundary of the transmembrane stem of S. cerevisiae subunit 8 leftwards from residue 35 (based on the GES algorithms and the fungal comparisons with subunit 8 and A6L). Second are the indications from Kyte-Doolittle (Fig. 2A) and Von Heijne [7] hydropathy plots suggesting that the N-terminal boundary of the transmembrane stem should not be placed much more upstream than residue 15. Further biochemical experiments will be needed to test this assignment more rigorously (but see discussion below of subunit 8 variants illustrated in Fig. 1D and E).

ROLE OF SUBUNIT 8 IN ASSEMBLY AND FUNCTION OF ATP SYNTHASE

The preliminary studies on the functional and assembly defects in aap1 mit⁻ mutants were made in parallel with the molecular genetic characterization of the aap1 gene [3]. The biochemical and bioenergetic phenotype of mutants was clarified by the definition of the nucleotide sequence of this gene and the sequence changes in a group of aap1 mit⁻ mutants and revertants [3]. Thus, the respiratory growth defect in the mit⁻ mutant M26-10, which is a complex frameshift mutant producing an 18 amino acid truncated polypeptide (Fig. 1B), was shown to result primarily from a total loss of ATP-Pi exchange activity in mitochondria [5]. As in the case for oli2 and oli1 mit⁻ mutants defective in synthesis of subunits 9 and 6, respectively, this aap1 mit⁻ mutant showed (compared to a wildtype control) substantial reduction in respiratory activity, NADH-cytochrome c reductase and cytochrome c oxidase activities. These data indicate that defective F_0 sector function has a deleterious effect on the activities of other respiratory enzyme systems; such pleiotropic effects are greatly accentuated in certain oli1 mit⁻ mutants that lack subunit 9, in which no significant assembly of cytochrome c oxidase occurs and which hence have negligible respiration [5,28].

A more detailed analysis was reported by Marzuki and colleagues [18] of the enzyme disturbances in three aap1 mit⁻ mutants, all containing truncated subunit 8 polypeptides, the extreme of which is M31 that retains only the first two residues (Fig. 1C). In all three mutants the total ATPase activity is about half that of wildtype, and this activity is inhibited only marginally by oligomycin; the ATP-Pi exchange activity is negligible. These results clearly indicate a defect in the function of the F_0 sector.

The earliest immunoprecipitation data, obtained using a polyclonal antibody raised against holo-mtATPase, suggested that in the aap1 mit⁻ mutant M26-10, neither subunit 8 nor subunit 6 were assembled into the complex [3]. More detailed assessments of mtATPase assembly, utilizing a monoclonal antibody recognizing the F_1-β subunit [29], confirmed the absence of both subunits 6 and 8 from the mtATPase complex of the aap1 mit⁻ mutants. The mtATPase did, however, contain subunit 9 associated with an assembled F_1 sector [17,18]. The critical role of subunit 8 in facilitating the assembly of subunit 6 was emphasized by the demonstration that in these aap1 mit⁻ mutants, subunit 6 continues to be produced inside mitochondria although it is not assembled into mtATPase [18]. Nonetheless, the conclusion could not be drawn from these studies that subunit 8 has a specific functional role in the F_0 sector, since subunit 6 itself is known to be required for a functional F_0 sector [30].

The sequence changes in revertants of the M26-10 frameshift mutation were reported by Macreadie and colleagues [4]. One of these, M26-10R1, restores the reading frame at the site of the original frameshift, leading to the amino acid substitution Leu18→Thr, in addition to a second substitution (Thr14→Met) which results from a further base substitution in M26-10 (see Fig. 1B). The revertants M26-10R7 and M26-10R13 both undergo compensatory base insertions about 25 nucleotides upstream of the original frameshift (itself a deletion of a thymine residue), resulting in both cases in a full length subunit 8 containing a stretch of amino acid replacements, as illustrated for M26-10R7 in Fig. 1D. Incidentally, nucleotide sequence changes in the revertants M26-10R7 and M26-10R13 were crucial to the assignment, on molecular genetic grounds, of a 48 amino acid polypeptide as being the authentic product of the aap1 gene; only one of the two candidate reading frames in this mtDNA region was found to be appropriately restored by the reversion events [4].

Measurement of growth properties and biochemical functions of the M26-10 revertants has provided some evidence for a possible functional role of subunit 8 in mitochondrial energy transduction. In spite of the two substitutions in its revertant subunit 8 protein, M26-10R1 grows as well as the wildtype parent J69-1B, and shows barely perceptible changes in its respiratory energy coupling and ATPase function [18]. By contrast, M26-10R7 and M26-10R13 are temperature-sensitive, showing reduced growth rates at both 18°C and 36°C [4]. Significantly, both revertants display definitive lesions in the respiratory coupling ratio, and in the ATP-Pi exchange activity [18]. M26-10R13 is more severely affected than M26-10R7 and shows a reduction in the oligomycin-sensitivity of its mtATPase, which correlates with an extra amino acid substitution (Tyr8→Ile) additional to the other 8 altered residues in these revertants (cf. Fig. 1D). Since subunit 6 is apparently assembled in all three revertants studied [18], the partial bioenergetic defects in M26-10R7 and M26-10R13 suggest a specific functional role for subunit 8 in the F_0 sector.

In terms of the positioning of the putative transmembrane stem of yeast subunit 8 (Fig. 1) it may be noted that the amino acid changes in revertant M26-10R7, which has a normal oxidative growth rate at 28°C, are centred on the N-terminal membrane boundary (Fig. 1D). A negatively charged glutamic acid (Asn11→Glu) is predicted to lie outside the membrane, whilst the polar residue asparagine (Thr14→Asn) lies directly at the membrane interface. It is thus unlikely that the boundary of the transmembrane stem lies much further towards the N-terminus of the protein than shown in Fig. 1D, applying the principle that highly polar or charged residues are unlikely to be buried in the hydrophobic lipid bilayer.

The conclusion that subunit 8 in yeast does itself play a role in the bioenergetic functions of mtATPase, in addition to its role in F_0 sector assembly, is supported by the studies of Higuti and colleagues on the A6L subunit of mammalian mtATPase. Studies on rat liver mitochondria photoaffinity labelled with the anisotropic inhibitor, monoazide ethidium, have demonstrated a specific binding of the inhibitor to proteins of mitochondrial membranes [31]. Purification of these mitochondrial components yielded two hydrophobic proteins of apparent sizes 8 and 13 kDa, named chargerin I and II respectively, whose affinity for the monoazide ethidium *in situ* was dependent on the state of mitochondrial energization [31]. Amino acid sequencing of a peptide derived from chargerin II showed it to have a sequence identical to a run of 12 amino acids from the C-terminal portion of rat A6L protein, in a region relatively well conserved among mammals [32]. Antibodies raised against purified chargerin II [31] were shown to inhibit ATP synthesis in isolated mitochondria. Moreover, ATP-Pi exchange activity was inhibited in an energy-dependent fashion [33], supporting the conclusion that energy transduction affects the conformational state of chargerin II. The findings are consistent with a role for subunit A6L in the energy transduction function of mammalian mtATPase.

Historically, an inorganic phosphate-binding property has been ascribed to certain proteolipids from yeast [34] and bovine [35] mitochondria, that were subsequently shown to be subunit 8 and A6L, respectively [7,26]. However, the physiological significance of this binding has not yet been established and it may represent an adventitious interaction of phosphate ions with clusters of the positively charged residues present in C-terminal regions of each proteolipid. There is no direct evidence for a role of either subunit 8 or A6L in the transport of inorganic phosphate across the mitochondrial membrane.

MANIPULATION OF SUBUNIT 8 BY ALLOTOPIC EXPRESSION IN THE NUCLEUS

The investigations on subunit 8 from S. cerevisiae described above have relied upon naturally occurring mutations in mitochondrial DNA, analysis of which led to the assessment of their impact on the structure of subunit 8 and their consequences for the assembly and function of mtATPase. It would indeed be of great utility to be able to manipulate the structure and expression of subunit 8 by directed mutagenesis of the mitochondrial aap1 gene, but various technical characteristics of yeast mitochondria in practice prevent the manipulation and expression of mtDNA for this purpose, either in vivo or in vitro. Thus, there is no efficient system for transforming mtDNA into the mitochondria of host yeast cells [36], although preliminary data are now being obtained with a novel approach utilizing microscopic projectiles, coated with DNA, shot directly at target cells [37, 38]. Moreover, it has not been possible to undertake studies on mitochondrial proteins based on the expression of yeast mtDNA sequences into the encoded protein products by transcription and translation in vitro. This is primarily because of the very low yields of such products, either using homologous or heterologous protein synthesizing systems.

It was therefore realized [36] that the optimal approach for studying mitochondrially encoded proteins, such as subunit 8 of mtATPase, was to restructure these genes for expression in the nucleocytosolic system which is well established for efficient and specific protein output both in vitro and in vivo (cf. Fig. 5). This strategy frees manipulation of the protein from constraints imposed by the organellar location of the gene. This process of relocating organellar genes to the nucleus, and redirecting the encoded proteins back to the organelle in order to explore the natural cellular protein targeting and import mechanisms, is termed allotopic expression [19]. As described below, allotopic expression of subunit 8 has permitted the implementation of a range of gene manipulation and expression strategies, which, when coupled with import and enzyme assembly technologies, provides new paradigms for analyzing the assembly and function of subunit 8.

The key to allotopic expression of subunit 8 was the construction of a chemically synthesized gene [40] that was designed for optimal expression in the nucleocytosolic system of yeast. Factors taken into account in gene design were differences in codon dictionary and preferred codon usage between the nucleocytosolic and mitochondrial translation systems of yeast. The synthetic gene containing 31 such modified codons directed the expression in vitro of a polypeptide which was identical in properties to authentic subunit 8 [36].

In order to facilitate import of the relocated subunit 8 protein, it was necessary to furnish the protein with a cleavable N-terminal leader sequence (Fig. 5A). The sequence chosen on the basis of its ability to direct import of a hydrophobic protein into yeast mitochondria was the 66 amino acid leader (N9L) of Neurospora crassa mtATPase subunit 9 (N9). The precursor to N. crassa subunit 9 (pN9) is encoded in the nucleus; the long hydrophilic positively charged N9L leader is responsible for the delivery of the hydrophobic 81 amino acid proteolipid subunit 9 into

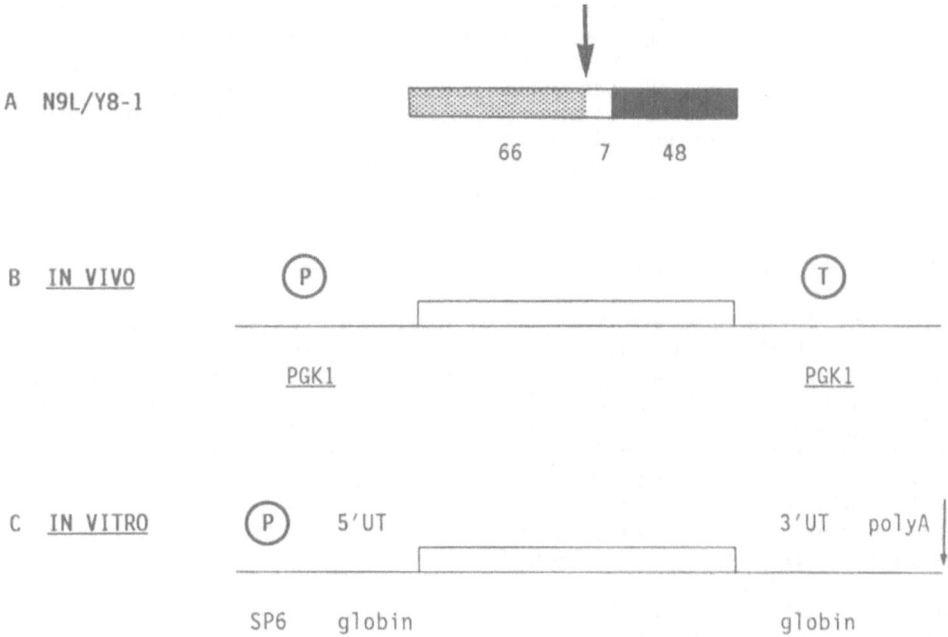

Fig. 5. Gene expression to produce yeast mtATPase subunit 8 in the nucleocytosolic compartment in a form suitable for import into mitochondria. A, gene encoding chimaeric precursor N9L/Y8-1 whose Y8 component was recoded for nucleocytosolic expression. Hatched box, N9L encoding sequence (specifies 66 amino acids); open box, sequence encoding the first 5 amino acids of mature subunit 9 of N. crassa plus two serine residues created by fusion of N9L to Y8 (specifies a total of 7 amino acids); filled box, Y8 coding sequence (specifies 48 amino acids); arrow indicates matrix protease cleavage site. B, allotopic expression of N9L/Y8-1 in vivo. The gene in context of yeast expression vector pLF1 [6] is shown under control of transcriptional signals from the yeast PGK1 gene: P, promoter and T transcription terminator. C, expression of N9L/Y8-1 in vitro. The gene in context of pSP64T expression plasmid [39] is shown under control of the SP6 transcriptional promoter P, allowing efficient transcription in vitro using the prokaryotic SP6 RNA polymerase; run off transcription proceeds as far as the point at which the plasmid is linearized by restriction endonuclease cleavage (arrow). Efficient translation of this mRNA in a rabbit reticulocyte lysate system is facilitated by the presence of translation control signals from the 5' and 3' untranslated (UT) sequences flanking the Xenopus laevis β-globin gene; poly A sequences are also represented in the in vitro transcript.

mitochondria [41,42]. Moreover, the pN9 precursor is able to be imported efficiently into mitochondria of yeast S. cerevisiae [43].

Fusion in-frame of DNA encoding the N9L sequence to the artificial subunit 8 gene [44] created a gene construct encoding the chimaeric precursor, N9L/Y8-1, that comprises the N-terminal leader peptide of 66 amino acids, the first 5 amino acids of mature N. crassa subunit 9, 2 additional serine residues resulting from the DNA sequence at the fusion point, and the 48 amino acids of subunit 8 (Fig. 5A). This chimaeric precursor is able to be imported into yeast mitochondria both <u>in vivo</u> [6] and <u>in vitro</u> [44], where it is processed at the natural cleavage point of pN9 (where the N9L and N9 sequences abut) recognized by the chelator-sensitive matrix protease enzyme [45]. The <u>in vivo</u> data concerning incorporation of imported subunit 8 into functional mtATPase will be first considered below, followed by a discussion of the ability of subunit 8 to be imported and assembled into mtATPase in isolated mitochondria.

ASSEMBLY OF NUCLEARLY ENCODED SUBUNIT 8 INTO mtATPase <u>IN VIVO</u>

In order to demonstrate <u>in vivo</u> allotopic expression of subunit 8, use was made of the <u>aap1</u> mutants M26-10 and M31 (Fig. 1B and C) that are unable to synthesize subunit 8 naturally inside mitochondria and therefore are unable to utilize the respiratory substrate ethanol for cellular growth. Expression of N9L/Y8-1 in these mutant cells was achieved utilizing a multicopy expression vector (pLF1) that is propagated in the nucleus of yeast cells under 2μ plasmid replicon control. This vector [6] carries the artificial gene construct under control of the promoter derived from the yeast gene <u>PGK1</u>, as well as the transcription termination sequence from the same gene (Fig. 5B). This transcriptional control unit in the yeast nucleus results in effectively constitutive expression of the N9L/Y8-1 gene, although the activity of the <u>PGK1</u> promoter is somewhat higher in the presence of glucose compared to ethanol, reflecting the fact that phosphoglycerate kinase is a glycolytic enzyme [46].

In <u>aap1</u> <u>mit$^-$</u> cells transformed with pLF1-N9L/Y8-1, the successful incorporation of the imported subunit 8 into a functional mtATPase complex was indicated by the efficient restoration of the cellular ability to grow on ethanol [6]. Further genetic analysis of the transformants showed that the presence of the vector directing expression of N9L/Y8-1 was required for the acquired ethanol-positive phenotype. Loss of the expression vector (by vegetative segregation during growth on non-selective medium) resulted in yeast cells with cellular growth and genetic properties of the original <u>aap1</u> <u>mit$^-$</u> mutants. Importantly, transformation of the <u>aap1</u> <u>mit$^-$</u> strain M26-10 with the expression vector pLF1 carrying the gene coding for subunit 8 not fused to the N9L leader did not give rise to rescue of the respiratory growth defect of the mutant strain [6]. The conclusion from this experiment is that genetic reconstitution of the mtATPase complex by a nuclearly encoded version of subunit 8 requires the presence of an N-terminal leader sequence to target the subunit 8 into mitochondria.

Assessment of the functional activities of mitochondria isolated from the archetypal transformant strain T2-1, in which the <u>aap1</u> mutation of strain M31 has been complemented, revealed levels of respiratory function, ATPase and ATP-Pi exchange activities generally comparable to those measured for the reference wildtype strain, J69-1B [6]. The transformant T2-1 did not, however, show as much inhibition by oligomycin of mitochondrial respiration and ATPase activity as the wildtype control. It was inferred that there is a minor perturbation in the proton channel function or coupling to the F_1 sector of the mtATPase complex in the transformant T2-1.

The assembly of imported subunit 8 into the mtATPase complex was demonstrated by immunoadsorption to an F_1-β antibody of mtATPase proteins radiolabelled with [^{35}S]-sulphate [6]. The immunopurified mtATPase complex of transformant T2-1 was shown to contain an imported version of subunit 8 that was slightly larger than the natural protein, by virtue of the extra seven amino acids present at the N-terminus of the allotopically expressed subunit 8. A further example of the data obtained demonstrating the assembly of mtATPase from imported subunit 8 in vivo is shown in Fig. 6, in this case involving allotopic expression of the chimaeric precursor N9L/Y8-2. In contrast to N9L/Y8-1 (Fig. 5A), the N9L leader sequence was directly fused to Y8 to generate N9L/Y8-2 [48]. Mitochondria were prepared from both wildtype J69-1B cells, and a transformant strain T10-1 (the aap1 mutant M31 cells transformed with the expression vector pLF1 carrying N9L/Y8-2) that had been pulse-labelled with [^{35}S]-sulphate in the absence of cycloheximide. The gel electrophoretic display of radiolabelled proteins immunoadsorbed to the F$_1$-β antibody, after its incubation with the cholate extract of isolated mitochondria, reveals the presence of subunit 8 in both strains, as expected (Fig. 6A and B, lane 1 in each case). Significantly, when cells were labelled in the presence of cycloheximide, which results in the suppression of labelling of cytosolically synthesized proteins, radiolabelled subunit 8 is recovered in the immunoprecipitate made from cholate extracts of wildtype mitochondria (J69-1B) (Fig. 6A, lane 2), but not in that of the transformant T10-1 (Fig. 6B, lane 2). The suppression of labelling of subunit 8 in the transformant supports the conclusion that in this strain subunit 8 is synthesized on cytosolic ribosomes and is imported before assembling with the other subunits into the mtATPase complex. The biochemical characterization of strain T10-1 is in progress.

IMPORT OF SUBUNIT 8 INTO MITOCHONDRIA IN VITRO

Expression of genes by transcription and translation in vitro allows the production of radiolabelled proteins of high specific activity and radiochemical purity, although in very small molar quantities. These products can then be used to characterize physically the proteins themselves and, when using chimaeric precursors, to study import of the proteins into isolated mitochondria. To achieve in vitro expression of N9L/Y8-1, the gene construct specifying this chimaeric precursor was inserted into the recombinant E. coli plasmid pSP64T, carrying the expression unit depicted in Fig. 5C. Transcripts consisting of the N9L/Y8-1 coding region, flanked by 5' and 3' untranslated sequences of the Xenopus laevis β-globin gene, were generated in vitro using SP6 RNA polymerase. These flanking sequences were included [39] to enhance the efficiency of translation of the RNA in a rabbit reticulocyte lysate. When N9L/Y8-1, labelled with [^{35}S]-methionine during translation in this system, was incubated with isolated wildtype mitochondria [44] import and cleavage of the precursor occurred to generate a product of slightly slower mobility than that of subunit 8. The imported subunit 8 was resistant to externally added protease because it was in a membrane-enclosed space. Radiosequencing of this protease-resistant product revealed that the predicted cleavage site of the matrix protease on the chimaeric precursor had indeed been recognized [44].

In contrast to these results, when subunit 8 was fused to the leader sequence from cytochrome c oxidase VI [36] (denoted here COX6L), import of subunit 8 into isolated wildtype mitochondria was not observed [44]. Moreover, aap1 mit⁻ cells transformed with pLF1 carrying COX6L/Y8 were not rescued to an ethanol-positive phenotype (R. H. P. Law, unpublished data). It is considered that the conformation of the chimaeric COX6L/Y8 protein may be such that it cannot unfold [49] to allow its translocation

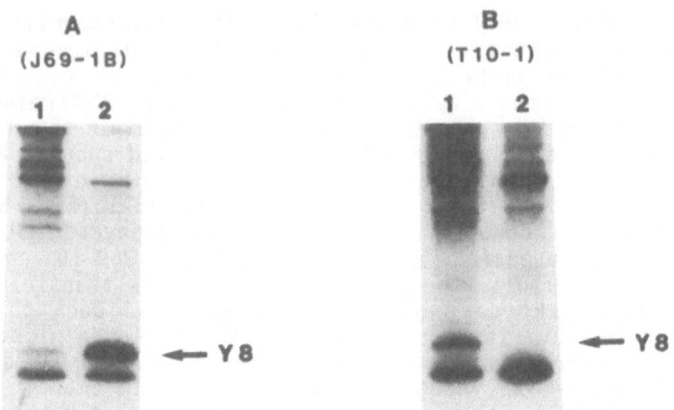

Fig. 6. Assembly of mtATPase from imported subunit 8 <u>in vivo</u>.
Mitochondria were prepared from J69-1B [<u>rho</u>+] cells (panel A) and
M31 cells transformed with plasmid pLF1-N9L/Y8-2, strain denoted
T10-1, (panel B). Cells were pulse-labelled with [^{35}S]-sulphate
in the absence of cycloheximide (lane 1) or in the presence of
this inhibitor of cytosolic ribosomes (lane 2). Isolated
mitochondria were solubilized in cholate extraction buffer [47].
Following centrifugation of this extract, assembled mtATPase
proteins were collected from the supernatant by immunoadsorption
to immobilized F_1-β antibody, and subjected to SDS-
polyacrylaminde gel electrophoresis. The position of subunit 8
(Y8) on the fluorogram is shown. N9L/Y8-2 is a chimaeric
precursor in which the N9L leader is directly fused to Y8 [48],
omitting the 7 supernumerary amino acids illustrated in Fig. 5A.

across the mitochondrial membranes; alternatively its leader may be
constrained from directly interacting with the import apparatus [50,51].
Note that the COX6L/Y8 exposed to mitochondria <u>in vitro</u> exhibited a
strong tendency to embed itself in mitochondrial membranes, which may
have interfered with its ability to be properly imported [44]. This
affinity for membranes of the Y8 moiety of N9L/Y8-1 was also recognized
by Gearing and Nagley [44]. It was indicated particularly by the recovery
of proteinase K-resistant material of size slightly smaller than subunit
8 and of hydrophobic character, that had been generated by interaction of
N9L/Y8-1 with inactivated mitochondria whose energization had been
collapsed and which were thus no longer competent for protein import.
Radiosequencing of the N-terminus of this material [44] suggested that it
was formed by embedding of all but the first 7 amino acids of subunit 8
in the outer mitochondrial membrane.

This property of hydrophobic passenger proteins such as subunit 8
(see above) or subunit 9 [10] to embed in the mitochondrial membranes can
be seen in a remarkably exaggerated manner if the chimaeric precursors
are unfolded with urea prior to their incubation with isolated
mitochondria. Thus, after urea pretreatment N9L/Y8-1 and N9L/Y9-2 (a
chimaeric precursor containing the N9L leader fused directly to yeast
subunit 9 with no intervening amino acids [48]) bind avidly to
mitochondria (either active, or treated with inhibitors to collapse
energization), but neither precursor is imported [52]. This property of

non-productive membrane-association is in direct contrast with the behaviour of the hydrophilic chimaeric precursor pCOXIV-DHFR, that is used widely as a 'model' import precursor [49]. In this case, urea pretreatment leads to enhanced binding and accelerated import of the mouse DHFR moiety into isolated yeast mitochondria, as confirmed by Law and Nagley [52].

Our findings that urea unfolding of chimaeric precursors containing hydrophobic domains inhibits import illustrates the delicate balance that must be maintained by the import system in dealing with hydrophobic precursors. Pfanner and colleagues [53] showed that precursor proteins with hydrophobic domains have an enhanced import efficiency relative to purely hydrophilic precursors, because hydrophobic interactions between the passenger protein and the membrane facilitate delivery of the precursor into the import apparatus itself, at sites of contact between the inner and outer membrane [51]. This delivery is probably dependent on the leader being able to interact with specific receptor proteins [51] or even with the phospholipids of the outer membrane [54]. Presumably, N9L/Y8-1 also may utilize such hydrophobic membrane interactions during its own import process, although some residual irreversible embedding of the subunit 8 passenger is noticeable [44]. However, import of N9L/Y8-1 is very sensitive even to relatively small concentrations of urea (down to 0.4 M urea), presumably because even a moderate degree of unfolding exposes hydrophobic regions which have a high intrinsic membrane affinity and lead to non-productive membrane embedding [52].

ASSEMBLY OF IMPORTED SUBUNIT 8 INTO mtATPase COMPLEX OF ISOLATED MITOCHONDRIA

As indicated above, N9L/Y8-1 can be efficiently imported in vitro into isolated mitochondria [44]. The question arises as to whether the imported, processed subunit 8 is merely entrapped in a compartment of mitochondria (that renders it resistant to proteolytic digestion by externally added proteinase K) or whether it assembles into a functional ATPase complex. It becomes important to answer this question, because if conditions can be found under which imported subunit 8 is assembled into mtATPase in vitro there is opened up the possibility of dissecting the membrane assembly process in some detail (especially concerning F_0). In addition, a powerful tool is provided for the assessment of assembly properties of N9L/Y8 variants made by site-directed or random mutagenesis.

Our approach has been to use the F_1-β monoclonal antibody as a tool to ascertain whether the radiolabelled imported subunit 8 becomes associated with the other subunits of mtATPase. To do this, the mitochondria were processed by the fractionation scheme outlined in Fig. 7A. This involves lysis of mitochondria using the detergents sodium cholate and N-octyl glucoside to solubilize the mtATPase [55]; more than 70% of subunit 8 pulse-labelled in vivo in wildtype cells is solubilized into fraction (b) (data not shown). To this cholate lysate are added sepharose beads on which the F_1-β monoclonal antibody has been immobilized [29]; more than 90% of the cholate-soluble label in subunit 8 is recovered in the immunoprecipitate (c), with only a trace remaining in the supernatant (d) (data not shown).

Fig. 7B shows the distribution of radiolabelled subunit 8 derived from N9L/Y8-1 imported in vitro into mitochondria from wildtype strain J69-1B [rho+]. The left panel shows that about one third of the bound

N9L/Y8-1 was processed to subunit 8; this was resistant to externally added proteinase K, unlike the N9L/Y8-1 precursor (data not shown). Fractionation of these mitochondria with cholate shows that a portion (about 10%) of processed subunit 8 is solubilized (b); the remainder remains cholate-insoluble (a) along with all of the unprocessed precursor which presumably is attached to the outside of the organellar membranes. It is unlikely that the cholate-insoluble subunit 8 is assembled into mtATPase. Moreover, only a small proportion of the cholate-soluble processed subunit 8 is able to be immunoprecipitated (c); note that the material in lane (c) is concentrated four-fold relative to that in lanes (a) (b) and (d) [see legend to Fig. 7]. By these criteria the assembly of imported subunit 8 into mitochondria is indeed inefficient [47]; the amount of labelled material recovered in lane (c) is barely suitable for detailed analysis of the assembly process. Presumably, the presence of fully assembled mtATPase complexes within the inner mitochondrial membrane of rho$^+$ cells precluded assembly of the imported subunit 8 into the complex. Taking this explanation as our starting premise we reasoned that if mitochondria could be depleted of subunit 8 then assembly of exogenous, imported subunit 8 might be facilitated by the availability within the membrane of partially assembled mtATPase complexes whose assembly is arrested at a stage dependent on the presence of subunit 8.

In our first experiments to investigate this route to achieving assembly of subunit 8 in vitro we prepared mitochondria from the aap1 mit⁻ strain M31 that lacks endogenous intra-mitochondrial subunit 8. However, these mit⁻ mitochondria were unsuitable for assembly studies because the physiological and structural consequences of their mitochondrial mutation renders them incompetent for import of the N9L/Y8-1 precursor [47]. We next sought to establish a controlled depletion of subunit 8 by placing the nuclear gene encoding N9L/Y8-1 under the control of the strictly inducible promoter GAL1. In the presence of galactose the activity of this promoter leads to high levels of expression of the associated N9L/Y8-1 reading frame. A yeast expression vector pED121 with single-copy replication control [47], carrying the regulated gene encoding N9L/Y8-1, was introduced into mit⁻ strain M31 to generate the transformant strain denoted YGL-1. These transformant cells are able to grow on galactose, but not on ethanol alone. The YGL-1 cells were then grown under a regime where the inducer galactose was first present to allow full development of mtATPase function. The galactose is then removed, and cells continue to grow on ethanol resulting in the depletion of subunit 8. Mitochondria isolated from YGL-1 cells that had undergone a shift from galactose to ethanol growth medium were shown to be depleted of subunit 8 based on biochemical assessment of F_0 sector function. The extent of continued growth in the absence of galactose influenced the degree of disruption of mtATPase function overall, which reflected an increasing proportion of non-functional mtATPase complexes arrested at the subunit 8-dependent stage of assembly alongside the residual population of preformed functional mtATPase complexes [47].

Mitochondria prepared from YGL-1 cells depleted of subunit 8 by the regulated allotopic expression of subunit 8 were able efficiently to import N9L/Y8-1 and thus to generate processed subunit 8 (Fig. 7C, left panel), to an extent noticeably greater than in the case of J69-1B (Fig. 7B). Fractionation of the YGL-1 mitochondria showed enhanced release of precessed subunit 8 into the cholate extraction buffer (b), and this material is almost quantitatively able to be immunoprecipitated. This is evidenced by the recovery of radioactivity in the material concentrated in lane (c) and, conversely by the barely perceptible amounts of subunit

A Approach:

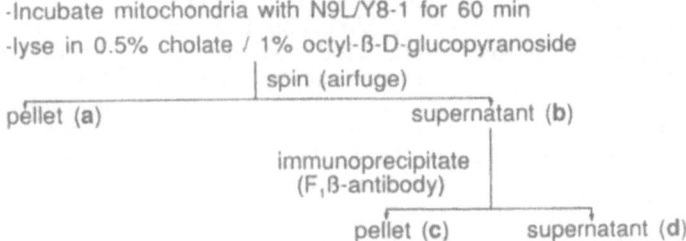

B Mitochondria of J69-1B (*rho+*)

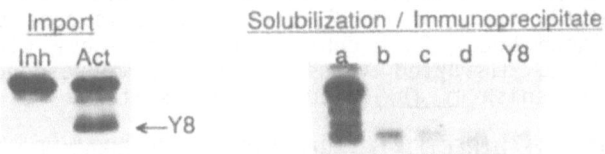

C Mitochondria of subunit 8-depleted transformant

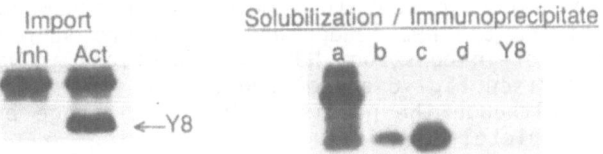

Fig. 7. Assembly of imported subunit 8 into mtATPase of mitochondria depleted of endogenous subunit 8. In panel A is given a summary of the experimental steps for tests of assembly into mtATPase of subunit 8 imported into mitochondria in vitro. Stages (a) through (d) at which protein samples were taken for analysis by gel electrophoresis are indicated. Detailed methods are described by Law and colleagues [47]. J69-1B and YGL-1 cells were pregrown in rich medium containing galactose and ethanol to mid-logarithmic phase, then shifted to rich medium containing ethanol as sole carbon source. These cells were grown at 28°C with aeration and harvested for mitochondrial preparation at the late logarithmic phase. Radiolabelled N9L/Y8-1 was incubated under import conditions with mitochondria from J69-1B (panel B) or from YGL-1 which had undergone depletion of subunit 8 under those conditions outlined in the text (panel C). Proteins were analyzed by gel electrophoresis directly from mitochondria pretreated with an inhibitor cocktail (Import, Inh) [10] or from untreated mitochondria (Import, Act). The fate of the radiolabelled Y8 imported into untreated mitochondria (Import, Act) through the fractionation scheme outlined in panel A is indicated in lanes (a) through (d). Note that the amount of mitochondrial-equivalent material recovered by immunoadsorption loaded in lane (c) was four times that loaded in lanes (a), (b) and (d). Radiolabelled subunit 8 was used as size marker (Y8).

8 remaining in the supernatant (d). Nevertheless, not all the imported subunit 8 is cholate-soluble (a); this perhaps indicates that there is a substantial proportion of non-assembled imported subunit 8 (Fig. 7C). This may reflect an inefficient throughput of imported subunit 8 because isolated mitochondria are being used. On the other hand, the process of importing and assembling subunit 8 may be intrinsically inefficient. Even in vivo, there may be a limitation on the rate of delivery of subunit 8 to the site of mtATPase assembly (in spite of the nuclear over-production of N9L/Y8-1 in strain T2-1), and this may be one cause of the slight aberration in F_O sector function (see above).

The ability to generate an intense band of radiolabelled immunoprecipitated subunit 8 in Fig. 7C, lane (c), provides a convenient means of assessing assembly of subunit 8 into mtATPase in vitro. We are exploring the generality of this subunit-depletion procedure to study the import and assembly properties of the nuclearly encoded subunits of mtATPase, notably those of the F_A sector. In this case the natural genes are being placed under GAL1 control on single-copy plasmids in mutant cells containing disrupted versions of the corresponding chromosomal genes (R. J. Devenish et al., unpublished results).

PROSPECTS FOR SITE-DIRECTED MUTAGENESIS OF THE SUBUNIT 8 GENE

The allotopic expression strategy provides a means of systematically analyzing structure-function aspects of subunit 8. In vitro mutagenesis has already been carried out to remove the supernumerary 7 amino acids interposed between the N9L leader and the mature subunit 8 moiety [48]. The resulting protein construct N9L/Y8-2 functions in vivo much like its N9L/Y8-1 parent (see Fig. 6 above) and can be imported into mitochondria in vitro [48] although the precise point of cleavage by matrix protease has not been completely clarified.

Our current program of site-directed mutagenesis on subunit 8 is focussed on three aspects of this hydrophobic polypeptide: (a) the N-terminal boundary of the transmembrane stem; (b) the C-terminal boundary of the transmembrane stem; (c) the C-terminal positively charged tail. Our studies on the N-terminal boundary focus on residue Gln12. Preliminary data have been obtained that one substitution at this position (Gln12→Leu) leads to a temperature-sensitive ethanol growth phenotype when the variant N9L/Y8-2 construct is expressed allotopically in the M31 mit⁻ host cells. It will be important to deduce the functional significance of a polar amino acid at the position of Gln12.

Consideration of the hydropathy plot using the GES algorithm of yeast subunit 8 suggested that the transmembrane stem of the protein covers amino acid residues 14 to 32 (see above). This same assignment for the fungi A. nidulans and N. crassa would place a charged residue within the transmembrane stem, namely at position 29 (Fig. 3B). We asked whether subunit 8 of S. cerevisiae could tolerate a similar positively charged residue at this position. The effects of converting the glutamine at position 29 to lysine (Fig. 1E) were assessed both in vivo and in vitro. The variant protein was expressed as a precursor protein with the N9L leader in M31 aap1 mit⁻ cells. Such cells were functionally rescued and grew vigorously on ethanol (D. Nero et al., manuscript in preparation). Assembly of the variant subunit 8 into the mtATPase complex was demonstrated after in vivo pulse labelling using the monoclonal F_1-β antibody to examine those subunits assembled into the complex, as already illustrated for N9L/Y8-2 itself in Fig. 6B. In vitro studies showed that the same N9L/Y8-2 (Gln29→Lys) variant can be imported into isolated mitochondria, and that the processed variant subunit 8 assembled into the mtATPase complex in YGL-1 target mitochondria. These data clearly show that subunit 8 can function with the substitution of Gln29→Lys. The lysine residue introduced at this position, being located perhaps 1 turn of the α-helix away from the oil/water interface itself, may protrude its side chain towards the more hydrophilic head-groups of the matrix-face of the lipid bilayer. Alternatively the transmembrane stem may involve fewer residues of subunit 8 than predicted above, the lysine in fact being outside of the hydrophobic lipid bilayer but closely juxtaposed to the matrix-surface of the membrane. We are currently changing residues further into the predicted transmembrane stem (e.g. Ile25→Lys), or at the other end of the stem (around residue 12), in order to shed further light on the location of the transmembrane stem.

Other mutants on which ongoing work is being undertaken concern the positively charged tail of subunit 8. A series of variants has been made in which this tail has been systematically truncated to convert each of the three positively charged residues to a STOP codon. Our analysis of these N9L/Y8-1 variants is so far limited to import experiments carried out _in vitro_. We have found that the variant Lys47→STOP is imported, but the corresponding Arg42→STOP is not able to be imported (D. Nero _et al_., manuscript in preparation). This identifies a further aspect of subunit 8 structure and topogenesis that will have to be taken into account, namely the ability of this hydrophobic protein to act as a satisfactory passenger in chimaeric precursors. This emphasizes the need for a full study of the _in vitro_ import and assembly properties of each of the variants made by _in vitro_ mutagenesis in parallel with functional and assembly studies _in vivo_, in order to achieve a proper appreciation of the relevant consequences of each mutation.

PERSPECTIVES

The introduction of the allotopic expression strategy has allowed this new range of investigations to be undertaken concerning the molecular biology, topogenesis and function of intrinsic membrane proteins normally encoded within the mitochondrial genome. The delivery of subunit 8 into yeast mitochondria, in regulated amounts, and of predetermined structure, has already been achieved. The systematic analysis of the assembly and function of subunit 8 variants and significantly, of intragenic and extragenic revertants of non-functional subunit 8 variants expressed allotopically, opens prospects for detailed biochemical genetic studies of subunit 8 structure-function and the interaction of subunit 8 with the other F_0 subunits 9 and 6. Additionally, it may be possible in the future to harness the 'biolistic' micro-projectile method for introducing functional mtDNA sequences into yeast mitochondria _in vivo_ [37,38], so as to control precisely the expression of individual mitochondrial genes. Coupled with increasingly powerful techniques of studying membrane assembly and bioenergetic functions, these molecular genetic advances underline our optimism that the studies of subunit 8 will play important role in defining in detail the process whereby a functional multisubunit enzyme complex is assembled in the inner mitochondrial membrane.

ACKNOWLEDGEMENTS

Work in our laboratory reviewed here is supported by the Australian Research Council. We thank Mr. D. Harari for assistance with the comparative analyses of subunit 8 and A6L, and Mrs. S. Ekkel for excellent technical assistance.

REFERENCES

1. Pedersen, P. L. and Carafoli, E. (1987) Trends Biochem. Sci. **12**, 146-150.
2. Nagley, P. (1988) Trends Genet. **4**, 46-52.
3. Macreadie, I. G. , Choo, W. M., Novitski, C. E., Marzuki, S., Nagley, P., Linnane, A. W. and Lukins, H. B. (1982) Biochem. Int. **5**, 129-136.
4. Macreadie, I. G., Novitski, C. E., Maxwell, R. J., John, U. P., Ooi, B. G., McMullen, G. L., Lukins, H. B., Linnane, A. W. and Nagley, P. (1983) Nucleic Acids Res. **11**, 4435-4451.
5. Linnane, A. W., Lukins, H. B., Nagley, P., Marzuki, S., Hadikusumo, R. G., Jean-Francois, M. J. B., John, U. P., Ooi, B. G., Watkins, L., Willson, T. A., Wright, J. and Meltzer, S. (1985) in Achivements and Perspectives of Mitochondrial Research (Quagliarello, E., Slater, E. C., Palmieri, F., Saccone, C. and Kroon, A. M., eds) vol. 1 Bioenergetics, pp. 211-222, Elsevier Science Publishers, Amsterdam.
6. Nagley, P., Farrell, L. B., Gearing, D. P., Nero, D., Meltzer, S. and Devenish, R. J. (1988) Proc. Natl. Acad. Sci. USA **85**, 2091-2095.
7. Velours, J., Esparza, M., Hoppe, J., Sebald, W. and Guerin, B. (1984) EMBO J. **3**, 207-212.
8. Velours, J., Esparza, M. and Guerin, B. (1982) Biochem. Biophys. Res. Commun. **109**, 1192-1199.
9. Sebald, W. and Hoppe, J. (1981) Curr. Top. Bioenerg. **12**, 1-64.
10. Farrell, L. B., Gearing, D. P. and Nagley, P. (1988) Eur. J. Biochem. **173**, 131-137.
11. Novitski, C. E., Macreadie, I. C., Maxwell, R. J., Lukins, H. B., Linnane, A. W. and Nagley, P. (1984) Curr. Genet. **8**, 135-146.
12. Cobon, G. S., Beilharz, M. W., Linnane, A. W. and Nagley, P. (1982) Curr. Genet. **5**, 97-107.
13. Osinga, K. A., De Vries, E., Van der Horst, G. and Tabak, H. F. (1984) EMBO J. **3**, 829-834.
14. Simon, M. and Faye, G. (1984) Mol. Gen. Genet. **196**, 266-274.
15. Grivell, L. (1989) Eur. J. Biochem. **182**, 477-493.
16. Fearnley, I. M. and Walker, J. E. (1986) EMBO J. **5**, 2003-2008.
17. Hadikusumo, R. G., Meltzer, S., Choo, W. M., Jean-Francois, M. J. B., Linnane, A. W. and Marzuki, S. (1988) Biochim. Biophys. Acta **933**, 212-222.
18. Marzuki, S., Watkins, L. C. and Choo, W. M. (1989) Biochim. Biophys. Acta **975**, 222-230.
19. Nagley, P. and Devenish, R. J. (1989) Trends Biochem. Sci. **14**, 31-35.
20. Von Heijne, G. (1981) Eur. J. Biochem. **120**, 275-278.
21. Velours, J. and Guerin, B. (1986) Biochem. Biophys. Res. Commun. **138**, 78-86.
22. Kyte, J. and Doolittle, R. F. (1982) J. Mol. Biol. **157**, 105-120.
23. Engelman, D. M., Steitz, T. A. and Goldman, A. (1986) Ann. Rev. Biophys. Biophys. Chem. **15**, 321-353.
24. Grisi, E., Brown, T. A. Waring, R. B., Scazzochio, C. and Davies, R. W. (1982) Nucleic Acids Res. **10**, 3531-3539.
25. Morelli, G. and Macino, G. (1984) J. Mol. Biol. **178**, 491-507.
26. Anderson, S., De Bruijn, M. H. L., Coulson, A. R., Eperon, I. C., Sanger, F. and Young, I. G. (1982) J. Mol. Biol. **156**, 683-717.
27. Walker, J. E., Tybulewicz, V. L. J., Falk, G., Gay, N. J. and Hampe, A. (1984) in H^+-ATPase (ATP synthase) Structure, Function, Biogenesis (Papa, S., Altendorf, K., Ernster, L. and Packer, L., eds) pp. 1-14, Adriatica Editrice, Bari.

28. Jean-Francois, M. J. B., Hadikusumo, R. G., Watkins, L. C., Lukins, H. B., Linnane, A. W. and Marzuki, S. (1986) Biochim. Biophys. Acta **852**, 133-143.
29. Hadikusumo, R. G., Hertzog, P. J. and Marzuki, S. (1984) Biochim. Biophys. Acta **765**, 258-267.
30. Choo, W. M., Hadikusumo, R. G. and Marzuki, S. (1985) Biochim. Biophys. Acta **806**, 290-304.
31. Higuti, T., Takigawa, M., Kotera, Y., Oka, H., Uchida, J., Arakaki, R., Fujita, T. and Ogawa, T (1985) Proc. Natl. Acad. Sci. USA **82**, 1331-1335.
32. Higuti, T., Negama, T., Takigawa, M., Uchida, J., Yamane, T., Asai, T., Tani, I., Oeda, K., Shimizu, M., Nakamura, K. and Ohkava, H. (1988) J. Biol. Chem. **263**, 6772-6776.
33. Uchida, J., Takigawa, M., Yamane, T., and Negama, Y., Tani, I. and Higuti, T. (1987) Biochem. Biophys. Res. Commun. **146**, 953-958.
34. Guerin, B. and Napias, C. (1978) Biochemistry **17**, 2510-2516.
35. Blondin, G. A. (1979) Biochem. Biophys. Res. Commun. **87**, 1087-1094.
36. Nagley, P., Willson, T. A., Tymms, M. J., Devenish, R. J. and Gearing, D. P. (1985) in Achievements and Perspectives of Mitochondrial Research (Quagliarello, E., Slater, E. C., Palmieri, F., Saccone, C. and Kroon, A. M., eds) vol. 2 Biogenesis, pp. 405-414, Elsevier Science Publishers, Amsterdam.
37. Johnston, S. A., Anziano, P. Q., Shark, K., Sanford, J. C. and Butow, R. A. (1988) Science **240**, 1538-1541.
38. Fox. T. D., Sanford, J. C. and McMullin, T. W. (1988) Proc. Natl. Acad. Sci. USA **85**, 7288-7292.
39. Krieg, P. A. and Melton, D. A. (1984) Nucleic Acids Res., **12**, 7057-7070.
40. Gearing, D. P., McMullen, G. L. and Nagley, P. (1985) Biochem. Int. **10**, 907-915.
41. Viebrock, A., Perz, A. and Sebald, W. (1982) EMBO J. **1**, 565-571.
42. Schmidt, B., Wachter, E., Sebald, W. and Neupert, W. (1984) Eur. J. Biochem. **144**, 581-588.
43. Schmidt, B., Henning, B., Zimmerman, R. and Neupert, W. (1983b) J. Cell Biol. **96**, 248-255.
44. Gearing, D. P. and Nagley, P (1986) EMBO J. **5**, 3651-3655.
45. Schmidt, B., Henning, B., Kohler, H. and Neupert, W. (1983a) J. Biol. Chem. **258**, 4687-4689.
46. Kingsman S. M. and Kingsman, A. J. (1983) in Interferons: from Molecular Biology to Clinical Applications (Burke, D. C. and Morris, A. G., eds) pp. 211-254, Cambridge University Press, Cambridge.
47. Law, R. H. P., Devenish, R. J. and Nagley, P. (1989) Eur. J. Biochem. (in press).
48. Law, R. H. P., Farrell, L. B., Nero, D., Devenish, R. J. and Nagley, P. (1988) FEBS Lett. **236**, 501-505.
49. Eilers, M. and Schatz, G. (1988) Cell **52**, 481-483.
50. Verner, K. and Lemire, B. D. (1989) EMBO J. **8**, 1491-1495.
51. Hartl, F. U., Pfanner, N., Nicholson, D. W. and Neupert, W. (1989) Biochim. Biophys. Acta **988**, 1-45.
52. Law, R. H. P. and Nagley, P. (1989) Biochim. Biophys. Acta (submitted).
53. Pfanner, N., Muller, H. K., Harmey, M. A. and Neupert, W. (1987) EMBO J. **6**, 3449-3454.
54. Endo, T., Eilers, M. and Schatz, G. (1989) J. Biol. Chem. **264**, 2951-2956.
55. Rott, R. and Nelson, N. (1981) J. Biol. Chem. **256**, 9224-9228.

GENE STRUCTURE OF HUMAN ATP SYNTHASE BETA SUBUNIT

Yasuo Kagawa and Shigeo Ohta

Department of Biochemistry
Jichi Medical School
Minamikawachi, Tochigi-ken, Japan 329-04

SUMMARY

The β subunit of ATP synthase is a catalytic subunit that is highly homologous among different species. Human nuclear genes for the mitochondrial enzymes including the β subunits and 8 related proteins were cloned to elucidate the coordinated transcriptional control of energy transduction.

The human β gene (8kbp) contained 10 exons and 8 Alu repeating sequences, without any TATA box. An S1 nuclease protection experiment revealed two initiation sites for the transcription. The promoter activity of the upstream region was examined by transfecting HeLa cells with a chloramphenicol acetyltransferase (CAT) gene ligated with various lengths of the 5' upstream region of the β gene. The regions with promoter activity were analyzed by DNase I foot printing and the gel retardation assay. We found a special enhancer with the motif GAGACAAGGTTTCACC and the transcriptional factor(s) to coordinate the expression of some mitochondrial energy transducing enzymes.

INTRODUCTION*

ATP synthase (FoF_1) is a central enzyme supplying energy to cells by respiration and photooxidoreduction (1,2). Recently, it was shown that only the αβ subunit complex of FoF_1 catalyzes ATP synthesis (3), the β subunit acts as the catalytic center, and the remaining subunits are essential for chemiosmotic coupling between ATP synthesis and proton translocation through biomembranes (2,4). The velocity of ATP synthesis fluctuates greatly depending on the cell activity, and it is regulated by both rapid respiratory control (5) and slow synthesis of mitochondrial enzyme complexes (6). Respiratory control is the regulation of ATP synthase (FoF_1) by the electrochemical potential of protons across membrane (1), which is established by the electron transport system (7). In fact, the Vmax of FoF_1 was shown to be controlled by the potential across a planar lipid bilayer containing FoF_1 (8). Because the maximal velocity of ATP synthesis is limited by the amount of either FoF_1 or enzyme complexes supplying energy to FoF_1 (5), the biosynthesis of these proteins is enhanced: for

---

*I would like to dedicate this article to Amenomori Houshuu (1666-1755), a scholar who respected Korean science and opened a long peaceful era of interaction between Korea and Japan.

Bioenergetics, Edited by C. H. Kim and
T. Ozawa, Plenum Press, New York, 1990

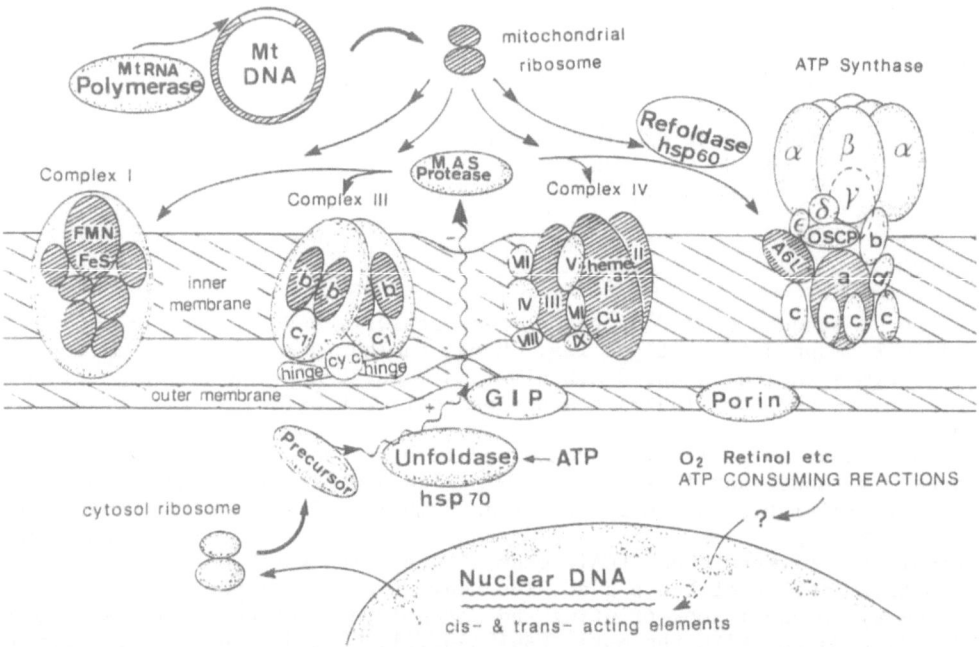

Fig. 1. Biosynthesis of complexes of oxidative phosphorylation.

Among all the mammalian enzymes, only the four enzyme complexes of oxidative phosphorylation are encoded by both mitochondrial and nuclear DNA. These are FoF_1, complex I (NADH:ubiquinol oxidoreductase), complex III (ubiquinol: ferricytochrome c oxidoreductase) and complex IV (ferrocytochrome c : oxygen oxidoreductase). The mitochondrion has its own genetic system for synthesizing some subunits of these complexes. The mammalian mitochondrial DNA encodes only 13 polypeptides, 22 tRNA and 2 ribosomal RNA (hatched by oblique lines). Seven, one, three and two subunits are found in complexes I, III, and IV, and FoF_1, respectively. The remainder of the subunits are nuclear gene products that are imported into the organelle. These proteins are translated as precursors, unfolded by unfoldase (heat shock protein 70), translocated by putative general import protein (GIP), their presequence is removed by MAS protease, and then assembled by refoldase (heat shock protein 60), as described in DISCUSSION. (Original figures in ref. 11 was modified).

example, this occurs during continuous muscle contraction (9) and in the stimulated dendrites of neurons (10). However, the mechanisms of this phenomenon have yet been elucidated.

Among all mammalian enzymes, only four enzyme complexes of oxidative phosphorylation (FoF$_1$ and Complexes I, III and IV) are encoded by both mitochondrial and nuclear DNA (Fig. 1)(11). Although complexes for energy transduction are encoded by typical housekeeping genes, the relative amounts of these subunits in mitochondria are kept nearly constant (7), because of subunit stoichiometry in oligomers and efficient ratios among oligomers. Thus, there must be a mechanism that coordinates the biosynthesis of the β subunit and other energy transducing enzymes. The primary structure of the β subunit of human FoF$_1$ (12) is highly homologous to those of other species (12-15), perhaps because of its universal catalytic role in ATP synthesis. However, in E. coli (13,14) and thermophilic bacteria (15,16), an operon consisting of all nine genes for subunits of FoF$_1$ is regulated by a single promoter. Mammalian mitochondrial DNA encodes 13 subunits of these complexes, which are also expressed as a single polycistronic transcript (6) (Fig.1). However, the nuclear DNA encodes most of the mitochondrial proteins, including the remaining subunits of these complexes, via many transcripts of different chromosomes. Thus, the mechanism coordinating the expression of the energy transducing proteins was studied using the eukaryotic nuclear genes.

In order to elucidate this transcriptional control, the human genes of the key subunits of the energy transducing system were cloned and sequenced. They are the catalytic subunit of FoF$_1$; the β subunit (12,17), cytochrome c1 (18, 19) and hinge protein (20), ATP/ADP translocator (21), pyruvate dehydrogenase E1α (22) and E1β subunits (23), complex IV (cytochrome oxidase) subunit VIc (24), and phospholipase C (25). The coordination of the expresseion of genes involved in mitochondrial biogenesis has been described in yeast (26). Here, we show the coordinated expression of these genes under several conditions and identified a regulatory sequence within the 5'-flanking regions of the FoF$_1$ β subunit gene that may be involved in the coordinated expression of some mitochondrial genes for energy transduction.

MATERIALS AND METHODS

Molecular Cloning, Plasmid Construction and DNA Sequencing—A human genomic gene library constructed in Charon 4A was screened with the cDNA of the human F$_1$ β subunit cloned in this laboratory (12). The positive plaques were purified by successive rounds of plaque hybridization. DNA fragments were subcloned into the plasmid pUC18 or pUC19. The nucleotide sequence was determined by the dideoxy nucleotide chain termination method using M13 phage single-stranded DNA as the template under the sequence strategy reported by us (17).

Coordinated Expression of Energy Transducing Genes—HeLa cells (human uterine cancer cell strain)(17) and HL60 cells (human myelogenous leukemia cell strain) (21) were cultured in RPMI-1640 containing 10% calf serum as described previously. Northern blotting was performed as described in the legend of Fig. 4. The hybridization probes for the Northern blot analysis were prepared by universal labeling of M13 single-strand DNA harboring cDNA's of F$_1$ β (12), ATP/ADP translocator (21), cytochrome c1 (19), hinge protein (20), pyruvate dehydrogenase E1α(22) and E1β (23), and cytochrome oxidase VIc (25).

Assay of Promoter and Enhancer Activities—We have used a linked reporter function, the bacterial enzyme, chloramphenicol acetyltransferase (CAT) to conveniently distinguish expression of exogenous human genes from that of the endogenous cognate gene in transient transfection assays in HeLa cells.

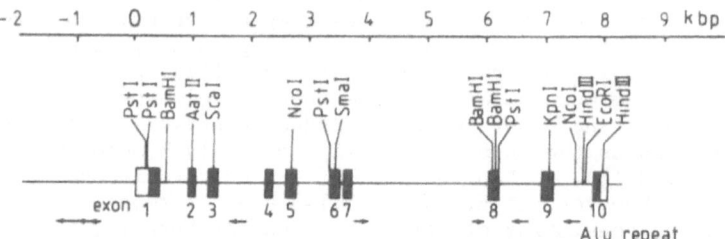

Fig. 2. The gene structure of human ATP synthase β subunit.
The exons are shown by boxes, and the encoding regions are shown by closed boxes. The location of the Alu repeats are shown by arrows.

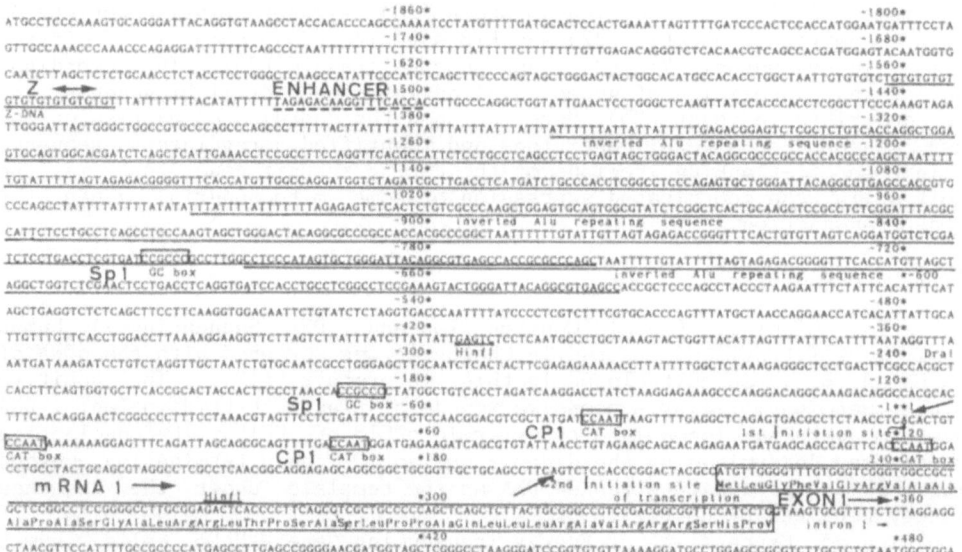

Fig. 3. Nucleotide sequence of the 5' upstream region of the human ATP synthase β subunit gene. The nucleotides are numbered from the upstream initiation site for transcription.

The CAT gene and the 5'-upstream region of gene of the FoF$_1$ β subunit were fused as follows: an EcoRI–BamHI fragment (the nucleotide position –1919 to +432 of the first transcription site) of the β gene was digested with several restriction enzymes. The ends of the digested fragments were filled-in with T4-polymerase, ligated with HindIII linker (dCAAGCTTG); and after digestion, these fragments were ligated to HindIII-digested pMLCAT (27). pSV2CAT was used as a positive control (28). Plasmid DNA (15 μg) was transfected into HeLa cells (1 x 10^6 cells in 9–cm dishes)(29). The CAT assay was performed as described previously (17). Introduction of double-stranded synthetic oligonucleotide into the CAT plasmid was performed by digesting the CAT-fusion gene with Th111I (at –1711 and –1339), filling by T4-polymerase, dephosphorylating by calf intestinal alkaline phosphatase, and then ligating with the oligonucleotide.

DNase I Foot-printing--Nuclear extract (5mg/ml) was prepared from HeLa cells (8 x 10^7 cells per 5 mg nuclear extract) by the method described by Dignam et al. (30). The foot-printing assay was performed by the method of Tamura et al. (31). An Nsp (7524)I–HaeII fragment was end-labelled by T4-polynucleotide kinase and [γ^{32}P]ATP and then digested with Tth111I. The Tth111I–Nsp(7524)I fragment was purified by gel electrophoresis. For 10 min in an ice bath, the labeled probe DNA (1–3 ng) was incubated with the nuclear extract in 20 μl of a solution containing 2 mM spermidine, 200 ng of pBR322 DNA, 0.136 mM EDTA and 10.2 % (v/v) glycerol. Then, the DNA fragment was digested by the addition of 100 ng DNase I and 3 mM CaCl$_2$. The resulting DNA fragments were extracted with phenol and then analyzed by 8% polyacrylamide 7 M urea gel electrophoresis followed by autoradiography.

Gel Retardation Assay--The 5' upstream region of the β gene was digested into fragments, which were labeled with T4-polynucleotide kinase. Each of the labeled DNA fragments was incubated in 25 μl of a solution containing the HeLa cell nuclear extract, 10 mM Tris–HCl, pH 7.5, 50 mM NaCl, 5 % (v/v) glycerol, 1 mM DTT, 1 mM EDTA, and 2μg poly(dI, dC). Protein–oligonucleotide complexes were separated from unbound DNA fragment on 5 % polyacrylamide gels. Samples were electrophoresed at 40 V at room temperature, then the gel was dried and autoradiographed according to the procedure of Carthew (32). For the competition experiments , unlabeled oligonucleotide (1:1 molar ratio of the corresponding labeled oligonucleotide) was added to the assay mixture.

RESULTS

Structure of Human F$_1$ β Gene--The organization of the human F$_1$ β gene is shown in Fig. 2. The entire length of the F$_1$ β gene is 11 kb. The intron/exon junctions were determined by direct comparison of the nucleotide sequences with the cloned cDNA. There are 10 exons of relatively uniform size (122 bp - 213 bp) which are separated by 9 introns of higly variable length containing 8 Alu repeats. The first exon codes for most of the presequence. The nucleotide sequence of the 5' upstream region is shown in Fig. 3. The stretch of 29 alternating Ts and Gs that begins at nucleotide –1558 has the potential to form a left-handed helical structure or Z-DNA. There are two initiation sites as determined by S1 nuclease mapping, and the first site is at +1 and the second, at +197. There were no putative TATA boxes found around 30 nucleotide upstream from either site. Three CAT boxes (CCAAT) were found between the two sites, and two GC boxes in the 5'-upstream region. The enhancer and silencer will be described in the later sections.

Coodinated Expression of Energy Transducing Enzymes--When the oxygen content for culturing HeLa cells was reduced from 20 % to 2 %, Northern blot analysis revealed a coodinated transient decrease of expression of F1 β and pyruvate dehydrogenase E1α genes without any change of the expression

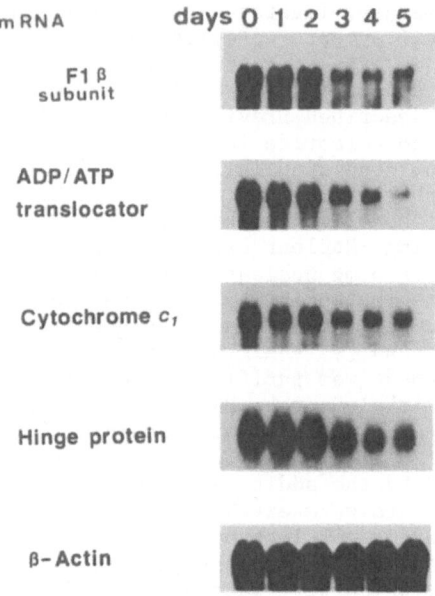

Fig. 4. Coordinated expression of F_1 β and energy transducing proteins in the HL60 cells treated with retinol: Northern blott analysis.

The cells were cultured for the indicated days (each lane) with 1 μM retinol. Total RNA was extracted from these cells by the guanidine thiocyanate–hot phenol method. Poly A RNA was purified from the total cellular RNA by oligo dT cellulose column chromatography. Samples of 5 μg of poly A RNA were applied to each lane of the 1 % agarose gel electrophoresis after denaturation and transfered to a nylon membrane (Hybond–N, Amersham). Hybridization probes were prepared by universal labeling of M 13 single-strand DNA corresponding to cDNAs of the proteins indicated in this Figure. Poly A tail of each cDNA had been removed by digestion with Bal 31 endonuclease. The cDNA of β–actin was labeled by nick-translation.

of the β-actin gene (21). As shown in Fig. 4, retinol decreased coodinated expressions of genes for the energy transducing enzymes without changing the expression of the β-actin gene. Retinol did not affect growth, therefore mitochondria became much smaller. Retinoic acid caused differentiation of HL60 cells including H_2O_2 production by phorbolester (8-fold increase) and reduction of cell proliferationt. However, both agents caused a similar expression pattern.

Promoter and Enhancer Activities of the FoF_1 β gene--The promoter activity of the 5'-upstream region of the β subunit was measured by the expressed CAT activity (Fig. 5), For a quantitative analysis, the efficiency of the transfection was normalized to the incorporated DNA content measured by Southern blotting. The enhancing elements were located in the fragments from the -1913th to -1339th nucleotides (3.5-fold stimulation) and from the -1190th to -544th nucleotides (4.5-fold stimulation). The region from the -1339th to -1190th nucleotides functioned as a silencer that repressed transcription down to one 3.5th the contol level (Fig. 5).

Binding of Nuclear Factors to the Enhancing DNA Fragments--The various restriction fragments were incubated with the nuclear extract and electrophoresed. The fragments containing the GC box (Spl binding site)(33,34) were shifted. An Nsp(7524)-Tth111I fragment (-1585th to -1339th) was shifted by incubation with the extract as shown in Fig. 6. The shift was restored by adding an excess amount of unlabeled fragment (Fig. 6, lane 3), which suggests the specific binding of a nuclear protein. The foot-printing assay located the binding sites (protected site) in the 246 bp long fragment between positions 165 and 173 from the Tth111I site. The sequence of this cis-element was 5'-GGTGAAACCC-TTGTCTC. This specific DNA-protein interaction was also confirmed with the synthetic oligonucleotide with or without mutated sites.

Properties of the Enhancing Element--The introduction of the above mentioned double stranded oligonucleotide into the plasmid containing the CAT gene fully recovered the promoter activity, and the effect was independent of the direction and position (even behind the CAT gene). The mutated oligonucleotide did not recover any of the promoter activity (Fig. 7). Therefore, the enhancing element of the F_1 β subunit gene is a typical enhancer.

DISCUSSION

Transcriptional Control of Housekeeping F_1 β Gene--The F1 β gene lacks TATA box, but instead has GC boxes, and generates transcripts with multiple 5'-ends (17). These are characteristic features of housekeeping genes that are constitutively expressed (35). Yet, there are controlling mechanisms in these genes (9,10, 36). In yeast, for example, when glucose is added to the medium, mRNAs of the F_1 α, β and γ subunits and the ATP/ADP translocator are reduced to 25 % of the amount present in the absence of glucose (36). In mammalian cells, too, the coordinated inhibition of expression of energy transducing genes was shown under hypoxia, retinol (Fig. 4) and several other conditions by Northern blotting analysis. The cis-elements were detected by virtue of its effects on transient expression of a linked reporter function (CAT) in transfected HeLa cells (Fig. 5). The efficiency of the transfection was normalized by estimating the amount of incorporated DNA by Southern blotting analysis.

New Enhancer for Energy Transduction System--A novel enhancer of the F_1 β gene has been discovered (Fig. 7). This new cis-element was also found in the 5'-upstream region of the genes for cytochrome c1 (18) and pyruvate dehydrogenase Eα subunit (21) (Fig. 8). The essential role of cytochrome c1-hinge protein in electron transport was established by Kim et al.(37) and pyruvate dehydrogenase, by Koike (23). These results suggest that the

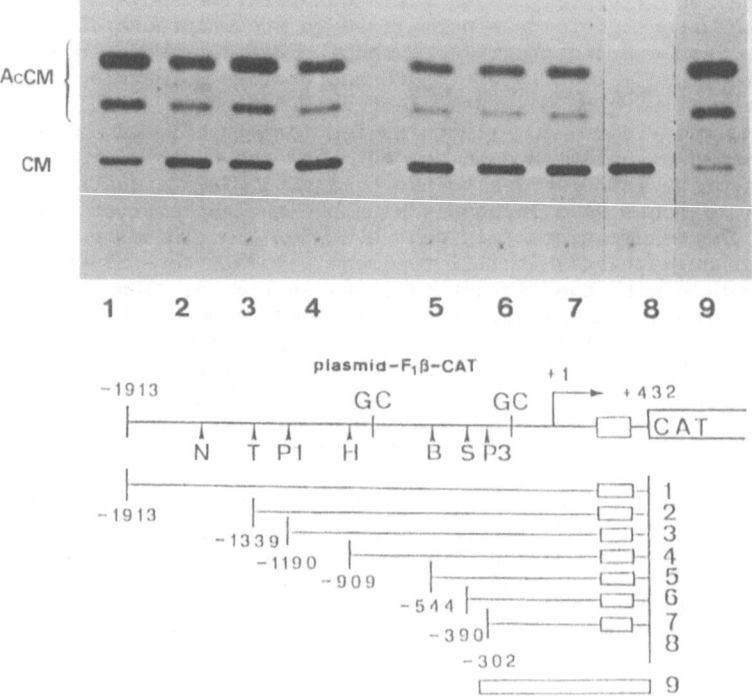

Fig. 5. CAT activity directed by various lengths of the 5'
flanking sequence of the human F_1 β gene.
Upper panel: Autoradiogram of a thin layer chromatogram showing
chloramphenicol acetylated by the expressed CAT activity. Lanes 1–9
show results of assays with plasmids (1 to 9) illustrated in the
lower panel. CM, chrloramphenicol; AcCM, acetylated products of
chloramphenicol.
Lower panel: Diagram of construction of the CAT-plasmids. The
restriction sites used are as follows: N; Nsp (7524)I, T; Tth111I:
H: Hae II, B; BstEII, and S: Sca I. P1 and P3 indicate pHB1 and
pHB3, respectively (17). GC indicates a GC-box. The open box of SV
40 at the bottom shows an early promoter region of SV 40.

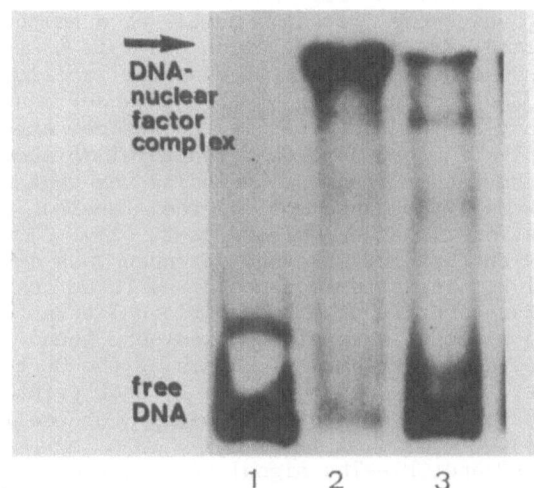

Fig. 6. Gel retardation assay and competition experiments of a Nsp(7524)I-Tth111I fragment of the human F_1 β gene.
A Nsp(7524)I-Tth111I fragment (246 bp, -1685 to -1339 from the first position of the transcription of F_1 β gene) was end-labeled. Lane 1, the labeled fragment with 2μg of poly(dI,dC); Lane 2, the components of Lane 1 plus the components of Lane 2 plus 10ng of the unlabeled fragment.

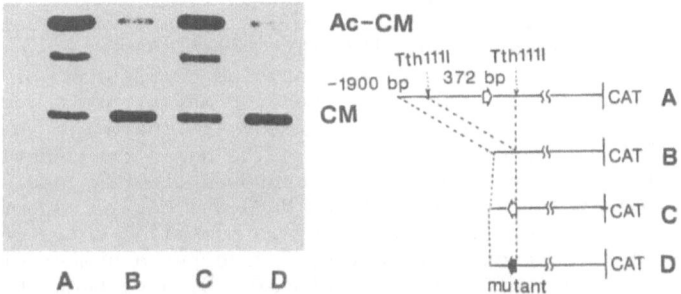

Fig. 7 The enhancer activity of a synthetic double stranded oligonucleotide shown by the CAT assay. The oligonucleotide 5'-TTTTTAGAGACAAGGTTTCACCACGTTGCCCAGGC-3' and its complementary strand were annealed and ligated into a Tth111I site, or a Sac I site. A; CAT gene with normal 5' flanking region of F_1 β gene. B. with out the flanking region. C. with an insert in the opposite direction at the Tth111I site. D. with an mutated insert 5'-TTTTTAGAGTCAAGGTTTCTCCACGTTGCCCAGGC-3'in the Tth111I site.

expression of specific mitochondrial energy transducing genes is regulated by a common mechanism involving enhancer elements. The consensus sequence has not been found in similar regions of the other mitochondrial genes such as those for the urea cycle (38) or a number of nonmitochondrial genes that have been reported (39). A deletion analysis of the enhancer region indicated the sequence essential for enhancer function. Both gel shift assay (Fig. 6) and foot printing of this enhancer sequence revealed the presence of a trans-acting factor(s) in the nuclear extract. For the binding of this putative factor(s) to DNA, length of 35 nucleotides was sufficient. However, the longer fragment including the consensus sequence seemed to bind more strongly, because most of the labeled oligonucleotide was shifted by incubating with the nuclear lysate (Fig. 6). The foot printing experiment showed the binding site and dependency on divalent cations. The consensus CTF/NF1 binding site sequence among three constitutive enzymes of the malate-aspartate shuttle of mitochondria has been reported (38). As a common cis-acting element, several GC boxes were found in these three genes in Fig. 8. By CAT assay, fragments containing the GC box (33) enhanced the promoter activity. This was confirmed by foot printing and gel shift assays. The GC box serves as binding sites for cellular transcription factor Spl (33,34).

Control by HAP and CP1--The signal transduction in the expression of the mammalian energy transducing system (Fig. 1) is still unknown. However, in yeast, the signals from oxygen (heme control) and carbon source (catabolite repression) to the cytochrome gene via trans-acting elements that bind to the cis-acting elements have been extensively studied (40-42). The cis-acting elements of the iso-1-cytochrome c gene (CYC 1) are called upstream activation sites 1 (UAS1) and 2 (UAS2). UAS1 and UAS2 are controlled by trans-elements called RAF+HAP1 (40) and HAP2+HAP3 (41), respectively. A pet mutant lacking heme synthesis (δ aminolevulinate dehydratase-less) is coordinately devoid of cytochrome c as well as none heme subunits of complexes III and IV (43). Transcription is activated by the binding of the HAP1 (heme activated protein 1) to UAS, and HAP1 is activated by binding of heme that is synthesized in the presence of oxygen. The expression of CYC 1 is coordinated with that of four nuclear subunits of complex III via general factors GI and GII (26), and with that of three nuclear subunits of complex IV via HAP2 plus heme (44).

In mammalian cells, anoxia also inhibits expression of genes for energy transduction enzymes, including the $F_1 \beta$ and pyruvate dehydrogenase (21). In the human $F_1 \beta$ gene (Fig. 3) and other enzymes, the HAP2+HAP3 binding site is the sequence motif CCAAT. The human trans-acting element CP1 binds to the CCAAT element, and is composed of CP1A and CP1B which could be functionally interchangeable for HAP3 and HAP2, respectively (45). However, the mechanism of signal transfer in mammalian cells may be quite different from that of yeast. Not only $F_1 \beta$ gene but a number of mammalian promoters, such as that of heat shock protein 70, have been shown to contain CCAAT element. Moreover, not only CP1 but CP2 and NF1 bind to CCAAT element.

Coordinated Expression of F_1 Gene and Differentiation--In yeast, many housekeeping genes are coordinately expressed by promoter binding factor TUF (46), GI, etc (26). But in mammalian cells, expression of genes for enzymes depends on differentiated tissues. The tissue specific expression of isoforms of F_1 and related enzymes suggests that the mechanism of these gene expression is under control of cell differentiation. The bovine $F_1 \alpha$ subunits of heart (47) and liver (48) are different in 10 amino acid residues and in 158 bases of the cDNA (47). There are other examples of multiple gene families subject to tissue-specific regulation for a number of other mitochondrial enzymes, including cytochrome oxidase minor subunits (49), ATP/ADP translocator (50,51) and the oligomycin conferral proteins

```
                          Mg-independent    Mg-dependent
                             factor            factor
  F1 beta  Z-DNA--AT- Rich --TAGAGACAAGGTTTCACCA-CCAGG
  Cyt.c1   Z-DNA--AT- Rich --TAGAGACAAGGTTTCACCA-CCAGG
PDH alpha  Z-DNA--AT---------AGAGACA--------CACCA--------
```

Fig. 8. Illustration of a consensus sequence among the 5'-flanking regions of
human nuclear genes for some energy transducing enzymes.
F1 beta: F_1 β gene, Cyt.c1: cytochrome c1
gene, PDH alpha: pyruvate dehydrogenase E1α
subunit gene.

(52). The tissue specificity and physiological meaning of multiple mRNAs
(17) and polymorphism (53) in the human genomic DNA for F_1 β subunit is
still obscure, but multiple mRNAs of yeast was shown to be differentlly
expressed (54).

In order to test the relationship between differentiation and coordination of expression of energy transducing enzymes, HL60 cells were differentiated into granulocytic and macrophagic cells by treatment with retinoic
acid and phorbor ester, respectively (Arai, Ohta and Kagawa, to be published). The expression of nuclear-coded mitochondrial proteins were decreased like in the case of retinol (Fig. 4). Although retinol did not
cause differentiation itself, retinoid family shares a common steroid
receptor superfamily. The expression of F_1 gene was sensitive to phorbolester, retinol and retinoic acid as shown in Fig. 4, and thus, the phospholipase C-protein kinase C system may be involved in the response. We have
cloned the human phospholipase C gene for this purpose (25), and the signal
transduction by phospholipase C is now intensively studied by Korean workers (55).

The coordinated expression of nine pancreatic exocrine enzymes by a
homologous cis-acting sequence via a cell specific trans-acting factor
called the "differentiator" was also proposed (56). The maternal mRNA of
the F_1 α subunit was shown to be one of the the most abundant mRNA (An2) in
the animal pole of Xenopus oocyte (57). This suggested that some energy
transforming enzyme may play an important role in differentiation. Differentiation of erythroleukemia cells was repressed by inhibiting mitochondrial protein synthesis with chloramphenicol (58).

Assembly of the F_1 β Subunit--The prokaryotic F_1 subunits assemble
automatically in vitro (2-4). However, as shown in Fig. 1., eukaryotic
F_1 subunits have presequence for post-translational peptide transport and
processing. This process require a) unfoldase (heat shock protein 70)(59),
b) general import protein (GIP)(60), c) MAS proteinase to remove the pre-
peptide (61) and d) refoldase to assemble subunits into complex (62). The
detailed mechanism of this post-translational process was studied on the F_1
β subunit that is highly homologous among species including man (12), rat
(63) and even prokaryotes (14,16).

Translational and Transcriptional Controls--Mitochondrial biogenesis
require coordination of mitochondrial genes that supply 13 subunits of
oxidative phosphorylation, and nuclear genes for the F_1 β subunit and other
component of the complexes. Modulation of mitochondrial gene expression of
subunits of complex IV in yeast occurs largely at the level of translation

by 3 nuclear genes for translation (64). In complex III, too, processing of the apocytochrome b pre-mRNA containing 3 exon portions, requires 11 nuclear gene products for tRNA processing, 5' processing, splicing, translation and maturation (65).

Therefore, the mitochondrial genes might be under the translational control of the nucleus-encoded proteins, while the nuclear genes for F_1 β and other energy transducing proteins might predominantly be under the transcriptional control during mitochondrial biogenesis.

ACKNOWLEDGEMENTS

We thank Dr. Chong H. Kim and Dr. Takayuki Ozawa for organizing this Symposium. We also thank Drs. Hitoshi Endo, Hajime Arai and Hideaki Tomura for their research, and Mrs. Toshiko Ohta, Miss Kakuko Matsuda and Miss Mamiko Hoshino for their assistance.
This work was supported in part by Grant #63617006 of Ministry of Education, Science and Culture, and Grant from Kyowa Hakko.

REFERENCES

1. Mitchell, P. (1979) Science 206, 1148-1159.
2. Kagawa, Y. (1984) in Bioenergetics (Ernster, L. ed.) pp. 149-186, Elsevier, Amsterdam.
3. Kagawa, Y., Ohta, S. and Otawara-Hamamoto, Y.(1989) FEBS Lett.49,67-69.
4. Yoshida, M., Okamoto, H., Sone, N., Hirata, H. and Kagawa, Y. (1977) Proc. Natl. Acad. Sci. U.S.A. 74, 936-940.
5. Chance, B. (1977) Ann. Rev. Biochem. 46, 967-980.
6. Tzagoloff, A. and Myers, A. M. (1986) Ann. Rev. Biochem. 55, 249-285.
7. Hatefi, Y. (1985) Ann. Rev. Biochem. 54, 1015-1069.
8. Muneyuki, E., Kagawa, Y. and Hirata, H. (1989) J. Biol. Chem. 264, 6092-6096.
9. Williams, R.S., Garcia-Moll, M., Mellor,J., Salmons,S., and Harlan, W. (1987) J. Biol. Chem. 262, 2764-2767.
10. Wong-Riley, M.T.T. (1989) Trends Nerosci. 12, 94-103.
11. Kagawa, Y. and Ohta, S. (1989) Japanese J. Biochem. 61, 75-88.
12. Ohta, S. and Kagawa, Y. (1986) J. Biochem. (Tokyo) 99, 135-141.
13. Walker, J.E., Eberle, A., Gay, N.J., Runswick, M.J. and Saraste, M. (1982) Biochem. Soc. Trans. 10, 203-206.
14. Futai, M., Noumi, T. and Maeda, M.(1989) Ann. Rev. Biochem. 58, 111-136.
15. Kagawa, Y., Ishizuka, M., Saishu, T. and Nakao, S. (1986) J. Biochem. 100, 923-936.
16. Ohta, S., Yohda, M., Ishizuka, M., Hirata, H., Hamamoto, T., Otawara-Hamamoto, Y., Matsuda, K. and Kagawa, Y. (1988) Biochim. Biophys. Acta 933, 141-155.
17. Ohta, S., Tomura, H., Matsuda, K. and Kagawa, Y. (1988) J. Biol. Chem. 263, 11257-11262.
18. Suzuki, H., Hosokawa, Y., Nishikimi, M. and Ozawa, T. (1989) J. Biol. Chem. 264, 1368-1374.
19. Nishikimi, M., Ohta, Suzuki, H., Tanaka, T., Kikkawa, F., Tanaka, M., Kagawa, Y., and Ozawa, T. (1988) Nucleic Acid R. 16, 3577.
20. Ohta, S., Goto, K., Arai, H. and Kagawa, Y. (1987) FEBS Lett. 226. 171-175.
21. Ohta, S., Endo, H., Matsuda, K. and Kagawa, Y. (1989) Ann. N.Y. Acad. Sci. in press.
22. Endo, H., Hasegawa, K., Narisawa, K., Tada, K. and Kagawa, Y. (1989) Am. J. Human Genet. 44, 358-364.

23. Koike, K., Ohta, S., Urata, Y., Kagawa, Y. and Koike, M. (1988) Proc. Natl. Acad. Sci. USA, 85, 41–45.
24. Ohtsuka, M., Mizuno, Y., Yoshida, M., Kagawa, Y. and Ohta, S. (1988) Nucleic Acid Res. 16, 10916.
25. Ohta, S., Matsui, A., Nozawa, Y. and Kagawa, Y. (1988) FEBS Lett. 242, 31–35.
26. Dorsman, J.C., van Heeswijk, W.C. and Grivell, L.A. (1988) Nucleic Acid Res. 16, 7287–7301.
27. Fujisawa-Sehara, A., Sogawa, K., Nishi, C. and Fujii-Kuriyama, Y. (1986) Nucleic Acid Res. 11, 1475–1489.
28. Laimins, L., Khoury, G., Gorman, C.M., Howard, B.H. and Gruss, P. (1982) Proc. Natl. Acad. Sci. USA, 79, 6453–6457.
29. Davis, L.G., Dibner, M.D. and Battey, J.F. (1986) Methods in Molecular Biology, Elsevier Scientific Publishing Co., Inc. Amsterdam.
30. Digman, J.D., Levovitz, R.M. and Roeder, R.G. (1983) Nucleic Acid Res. 11, 1475–1489.
31. Tamura, T., Inoue, T., Nagata, K., and Mikoshiba, K. (1988) Biochem. Biophys. Res. Commun. 157, 419–425.
32. Carthew, R.W., Chodosh, L.A., and Sharp, (1985) Cell, 43, 439–448.
33. Dynan, W.S. and Tjian, R. (1985) Nature 316, 774–778.
34. Jones, K.A., Kadonaga, J.T., Rosenfeld, P.J., Kelly, T.J. and Tjian, R. (1987) Cell 48, 79–89.
35. Kim, S. H., Moors, J.C., David, D., Respess, J.G., Jolly, D.J. and Freedman, T. (1986) Nucleic Acid Res. 14, 3103–3118.
36. Szekely, E. and Montogomery, D.L. (1984) Mol. Cell Biol. 4, 939–946.
37. Kim, C.H., Balny, C. and King, T.E. (1987) J. Biol. Chem. 262, 8103–8108.
38. Ohtake, A., Takiguchi, M., Shigeto, Y., Amaya, Y., Kawamoto, S. and Mori, M. (1988) J. Biol. Chem. 263, 2245–2249.
39. Wingender, E. (1988) Nucleic Acid Res. 16, 1879–1902.
40. Pfeifer, K., Kim, K.S., Kogan, S. and Guarente, L. (1989) Cell 56, 291–301.
41. Hahn, S. and Guarente, L. (1988) Science 240, 317–321.
42. Pfeifer, K., Arcangioli, B., and Guarente, L. (1987) Cell 49, 9–18.
43. Myers, A.M., Crivellone, M.D., Koerner, T.J., and Tzagoloff, A. (1987) J. Biol. Chem. 262, 16822–16829.
44. Trawick, J.D., Wright, R.M. and Poyton, R.O. (1989) J. Biol. Chem. 264, 7005–7008.
45. Chodosh, L.A., Olesen, J., Hahn, S., Baldwin, A.S., Guarente, L. and Sharp, P.A. (1988) Cell 53, 25–35.
46. Capieaux, E., Vignais, M–L., Sentenac, A. and Goffeau, A. (1989) J. Biol. Chem. 264, 7437–7446.
47. Walker, J. E., Powell, S.J., Vinas, O. and Runswick, M.J. (1989) Biochemistry 28, 4702–4708.
48. Breen, G.A.M. (1988) Biochem. Biophys. Res. Commun. 152, 264–269.
49. Kuhn-Nentwig, L. and Kadenbach, B. (1985) Europ. J. Biochem.149, 147–158.
50. Powell, S.J., Medd, S.M., Nyren, P., and Walker, J.E.(1989) Biochemistry 28, 866–873.
51. Battini, R., Ferrari, S., Kaczmarek, L., Calabretta, B., Chen, S.T. and Baserga, R. (1987) J. Biol. Chem. 262, 4355–4359.
52. Walker, J.E., Gay, N.J., Powell, S.J., Kostina, M., and Dyer,M.R. (1987) Biochemistry 26, 8613–8619.
53. Hasegawa, K., Ohta, S., Narisawa, K., Tada, K. and Kagawa, Y.(1987) Japanese J. Human Genet. 32, 275–281.
54. Wright, R.M., Rosenzweig, B. and Poyton, R. O. (1989) Nucleic Acid Res. 17, 1103–1120
55. Rhee, S.G., Suh, P.G., Ryu, S.H. and Lee, S.Y. (1989) Science 244, 546–550

56. Boulet, A.M., Erwin, C.R. and Rutter, W.J. (1986) Proc. Natl. Acad. Sci. U.S.A. 83, 3599–3603.
57. Weeks, D.L. and Melton, D.A. (1987) Proc. Natl. Acad. Sci. U.S.A. 84, 2798–2802.
58. Kaneko, T., Watanabe, T. and Oishi, M. (1988) Mol. Cell Biol. 8, 3311–3315.
59. Deshaies, R.J., Koch, B.D., Werner-Washburn, M., Craig., E., and Scheckman, R. (1988) Nature 332, 800–805.
60. Hartl, F-U., Pfanner, N., Nicholson, D.W. and Neupert, W. (1989) Bio chim. Biophys. Acta 988, 1–45.
61. Yang, M., Jensen, R.E., Yaffe, M.P., Oppliger, W. and Schatz, G. (1988) EMBO J. 7, 3857–3862.
62. Cheng, M.Y., Hartl, F-U., Martin, J., Pollock, R.A., Kalousek, F., Neupert, W., Hallberg, E.M., Hallberg, R.L. an Horwich, A.L. (1989) Nature 337, 620–625.
63. Lee, Y-M., Chu, L-P. and Lee, S-C.(1988) J. Formosan Med. Assoc. 87, 933–938.
64. Fox, T.D., Costanzo, M.C., Strick, C.A., Marykwas, D.L.,Seaver, E.C. and Rosenthal, J.K. (1988) Philos. Trans. R. Soc. London B 319, 97–105.
65. Tzagoloff, A., Crivellone, M.D., Gampel, A., Muroff, I., Nishikimi, M. and Wu, M. (1988) Philos. Trans. R. Soc. London B 319, 107–120.

Molecule and Gene of *Sulfolobus acidocaldarius* ATPase

Kimitoshi Denda, Jin Konishi, Kyoko Hajiro, Tairo Oshima,
Takayasu Date* and Masasuke Yoshida

Department of Life Science, Tokyo Institute of Technology, Nagatsuta, Yokohama, Japan 227
*Department of Biochemstry, Kanazawa Medical University, Uchinada, Ishikawa, Japan 920-02

SUMMARY

A novel ATPase (*Sul-* ATPase) was solubilized from membranes of an archaebacterium, *Sulfolobus acidocaldarius,* by washing with a buffer containing EDTA. Enzymatic characteristics of this ATPase are distinctly different from F_1-ATPase and resemble eukaryotic endomembrane H^+-ATPase. Attempt was made to isolate the whole ATPase complex from membranes using a detergent and the isolated complex showed several protein bands in SDS-PAGE including a band at the position around 40KD in addition to subunits of water-soluble *Sul-* ATPase. By the same procedure to isolate proteolipid subunit, or F_{oc} subunit, of F_oF_1-ATPase, a very hydrophobic protein was extracted from membranes of this bacterium by chloroform-methanol treatment. $[^{14}C]DCCD$ bound to this protein when the membrane was incubated with this reagent. Molecular cloning of the *Sul*-ATPase operon revealed that the α and β subunits of *Sul*-ATPase show a remarkable amino acid sequence homology to the 70K and 60K subunits of eukaryotic vacuolar H^+-ATPases, respectively. They also show significant, though less remarkable, homology to the α and β subunits of F_oF_1-ATPase. The operon also contains a gene encoding the proteolipid subunit above mentioned. Its sequence clearly shows that it is a *Sulfolobus* equivalence of F_{oc} subunit. From the analysis of the sequences, a phylogenetic tree of evolution of H^+-ATPases was constructed and we propose the monomeric H^+-ATPase as the most primordial H^+-ATPase. If it is really the case, cooperative kinetics observed for F_oF_1-ATPases are related to some regulatory functions which were added to the core catalytic function later in evolution. Close relation between archaebacterial ATPases and the eukaryotic endomembrane H^+-ATPases also implies that the eukaryotic endomembrane system has been originated from inclusion of the plasma membranes of an ancestral archaebacterial cell. At the standpoint of symbiotic theory of the origin of eukaryote, above proposition means that the host cell for symbionts was an ancestral archaebacterial cell.

INTRODUCTION

F_oF_1-ATPases are considered to be present in all cells, which obtain ATP from coupled proton flow (1-4) . F_1-ATPase, comprised of five kinds of subunits (α, β, γ, δ, and ε), is the water-soluble catalytic part of F_oF_1-ATPase and F_o, comprised of three kinds of subunits (a, b , and c), is a membraneous part which acts as a proton channel. F_oF_1-ATPases from various organisms are considered to constitute a genetically-related family on the basis of the homologous primary strucutures of their subunits, especially those of the α, β, and c subunits (5,6). However, the ATPases found in archaebacteria differ from typical F_1-ATPases with respect to subunit properties, sensitivity toward inhibitors, and pH optima (7-12). Furthermore, archaebacterial ATPases prepared from three species immunologically cross-reacted with *Saccharomyces cerevisiae* vacuolar H^+-ATPase, a member of another family of H^+-ATPase, V-ATPase, which includes variuos eukaryotic endomembrane H^+-ATPases (13). Here we describe the molecule and gene of ATPase from *Sulfolobus acidocaldarius*, an acidothermophilic archaebacterium, and discuss the evolution of H^+-ATPase.

1. Molecule of *S. acidocaldarius* ATPase

Soluble *Sul*-ATPase

The ATPase was solubilized easily with low ionic strength buffer containing EDTA from membranes prepared by French press treatment. The purified ATPase was a large multisubunit complex. The molecular weight was estimated to be 360 K. We previously reported that the *Sul*-ATPase seemed to consist of at least three kinds of polypeptides, α, β, and γ, the molecular weights of which were estimated to be 66K, 51K, and 28K, respectively. However, an additonal subunit (δ) with approximate Mr 12K was found later. Electron microscopic observation revealed that the *Sul*-ATPase was very similar in gross structure to the F_1-ATPase, exhibiting a hexagonal array with diameter of about 100 Å. The ATPase activity of the *Sul*-ATPase apparently obeyed simple Michaelis-Menten type kinetics. Its Km value, 1.6 mM, is significantly larger than that of an F_1-ATPase, which is usually in the range of 0.1-0.4 mM. Since the Km value for *H. halobium* is reported to be 1.4 mM, the apparent low affinity for ATP appears to be one of the common characteristics of archaebacterial ATPases. The enzyme had an optimum pH of about 5. Na_2SO_3 has a remarkable stimulatory effect on the ATPase activity. Vanadate, an inhibitor of ATPases that form E-P intermedeates, and azide, a specific inhibitor of F_1-ATPases, showed no effect on *Sul*-ATPase. One of covalent inhibitors of F_1-ATPase, 7-chloro-4-nitrobenzo-2-oxa-1,3-diazol (NBD-Cl), inactivated *Sul*-ATPase. However, inactivation was less efficient than F_1-ATPase and bound [^{14}C]NBD was almost equally distributed between the α and β subunits of *Sul*-ATPase whereas β subunit is exclusively labeled for F_1-ATPase (Fig.1). It is consistent to the fact that *Sul*-ATPase α subunit lacks the specific tyrosine residue which reacts with NBD-Cl in case of F_1-ATPase. The activity of *Sul*-ATPase is inhibitied by nitrate, an inhibitor of V-ATPases.

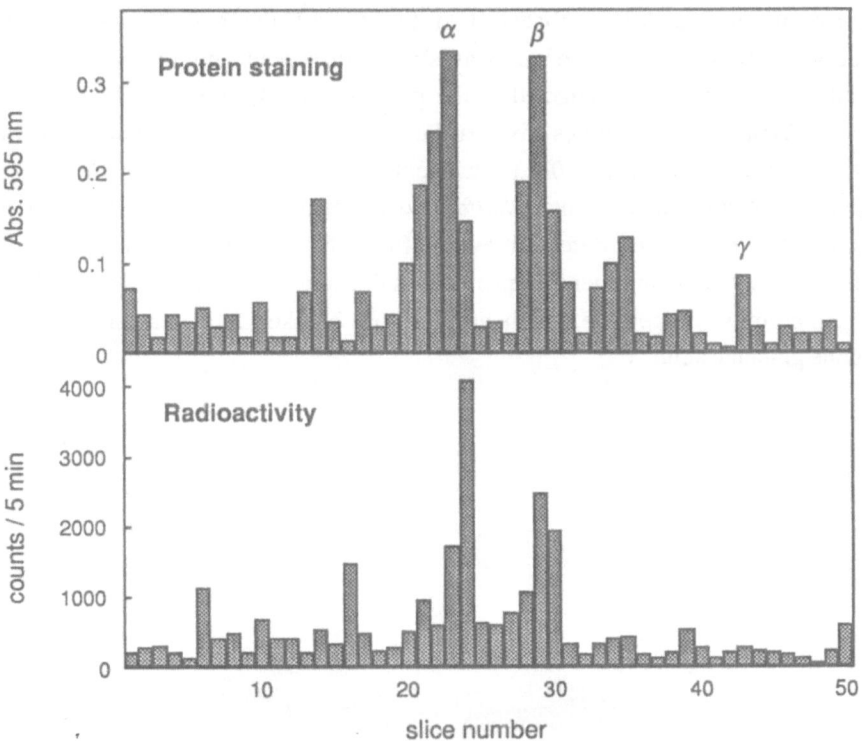

Fig.1. Binding of [^{14}C]NBD-Cl to *Sul*-ATPase. To a 100μl solution containing 0.5mg of purified ATPase, 10% glycerol, 20mM Tris-Cl, pH 8.0, 1mM EDTA, 1μl of ethanolic solution of 10mM [^{14}C] NBD-Cl (54 μCi/mmol) was added and incubated for 2 hr at room temperature. Unbound [^{14}C]NBD-Cl was removed by a Sephadex G50 column. Proteins were subjected to SDS-PAGE (12 %) and the gel was sliced into 1.5 mm sections. These pieces were dissolved with 20% H_2O_2 for overnight at 70°C. The dissolved gels were transferred into vials and radioactivities were measured. To indicate the position of subunits in the gel, another gel was stained with Coomassie brilliant blue R and the protein-bound dye of each cut slice was extracted and measured by the absorbance at 595 nm (9).

Proteolipid subunit of *Sulfolobus* ATPase complex

When F_oF_1-ATPases are incubated with N,N'-dicyclohexylcarbodiimide (DCCD), this reagent binds preferentially to a specific glutamic (or aspartic) acid residue of F_oc subunit with concomitant loss of activity of H^+-ATPase (14). F_oc subunit is also called as a proteolipid subunit since it is a very hydrophobic protein which can be extracted directly in organic solvent from membrane by washing with chloroform-mehtanol (2:1,v/v). This procedures were applied to *Sulfolobus* membranes preincubated with [^{14}C]DCCD. Extracted proteins were precipitated by the addition of diethyl ether and the brownish precipitate was analyzed with SDS-urea PAGE. Silver staining of the gel gave a smear band at the position of 10K. Without urea, an aggregated protein band was observed at the top of the gel. To examine the distribution of radioactivity, the gel was sliced, dissolved and radioactivity was measured. Radioactivity was detected at the position of Mr 10K (Fig.2). Sequence analysis of CNBr-cleaved peptides gave partial amino acid sequences similar to F_oc subunit. Thus, DCCD-reactive 10K protein extracted with organic solvent is most likely the proteolipid subunit of intact *Sulfolobus* ATPase complex. The NH_2-terminus of this protein has never detected even after CNBr cleavage. Therefore, a translational product of this gene is probably subjected to processing, producing a mature proteolipid polypeptide which has a stable blocked NH_2-terminus such as pyrrolidone glutamic acid.

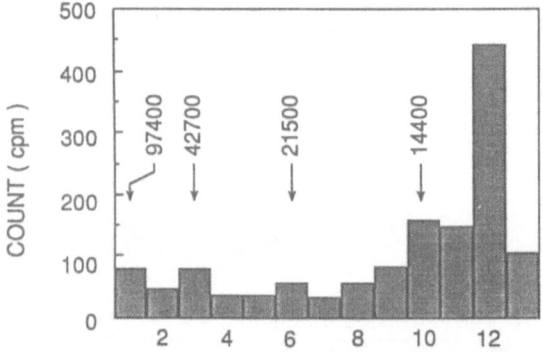

Fig.2. Distribution of the radioactivity in a SDS-PAGE of the proteins extracted by chloroform/methanol from *Sulfolobus* membranes incubated with [^{14}C]DCCD. Membranes were incubated with 200 μM of [^{14}C]DCCD (54 mCi/mmol, ethanol solution) in the reaction buffer (50mM HEPES-NaOH, 1mM ADP, pH 7.2) for 2 hr at room temperature. After the incubation, free [^{14}C]DCCD was removed by cetrifugation. The precipitated membranes were treated with two volumes of chloroform/methanol (2:1 V/V). Proteins extracted in organic solvent were subjected to SDS-PAGE (12%). The gel was sliced into 2mm sections and these pieces were dissolved with 20% H_2O_2 at 70°C for 12 hr. The dissolved gels were transferred into vials and radioactivities were measured.

Attempt to purify the intact ATPase complex

Water-soluble *Sul*-ATPase is probably a part of intact H^+-ATPase complex which exists in the membrane of this bacterium. Although *Sul*-ATPase seems to be very easily detached from the membranes upon disruption of the cells, about 20% of ATPase activity remain on the membranes. The attempt was made to purify the membraneous ATPase complex using a detergent. The membranes of *Sulfolobus acidocaldarius* were incubated with 1% octylglucoside with stirring at room temperature for 1 hr. After cetrifugation, the supernatant was applied to a DEAE-Toyopearl column. The column was eluted with a NaCl gradient and active fractions were pooled and concentrated. Then the solution was applied to a gel permeation HPLC column (G3000SW). The ATPase activity was eluted as a single peak at the position ahead of usual F_1-ATPase, indicating the molecular size of the ATPase complex was larger than 380KD. SDS-PAGE of this fraction showed that it contained 45K, 40K, and probably proteins less than 15K in addition to the subunits of water-soluble *Sul*-ATPase (Fig. 3). It is not known yet which are the subunits of ATPase complex and co-purified cotaminations.

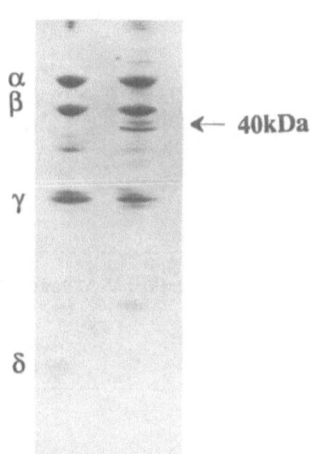

Fig.3. SDS-PAGE of the detergent-solubilized ATPase from *S.acido-caldarius* membranes. Electrophoresis was performed in a gel containing 12% acrylamide. The proteins in the gel were silver stained.
Lane 1, soluble ATPase.
Lane 2, detergent (1% octylglucoside) solubilized ATPase.

2. Molecular Cloning of *S. acidocaldarius* ATPase Complex

Prokaryotic V-ATPase

By using the oligonucleotide probes deduced from the partial amino acid sequence and pUC118 genomic library, we cloned a gene for each subunit of *Sul*-ATPase. Moderate homologies (about 25 %) between the α and β subunits of *Sul*-ATPase and those of F_oF_1-ATPases were found, indicating that they may have possibly evolved from a common ancestor of F_oF_1-ATPases. However, since it is known that the amino acid sequences of the α and β subunits of F_oF_1-ATPases have been highly conserved during evolution (more than 60% between human and *E.coli*), it was concluded that this archaebacterial ATPase appears to belong to a new group of ion-translocating ATPase that is only distantly related to F_oF_1-ATPases (Fig. 4). The sequences of several V-ATPases have been recently available (17-21). They show a certain similarity to the major subunits of F_oF_1-ATPases (about 25%). The most surprising finding was that they are highly homologous to the α, and β subunits of the *Sul*-ATPase (22). Overall, the 70K subunits of two eukaryotic V-ATPases are 62% homologous to each other, and about 50% homologous to the α subunits of *Sul*-ATPase. The 60K subunits of two eukaryotic V-ATPases are about 70% homologous each other, and about 55% to the β subunit of *Sul*-ATPase. Thus it is now evident that *Sulfolobus* ATPase is a prokaryotic V-ATPase. This conclusion has been further confirmed for other archaebacterial ATPases by recent reports from other laboratories (23, 24).

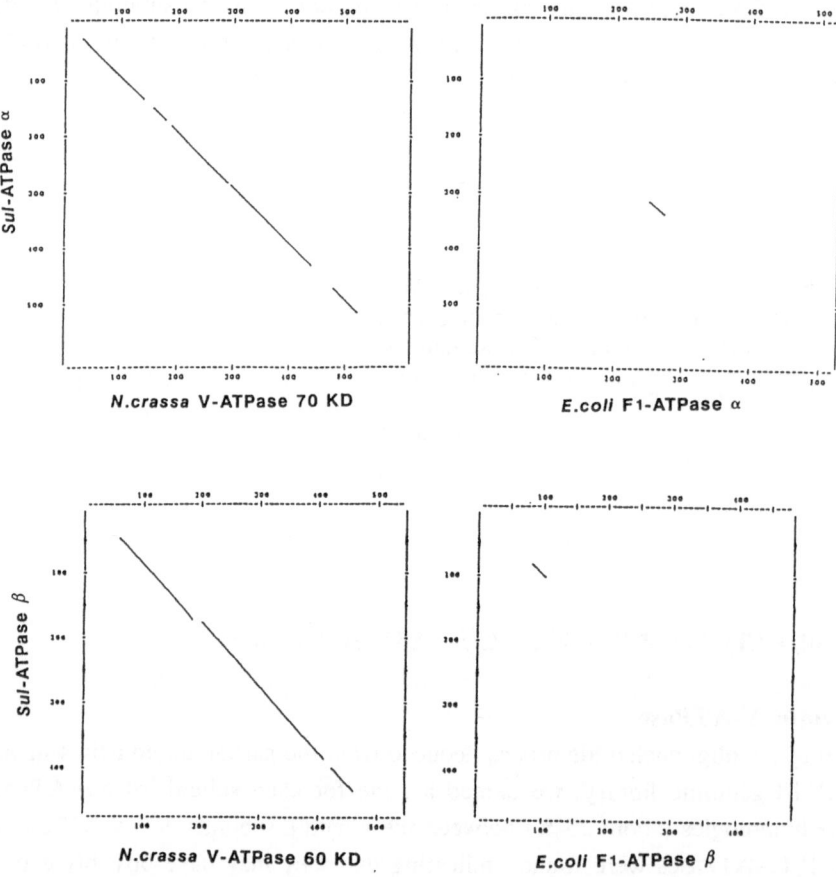

Fig.4. Comparison of the amino acid sequence of the α and β subunits of *Sul*-ATPase with 70K and 60K subunits of *N.crassa* vacuolar H⁺-ATPase (left) or *E.coli* F_0F_1-ATPase (right). A dot was generated if the number of identical amino acid residues an 11-residue segment was more than 50 %.

Sulfolobus proteolipid subunit is not double-sized

In the same operon, we found a gene encoding a 101 amino-acid protein (25). It contains the same amino acid sequences obtained from chloroform/methanol extracted protein above mentioned. Its homologous sequence and similar hydropathy profile to $F_o c$ subunit lead us to the conclusion that it is a *Sulfolobus* equivalence of $F_o c$ subunit of $F_o F_1$-ATPases. It is about 30 amino acid larger than the $F_o c$ subunit and has a small amino acid repeat sequence. However, it is distinct from the double-sized 16K proteolipid subunit of eukaryotic V-ATPases. This implies that gene duplication occurred in proteolipid subunit after ancestral V-ATPase had been differentiated from ancestral archaebacterial ATPase.

3. Molecular Evolution of H^+-ATPase

Proto-VAF ATPase

Thus, close similarity between archaebacterial ATPases and eukaryotic vacuolar H^+-ATPases have become evident. Both ATPases also show moderate similarity to $F_o F_1$-ATPases: the amino acid sequences of major two subunits and proteolipid subunit of each ATPase are more or less homologous and the $\alpha_3 \beta_3$-type hexagonal subunit structure is conserved for all ATPases. We can construct an evolutional tree of this superfamily of H^+-ATPases by comparing the sequence similarity (Fig. 5). Archaebacterial ATPases (A) and vacuolar H^+-ATPases (V) are the closest relatives and they have been evolved from a common ancestor, designated as VA in

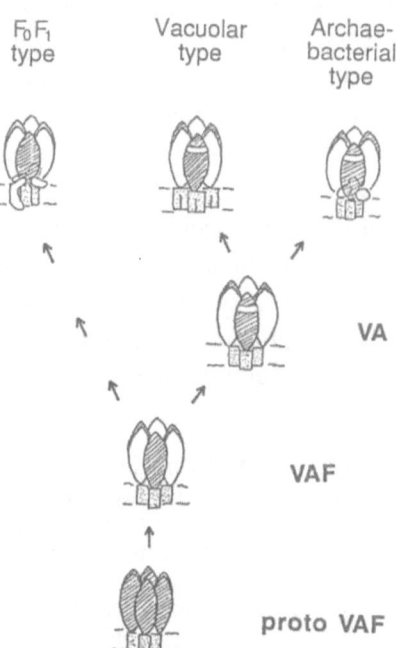

Fig.5. An evolutional tree of the H^+-ATPase superfamily. Shadowed subunits represent catalytic subunits; the β subunit of $F_o F_1$-ATPase, 70K subuint of vacuolar H^+-ATPase, and the α subunit of *Sul*-ATPase. Dotted subunits represent proteolipid subunits (or c subunit of $F_o F_1$-ATPase) which are specifically labeled by DCCD. A white band in the catalytic subunits of V, A, and VA ATPase shows the inserted ~95 amino acid sequence. The proto VAF ATPase is made up from six catalytic subunits and six proteolipid subunits.

the figure. The VA-ATPases and F_oF_1-ATPases (F) diverged from a common ancestor, shown as VAF in the figure. The VAF ATPase consisted of at least three kinds of subunits, three catalytic subunits, three non-catalytic subunits and six proton-channel proteolipids. During evolution from VAF to VA, the catalytic subunits gained a large inserted sequence near the amino terminal region. Since the significant sequence similarity has been found between catalytic and non-catalytic subunits of each ATPase discussed here, it is reasonable to conclude that they were created from the same gene by gene duplication. If it is really so, we can trace back to the more primordial ATPase (proto-VAF) which had six identical catalytic subunits and six proteolipid subunits.

Protomer H^+-ATPase

The proto-AVF ATPase is likely to be the oldest *hexagonal* H^+-ATPase, however, is it the oldest H^+-ATPase? It has been known that oligomeric proteins and enzymes composed of the identical or semi-identical subunits often have its monomeric counterpart, as seen for haemoglobin and myoglobin. Therefore, we should consider a possibility that monomeric H^+-ATPase can precede the appearance of hexameric one. This monomeric enzyme, designated as protomer H^+-ATPase in Fig. 6, could contain a catalytic moiety and a proton-channel moiety (proteolipid) in a single polypeptide or in two separarated polypeptides. From the analogy to the myoglobin, the protomer H^+-ATPase has neither regulatory functions, nor kinetic cooperativity caused from subunit interactions but it can still catalyze by itself the core function, that is, coupling of proton translocation and ATP synthesis/hydrolysis. The evolutional advantage of hexameric H^+-ATPase might be fine regulatory function which can meet variable energy demand of cells. This implies that the strong kinetic cooperativity observed for present-day H^+-ATPases should be related to some regulatory functions of the enzyme rather than the catalysis itself. The ATP hydrolysis catalyzed by a single catalytic site with a very high affinity to ATP, unisite catalysis, observed for mitochondrial F_1-ATPase can occur to the protomer H^+-ATPase but it is too slow to play a biological function in the primordial cells. Alternative binding site theory cannot be appricable to protomer-H^+-ATPase since it has only one adenine nucleotide binding site. If this protomer enzyme really exist, it also seems to be in conflict with the idea that the catalytic site(s) resides at the interface(s) of the two major subunits.

Fig.6. Hypothetical protomer H^+-ATPase. It may catalyze ATP synthesis coupled with proton flow but the functions may be no longer cooperative. Remarkable cooperative natures of present-day H^+-ATPases may be related to regulatory functions which have evolved later.

The origin of eukaryotic cells

According to the endosymbiotic theory of the origin of eukaryitc cells, the evolution of symbioses has been the origin of mitochondria and chloroplasts. Eukaryotic endomembrane system, including endoplasmic reticulum, Golgi bodies, lysosomes, vacuoles and secretory vesicles, are proposed to be originated from invagination of the host cell plasma membrane rather than endosymbionts (Fig. 7). Therefore, information about the origin of the endomembrane system may provide insight into the nature of the host cell. In this respect, the finding of close similarity between archaebacterial ATPases and vacuolar H^+-ATPases suggests a possibility that the host cell could be an prokaryote whose ATPase was V type. Although it is not established that distribution of prokaryotic V-ATPase is really restricted in archaebacteria, the primordial prokaryote from which archaebacteria have been evolved is the candidate of the host cell that has been searched for.

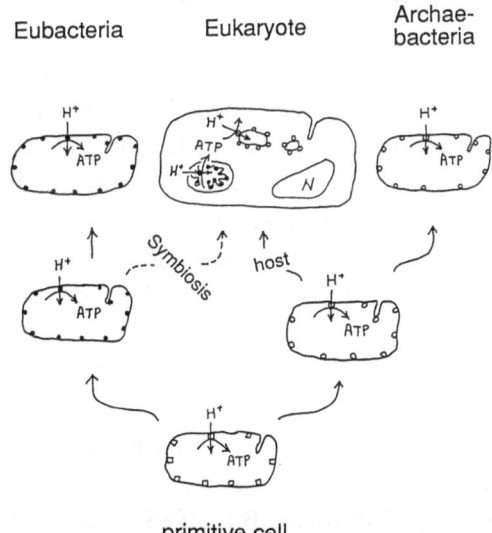

primitive cell

Fig.7. Was an archaebacteria-like cell the host cell for symbionts which later have evoluved into eukaryotic cells ? Eukaryotic vacuolar systems were originated from invagination of plasma membranes of an ancient archaebacteria-like cell. After separation from the plasma membranes, H^+-ATPase changed its physiological function from ATP synthesis to ATP-driven proton pump because mitochondria, descendants of symbionts, were more effective to produce ATP.

ACKNOWLEDGEMENTS

We thank Mr. Masato Sumi for his assistence with preparing our manuscript.

References

1. Hatefi, Y. (1985) *Ann. Rev. Biochem.* **54**, 1015-1069
2. Boyer, P.D. (1987) *Biochemistry* **26**, 8503-8507
3. Futai, M., Noumi, T., and Maeda, M. (1989) *Ann. Rev. Bichem.* **58**, 111-136
4. Senior, A.E. (1990) *Ann. Rev. Biophys.* in press
5. Walker, J.E., Fearnly, I.M., Gay, N.J., Gibson, B.W., Northrop, F.D., Powell, S.J., Runswick, M.J., Saraste, M., and Tybulewicz, V.L.J. (1985) *J. Mol. Biol.* **184**, 677-701
6. Walker, J.E., Cozens, A.L., Dyer, M.R., Fearnley, I.M., Powell, S.J., and Runswick, M.J. (1987) *Biochem. Soc. Trans.* **15**, 104-106
7. Inatomi, K. (1986) *J. Bacteriol.* **167**, 837-841
8. Mukohata, Y. and Yoshida, M. (1987) *J. Biochem.* **102**, 797-802
9. Konishi, J., Wakagi, T., Oshima, T., and Yoshida, M. (1987) *J. Biochem.* **102**, 1379- 1387
10. Lübben, M., Lünsdorf, H., Schäfer, G. (1988) *Biol. Chem. Hoppe-Seyler* **369**, 1259-1266
11. Stan-Lotter, H. and Hochstein, L.I. (1989) *Eur. J. Biochem.* **179**, 155-160
12. Schobert, B. and Lanyi, J.K. (1989) *J. Biol. Chem.* **264**, 12805-12812
13. Konishi, J., Denda, K., Oshima, T., Wakagi, T., Uchida, E., Ohsumi, Y., Anraku, Y., Matsumoto, T., Wakabayashi, T., Mukohata, Y., Ihara, K., Inatomi, K., Kato, K., Ohta, T., Allison, W.S., and Yoshida, M. (1989) *J. Biochem.* in press
14. Sebald, W. and Hoppe, J. (1981) *Curr. Top. Bioenerg.* **12**, 1-64
15. Denda, K., Konishi, J., Oshima, T., Date, T., and Yoshida, M. (1988) *J. Biol. Chem.* **263**, 6012-6015
16. Denda, K., Konishi, J., Oshima, T., Date, T., and Yoshida, M. (1988) *J. Biol. Chem.* **263**, 17251-17254
17. Zimniak, L., Dittrich, P., Gogarten, J. P., Kibak, H., and Taiz, L. (1988) *J. Biol.Chem.* **263**, 9102-9112
18. Bowman, E.J., Tenney, K., and Bowman, B.J. (1988) *J. Biol. Chem.* **263**, 13994-14001
19. Bowman, B.J., Allen, R., Wechser, M.A., and Bowman, E.J. (1988) *J. Biol. Chem.* **263**, 14002-14007
20. Manolson, M.F., Ouellette, B.F.F., Filion, M., and Poole, R.J. (1988) *J. Biol. Chem.* **263**, 17987-17994
21. Nelson, H., Mandiyan, S., and Nelson, N. (1989) *J. Biol. Chem.* **264**, 1775-1778
22. Gogarten, J.P., Kibak, H., Dittrich, P., Taiz, L., Bowman, E.J., Bowman, B.J., Manolson, M.F., Poole, R.J., Date, T., Oshima, T., Konishi, J., Denda, K., and Yoshida, M. (1989) *Proc. Natl. Acad. Sci. U. S. A.* **86**, 6661-6665
23. Inatomi, K., Eya, S., Maeda, M., and Futai, M. (1989) *J. Biol. Chem.* **264**, 10954-10959

24. Bernasconi, P., Rausch, T., Gogarten, J.P., and Taiz, L. (1989) *FEBS Lett.* **251**, 132-136

25. Denda, K., Konishi, J., Oshima, T., Date,T., and Yoshida, M. (1989) *J. Biol. Chem.* **264**, 7119-7121

Nuclear Genes Encoding Two Subunits of Human Mitochondrial Cytochrome *bc*1 Complex, Cytochrome *c*1 and Ubiquinone-Binding Protein: THEIR STRUCTURAL ORGANIZATION OF THE 5'-FLANKING REGIONS AND CHROMOSOMAL LOCALIZATION

Hiroshi Suzuki[1], Yoshitaka Hosokawa[1], Haruo Toda[1], Morimitsu Nishikimi[1]*, Akio Matsukage[2], Michihiro C. Yoshida[3], and Takayuki Ozawa[1]

[1]Department of Biomedical Chemistry, Faculty of Medicine University of Nagoya, Nagoya 466, [2]Laboratory of Cell Biology Aichi Cancer Center Research Institute, Nagoya 464, and [3]Chromosome Research Unit, Faculty of Sciences, Hokkaido University, Sapporo 060, Japan

SUMMARY

We have isolated nuclear genes encoding human cytochrome *c*1 (C1) and ubiquinone-binding protein (QP) of cytochrome *bc*1 complex. The C1 and QP genes span 2.4 and 4.5 kilobase pairs, respectively. All intron/exon splice junctions in both genes follow the GT/AG rule. The 5'-flanking region of the C1 gene contains seven putative GC boxes as typical transcriptional regulatory sequence elements but lacks TATA and CCAAT boxes. The same region of the QP gene contains four putative CCAAT boxes and one putative NF-Y binding site but lacks TATA and GC boxes. In the 5'-flanking regions of both genes, there are homologous sequences to AP-1 recognition site and three common sequences, 5'-TATTCAGGT-3', 5'-ATCTGGCT-3', and 5'-TGGTGA(T/G)AG-3'. A homologous sequence to each of the three common sequences is also found in the 5'-flanking regions of the genes for the β subunit of human F0F1-ATPase and rat somatic cytochrome *c*. It may be that the common sequences play an important role in the coordinate expression of nuclear genes encoding mitochondrial proteins responsible for oxidative phosphorylation. Southern blot analyses showed that both genes are present in a single copy in the human genome and are located on chromosome 8.

INTRODUCTION

The mitochondrial oxidative phosphorylation system in mammals consists of five multisubunit complexes. Since all the complexes but Complex II comprise subunits encoded in both the mitochondrial and the nuclear genomes, a regulatory system must be continuously operating to coordinate the transcription of mitochondrial and nuclear genes under various conditions such as proliferation (1, 2), malignant transformation (3-5), and differentiation (1, 6) of the cells, and mitochondrial DNA mutations in mitochondrial cytopathies (7). Since most of the genetic information required for mitochondrial

---

* Present address : Institute of Applied Biochemistry, Yagi Memorial Park, Mitake, Gifu 501-01, Japan

biogenesis is provided by the nuclear genome, genetic and biochemical analyses of nuclear genes encoding mitochondrial proteins are essential to an understanding of the biogenesis and the molecular mechanism of communication between physically separated nuclear and mitochondrial genomes. Ozawa et al. (8) and Zeviani et al. (9) demonstrated the first examples of a maternal and an autosomal heritable way, respectively, associated with lessions of the human mitochondrial genome. Zeviani et al. (9) also suggested the existence of a nuclear-coded, *trans*-acting factor involved in the lessions of the genome. We have focused on the genetic information residing in the flanking regions of nuclear genes encoding mitochondrial respiratory components, and found two common sequence elements (designated Mt1 and Mt2 in Ref.10) in the 5'-flanking regions of nuclear genes encoding human cytochrome c_1 (C1), a subunit of cytochrome bc_1 (Complex III), the β subunit of human F_0F_1-ATPase (HATPase β) (11) and chicken δ-aminolevulinate synthase (12). It is very interesting to examine whether the same sequences occur in nuclear genes for other mammalian mitochondrial respiratory components and participate in the cooperative expresssion of many nuclear genes responsible for the respiration. Indeed, in yeast, two consensus sequence elements participating in the coordinate expression of nuclear genes for different subunits of Complex III have been reported (13).

We isolated and characterized a nuclear gene encoding human ubiquinone-binding protein (QP), a subunit of Complex III (14). Here we have determined the nucleotide sequence of the 5'-flanking region and the chromosomal localization of the QP gene, and found that the three common sequences exist in the 5'-flanking regions of the QP and the C1 genes. We will discuss the possibility of the involvement of the common sequences in the coordinate expression of the nuclear genes encoding mitochondrial proteins responsible for oxidative phosphorylation.

EXPERIMENTAL PROCEDURES

Materials - Enzymes and chemicals were purchased from the following sources:restriction enzymes and T4 DNA ligase from Toyobo (Osaka, Japan);Klenow fragment of DNA polymerase I, and a 7-deaza sequencing kit from Takara Shuzo (Kyoto, Japan); calf intestinal alkaline posphatase from Boehringer Mannheim; and a multiprime DNA labeling system and [α-^{32}P]dCTP (3,000 Ci/mmol) from Amersham. GeneScreen *Plus* nylon membranes were obtained from Du Pont. The human and mouse parental cells and karyotypical characterization of human-mouse somatic cell hybrids were described elsewhere (15).

Isolation of Phage Clones Containing the Human C1 and QP Genes - A genomic DNA library of about 1.6 x 10^6 EMBL3 phages containing a partial *Mbo*I digest of human leukocyte DNA was screened by plaque hybridization as previously described in (10) using human C1 (16) and QP (17) cDNAs as probes, respectively. The genomic DNA inserts in cloned phages were excised by digestion with *Sal*I for subsequent restriction mapping and sucloning to sequence.

DNA Fragments Used as Probes-The following DNA fragments were prepared from the recombinant DNA containing the 5'-flanking regions of the C1 (10) and the QP genes.The 154-bp *Sal*I-*Pst*I fragment (nucleotide position -1405 to -1125 relative to the translation initiation site) from the C1 gene and the 280-bp *Bam*HI-*Hind*III fragment (see Fig. 1) from the QP gene were used as hybridization probes. These probes were radioactively labeled with ^{32}P-dCTP by multiple priming as described in (10).

Southern Hybridization - Southern blot analysis was performed at high sringency as described in (10).

DNA Sequencing Analysis - The 1.3-kb *Pst*I-*Bam*HI, 0.3-kb *Bam*HI-*Hind*III, and 0.6-kb *Hind*III-*Pst*I fragments from the 5'-flanking region of the QP gene (see Fig. 1) were subcloned into pUC18 plasmid. DNA sequencing was carried out as described in (10).

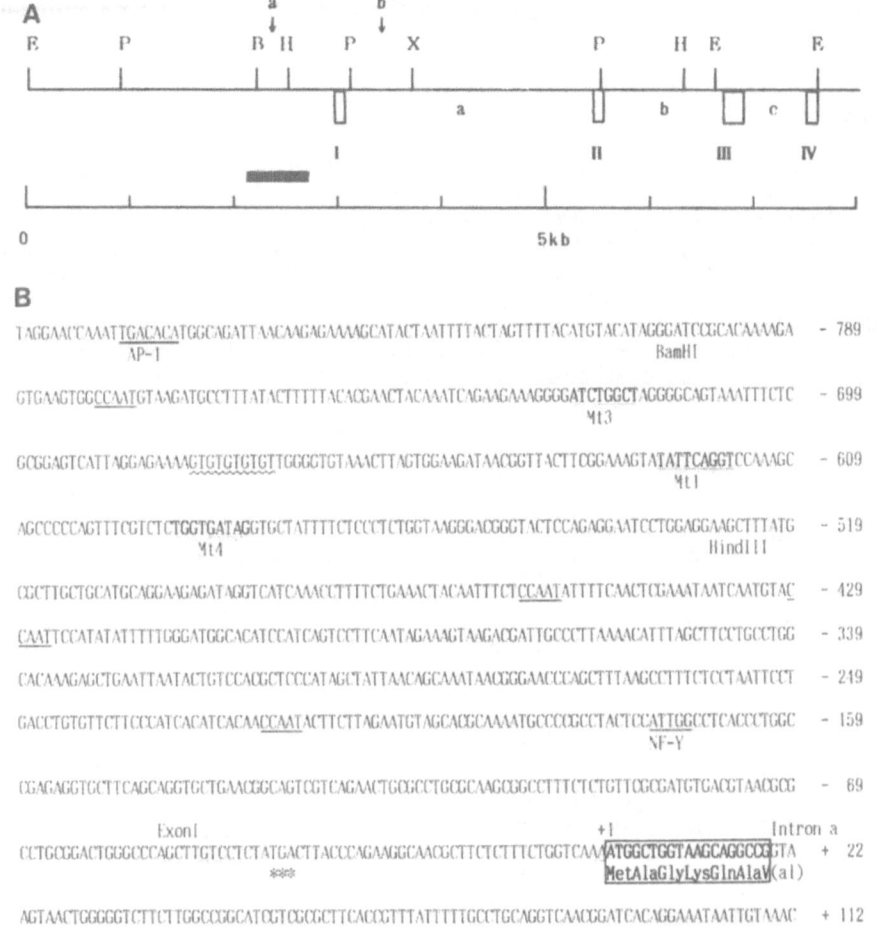

Fig. 1. Structural organization and nucleotide sequence of 5'-flanking region of the human QP gene. A, structural organization of the QP gene. Coding regions (exons) are represented by open boxes and are designated by Roman numerals. Introns are named alphabetically as indicated. The direction of transcription is from left to right. E, P, B, H, and X represnt *Eco*RI, *Pst*I, *Bam*HI, *Hind*III, and *Xba*I, respectively. A thick bar represents the sequenced region in this study. Vertical arrows a and b indicate the fragments used as probes for the hybridization with human genomic DNAs (Fig. 3) and with human-mouse somatic cell hybrid DNAs (Fig, 4), respectively. B, nucleotide sequences of the 5'-flanking region of the human QP gene. Nucleotide +1 denotes the A of the translation initiation codon ATG and residues preceding it are indicated by negative numbers. Mt1, Mt3, and Mt4 show the common sequences found in the 5'-flanking regions of the QP and C1 genes. The nucleotide sequence and the deduced amino acid sequence of the first coding region are boxed. Four putative CCAAT boxes and one putative NF-Y binding site are underlined. The putative AP-1 recognition sequence is indicated by thick underline, and a potential Z-DNA-forming sequence by wavy line. *** represents a stop codon in frame.

Computor Analysis - Computor analysis was performed with a GENETYX program (SDC, Tokyo).

RESULTS

Nucleotide Sequence of the 5'-Flanking Region of the Human QP Gene - The single nuclear gene encoding human QP of Complex III was isolated and characterized (14). The nucleotide sequence of 878 bp upstream from the translation initiation site revealed two structural features (Fig. 1). Fisrt, there are four putative CCAAT boxes, but no typical TATA boxes exist at normal distances downstream from the CCAAT boxes. Second, no GC boxes exist in the region. We found one putative NF-Y binding site (ATTGG) , the opposite orientation of the CCAAT motif, and a homologous sequence (TGACACA) to GCN4 or AP-1 recognition site (TGACTCA)(18,19). The 5'-flanking region of the C1 gene (10) has seven putative GC boxes but neither a TATA box- nor CCAAT box-like

```
                           Mt1

HQP          -624    TATTCAGGT    -616     sense strand
HC1          -1304   TATTCAGGT    -1296    sence strand
HATPase β    -1790   TAGCCAGGT    -1798    antisense strand
RC           -660    TAATCAGGC    -652     sense strand

     Common
     sequence        TATTCAGGT


                           Mt3

HQP          -724    ATCTGGCT     -717     sense strand
HC1          -1381   ATCTGGCT     -1374    sense strand
HATPase β    -1798   ACCTGGCT     -1791    sence strand
RC           -685    ATCTAGCA     -678     sence strand

     Common          ATCTGGCT
     sequence


                           Mt4

HQP          -590    TGGTGATAG    -582     sense strand
HC1          -1284   TGGTGAGAG    -1292    antisense strand
HATPase β    -1718   TGGTGAAAC    -1726    antisense strand
RC           -539    TGGGGATGG    -547     antisense strand

     Common              T
     sequence        TGGTGA AG
                          G
```

Fig. 2. Three common sequences, Mt1, Mt3 and Mt4, found in the 5'-flanking regions of the genes for human QP (HQP), human C1 (HC1) (10), the β subunit of human F0F1-ATPase (HATPase β) (11), and rat somatic cytochrome *c* (RC) (21). Nucleotide positions relative to the translation initiation site for the HQP, HC1, and HATPase β genes and relative to the transcription initiation site for the RC gene are shown.

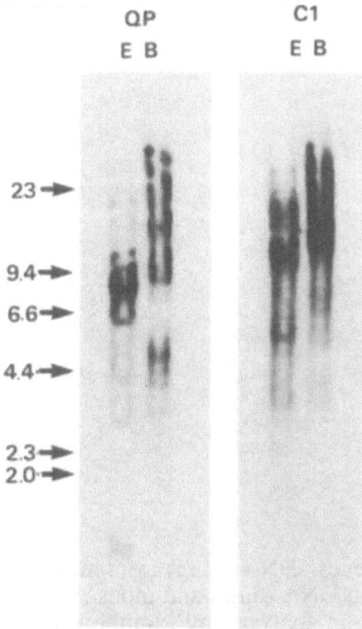

Fig. 3. Existence of multiple nucleotide sequences in the human genome which hybridize to the *Bam*HI-*Hin*dIII or the *Sal*I-*Pst*I fragment. Human placenta DNAs (5µg each) were digested with *Eco*RI (E) or *Bam*HI (B), and Southern blot hybridization was carried out using the 280-bp *Bam*HI-*Hin*dIII (QP) (see Fig. 1) or the 154-bp *Sal*I-*Pst*I fragment (C1) was used as a probe. The two fragments contain Mt1, Mt3, and Mt4 sequences.

sequence. Two of them display the same decanucleotide GCCCCGCCCC motif, which has been shown to exhibit a relatively high affinity for Sp1 (20). We also found three sequences, TGACTCA (on the antisense strand), TGACTGA, and TGACTCG homologous to the GCN4 or AP-1 recognition site.

Three Common Sequences Found in the 5'-Flanking Regions of the QP and C1 Genes - As shown in Fig. 2, three common sequences, 5'-TATTCAGGT-3', 5'-ATCTGGCT-3', and 5'-TGGTGA(T/G)AG-3', designated Mt1, Mt3, and Mt4, respectivey, were found in the 5'-flanking regions of the C1 and QP genes by computor-assisted homology search. Highly homologous sequences to the three sequences were also found in the corresponding regions of two other nuclear genes for HATPase β (11) and rat somatic cytochrome *c* (RC) (21). These three sequences are located closely on the 154-bp *Sal*I-*Pst*I fragment of the C1 gene and in the 280-bp *Bam*HI-*Hin*dIII fragment of the QP gene (Figs 1 and 2). As shown in Fig. 3, Southern blot analyses using the two fragments as probes showed that a number of the nucleotide sequences which hybridize to the two fragments at high stringency, exist in the human genes.

Chromosomal Localization of the Human C1 and QP Genes - DNAs extracted from 14 human-mouse somatic cell hybrids were examined to determine the chromosome on which the human QP gene is located. The *Eco*RI-digested human genomic DNA on a membrane was hybridized with the 0.6-kb *Pst*I-*Xba*I fragment of the QP gene. The fragment recognizes a single human *Eco*RI fragment of 7 kb (14). As shown in Fig. 4, DNAs from four hybrid clones, 1-4, 3A6, 7D4, and 6-3, gave a definitely positive band. Although hybrid clone 3B5 gave a weakly positive band at the position of the human specific DNA, this is due to a small number of cells retaining chromosome 8. As shown in Table 1, chromosome 8 is the only human chromosome that showed the most concordance with the presence of human QP gene DNA. The DNAs from the same clones also gave positive bands when the C1 cDNA or the DNA polymerase β cDNA (10SL), for which genes have been assigned to chromosome 8 (22, 23), were used as a probe. Thus, we conclude unambiguously that the QP gene is also located on chromosome 8.

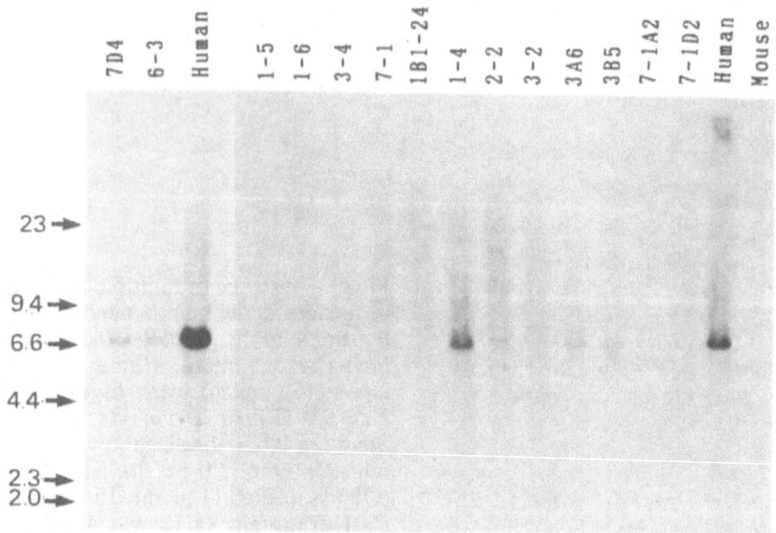

Fig. 4. Chromosomal localization of QP genes. DNAs (5μg) from human-
mouse somatic cell hybrids, human DNA (5μg), and mouse DNA
(5μg) were digested with *Eco*RI and analyzed by tSouthern blot
hybridization using the 0.6-kb *Pst*I-*Xba*I fragment of the QP gene
corresponding to a part of intron a (see Fig. 1). Indicated size
markers are in kb.

Table 1. Segregation of the QP gene with human chromosomes in human-mouse hybrids.

Hybrid clone	\multicolumn{24}{c}{Human chromosome}																							QP	C1[a]	β[b]
	1	2	3	4	5	6	7	8	9	10	11	12	13	14	15	16	17	18	19	20	21	22	X	Y			
1-5	-	-	-	+	-	-	-	-	-	-	+	-	(+)	-	-	+	-	+	+	-	-	+	-		-	-	-
1-6	-	-	+	+	-	+	-	-	+	+	+	-	-	+	-	+	+	(+)	-	-	-	+	-		-	-	-
3-4	-	-	-	-	(+)	-	-	-	-	-	-	-	+	+	-	-	-	-	-	+	(+)	(+)	-		-	-	-
7-1	-	-	+	-	-	-	-	-	-	-	+	+	+	-	+	-	-	-	+	-	+	-	-		-	-	-
1B1-24	-	-	-	-	-	-	+	-	-	-	-	-	+	-	+	-	-	-	-	-	-	-	-		-	-	-
1-4	-	-	-	(+)	-	-	(+)	(+)	-	-	-	+	+	-	-	+	+	-	-	+	+	-	-		+	+	+
2-2	-	-	-	-	-	-	+	-	-	-	+	+	-	-	(+)	+	-	-	-	-	-	-	-		-	-	-
3-2	-	-	-	-	-	-	+	-	-	-	+	+	-	-	(+)	+	-	-	-	-	-	-	-		-	-	-
3A6	-	-	(+)	(+)	-	(+)	-	(+)	-	-	-	-	-	-	-	(+)	+	-	+	+	-	-	-		+	+	+
3B5	-	-	-	+	+	-	-	-	-	-	-	-	+	-	-	-	+	-	(+)	(+)	-	-	-		[+]	[+]	[+]
7-1A2	+	-	+	-	-	-	+	-	-	-	+	+	+	-	+	-	(+)	+	-	-	-	-	-		-	-	-
7-1D2	-	-	-	-	-	-	-	-	-	-	+	+	-	-	-	-	-	-	-	-	-	-	-		-	-	-
7D4	-	-	-	+	+	-	+	+	-	+	-	+	-	-	-	+	-	-	+	-	+	-		+	+	+
6-3	-	-	-	+	+	(+)	+	+	-	-	+	+	-	+	-	-	+	-	+	-	+	-	-		+	+	+

+, (+), and -, more than 30%, 20-30%, and less than 20% of cells retained the
chromosome, respectively. The presence of the diagnostic band in Southern hybridization
was shown by a + designation in the column designated ubiquinone-binding protein (QP),
cytochrome *c*1 (C1), and DNA polymerase β (β). [+] represents the presence of the
diagnostic band resulting from existence of a small number of cells retaining chromosome
8. a and b from ref. 22.

DISCUSSION

Coordinate Expression of the Nuclear and Mitochondrial Genes - As summarized in Table 2, the two genomic systems are coordinately regulated. In the expression of the mitochondrial genome, two different regulatory mechanisms are likely to operate in the short and long terms. The long-term expression is caused by the increase in the transcriptional activity, and the long-term expression is supported by the increase in the copy number of the genes resulting from the increase in the number of mitochondrion per cell (24). In the case of the short-term expression, several factors affect the coordinate regulation in expression of the nuclear and mitochondrial genes. Expression of nuclear genes for ADP/ATP carrier and c-*myc* is cell-cycle dependent in phytohemagglutinin-stimulated human lymphocytes (2). When cells are transformed by polyoma virus DNA, the polyoma large T, adenovirus E1A , and c-*myc* oncogene, steady-state level of the mitochondrial mRNA encoding COII (3) or the nuclear mRNA for ADP/ATP carrier (5) was increased. The protein products of oncogenes (for example, c-*jun* and c-*myc*) act as *trans*-acting factors in RNA and/or DNA synthesis (25). These facts suggest the

Table 2. Coordinate expression of the mitochondrial gene and the nuclear genes encoding mitochondrial proteins responsible for oxidative phosphorylation.

			Copy number of Mt DNA	Tested genes	mRNA	Ref.
I. Short-term expression(5~48hr)						
a. Cell proliferation	1.	Serum	N.D.	N	ADP/ATP↑ carrier	(1)
	2.	Growth factors	N.D.	N	ADP/ATP↑ carrier	(1)
	3.	Phytohemagglutinin	N.D.	N	ADP/ATP↑ carrier	(2)
	4.	Transformation by oncogenes or virus	→	Mt	COII ↑	(3)
			N.D.	N	ADP/ATP↑ carrier	(4)
	5.	Nuclear gene lessions	→	Mt	ND1, ↑ 16SrRNA	(5)
b. Cell differentiation	1.	Phorbol esters	N.D.	N	ADP/ATP↓ carrier	(1)
	2.	Retinoic acids	N.D.	N	ADP/ATP↓ carrier	(1)
			N.D.	N	Hinge ↓ protein	(6)
c. Hormonal regulation	1.	Thyroid hormone	N.D.	Mt	All Mt ↑ mRNA rRNA →	(25)
II. Long-term expression(5~30 days)						
a. Chronic stimulation of muscles			↑	Mt	Cyt b ↑ rRNA ↑	(26)
			↑	N	ATPaseβ ↑	(27)
			N.D.	Mt N	COIII ↑ COVIc ↑	(28)

N and Mt represent the nuclear and mitochondrial genes, respectively. →, ↑, and ↓ indicate no change, increase, and decrease, respectively. N.D.; not determined.

possibility that the oncogene products play an important role in the coordinate expression. On the other hand, mRNA levels of the ADP/ATP carrier (1) and the hinge protein (6) were decreased when growing HL-60 cells are induced to differentiate by either phorbol ester or retinoic acid. It seems that a regularory system operates to maintain or modulate the delicate balance of the coordinate expression of the mitochondrial and nuclear genes during the cell cycle and differentiation.

Zeviani et al. (9) reported an autosomally-inherited case of mitochondrial myopathy and suggested the existence of a nuclear-coded, *trans*-acting factor involved in the lessions of the genome. In polyps of familial polyposis coli (FPC) patients, the level of mitochondrial mRNA was increased (5). FPC is inherited as an autosomal dominant trait suggesting that the increased level of the mRNA is associated with a lession of the nuclear genome. These findings indicate that mitochondrial biogenesis is regulated by the tight communication between physically separated nuclear and mitochondrial genes, and that the nuclear-coded *trans*-acting protein factors may play important roles in the communication as well as in the transcription. The *trans*-acting proteins which specifically bind to the the common sequences discovered in the present study, therefore, could be candidates for the communicators between the nuclear and mitochondrial genes.

Structural Features of the 5'-Flanking Regions of the Nuclear Genes Encoding Mitochondrial Proteins Responsible for Oxidative Phosphorylation - The structures of the 5'-flanking regions of the four nuclear genes encoding mitochondrial proteins responsible for oxidative phosphorylation were summarized schematically in Fig. 5. They consist of GC and/or CCAAT boxes as *cis*-acting promoter elements regulating the initiation of

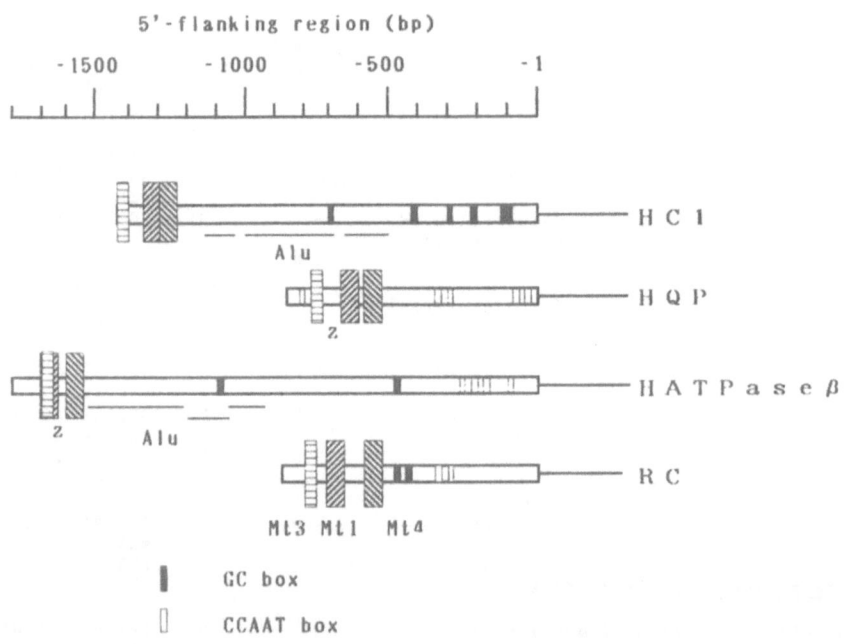

Fig. 5. Location of Mt1, Mt3, and Mt4 in the 5'-flanking regions of four nuclear genes responsible for mitochondrial respiration. Alu and Z indicate the *Alu* sequences and potential Z-DNA-forming sequences, respectively. HC1, the human cytochrome c_1 gene; HQP, the human ubiquinone-binding protein gene; HATPase β, the β subunit gene of human F_0F_1-ATPase; RC, the rat somatic cytochrome *c*. Nucleotide positions are the same as described in the legend of Fig. 2.

transcription. The C1 and QP genes use only GC and CCAAT boxes, respectively. On the other hand, the HATPase β and RC genes use both GC and CCAAT boxes. Absence of the typical TATA box is characteritic of all these genes. They, therefore, belong to a housekeeping type of genes. Another feature of the 5'-flanking regions is that three common sequences, Mt3, Mt1, and Mt4 (Fig. 2), are located in this order in the limited region , although Mt1 overlaps Mt3 in the HATPase gene (Fig. 5). Highly homologous nucleotide sequences to the 280-bp *Bam*HI-*Hind*III or the 154-bp *Sal*I-*Pst*I fragments which contain the three common sequences, occur in other nuclear genes (Fig. 3), suggesting the possibility that these sequences function as *cis*-acting elements in the transcription of nuclear genes.

Employing gel retardation assays, we tried to examine the association of whole cell extract from HeLa cells with fragments containing the common sequences A few proteins in the extract bind specifically to the 280-bp *Bam*HI-*Hind*III fragment from the QP gene and the 154-bp *Sal*I-*Pst*I fragment from the C1 gene. It was also found that one of them specifically binds to both fragments, demonstrated by the competition experiment. This result strongly suggests that at least one of the three common sequences in the 5'-flanking regions of the QP and C1 genes is a *cis*-acting element responsible for the coordinate regulation of the transcription. Analysis of the nucleotide sequence of the binding site is in progress with DNase I foot printing.

ACKNOWLEDGEMENTS

DNAs extracted from some of human-mouse somatic cell hybrids were supplied by the Cancer Research Program.   This work was supported in part by Grants-in-Aids for Scientific Research on Priority Areas (62617002) from the Ministry of Education, Sience and Culture of Japan.

REFERENCES

1. Battini, R., Ferrari, S., Kaczmarek, L., Calabretta, B., Chen, S-T., and Baserga, R. (1987) *J. Biol. Chem.* 262, 4355-4359.
2. Kaczmarek, L., Calabretta, B., and Baserga, R. (1985) *Proc. Natl. Acad. Sci. U. S. A.* 82, 5375-5379.
3. Glaichenhaus, N., Leopold, P., and Cuzin, F. (1986) EMBO J. 5, 1261- 1265.
4. Liu, H.-T., Baserga, R., and Mercer, W. E. (1985) *Mol. Cell. Biol.* 5, 2936-2942.
5. Yamamoto, A., Horai, S., and Yuasa, Y. (1989) *Biochem. Biophys. Res. Commun.* 159, 1100-1106.
6. Ohta, S., Goto, K., Arai, H., and Kagawa, Y. (1987) *FEBS Lett.* 226, 171-175.
7. Tanaka, M., Miyabayashi, S., Nishikimi, M., Suzuki, H., Shimomura, Y., Ito, K., Narisawa, K., Tada, K., and Ozawa, T. (1988) *Pediatric Res.* 24, 447-454.
8. Ozawa, T., Yoneda, M., Tanaka, M., Ohno, K., Sato, W., Suzuki, H., Nishikimi, M., Yamamoto, M., Nonaka, I., and Horai, S. (1988) *Biochem. Biophys. Res. Commun.* 154, 1240-1247.
9. Zeviani, M., Servidei, S., Gellera, C., Bertini, E., DiMauro, S., and DiDonato, S. (1989) *Nature* 339, 309-311.
10. Suzuki, H., Hosokawa, Y., Nishikimi, M., and Ozawa, T. (1989) *J. Biol. Chem.* 264, 1368-1374.
11. Ohta, S., Tomura, H., Matsuda, K., and Kagawa, Y. (1988) *J. Biol. Chem.* 263, 11257-11262.
12. Maguire, D. J., Day, A. R., Borthwick, I. A., Srivastava, G., Wigley, P. L., May, B. K., and Elliott, W. H. (1986) *Nucleic Acids Res.* 14, 1379-1391.

Suzuki et al.

13. Dorsman, J. C., van Heeswijk, W. C., and Grivell, L. A. (1988) *Nucleic Acids Res.* 16, 7287-7301.
14. Suzuki, H., Hosokawa, Y., Toda, H., Nishikimi, M., and Ozawa, T. (1989) *Biochem. Biophys. Res. commun.* 161, 371-378.
15. Semba, K., Yamanashi, Y. Nishizawa, M., Sukegawa, J., Yoshida, M., Sasaki, M., Yamamoto, T., and Toyoshima, K. (1985) *Science* 227, 1038-1040.
16. Nishikimi, M., Ohta, S., Suzuki, H., Tanaka, T., Kikkawa, F., Tanaka, M., Kagawa, Y., and Ozawa, T. (1988) Nucleic Acids Res. 16, 3577.
17. Suzuki, H., Hosokawa, Y., Toda, H., Nishikimi, M., and Ozawa, T.(1988) *Biochem. Biophys. Res. Commun.* 156, 987-994.
18. Arndt, K., and Fink, G. R. (1986) *Proc. Natl. Acad. Sci. U. S. A.* 83, 8516-8520.
19. Bohmann, D., Bos, T. J., Admon, A., Nishimura, T., Vogt, P. K., and Tjian, R. (1987) *Science* 238, 1386-1392.
20. Kadonaga, J. T., Jones, K. A., and Tijan. R. (1986) *Trends Biochem. Sci.* 11, 20-23.
21. Evans, M. J., and Scarpulla, R. C. (1988) *Mol. Cell. Biol.* 8, 35-41.
22. Nishikimi, M., Suzuki, H., Yamaguchi, M., Matsukage, A., Yoshida, M.C., and Ozawa, T. (1988) *Biochem. Intl.* 16, 655-660.
23. Matsukage, A., Yamaguchi, M., Utsumi, K. R., Hayashi, Y., Ueda, R., and Yoshida, M. C. (1986) *Jpn. J. Cancer Res.* (Gann) 77, 330-333.
24. Robin, E. D., and Wong, R. (1988) *J. Cell. Physiol.* 136, 507-513.
25. Kingston, R. E., Baldwin, A. S., and Sharp, P. A. (1985) *Cell* 41, 3- 5.
26. Mutvei, A., Kuzela, S., and Nelson, B. D. (1989) *Eur. J. Biochem.*180, 235-240.
27. Williams, R. S. (1986) *J. Biol. Chem.* 261, 12390-12394.
28. Williams, R. S., Garcia-Moll, M., Mellor, J., Salmons, S., and Harlan, W. (1987) *J. Biol. Chem.* 262, 2764-2767.
29. Hood, D. A., Zak, R., and Pette, D. (1989) *Eur. J. Biochem.* 179, 275- 280.

STUDIES ON THE FUNCTION OF THE MITOCHONDRIAL HINGE PROTEIN:

MOLECULAR GENETIC APPROACH USING YEAST AS A MODEL SYSTEM

Chong H. Kim[1,2] and Richard S. Zitomer[2]

[1]Department of Biology, Rensselaer Polytechnic Institute, Troy, New York 12180, and [2]Department of Biological Sciences, State University of New York Albany, New York 12222, U.S.A.

SUMMARY

This chapter summarizes our recent studies on the 17 kDa protein, subunit VI of yeast mitochondrial cytochrome bc_1 complex, as a model system for the investigation of the function of the bovine heart mitochondrial hinge protein. The yeast 17 kDa protein shows an extensive homology to the hinge protein from the bovine heart as well as the human mitochondria. A double deletion mutant of *Saccahromyces cerevisiae*, which lacks the 17 kDa protein and *iso*-1-cytochrome *c*, was constructed. This double deletion mutant *(cyc1-1,17kDa^{-1})* cannot grow on nonfermentable carbon source (glycerol$^-$), but can be complemented to normal growth, when it is transformed by yeast vectors containing the wild type 17 kDa protein gene or the *CYC1* gene. This double mutant showed a growth rate and a oxygen consumption rate three times slower than those of the wild type. These results suggest that the 17 kDa protein may be essential for cell respiration of a yeast strain whose cytochrome *c* level is limited (only 5%). This glycerol phenotype allows genetic studies on the function of the 17 kDa protein in yeast mitochondria and those information on the function of the 17 kDa protein can lead us to elucidate the role of the hinge protein in mammalian mitochondria.

INTRODUCTION

The ubiquinol cytochrome *c* reductase, also called the cytochrome bc_1 complex, is a mitochondrial respiratory chain enzyme responsible for catalyzing the electron transfer from ubiquinol to the cytochrome *c* and is common enzyme complex to eukaryotic energy tranducing systems or to some bacterial systems. This process is coupled with a proton translocation, thus generating ATP (1-4). The number of subunits present in eukaryotes is larger (11 subunits in bovine heart bc_1 complex (5) and 8-10 subunits in yeast bc_1 complex (1,2)) than in prokaryotes, such as *Paracoccus denitrificans* (4) that contains only 3 subunits (see Table 1). Therefore, there have been questions raised regarding the function of those

---

[1]The gene for the 17 kDa protein in yeast bc_1 complex is expressed in italic, *17 kDa*, since this gene has not been named yet.

Table 1. Subunit composition of beef heart, *S. cerevisiae*, and *P. denitrificans* cytochrome bc_1 complex.

Subunit		Beef heart Mr (kDa)[a]	*S. cerevisiae* Mr (kDa)[b]	*P. denitrificans* Mr (kDa) by SDS-PAGE[c]
Core I		49	44	
Core II		47	40	
Cyt. b		42.6	43.7	39
Cyt. c_1		27.3	27.5	62
Fe-S		21.6	20.1	20
QP	VI	13.4	14.5 (VII) (14 kDa)	?
	VII	9.5	12.3 (VIII) (11 kDa)	
Hp	VIII	9.2	14.6 (VI) (17 kDa)	?
	IX	8.0	—	
	X	7.2	7.3 (IX) (7.2 kDa)	
	XI	6.4	—	

QP, Q binding protein; Hp, Hinge protein. [a]from ref. 5, [b]from ref. 1 and [c]from ref. 4.

subunits which do not have the prosthetic groups. However, the studies on the role of these subunits are also important to understand the activity and its mechanism of the cytochrome bc_1 complex in the respiratory chain. Among these, the hinge protein, subunit VIII of the bovine heart cytochrome bc_1 complex has been isolated to a purified form and most extensively studied, and a possible important role of the hinge protein in the mitochondrial respiratory chain has been proposed (6-10). Although the cytochrome bc_1 complex of *P. denitrificans* consists of only 3 polypeptides with which a respiratory chain appears to be functional (4) as in its mitochondrial counterpart, it cannot be ruled out that some smaller subunits are not identified by SDS-PAGE, or lost during purification of the complex. It is also interesting to identify the functional role of the highly acidic nature of 150 amino acids long segment (11) in the sequence of its cytochrome c_1, of which the overall amino acid composition is analogous to the hinge protein of the mitochondrial bc_1 complex.

The hinge protein is originally identified in the beef heart mitochondria. The beef heart mitochondrial hinge protein is very tightly associated with cytochrome c_1 and it is not easily dissociable from cytochrome c_1 except under certain conditions (9). The unique structural characteristic of the hinge protein and its necessity for the formation of the cytochrome c_1-Hp-c complex led us to conceive that the hinge protein or c_1-Hp-c complex may have a certain role in the mitochondrial respiratory chain. After extensive kinetics as well as spectroscopic studies on this protein, we have postulated that the hinge protein or c_1-Hp-c complex probably functions as a regulator for the reaction of electron transfer from cytochrome c_1 to cytochrome c and the mechanism of regulation by the hinge protein can be formulated in so that the hinge protein induces a special conformational arrangement of cytochrome c_1 through which the electron transfer from cytochrome c_1 to cytochrome c is favored. However, the difficulty to modify or remove a single subunit from the cytochrome bc_1 complex without affecting the rest of its components have not allowed us to elucidate a genuine physiological role of the hinge protein in the mitochondrial respiratory chain. Therefore, we have initiated the studies with yeast mitochondrial bc_1 complex, which is genetically manipulable and whose components are very similar to those of the beef heart mitochondrial bc_1 complex.

The 17 kDa protein of the yeast bc_1 complex, also called yeast hinge protein, was proved to be the same as the 18.5 kDa protein (13) which has been found to be associated with isolated yeast cytochrome c_1. Based on the gene deletion and disruption experiments, Berden's group (14) and Trumpower's group (15) have independently reported that the 17 kDa protein appeared to be required for the synthesis of both cytochrome bc_1 complex and cytochrome c oxidase in yeast mitochondria. However, Schoppink *et al.*(16) later reported that the 17 kDa deletion mutant, 17 kDa⁻, did not show any defect in the assembly of the bc_1 complex or that of cytochrome c oxidase, but its cytochrome c reductase activity was reduced to one half of that of the wild type. However, they found that an additional mutation was frequently occurred in the 17 kDa mutants resulting a respiratory deficient phenotype and low cytochromes b and c_1 levels in the bc_1 complex. The site or the significance of this secondary mutation has not been identified so far.

To establish a phenotype for the 17 kDa protein gene mutations with which genetic approaches on the structure-function studies of the 17 kDa protein or the hinge protein can be carried out, we constructed a double mutant in which the 17 kDa protein gene and CYC1 gene (*iso*-l-cytochrome c gene) were deleted. This double mutant *(cyc1-1,17kDa⁻)* could not grow on nonfermentable carbon source, but can be complemented to normal growth, when it is transformed by yeast vectors containing the wild type 17 kDa protein gene or the CYC1 gene. These results demonstrate that the 17 kDa protein may be essential for mitochondrial respiration when the level of the cytochrome c is limited (only 5%).

In this chapter we briefly summarize the construction and characterization of the double deletion mutant of *S. cerevisiae, cyc1-1,17kDa⁻* and the observed results are compared with those reported (14-16) from the genetic studies on the 17 kDa protein of yeast mitochondria. The detailed experimental procedures as well as other results are appeared in the manuscript by Kim and Zitomer (17)

THE HOMOLOGY BETWEEN THE BEEF HEART HINGE PROTEIN AND THE YEAST 17 kDa PROTEIN

The cDNA of the yeast mitochondrial hinge protein, named the 17 kDa protein and the human mitochondrial hinge protein were cloned and sequenced by van Loon *et al.*(18) and Ohta *et al.*(19), respectively. The amino acid sequences, deduced from gene sequences, of the 17 kDa protein and the human hinge protein are indeed homologous to that of the beef heart hinge protein as shown in Fig. 1. The homology between the yeast 17 kDa protein and the beef heart hinge protein is shown to be 36%, while the mature form of the human mitochondrial hinge protein shows 95% homology with the beef heart hinge protein.

The 17 kDa protein in yeast mitochondria is extremely rich in charged residues (52%), most of which are acidic residues (38%). The continuous stretch of 25 acidic amino acid residues is a unique structural feature as the eight glutamic acids stretch in the beef heart hinge protein, and is considered to be the longest stretch of acidic residues found in a naturally occurring protein so far. Alignment of amino acid sequences of the beef heart hinge protein and the yeast 17 kDa protein reveals a homology of 35-40% at the carboxy-

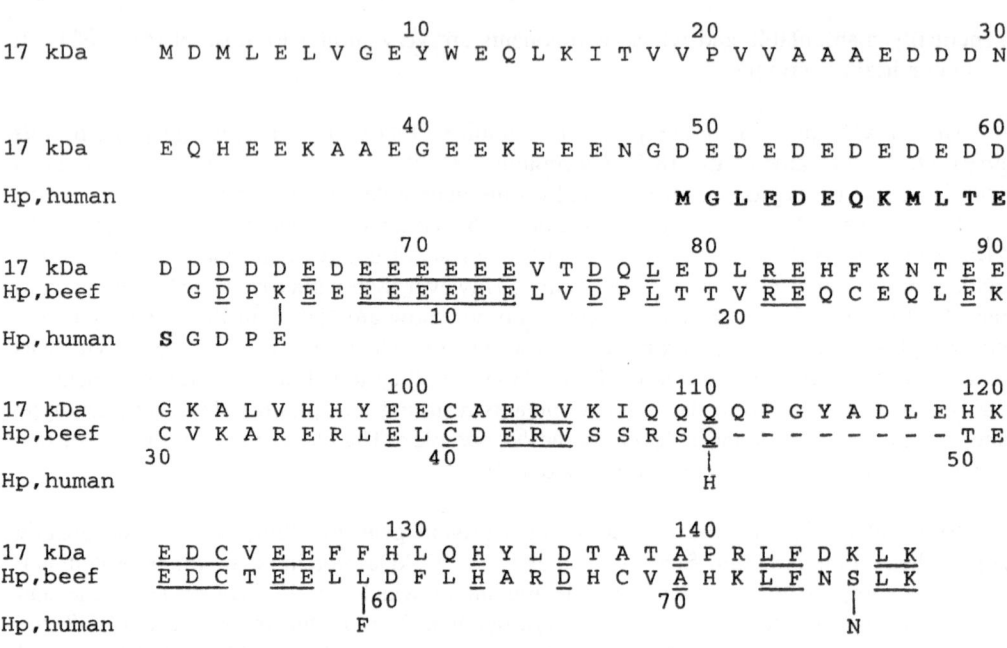

Fig. 1. Amino acid sequence comparison of the 17 kDa and hinge protein(Hp). They are aligned to give a most probable homology. The identical residues are underlined. In the sequence of the human hinge protein the only amino acids differing from those of the beef heart hinge protein are indicated. The presequence of 13 amino acids in the human hinge protein are expressed in the bold letter.

terminal side of the block of acidic amino acid. Compared with the beef heart hinge protein, the yeast protein extends 61 amino acids more at the amino terminal.

Recent studies on the 17 kDa protein by Schoppink *et al.*(20) suggested that the 17 kDa protein appears to play a stimulating role for the electron transport from the cytochrome bc_I complex to the cytochrome c in yeast mitochondria. They reported that the 17 kDa protein is responsible for the association of basic ferricytochrome c with the bc_I complex and this binding becomes rate limiting at the high ionic strength (>225 mM). From these results, it is conceivable that the role of the 17 kDa protein in the yeast mitochondrial bc_I complex may be similar to that of the hinge protein in beef heart mitochondrial bc_I complex. And the elucidation of the function of the 17 kDa protein in yeast mitochondria can help us to unravel the function of the hinge protein in mammalian mitochondria.

THE CONSTRUCTION OF A DOUBLE MUTANT

Since the beef heart hinge protein, the counter part of the yeast 17 kDa protein is believed to play a role in the interaction of the cytochrome bc_1 complex and cytochrome c, and a deletion of the 17 kDa protein gene in yeast results in the reduction of the cytochrome c reductase activity, it is interesting to see whether a reduction of cytochrome c level could amplify the effects of a 17 kDa protein gene deletion and bring a distinct phenotype which is measurable.

To conduct this investigation, *S. cerevisiae,* strain aZO60 *(MATa, trp1-1, leu2-3,-112, his4-519, his3-11,-15, cyc1-1),* which contains a leu2 mutation, a deletion of the *CYC1* gene, was used. The *CYC1* gene encodes the *iso*-1-cytochrome c protein (21) which constitutes 95% of the cytochrome c level in aerobically grown yeast cells, and the remaining 5% of cytochrome c is constituted by the *iso*-2-cytochrome c, which is encoded by the *CYC7* gene (22). Interestingly, this *cyc1-1* mutant (aZO60) cells can still grow at a rate of near wild type on various nonfermentable energy sources, such as glycerol, ethanol, and acetate, despite of this twenty fold reduction in cytochrome c content. However, their growth is impaired in lactate media (23). The growth condition of such cytochrome c mutants are shown in Table 2.

Table 2. Growth condition of cytochrome c mutants of *S. cerevisiae.*

	Growth Media	
Genotype	Glycerol	Lactate
CYC1[+], *CYC7*[+]	+	+
cyc1[-], *CYC7*[+]	+	−
cyc1[-], *cyc7*[-]	−	−
CYC1[+], *cyc7*[-]	+	+

The 17 kDa protein gene was deleted in yeast strain, aZO60 *(cyc1-1, CYC7),* by a one step gene replacement using a linear fragment of DNA containing *LEU2* disruption of the 17 kDa protein gene as shown in Fig. 2A. Transformation of yeast was carried out as described (24). Transformants showing a leu[+] phenotype were selected, and two classes of leu[+] transformants were obtained; the first class was a glycerol[+] phenotype and the other was a glycerol[-] phenotype. The transformants which contained the disruption of the 17 kDa protein were identified by a Southern analysis (25), using a chromosomal DNA from the transformants digested with *ClaI,* and the 1.3 kb *AccI* fragment containing the 17 kDa protein gene, as a probe (Fig. 2A). As shown in Fig. 2B, the yeast strain, aZO60 (the wild type for the 17 kDa protein) and a glycerol[+] transformant (lane 3) showed a hybridization band of 2.8 kb which is a characteristic of the wild type 17 kDa protein gene. The glycerol[-] transformants (lane 1 & 2) showed a hybridization band of 3.5 kb which is expected from the integration of the *LEU2* disruption at the 17 kDa protein gene locus (see Fig. 2A). Thus constructed double mutant, *cyc1-1,17kDa[-],* was designated, aZO60-17 and used for further investigation.

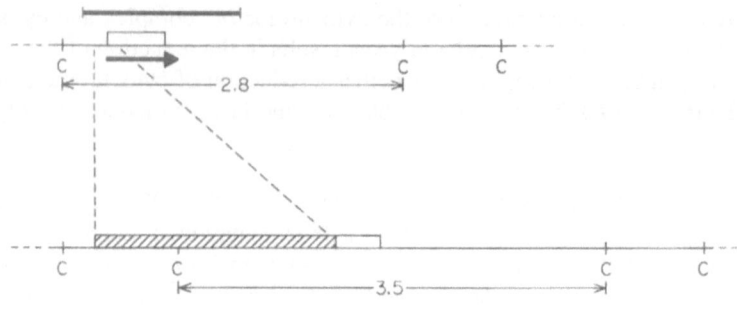

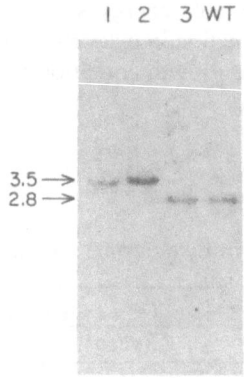

Fig. 2. Southern analysis. *A.* The predicted physical maps of the yeast genome at the wild type 17 kDa locus and at the *LEU2*-17 kDa gene disruption. C represents the restriction site for *ClaI*. The 17 kDa protein gene locus is marked in a rectangular box and *LEU2* gene is shown in a shaded box. A bold bar above the 17 kDa protein gene indicates an *AccI* fragment (1.3 kb) which was used as a DNA probe. *B.* Genomic DNA isolated from leu+ transformants of aZO60, glycerol‾ phenotype (lane 1 & 2), glycerol+ phenotype (lane 3) and aZO60 wild type (WT), were digested with *ClaI* and transferred to a nylon membrane. The autoradiogram was obtained by hybridization of the bound DNA to the 32P labeled *AccI* fragment. (from Ref. 17)

THE DOUBLE MUTANT CAN BE COMPLEMENTED WITH PLASMIDS CARRYING *CYC1* OR 17 kDa PROTEIN GENE

It is important to confirm the glycerol‾ phenotype was indeed due to the deletion of the 17 kDa protein gene. Therefore, the double mutant, aZO60-17, cells were transformed with the centrometric plasmids carrying either the 17 kDa gene (YCp*17kDa*) (17), or the *CYC1* gene (YCp*CYC1(2.4)*) (26) to complement the respective deleted gene. These transformations could confirm whether the glycerol‾ phenotype is not due to the third site mutation, but is indeed due to the combined deletions of those two genes. Therefore, the complementation of either the *CYC1* or 17 kDa protein gene into the double mutant should convert the glycerol‾ phenotype to the glycerol+ phenotype. Transformants selected via the *TRP1* marker on the plasmids were tested for the growth on glycerol‾ and the representing results are shown in Fig. 3. The introduction of either the wild type *CYC1* or the 17 kDa protein gene

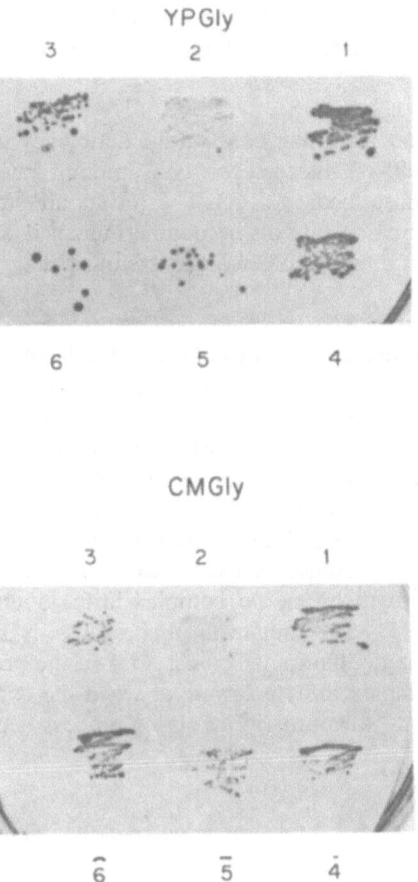

Fig. 3. Comparison of growth and respiratory capacity of aZO60, wild-type (1), aZO60-17(2),and aZO60-17 transformants with plasmids carrying *CYC1*(3 & 4) or *17kDa*(5 & 6) **gene.** Cells were plated on YPGly (3% glycerol) and CMGly (3% glycerol) media (from Ref. 17).

restored the growth ability on glycerol, while the transformation of cells with the yeast vectors without either of the gene (as a control) did not restore the respiratory capacity.

RESPIRATORY CAPACITY AND CYTOCHROME c REDUCTASE ACTIVITY OF THE DOUBLE MUTANT, aZO60-17

Growth rate. In order to determine the respiratory capacity of the double deletion mutant, aZO60-17, which shows a glycerol⁻ phenotype, the growth rates of aZO60 (WT for the 17kDa protein), aZO60-17, aZO60-17 transformants with plasmid YCp*CYC1(2.4)* or YCp*17kDa* as well as HR2, wild type and *17kDa⁻* mutant (16) in liquid glycerol media were monitored for 48 hrs. The doubling time of the cells are summarized in Table 3. The growth rate of the double mutant was about three times slower than that of cells carrying either the *CYC1* gene or the 17kDa protein gene. It is obvious that the deletion of the 17kDa protein together with a reduction of cytochrome c level has a drastic effect on the cell respiratory capacity.

Oxygen consumption rate. To further confirm the deficiency in the respiratory capacity of the double mutant, aZO60-17, the oxygen consumption rate of those mutants were measured in the liquid raffinose media (YPR). The results are summarized in Table 4. As predicted, the oxygen consumption rate of the double mutant is about 3 times slower than that of cells carrying the *CYC1* gene and two times slower than that of the cells carrying the 17 kDa protein gene.

Cytochrome c reductase activity. The cytochrome c reductase activity of mitochondria prepared from the double mutant, aZO60-17, was measured to be only 45% of that of the wild type, aZO60, which contains the 17 kDa protein. However, the double mutant complemented with the 17 kDa protein gene, aZO60-17 containing the plasmid YCp*17kDa,* showed the cytochrome c reductase activity to the level of the wild type. This confirms that the 17 kDa protein indeed affects the cytochrome c reductase activity in yeast mitochondria as reported by Schoppink *et al.*(20) and also confirmed in our laboratory (17). Since these assays were carried out in the presence of excess exogenous cytochrome c, the observed results demonstrated that the cytochrome bc_1 complex in the double mutant is functionally identical to that in the single *17 kDa⁻* mutant. In this regards, it is feasible to consider that the glycerol⁻ phenotype in the double mutant must arise in the process of electron transfer from the cytochrome bc_1 complex to cytochrome c; when the cytochrome c level in yeast mitochondria is reduced, the 17 kDa protein involving process becomes a rate limiting step in respiration.

Table 3. Doubling time of yeast strains cultured in liquid media containing a nonfermentable carbon source (YPGly). Culture temperature was 30°C.

Strain	Genotype	Plasmid	Doubling Time (hrs)
HR2	*CYC1,17kDa⁺*	–	5.0
	CYC1,17kDa⁻	–	6.5
aZO60	*cyc1-1,17kDa⁺*	–	6.5
aZO60-17	*cyc1-1,17kDa⁻*	–	17.5
	cyc1-1,17kDa⁻	YCp*CYC1(2.4)*	5.5
	cyc1-1,17kDa⁻	YCp*17kDa*	6.5

Table 4. The oxygen consumption rate of yeast cells grown to mid-log phase in liquid raffinose media (YPR).

Strain	Plasmid	$\Delta O_2/min, \%^a$
aZO60	–	11.0
aZO60-17	–	3.7
aZO60-17	YCp*CYC1(2.4)*	10.5
aZO60-17	YCp*17kDa*	8.0

[a]Expressed as the decrease of oxygen concentration (%) per minute per cell suspension of A_{550}=0.5. ΔO_2 is average of six measurements for three different cultures for each strain. Cells were grown at 30°C and oxygen measurement was done at 28°C.

CONCLUSION

The present studies demonstrate that the disruption of the gene encoding the 17 kDa protein in the cytochrome bc_1 complex affects the cell respiration in a yeast strain with the decreased cytochrome c level (5%); *S. cerevisiae* which lacks the 17 kDa protein, subunit VI of the bc_1 complex and *iso*-1-cytochrome c showed a growth rate and oxygen consumption rate three times slower than those of the wild type. This double deletion mutant, aZO60-17, which shows glycerol⁻ phenotype, would allow the further genetic analysis to investigate the role of the 17 kDa protein in the electron transfer and assembly of the cytochrome bc_1 complex in yeast mitochondria. And it would provide a clue to elucidate the physiological function of the hinge protein in beef heart mitochondria.

ACKNOWLEDGEMENT

This work was supported by grants from American Heart Association and National Institutes of Health.

We would like to thank Drs. L. Grivell and J.A. Berden for their kind gifts of yeast strain HR2 (wild type and 17 kDa⁻ mutant) and plasmid carrying 17 kDa gene.

REFERENCES

1. Katan, M.B., Pool, L., and Groot, G.S.P. (1976) *Eur. J. Biochem.* **65,** 95-105
2. Siedow, J.N., Power, S., De La Rosa, F.F., and Palmer, G. (1978) *J. Biol. Chem.* **253,** 2392-2399
3. Hatefi, Y. (1985) *Annu. Rev. Biochem.* **54,** 1105-1169
4. Yang, X., and Trumpower, B. L. (1986) *J. Biol. Chem.* **261,** 12282-12289
5. Schägger, H., Link, Th.A., Engel, W.D., and von Jagow, G. (1986) *Methods Enzymol.* **126,** 224-237

6. Kim, C.H., and King, T.E.(1981) *Biochem. Biophys. Res. Commun.* **101**, 29-36
7. Wakabayashi, S., Takeda, H., Matsubara, H., Kim, C.H. and King, T.E. (1982) *J. Biochem.* **91**, 2077-2085
8. Kim, C.H., and King, T.E. (1983) *J. Biol. Chem.* **258**, 13543-13551
9. Kim, C.H. (1987) *in* Advances in Membrane Biochemistry and Bioenergetics (Kim, C.H., Tedeschi, H., Diwan, J.J. and Salerno, J.C., eds.) Plenum Press, New York, pp. 167-178
10. Kim, C.H., Balny, C., and King, T.E. (1987) *J. Biol. Chem.* **212**, 8103-8108
11. Ludwig, B., Kurowski, B., Panskus, G. and Steinrucke, P. (1987) *in* Cytochromes; Molecular Biology and Bioenergetics (Papa, S., Chance, B. and Ernster, L., eds.) Plenum Press, New York, pp. 41-47
12. Kim, C.H., Balny, C. and King, T.E. (1987) *in* Cytochromes; Molecular Biology and Bioenergetics (Papa, S., Chance, B., and Ernster, L., eds.) Plenum Press, New York, pp. 415-422
13. Ross, E., and Schatz, G. (1976) *J. Biol. Chem.* **251**, 1991-1996
14. Schoppink, P.J., Grivell, L.A., and Berden, J.A. (1987) *in* Advances in Membrane Biochemistry and Bioenergetics (Kim, C.H., Tedeschi, H., Diwan, J.J., and Salerno, J.C., eds.) Plenum Press, New York, pp.129-139
15. Schmitt, M.E., and Trumpower, B. (1987) *in* Cytochromes; Molecular Biology and Bioenergetics (Papa, S., Chance, B., and Ernster, L., eds.) Plenum Press, New York, pp. 177-187
16. Schoppink, P.J., Hemrika, W., Reyner, J.M., Grivell, L.A., and Berden, J.A.(1988) *Eur. J. Biochem.* **173**, 115-122
17. Kim, C.H., and Zitomer, R.S. (1990) submitted
18. Van Loon, A.P.G.M., de Groot, R.J., de Haan, M., Dekker, A., and Grivell, L.A. (1984) *EMBO J.* **3**, 1039-1043
19. Ohta, S., Goto, K., Arai, H., and Kagawa, Y. (1987) *FEBS Lett.* **226**, 171-175
20. Schoppink, P.J., Hemrika, W., and Berden, J.A. (1989) *Biochim. Biophys. Acta* **974**, 192-201
21. Sherman, F., Stewart, J.W., Margoliash, E., Parker, J., and Campbell, W.(1966) *Proc. Natl. Acad. Sci.,* U.S.A. **55**, 1498-1504
22. Downie, J.A., Stewart, J.W., Brockman, N., Schweingruber, A., and Sherman, F. (1977) *J. Biol. Chem.* **113**, 369-384
23. Sherman, F., Stewart, J. W., Jackson, M., Gilmore, R. A., and Parker, J.H. (1974) *Genetics* **77**, 255-284
24. Klebe, R. J., Harris, J. V., Sharp, Z. D., and Douglas, M. G. (1983) *Gene* **25**, 333-341
25. Southern, E.M. (1975) *J. Mol. Biol.* **98**, 503-517
26. Lowry, C. V., Weiss, J. L., Walthall, D.A., and Zitomer, R.S. (1983) *Proc. Natl. Acad. Sci., U.S.A.* **80**, 151-155

Molecular evolution and biology of human mitochondrial DNA

Satoshi Horai and Kenji Hayasaka

Department of Human Genetics
National Institute of Genetics
Mishima, Shizuoka, Japan

Summary

Nucleotide sequences of the major noncoding region of human mitochondrial DNA (mtDNA) from 95 human placentas have been determined. These sequences include 482 base-pair (bp) long region encompassing most part of the D-loop forming region. Comparisons of these sequences with those previously determined have revealed a remarkable features of nucleotide substitutions and insertion/deletion events. Average nucleotide diversity among the sequences is estimated as 1.45%, which is three to four-fold higher than the corresponding value estimated from restriction-enzyme analysis of whole mtDNA genome. This large variability in the noncoding region is consistent with the weak functional constraints on the region. A hypervariable domain has also been defined. In a 14 bp stretch, at least 17 different sequences were detected in this hypervariable domain. More than 97% of the base changes are transitions. Length differences have occurred exclusively as insertions of cytosine(s) or deletions of adenine(s) within stretch of adenines or cytocines, respectively. A significantly nonrandom distribution of nucleotide substitutions and sequence length variations were also noted. The results in the present study confirmed several features of the noncoding region of mtDNA which previously reported by Greenberg et al. (1) and Aquadro and Greenberg (2), though only seven sequences were determined and compared by these authors.

The phylogenetic analysis indicates that diversity among the Negroids is much larger than that among the Caucasoids or the Mongoloids. In fact, part of Negroids first diverged from the rest of individuals in the phylogenetic tree. A striking finding in the phylogenetic analysis is that the Mongoloids can be separated into two distinct groups. Divergence of part of the Mongoloids follows the earliest divergence of the Negroids. Remainder of the Mongoloids subsequently diverged together with the Caucasoids. This observation confirmed our earlier study which clearly demonstrated existence of two distinct groups in the Japanese by the restriction enzyme analysis (3).

Introduction

The mammalian mitochondrial DNA (mtDNA) is a circular genome of approx. 16.5 kilo-base-pairs (kb) in length, and encodes 13 subunits of the inner membrane respiratory complexes. The complete nucleotide sequences of human (4), mouse (5), cow (6) and rat mtDNAs (7) have been reported. The gross genetic arrangement of these genomes is remarkably conserved. They can be divided into two domains—a coding region constituting over 90% of the genome, and a noncoding region which contains the origin of H-strand replication (4) and the origins of transcription of both strands (8).

Since the original studies which confirmed the maternal inheritance and predominantly uniclonal nature of mammalian mtDNA within an individual (9 - 11), numerous reports have indicated that its nucleotide sequence is evolving much faster than that of single copy nuclear genes (12, 13).

Since there are substantial sequence variations among individuals (14), restriction enzyme analysis of mtDNAs has become a powerful tool in elucidating evolutionary relationships among human ethnic groups (15 - 23). Results obtained from these studies suggest that there is a high correlation between mtDNA restriction types and ethnic origins of individuals. The phylogenetic analysis in the Japanese indicated that they could be separated into two major groups (groups I and II; 3). This grouping may also apply to other Mongoloid populations, because a 9-bp deletion in region V (24), which characterizes the group I Japanese, was also observed in non-Japanese Mongoloid populations (25 - 27). Therefore, we extended our analysis to include a phylogenetic analysis of the sequence in a particular region of mtDNA.

To this end we have sequenced part of the major noncoding region encompassing the D-loop region of mtDNAs isolated from 95 human placentas. Comparisons of these sequences with others determined previously (1, 4) have revealed a striking features of nucleotide substitutions, insertions, and deletions. These results are discussed in light of sequence evolution and functional constraints on this region. We have also examined the evolution of mtDNA sequences at a gene level as well as a population level in detail.

Experimental Procedures

Samples

MtDNA was purified to homogeneity from placentas of 95 individuals from three different racial origins as described by Brown et al. (12) and Horai et al. (19). Sixty-one of these people were from the central part of Japan (3, 19); ten were non-Japanese Mongoloids which consisted of three Koreans, four Chinese, an Indonesian, a Philippine and a Papua New Guinean. We have also collected placenta samples from 17 European or American Caucasoids and seven African Negroids, whose mothers gave birth to their babies in Tokyo, Japan.

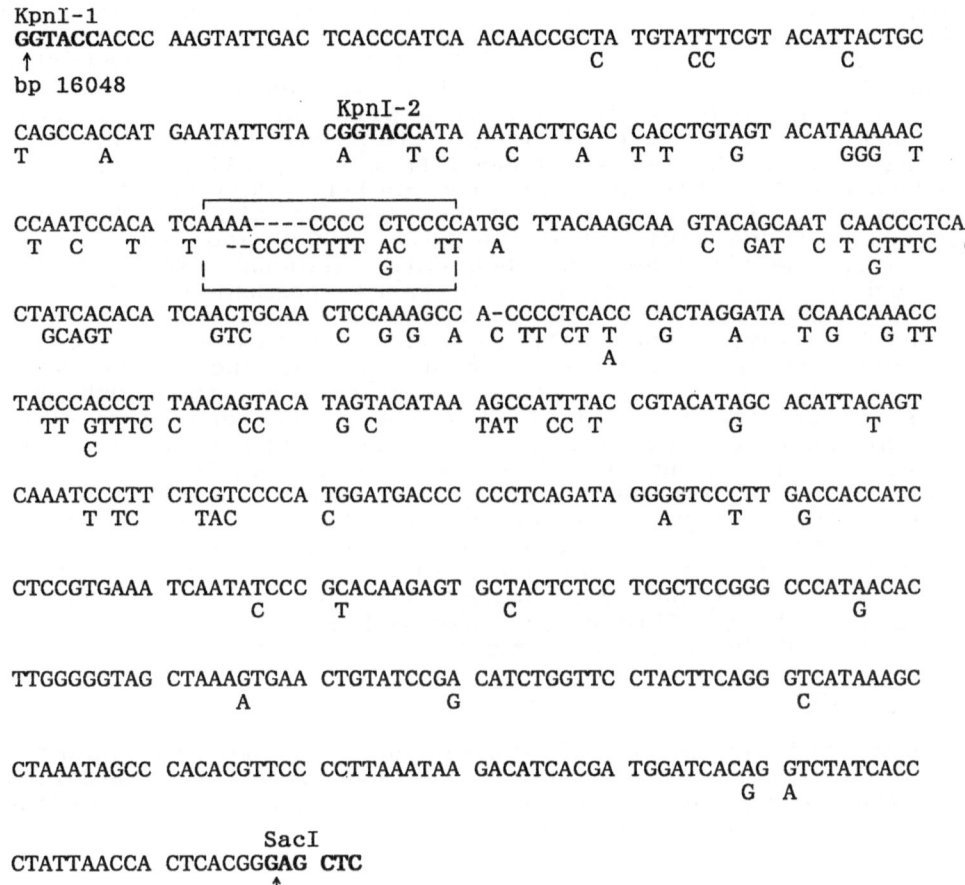

KpnI-1
GGTACCACCC AAGTATTGAC TCACCCATCA ACAACCGCTA TGTATTTCGT ACATTACTGC
↑   C   CC   C
bp 16048
                              KpnI-2
CAGCCACCAT GAATATTGTA C**GGTACC**ATA AATACTTGAC CACCTGTAGT ACATAAAAAC
T   A   A   T C   C   A T T   G   GGG T

CCAATCCACA TCAAAA----CCCC CTCCCCATGC TTACAAGCAA GTACAGCAAT CAACCCTCAA
T   C   T   T   --CCCCTTTT AC TT A   C   GAT C T CTTTC G
                            G   G

CTATCACACA TCAACTGCAA CTCCAAAGCC A-CCCCTCACC CACTAGGATA CCAACAAACC
GCAGT   GTC   C G G A C TT CT T   G   A   T G   G TT
                                            A

TACCCACCCT TAACAGTACA TAGTACATAA AGCCATTTAC CGTACATAGC ACATTACAGT
TT GTTTC C   CC   G C   TAT CC T   G   T
C

CAAATCCCTT CTCGTCCCCA TGGATGACCC CCCTCAGATA GGGGTCCCTT GACCACCATC
T TC   TAC   C   A   T   G

CTCCGTGAAA TCAATATCCC GCACAAGAGT GCTACTCTCC TCGCTCCGGG CCCATAACAC
C   T   C   G

TTGGGGGTAG CTAAAGTGAA CTGTATCCGA CATCTGGTTC CTACTTCAGG GTCATAAAGC
A   G   C

CTAAATAGCC CACACGTTCC CCTTAAATAA GACATCACGA TGGATCACAG GTCTATCACC
                                                       G A

        SacI
CTATTAACCA CTCACGG**GAG CTC**
              ↑
              bp 36

Fig. 1.   **Nucleotide sequences of the major noncoding region.**
           The sequence reported by Anderson et al. (1981) is
           shown in the upper lines with the base numbers.
           Mutations observed at least in one individuals are
           shown in the lower lines.   -'s stand for
           deletions/insertions and letters represent base
           substitutions.   Cloning sites (SacI and KpnI-1 or
           KpnI-2) are also indicated by bold face letters.   A
           hypervariable domain is boxed.

Cloning of mt DNA

Purified mtDNAs were doubly digested with KpnI and SacI. Digestion products were extracted three times with phenol/chloroform (1/1), precipitated with ethanol and dissolved in TE(pH 8.0). Then the fragments were inserted into pBluescript(KS-, Stratagene) which also doubly digested with KpnI and SacI, and purified by gel electrophoresis.

When human mtDNAs were digested with SacI, two fragments (9.6 and 7.0 kb) were usually observed. These fragment sizes agree with the sizes deduced from the published sequence of Anderson et al. (4) which has two SacI recognition sites at bp 36 and 9643 (the notations of Anderson et al. (4) for base numbers). Three KpnI sites (bp 2573, 16048 and 16129) are observed in the published sequence data. Although two fragments (13.5 and 3.0 kb) are visible, the third fragment of 81 bp is invisible under the usual electrophoretic condition. Through the careful examinations with prolonged gel electrophoresis, slight differences in mobility were observed in the 3.0 kb bands among individual samples. In the variant pattern, the 3.0kb band moves slightly slower than that of the usual pattern, suggesting that contiguous two fragments (3013 and 81 bp) were fused to generate a larger fragment of 3.1 kb. This mutation is rather frequent in the Japanese population (Horai unpublished). Therefore, in the double digestions with KpnI and SacI, four or five fragments were produced in each sample due to the polymorphism of one of the KpnI sites. Thus, the smallest fragment with SacI and KpnI digestion is either 482 or 563 bp in length, and is efficiently inserted into vectors without purification of the desired fragment.

DNA Sequencing

Nucleotide sequences of 482 or 563 bp fragments inserted into pBluescript were determined by the dideoxy-chain termination method (28) using the Sequenase Kit(U.S. Biochemical) according to the manufacturer's directions.

Data Analysis

We aligned the nucleotide sequences of the 482 bp fragments, which were common for 95 individuals, with those of six individuals reported previously (1, 4). We estimated the number of nucleotide substitution per site between individual sequences by using the six-parameter method of nucleotide substitution (29). On the basis of the estimated numbers, a phylogenetic tree was constructed by the unweighted pair grouping (UPG) method (30).

Results and Discussion

Fig.1 shows the nucleotide sequence of the light strand of the noncoding region which originally reported by Anderson et al. (4). KpnI-site1 and SacI site are conserved among all individuals, however KpnI-site2 is polymorphic. Therefore, we cloned and sequenced either 563 base pair or 482 base pair fragment for each individual. A total of 100 sequences were

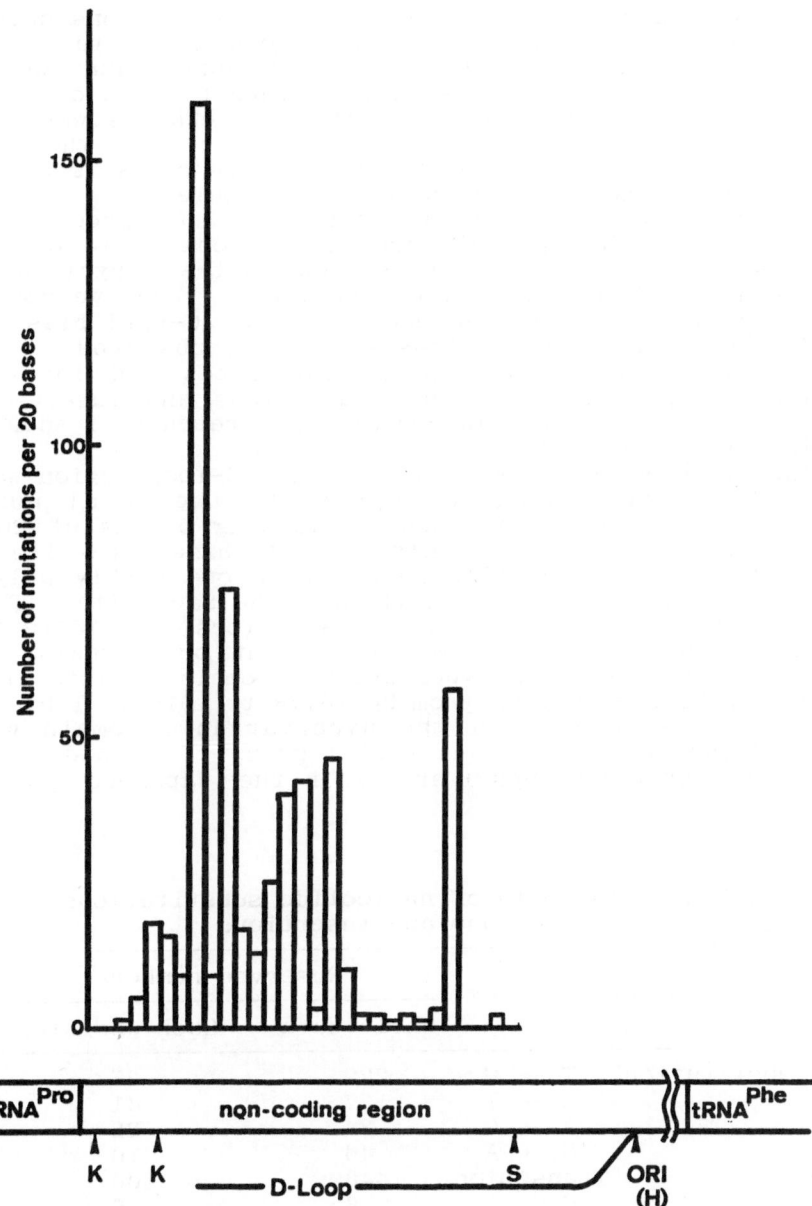

Fig. 2.   Distribution of mutations
The histogram shows the total number of base substitutions, deletions and insertions within contiguous non-overlapping blocks of 20 bases. A hypervariable domain is apparent near one of the KpnI sites within the D-loop region. K and S stand for KpnI and SacI recognition sites, respectively, used for cloning.

compared with the sequence reported by Anderson et al. (4). In Fig 1, base substitutions, deletions and insertions detected at least in one individual were also shown below the nucleotide sequence of Anderson et al. (4). We found mutations at 113 sites in total. At four sites two different kinds of nucleotide substitutions were observed. Nucleotide changes were observed more frequently in the 5' half of the region than the 3' half. The observed numbers of mutations which were classified according to the types of mutations were shown in Table 1. In the whole data, transition types of substitution are more predominant than transversions. Namely, 97% are transitions while only 3% are transversions. Moreover transitions between pyrimidines are more prevalent than those between purines. When we counted the nucleotide substitutions for each site, a 10-fold bias favoring transitions over transversions was still observed. We also observed deletion of adenine at two sites and insertion of cytocine at five sites. These deletions and insertions were found in a particular domain containing stretches of adenines and cytocines (see Table 2).

The distribution of mutations in the D-loop region was shown in Fig 2. The histogram represents the total number of mutations within contiguous non-overlapping blocks of 20 bases. Fig. 2 shows that most of highly variable base sites lies in the blocks near the tRNA-Proline gene, while one highly polymorphic base site was observed in a block near the SacI site. This site is polymorphic for all three major races (position 16519 in Table 4) and probably derived from the ancient polymorphism of this site. In this Figure a hypervariable domain exists in one of the 20 base blocks ranging from bp 16178 to 16197. Table 2 shows 17 different sequences in the hypervariable domain with the number of individuals in each racial group. J1 and J2 in this table represent the Groups I and II in the Japanese population,

Table 1. Analysis of nucleotide substitutions, deletions and insertions

Type of mutation			Observed number	
			Whole data	Each position
Substitution	T	→ C	229	27
	C	→ T	137	41
	A	→ G	34	22
	G	→ A	46	10
	Transition		446	100
	C	→ A	6	5
	C	→ G	1	1
	A	→ C	2	2
	A	→ T	2	1
	G	→ C	1	1
	Transversion		12	10
% Transition			97%	91%
Deletion	A	→ -	32	2
Insertion	-	→ C	69	5

Table 2. Mitochondrial DNA sequences in a hypervariable domain and observed numbers in the three racial groups

	mtDNA sequences	C	N	M	J1	J2	Total
1.	AAAA----CCCCCTCCCC	15	5	7	1	33	61
2.	AAAA----CCCCCTCCCT	1	-	-	-	-	1
3.	AAAA----CCCCCTCCTC	-	-	1	-	-	1
4.	AAAA----CTCCCTCCCC	-	-	-	-	1	1
5.	AAAA----CCCTCTCCCC	-	-	-	-	3	3
6.	AAAA----CCCTCCCCCC	-	2	-	1	-	3
7.	AAAA----CCCTGCCCCC	-	1	-	-	-	1
8.	AAAA----TCCTACCCCC	-	1	-	-	-	1
9.	AAAA----CCTCCCCCCC	1	-	-	-	-	1
10.	AAAA---CCCCCCCCCCC	1	-	1	-	3	5
11.	AAAA--CCCCCCCCCCCC	1	-	-	-	-	1
12.	AAA-CCCCCCCCCCCCCC	-	-	-	1	-	1
13.	AAA--CCCCCCCCCCCCC	-	-	-	1	-	1
14.	AAA---CCCCCCCCCCCC	1	-	-	4	4	9
15.	AAA-----CCCCCCCCCC	-	-	-	-	1	1
16.	AA--CCCCCCCCCCCCCC	-	-	-	4	2	6
17.	AA---CCCCCCCCCCCCC	-	1	1	2	-	4
	Total	20	10	10	14	47	101

N, C, M, J1 and J2 stand for Negroids, Caucasoids, non-Japanese Mongoloids, Japanese group 1 and Japanese group 2 , respectively. (For the division of Japanese population, see text.)

respectively, which were classified in our earlier study (3; also see below). About 60% of individuals exhibit the sequence which contains a 14 base stretch of four adenines, five cytocines, one thymine and four cytocines. However, the remainder showed a variety of sequences in this region. From No.2 to No.9 there are base substitution in the cytocine stretch. From No.10 to No.17 sequence differences were resulted from elongation of the cytocine stretch and shortening of the adenine stretch. Once the thymine at bp 16189 is replaced by cytocine, the number of adenines and cytocines becomes flexible, probably due to the replication error. This thymine to cytocine transition has independently occurred several times in the different lineages, which was confirmed by the phylogenetic analysis as mentioned later. We aligned and compared the 482 base pair sequences from 101 individuals. We also estimated the numbers of nucleotide substitutions between each pair of sequences. Average number of nucleotide diversity among 101 individual is 1.45%, which is three to four-fold higher than estimates from restriction enzyme analysis of mtDNA previously reported in human population (3). On the bases of the estimated numbers of nucleotide substitutions between individual sequences we constructed a phylogenetic tree as shown in Fig 3. In this tree solid circles present Negroids, open circles indicate Mongoloids and Caucasoids are shown by solid triangles. From the clustering patterns, we could classify all individuals into at least 10 clusters, designated as C1 to C10. Among these clusters, C5, C9 and C10 exhibit interminglings of individuals

Fig. 3.   A phylogenetic tree showing the 101 mtDNA lineages from
the three racial groups
All lineages were tentatively classified into ten
clusters disignated as C1 to C10.

Table 3. Racial distribution of polymorphic sites

Racial group			No. of mutated sites
Caucasoid N=20	Negroid N=10	Mongoloid N=71	
+	+	+	12
–	+	+	10
+	+	–	2
+	–	+	2
–	–	+	55 (0.77)
–	+	–	21 (2.10)
+	–	–	15 (0.75)
	Total		117

+ and – represent the presence and absence of polymorphism, respectively. Numbers in parenthesis are the numbers of mutated site proportioned to the numbers of individuals (N).

from different racial groups. However, the remaining seven consist of individuals from single races. C1 and C4 are cluster for Negroids. C2, C3 and C6 to C8 consist only of Mongoloids. Although no cluster specific for Caucasoids was observed in this tree, C9 mainly consists of Caucasoids. In this tree four Negroid individuals belonging to C1 first diverged from the rest of the clusters, subsequently two Mongoloid specific clusters (C2 and C3) diverged from the second Negroid cluster (C4) and other clusters. It is relevant that some Mongoloids occupy a phylogenetic site distinct from other Mongoloids. This observation confirms our earlier study of restriction enzyme analysis, where part of Negroids and part of Mongoloids (Japanese) first diverged from the rest of the individuals (31, 32).

In this sequence analysis, there are 117 base substitutions in 113 sites as shown in Table 3 (see also Fig.1). Of these, 12 sites are shared by the three racial groups. Ten sites are found in common between Negroids and Mongoloids. Whereas Caucasoids and Negroids shared two polymorphic sites and Caucasoids and Mongoloids shared two sites. The presence of polymorphic sites shared by two or three racial groups is probably due to the ancient polymorphism. In other words, mtDNAs have already been polymorphic at these sites before the divergence of racial groups. On the other hand, we observed many polymorphic sites which are specific for each race, 55 for Mongoloids, 21 for Negroids and 15 for Caucasoids. When these numbers are proportioned to the numbers of individuals tested, racial specific polymorphic sites are predominantly found in Negroids (Table 3). This indicates that Negroids are much more diverse than Caucasoids or Mongoloids.

How can the evolutionary inference of mtDNA divergences be drawn from the phylogenetic tree ? What else are evolutionary implications of polymorphic sites shared by three racial groups ?

When we arbitrarily chose three sites which were highly polymorphic among three racial groups, all individuals could be

Table 4. Combinations of nucleotides at three typical polymorphic
         sites and observed numbers of individuals from three
         racial groups.

Mt type	Position			Total	N	C	M	J1	J2
	16223	16519	16362						
M 1	T	C	C	13	1	–	1	–	11
M 2	T	C	T	11	4	1	–	2	4
M 3	T	T	C	13	1	–	–	–	12
M 4	T	T	T	19	2	–	1	–	16
M 5	C	C	T	30	1	10	7	12	–
M 6	C	C	C	4	1	2	–	–	1
M 7	C	T	C	2	–	–	–	–	2
M 8	C	T	T	9	–	7	1	–	1

N, C, M, J1 and J2 stand for Negroids, Caucasoids, non-Japanese
Mongoloids, Japanese group 1 and Japanese group 2, respectively.
(For the division of Japanese population, see text.) The sites
are presented by base numbers in Anderson et al. (1981).

classified into the eight combinations of nucleotides. We
tentatively call these combinations as mitochondrial types M1 to
M8. The phylogenetic tree shown in Fig. 4 is the same as Fig.
3, but each cluster is represented by the mitochondrial types.
For example, in the first cluster (C1 in Fig. 3) there are
mitochondrial types M2 and M3. Thus, at the earliest branching
point there already existed two types, M2 and M3. In the same
way, the second cluster (C2) contains only mitochondrial type M5,
so that this type is assumed to have first appeared around the
branching point indicated by "M5" in Fig. 4. Similarly, since
mitochondrial type M6 is observed only in the last cluster
(C10), the emergence of this type should be located around the
branching point which is indicated by "M6" in Fig. 4.

Fig. 5 shows a schematic presentation of the appearance of
each ancestral mitochondrial type. We assume that the mtDNA
divergence took place about 200 thousands years ago which is
deduced from the restriction enzyme analysis of mtDNAs (22, 31,
32). Although the most ancestral type could not be identified
from the present study, types M2 and M3 had already existed some
180 thousand years ago. Remaining types are subsequently
derived from either type M2 or type M3. Thus, at the present
time we can observe eight types in human populations.

To analyze the nucleotide diversities within and between the
racial groups the nucleotide sequences were compared
quantitatively.

The number of nucleotide differences per site were estimated
for every pair of individuals. Based on these estimates net
nucleotide differences (d) between two races was calculated using
the following equation (33),

$$d = d_{XY} - (d_X + d_Y) / 2,$$

where d_{XY} is the average number of nucleotide differences between
two races, X and Y, and d_X and d_Y are the average number of

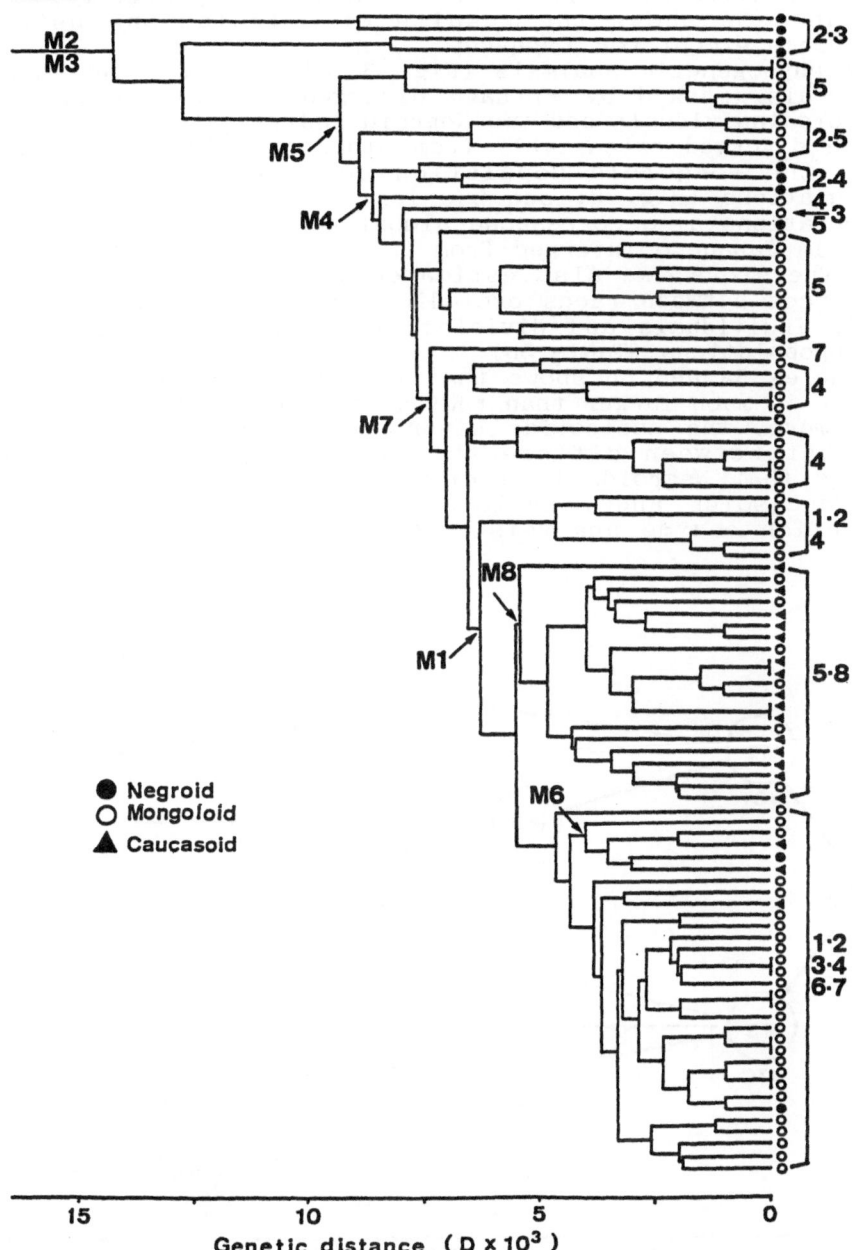

Fig. 4. A phylogenetic tree showing the 101 mtDNA lineages with mitochondrial types.
Each cluster is represented by the mitochondrial types 1 to 8 as designated in Table 4. M1 to M8 shown in the Figure indicate the "appearances" of these types in the phylogenetic tree.

nucleotide difference within races X and Y, respectively.

Table 5a shows the results of calculations using all individuals except for a Papuan.

A phylogenetic analysis (Fig. 3) showed that two Caucasoid individuals (marked by *) have diverged extensively from other Caucasoid individuals and two Negroid individuals (also marked by *) have diverged extensively from other Negroids. As mentioned above, the Mongoloid population could be divided into two subpopulations (Fig. 3; see the legend to Table 5). These two subpopulations also correspond almost completely to the groups I and II in Japanese inferred from the restriction enzyme analysis (3). Therefore, we also estimated the numbers of nucleotide differences between races excluding these two Caucasoid and two Negroid individuals and dividing the Mongoloid population into two subpopulations (Table 5b).

Tables 5a and 5b shows that the nucleotide diversity among Negroids is much larger than that among Caucasoids or Mongoloids. Furthermore, the nucleotide diversity among Negroids is larger than that between different racial groups. Thus, earlier findings that Negroids have highly diversified mtDNAs compared with Caucasoids and Mongoloids, which was deduced from the restriction enzyme analysis (22,31), was confirmed by the

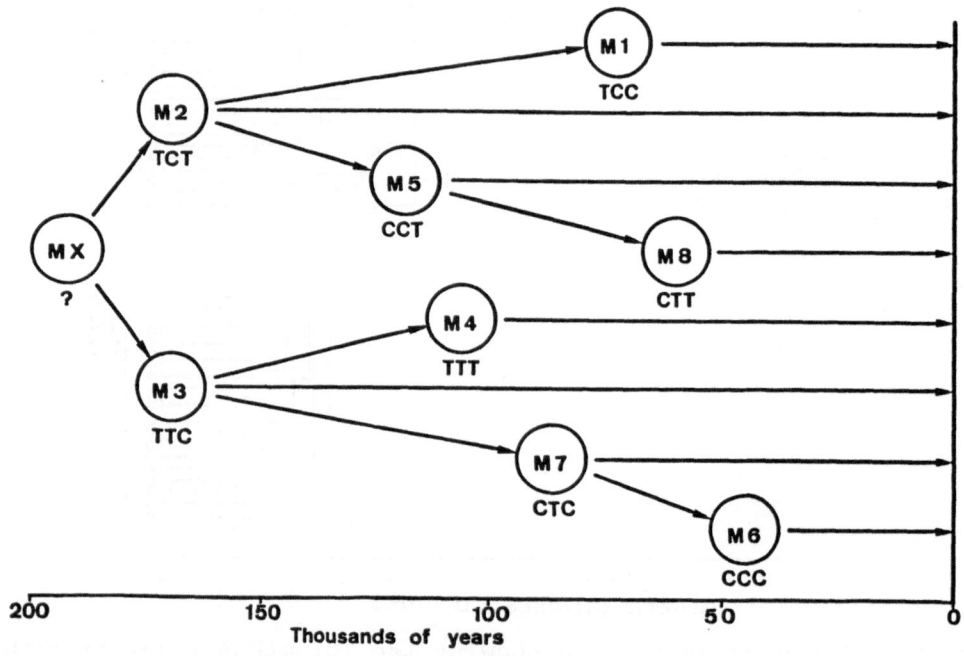

Fig. 5.   A schematic presentation of the appearances of each ancestral mitochondrial type

Time scale is based on the assumption that mitochondrial divergence took place about 200,000 years ago which is deduced from the restriction-enzyme analysis of mtDNAs (22,31).

Table 5. Estimates of the number of nucleotide differences per site among the three races (d_{XY}) and within each of the three races (d_X or d_Y), and net nucleotide differences among the races (d).

(a)

	Caucasoid N=20	Mongoloid N=70	Negroid N=10
Caucasoid	0.0094	0.0013	0.0028
Mongoloid	0.0128	0.0137	0.0015
Negroid	0.0194	0.0203	0.0238

(b)

	Caucasoid N=18	Mongoloid-2 N=51	Mongoloid-1 N=19	Negroid N=8
Caucasoid	0.0084	0.0019	0.0019	0.0036
Mongoloid-2	0.0115	0.0107	0.0032	0.0026
Mongoloid-1	0.0145	0.0170	0.0169	0.0014
Negroid	0.0212	0.0213	0.0232	0.0268

Figures on the diagonal represents d_X or d_Y. Those below and above diagonal represents d_{XY} and d, respectively. Mongoloid-2 includes those belonging to C6, C7, C8, C9 and C10 in Fig. 3, while the other Mongoloids are included in Mongoloid-1.

quantitative analysis of the nucleotide sequences.

In Table 5b the nucleotide differences between Mongoloids-1 and Mongoloids-2 (d_{XY}=0.0170) was much larger than that between Caucasoids and Mongoloids-2 (d_{XY}=0.0115). This table also shows that the nucleotide diverstiy among Mongoloids-1 (d_X=0.0169) is much larger than that among Mongoloids-2 (d_X=0.0107). These two subpopulations of the Mongoloids correspond roughly to groups I and II of Japanese, respectively, inferred from the restriction enzyme analysis (3). The restriction enzyme analysis also revealed that nucleotide diversity within group I was larger than within group II (3), which agrees well with the results of the present study. Thus, it is evident that the Mongoloid population can be separated into two subpopulations.

Net nucleotide differneces (33) of mtDNAs are not effective in elucidating the relationships between human populations, as previously noted by one of these authors (34). This may be partly due to the highly diversified mtDNAs existed in the Negroid population and a subpopulation of Mongoloids which might be resulted from the ancient polymorphisms, as is seen in Fig. 3 and Table 5b.

Nowadays Mongoloid descendants live in North and South America, Oceania, South East Asia, East Asia and Siberia, adapting themselves to various environment on the earth. An extended study of mtDNA polymorphism and its genealogy will be required to confirm the origins and dispersal of the two subpopulations of Mongoloids.

Acknowledgments

This work was supported by Grants-in-aids for General Scientific Research and Scientific Research on Priority Areas of "Bioenergetics", "Development of Evolutionary and Population Genetics Incorporating Newer Molecular Findings" and "From Asia to America: Prehistoric Mongoloid Dispersal" to S.H. from the Ministry of Education, Science and Culture, Japan.

References

1. Greenberg, B. D., Newbold, J. E., and Sugino, A. (1983) Gene 21, 33-49
2. Aquadro, C. F., and Greenberg, B. D. (1983) Genetics 103, 287-312
3. Horai, S., and Matsunaga, E. (1986) Hum. Genet. 72, 105-117
4. Anderson, S., Bankier, A. T., Barrell, B. G., De Bruijn, M. H. L., Coulson, A. R., Drouin, J., Eperon, I. C., Nierlich, D. P., Roe, B. A., Sanger, F., Schreier, P. H., Smith, A. J. H., Staden, R., and Young, I. G. (1981) Nature 296, 457-465
5. Bibb, M. J., Van Etten, R. A., Wright, C. T., Walberg, M. W., and Clayton, D. A. (1981) Cell 26, 43-68
6. Anderson, S., De Bruijn, M. H. L., Coulson, A. R., Eperon, I. C., Sanger, F., and Young, I. G. (1982) J. Mol. Biol. 156, 683-717
7. Gadaleta, G., Pere, G., Candia, G. D., Quagliariello, C., Sbisa, E., and Saccone, C. (1989) J. Mol. Evol. 28, 497-516
8. Cantatore, P., and Attardi, G. (1980) Nucl. Acids. Res. 8, 2605-2625
9. Hutchison, C. A. III., Newbold, J. E., Potter, S. S., and Edgell, M. H. (1974) Nature 251, 536-538
10. Potter, S. S., Newbold, J. E., Hutchison, C. A. III., and Edgell, M. H. (1975) Proc. Natl. Acad. Sci. U.S.A. 72, 4496-4500
11. Giles, R. E., Blanc, H., Cann, H. M., Wallace, D. C. (1980) Proc. Natl. Acad. Sci. U.S.A. 77, 6715-6719
12. Brown, W. M., George, M. JR., and Wilson, A. C. (1979) Proc. Natl. Acad. Sci. U.S.A. 76, 1967-1971
13. Ferris, S., Brown, W. M., Davidson, W. S., and Wilson, A. C. (1981) Proc. Natl. Acad. Sci. U.S.A. 78, 6319-6323
14. Brown, W. M., and Goodman, H. M. (1979) in Extrachromosomal DNA, ICN-UCLA symposia (Cummings, D., Borst, P., Dawid, I., Weisman, S., and Fox, C. F., eds) pp 485-500, Academic. New York.
15. Brown, W. M. (1980) Proc. Natl. Acad. Sci. U.S.A. 77, 3605-3609
16. Denaro, M., Blanc, H., Johnson, M. J., Chen, K. H., Wilmsen, E., Cavalli-Sforza, L. L., and Wallace, D. C. (1981) Proc. Natl. Acad. Sci. U.S.A. 78, 5768-5772

17. Blanc, H., Chen, K. H., D'Amore, M. A., and Wallace, D. C. (1983) Am. J. Hum. Genet. 35, 167-176
18. Johnson, M. J., Wallace, D. C., Ferris, S. D., Rattazzi, M. C., and Cavalli-Sforza, L. L. (1983) J. Mol. Evol. 19, 255-271
19. Horai, S., Gojobori, T., and Matsunaga, E. (1984) Hum. Genet. 68, 324-332
20. Wallace, D. C., Garrison, K., and Knowler, W. C. (1985) Am. J. Phys. Anthropol. 68, 149-155
21. Brega, A., Gardella, R., Semino, O., Morpurgo, G., Astaldi, G. B., Ricotti., Wallace, D. C., and Santachiara-Benerecetti, A. S. (1986) Am. J. Hum. Genet. 39, 502-512
22. Cann, R. L., Stoneking, M., and Wilson, A. C. (1987) Nature 325, 31-36
23. Harihara, S., Saitou, N., Hirai, M., Gojobori, T., Park, K. S., Misawa, S., Ellepola, S. B., Ishida, T., and Omoto, K. (1988) Am. J. Hum. Genet. 43, 134-143
24. Cann, R. L., and Wilson, A. C. (1983) Genetics 104, 699-711
25. Stoneking, M., and Wilson, A. C. (1989) in The colonization of the Pacific : a genetics trail (Hill, A., and Serjeantson, S., eds) Oxford University Press, Oxford, in press
26. Hertzberg, M., Mickleson, K. N. P., Serjeantson, S. W., Prior, J. F., and Trent, R. J. (1989) Am. J. Hum. Genet. 44, 504-510
27. Horai, S., Hayasaka, K., Hirayama, K., Takenaka, S., and I-Hung Pan. (1989) in Proceeding of Circum-Pacific Prehistory Conference, in Press
28. Sanger, F., Nicklen, S., and Coulson, A. R. (1977) Proc. Natl. Acad. Sci. U.S.A. 74, 5463-5467
29. Gojobori, T., Ishii, K., and Nei, M. (1982) J. Mol. Evol. 18, 414-423
30. Sokal, R. R., and Sneath, P. H. A. (1963) Principles of numerical taxonomy. Freeman, W. H., San Francisco.
31. Horai, S., Gojobori, T., and Matsunaga, E. (1986) Jpn. J. Genet. 61, 271-275
32. Horai, S., Gojobori, T., and Matsunaga, E. (1987) in HUMAN GENETICS, Proceeding of the 7th International Congress, Vogel F., and Sperling, K. eds Springer-Verlag, Heidelberg. pp, 177-181
33. Nei, M., and Li, W. H. (1979) proc. Natl. Acad. Sci. U.S.A. 76, 5269-5273
34. Nei, M. (1985) in Population Genetics and Molecular Evolution (Ohta, T., and Aoki K., eds) pp 41-64, Japan Sci. Soc Press, Tokyo/Springer-Verlag, Berlin

Part III. Mitochondrial Pathology

STABLE AND UNSTABLE BIOENERGETICS *in Vivo*

B. Chance

Department of Biochemistry and Biophysics
University of Pennsylvania
Philadelphia, PA 19104

SUMMARY

This symposium, up to this point has focused on the biochemistry and molecular biology of mitochondrial energy conservation with the aspiration that the principles would be applicable to medical and biological problems *in vivo*. To this point, this paper attempts to realize this aspiration and to establish the applicability of basic principles of biochemistry *in vivo* in normal and genetically disturbed/altered metabolic states.

Metabolic Control Principles

The basic principle of metabolic control that has even more force now than it did when it was enunciated by Claude Bernard in 1878 ("la *fixite du milieu interieur*") (1) is interpreted here that in life the internal environment be fixed. While Bernard did not understand the biochemical consequences of his ideas, we bioenergetisists now express this idea by the statement that the rate of ATP synthesis must equal the rate of ATP utilization. Otherwise the system cannot maintain itself in the required steady-state of life (see Fig. 1).

> La *fixité du milieu intérieur est la condition de la vie libre, indépendante* : le mécanisme qui la permet est celui qui assure dans le *milieu intérieur* le maintien de toutes les conditions nécessaires à la vie des éléments.
>
> *C.L. Bernard, 1878*

Fig. 1. Quotation from Claude Bernard, one of the first suggestions of the steady state (1) (our italics).

Criticality of Metabolic Control

There is no criticality of thermodynamic properties of biochemical systems, it is their kinetic properties that are keys to the maintenance of the steady state of equality of synthesis and breakdown of ATP. To this point Michaelis-Menten formulation for the control of mitochondria *in vitro* is applicable *in vivo* as indicated in Fig. 2 for control of oxidative metabolism specifically by ADP as indicated in the top line (2-4). We may substitute for ADP its value from the creatine kinase equilibrium on the assumption that the free creatine and free P_i are equal, that pH is constant, and that V/V_m is low. Thus, a direct relationship between NMR determinable cytosolic phospho-creatine (PCr) and inorganic phosphate (P_i) is established. In short the ratio of the oxidative metabolism rate in the steady state (which equals the rate of breakdown of ATP) is simply related to the maximal capability of oxidative phosphorylation, the maximum capability of mitochondria to produce ATP. Thus, NMR affords a method of determining the fraction of available oxidative capacity that is mobilized at any time. This is of great importance in cases where genetic and metabolic disease may have altered the capacity of the mitochondria to produce ATP, i.e., lower V_{max}. Alternatively one of the other reactants in the production of ATP, particularly oxygen may be limiting and thus lower V_{max}.

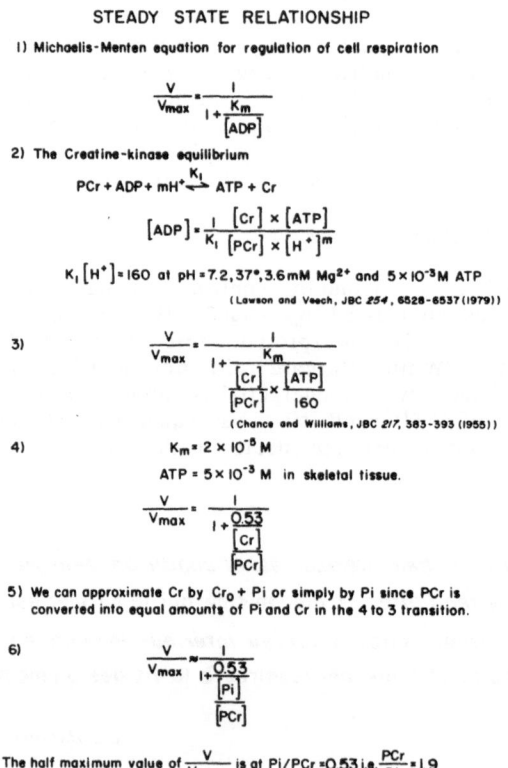

Fig. 2. Application of Michaelis-Menten formulation to metabolic control (5).

Experimental Verification

While at any one metabolic rate, the ratio V/V_{max} is readily obtained by Eqn. 6 of Fig. 2 above, the determination of V_{max} requires a "titration" which is most conveniently done by altering the work rate: in muscle, the external work (Fig. 3); in heart, the heart rate times pressure; and in other organs, the metabolic load, for example the breakdown of ATP in biosynthesis or ion transport, and in the brain neuronal function. Whatever the pertur-bation may be, it must produce quantifiable increments of metabolic activity.

Demonstrations of control the rate of oxidative metabolism in human skeletal muscle by systematic measurement of the external work had been first attempted by us in 1978 when a magnet large enough to accommodate the human limb at work on an ergometer was achieved (5,6). The protocol for this study is illustrated in Fig. 3 where increasing work rates would be expected to cause increasing values of P_i/PCr or ADP. The results obtained in these work/metabolic activity relationships are sufficiently accurate to allow plotting the data as a linearized reciprocal Hanes plot where V_{max} and K_m are readily determined (Fig. 4). The units of K_m are those of P_i/PCr as obtained through the creatine kinase equilibrium, i.e., ADP = 33 x P_i/PCr, near pH 7. The

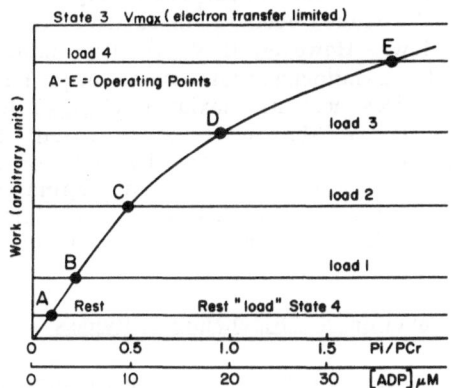

Fig. 3. A protocol for examining the V/V_m relationship to the control chemical, in this case as indicated as ADP or according to Fig. 2, P_i/PCr (5).

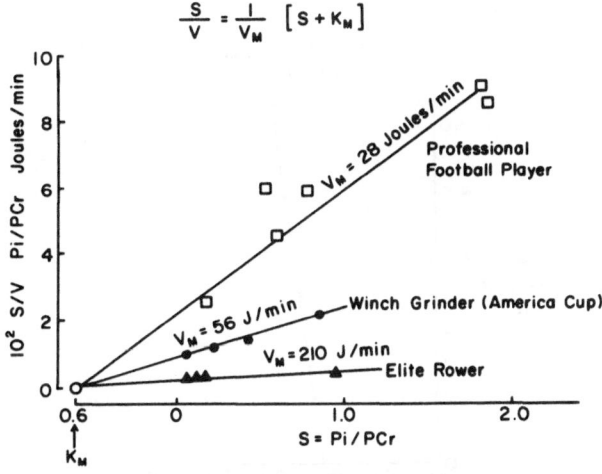

Fig. 4. Analysis of V/V_m, ADP relationships (calculated as P_i/PCr) for different individuals in which steady state exercise was obtained from the same muscle group over the working range and without a sig-nificant acidosis. Under these conditions, the hyperbolic relation-ship obtains as indicated by the straight line of the Hanes plot.

V_m's are those of mitochondrial oxidative phosphorylation and are given in work rates, watts, joules per min and represent the external manifestations of muscle work.

The Effect of Mitochondrial Concentration

While an increase of mitochondrial oxidative capacity is a usual response to exercise training adaptation, evidence of a much greater "adaptive response" can be obtained by comparing skeletal muscle with heart muscle, where continuous muscle function requires the hypertrophy of mitochondrial content in adult myocardium to the point of 40% of the heart volume occupied by mitochondria (6). In this case, V_{max} is so high that when the heart is perturbed by work, no detectable change of P_i/PCr is observed over the range available to ordinary work range of the myocardium (Fig. 5)(6). This is explained by an experimental and a theoretical chart (Fig. 6). The large V_{max} illustrates how the PCr/P_i value become insensitive to increases of work because V/V_{max} is very low. However, if V_{max} (the mitochondrial ability to produce ATP) is lowered by a mitochondrial disease or a lack of substrate, for example oxygen (Fig. 6), V rises near its maximum (V_{max}) leading to metabolic instability. If now the work of the heart is increased then V cannot be increased to meet the demand, metabolic control by ADP or any other substrate is no longer functional and lactic acidosis and angina result, with heart failure as an ultimate consequence. In this way, we demonstrate the great importance of an excess of functional mitochondria in maintaining the stable steady state.

Decreased Mitochondrial Function - "Biochemical Bypass"

Vastly exceeding the variation of mitochondrial content and effectiveness between heart and skeletal muscle is the effect of genetic and molecular disease. In one case we have been privileged to study a young lady lacking the cytochromebc_1 complex (7,8) and her pre-therapy transfer function with respect to a normal age-matched control is shown in Fig. 7. Her response to "biochemical by-pass therapy" vitamin K_3 plus C intake has been

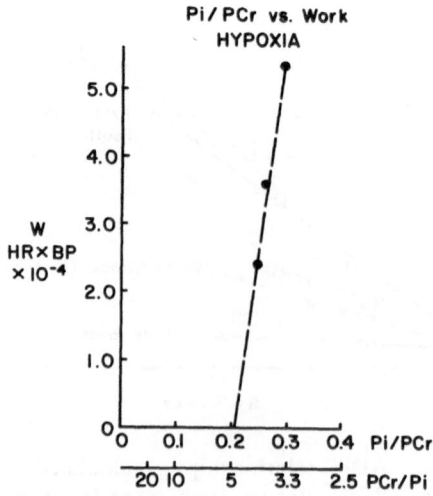

Fig. 5.   Illustrating the steeper transfer function of the myocardium due to the greatly increased concentration of mitochondria and to supplementary metabolic controls.

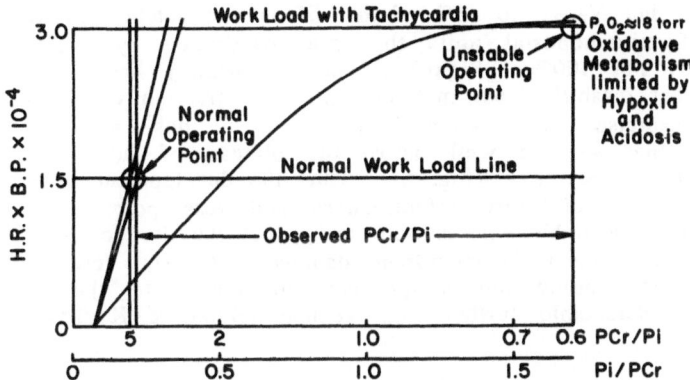

Fig. 6. A graphical explanation of the insensitivity of the P_i/PCr value of the heart to increased work load under normal conditions. An indication of the effect of hypoxia acidosis and increased work upon the transfer characteristic with consequent shift of the operating point to a higher ADP (P_i/PCr) level is also afforded.

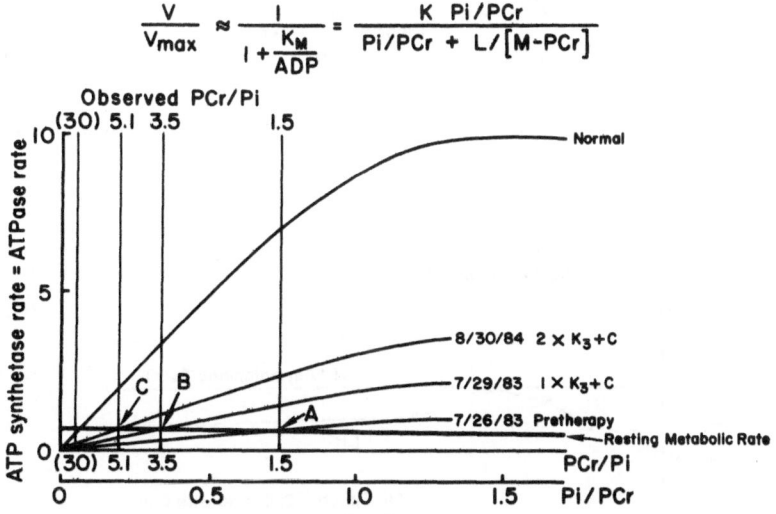

Fig. 7. Graphical analysis of the effect of therapeutic doses of vitamin K_3 and vitamin C upon the operating point for a patient lacking the cytochrome bc_1 complex, doses indicate a left shift of operating point from A to B to C in the indicated amounts, and the calculated transfer functions are illustrated with respect to those of an age and sex matched control ("normal").

monitored by NMR since the institution of therapy seven years ago, and she reports at this moment good health and function to the point of being able to do significant work. However, the best therapeutic treatment that we are able to provide raises her V_{max} to 40% of that of an age-matched control (Fig. 7). The number of mitochondrial units that are reactivated by the biochemical bypass is clearly not 100%. Significant muscle atrophy has occurred during the interval of her disability; our method scores the fraction of V/V_{max} that is evoked by her exercise. This is consistent with the dose response curve (8) which indicates that we are well below the plateau of full reactivation of electron transport by K_3 + C (Fig. 8). Thus the development of reagents for "biochem-ical bypass" of more effectiveness and less potential for radical generation is highly desirable. It is noteworthy that the dosage of K_3 is so far insufficient to cause detectable membrane damage. As yet, few variations of therapy have been studied but ubiquinone supplement to the therapy has resulted with no detectable further improvement (Buist & Kennaway, person. commun).

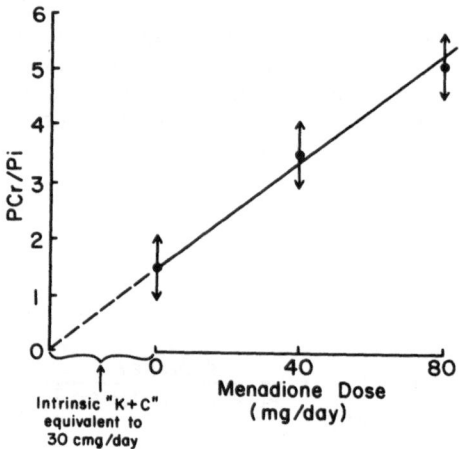

Fig. 8. Dose-response profile for K_3 titration of resting PCr/P_i of cytochrome *b* deficient subject.

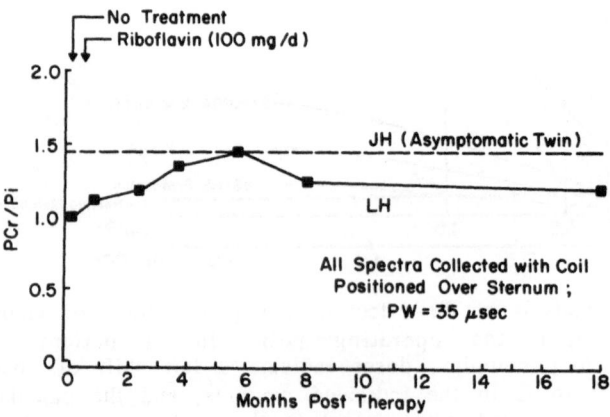

Fig. 9. Effect of riboflavin upon the PCr/P_i in the working myocardium of glutaric acid urea type II syndrome in one of two twins; ther other was asymptomatic.

A Flavin-Linked Genetic Deficiency

Finally, and as an acknowledgement of Drs. Yagi's and Massey's interest in flavins, we refer to our studies of genetic deficiency in twins, one of whom is normal and the other exhibited glutaric acid urea type II, presumably due to deficiency in the flavo enzyme itself (9). The course of attempted therapy by riboflavin as indicated by Fig. 9. The riboflavin indeed caused an increase of bioenergetic capacity which reached a maximum at 6 months. It is an appropriate question to ask whether or not a flavin derivative would be more successful in this therapy since the NMR study is an exemplary situation for evaluating the putative therapeutic procedure.

Discussion

Experimental results and the Michaelis-Menten metabolic control hypothesis afford the rationale quantifying the role of mitochondria in health and disease. The simplicity of the measure (PCr/P_i), and the applicability of the theorem for determining V_{max} explicitly have been demonstrated, and it is suggested that protocols in which variation of metabolic load on tissues in normal and diseased states are essential. However, in numerous cases the "rest work" gives a value of PCr/P_i that is sensitive to therapeutic procedures. Finally, the opportunity to administer bypass reagents which reactivate electron transfer in themselves may activate radical generation in tissues requires the design of accompanying radical traps for the bypass reagent that have radical generating properties.

It is expected that increasing use of NMR evaluation of tissue bioenergetics will lead to the evaluation of numerous therapeutic procedures in metabolic disease and evaluation of the limits of tolerance of biochemical bypasses of deletions in the respiratory change. Possibilities for a true evaluation of these potentialities under controlled conditions in human subjects is a novel and a much needed approach.

ACKNOWLEDGEMENTS

This work was supported in part by National Institutes of Health Grant HL 18708

REFERENCES

1. Bernard, C., Les Phenomenes de la Vie, Vols 1,2, Paris, J.B. Balliere et fils, 1878
2. Eleff, S., Leigh, J.S., McDonald, G.G. (1980) ^{31}P NMR Spectra of Human Limbs. Fed. Proc. 39:
3. Chance, B., Eleff, S. and Leigh, Jr., J.S. (1980) Noninvasive, Nondestructive Approaches to Cell Bioenergetics, Proc. Natl. Acad. Sci. USA 77:7430-7434.
4. Chance, B., Leigh, Jr., J.S., Clark, B.J. et al (1985) Control of Oxidative Metabolism and Oxygen Delivery in Human Skeletal Muscle: A Steady State Analysis of the Work/Energy Cost Transfer Function. Proc. Natl. Acad. Sci. USA 82:8384-8388.
5. Chance, B., Leigh, Jr., J.S., Kent, J. et al. (1986) Multiple Controls of Oxidative Metabolism of Living Tissues as Studied by ^{31}P MRS. Proc. Natl. Acad. Sci. USA 83:9458-9462.
6. Osbakken, M., Young, M., Closter, J., Huddewll, J. and Chance, B. (1987) Myocardial Bioenergetics Studied with 31P Nuclear Magnetic Resonance. Heart Failure 3:148-155.

7. Eleff, S., Kennaway, N.g., Buist, N.R.M., Darley-Usmar, V.M., Capaldi, R.S., Bank, W.J. and Chance, B. (1984) ^{31}P NMR Study of Improvement in Oxidative Phosphorylation by Vitamins K_3 and C in a Patient with a Defect in Electron Transport at Complex III in Skeletal Muscle. Proc. Natl. Acad. Sci. USA 81:3529-3533.

8. Argov, Z., Bank, W.J., Maris, J., Eleff, S., Kennaway, N.G., Olson, R.E. and Chance, B. (1986) Treatment of Mitochondrial Myopathy due to Complex III Deficiency with Vitamins K_3 and C: A ^{31}P NMR Follow-up Study. Ann. Neurol. 29:598-602.

9. Whitman, G.J., Chance, B., Bode, H., Maris, J., Haselgrove, J., Kelley, R., Clark, B.J. and Harken, A.H. (1985) Diagnosis and Therapeutic Evalution of a Pediatric Cardiomyopathy using ^{31}P NMR, J. Am. College of Cardiology. (JACC) 5:745-749.

REGULATION OF ACTIVITY AND TISSUE-SPECIFIC EXPRESSION OF CYTO-

CHROME C OXIDASE GENES UNDER NORMAL AND PATHOLOGICAL CONDITIONS

B. Kadenbach, U. Büge, F.-J. Hüther, C. Thiel,
M. Droste, A. Schlerf, B. Schneyder, L. Kuhn-Nentwig
and J. Müller-Höcker

Fachbereich Chemie der Philipps-Universität
Hans-Meerwein-Strasse, D-3550 Marburg, F.R.G.

INTRODUCTION

Life originated from anaerobic conditions, and also today some unicellular organisms can live without oxygen. But all multicellular organisms require mitochondrial respiration, which is coupled to phosphorylation of ADP to ATP, as an additional and essential source of energy. The rate of cellular respiration varies largely between different cells and tissues and under different endogenous or environmental conditions of an organism. Thus regulation of the rate of respiration represents an essential feature of eucaryotic cell metabolism. The terminal step of the mitochondrial respiratory chain, the formation of water from oxygen, electrons and protons, is catalyzed by cytochrome c oxidase (COX) and involves a large change of the free energy G. Therefore this enzyme represents a bottle-neck or pacemaker of oxidative metabolism. Three mechanisms are known which influence the rate of COX activity: 1) Respiratory control; according to the chemiosmotic hypothesis (1) the activity is controlled by the proton chemical potential across the inner mitochondrial membrane (2). 2) Regulation of activity by substrates (i.e. the concentrations of reduced cytochrome c and oxygen (3)) and allosteric effectors like phosphate, ADP and ATP (see below). 3) Variation of the regulatory response to allosteric effectors by expression of tissue- and developmental-specific isozymes (4-5). Regulation of COX activity by allosteric effectors is assumed to involve interaction with one of the nucleus-encoded subunits of the enzyme (subunits IV-VIII) (6). The three largest subunits (I-III), which are encoded on mitochondrial DNA, have a catalytic function, they contain the 5 redox centers, 3 copper ions (7,8) and two heme a groups. Subunits I-III occur also in COX from bacteria like Paracoccus denitrificans (9), the additional subunits occur only in eucaryotes at variable number, depending on the evolutionary stage of the organism (5). In mammals the largest number of 10 nucleus-encoded subunits are found as shown in fig. 1, where the SDS-PAGE gel patterns of the isolated enzymes from bovine, pig, rat and human liver are presented.

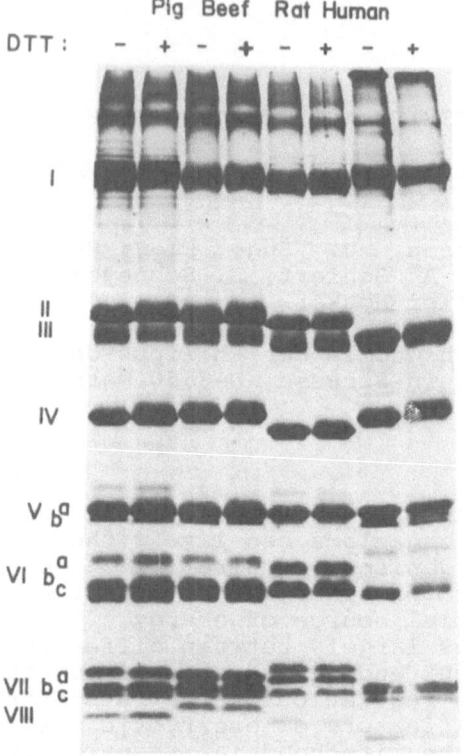

Fig.1. Sodium dodecylsulfate polyacrylamid gelelectrophoresis (SDS-PAGE) of isolated COX from the hearts of pig, bovine, rat and human. The SDS-dissociated samples were preincubated with 10 mM dithiothreitol (DTT) where indicated.

A model of COX emphasizing the regulatory function of nucleus-encoded subunits is shown in fig. 2. The model is based on experimental data on a) the nearest neighbour relationship of subunits by crosslinking experiments (10), b) the orientation of subunits in the membrane by protease digestion and antibody reactivity (11,12), c) the binding domain for cytochrome c by labelling of carboxylic groups with glycineethylester after activation with a carbodiimide (13,14), d) the occurrence of a third copper (7,8) and e) the occurrence of hydrophobic transmembraneous segments in the amino acid sequences of subunits (5,15). Recently Zhang et al. (16) identified subunit VIb only on the cytosolic side, using protease digestion and antibody detection. In the following we will summarize recent kinetic data on the regulation of activity and on the tissue-specific expression of nucleus-encoded genes of COX.

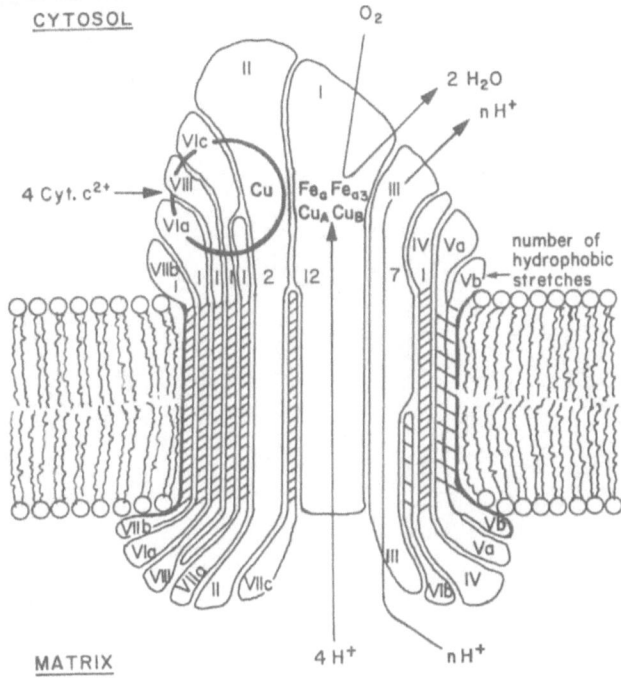

Fig. 2. Model of the mammalian COX complex as located in the inner mitochondrial membrane. The model is based on experimental data on the nearest neighbour relationship and membrane orientation of subunits, the binding domain for cytochrome c, the occurrence of transmembraneous hydrophobic segments in the amino acid sequence of subunits and on a third copper ion within the isolated enzyme.

RESULTS AND DISCUSSION

Regulation of COX activity by ATP and ADP

COX was isolated from bovine heart mitochondria and recon-stituted in asolectin-liposomes in the absence or presence of 10 mM ATP or ADP as described in (17). The kinetics of ferro-cytochrome c oxidation was measured by the photometric method of assay in the absence or presence of 10 mM ATP. The data are presented by a reversed Eadie-Hofstee plot (fig. 3). ATP, either present inside or outside of the liposomes increases the K_m for cytochrome c. In contrast, intraliposomal ADP decreases the K_m for cytochrome c (17). Because ferrocytochrome c interacts from the outside with COX, the effect of intraliposo-mal nucleotides must occur via binding at matrix domains of subunits and conformational change through the lipid bilayer. If the same proteoliposomes are used to measure their activity at various cytochrome c concentrations by the polarographic assay using ascorbate and TMPD as substrate, no influence of intraliposomal nucleotides is found (fig. 4). In contrast, extraliposomal ATP again increases the K_m for cytochrome c (17).

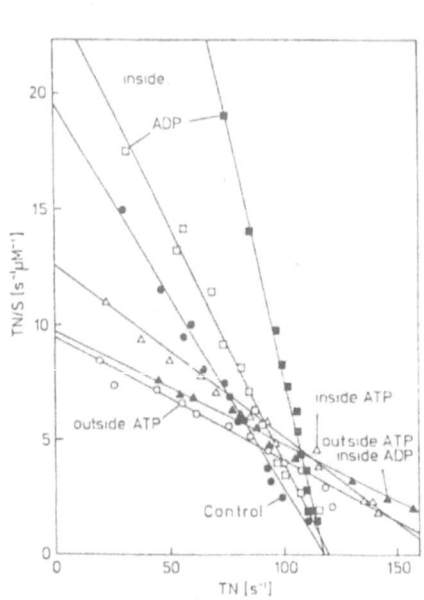

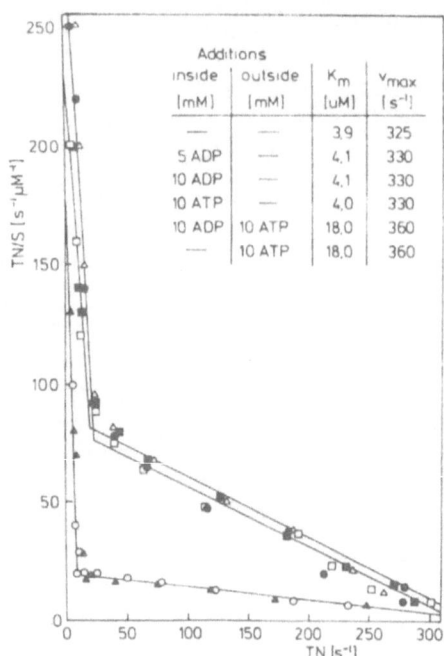

Fig. 3. Influence of intraliposomal ATP or ADP and
extraliposomal ATP (10 mM
each) on the kinetics of
ferrocytochrome c oxidation
by reconstituted COX from
bovine heart. The spectrophotometric assay of activity was measured at 1-80 μM
ferrocytochrome c concen-
trations in the presence of
uncoupler (17). The data are
presented in a reversed
Eadie-Hofstee plot.

Fig. 4. Polarographic assay
of the effect of nucleotides on
the kinetics of ferrocytochrome
c oxidation by reconstituted COX
from bovine heart. The conditions for measurement of activity were the same as given in
the legend to fig. 3 except that
25 mM ascorbate, 0.7 mM TMPD and
0.02-30 μM ferricytochrome c were
used as substrates (17). Symbols
as in Fig. 3.

The difference between the two methods of assay is illustrated in fig. 5. The photometric assay represents the more
physiological method, because each oxidized cytochrome c molecular must dissociate from the COX before the next ferrocytochrome c molecule can bind, (as it occurs in mitochondria). In
the polarographic assay, in contrast, cytochrome c is assumed
to remain bound to COX, while being reduced by ascorbate via
TMPD. We therefore suggest that intraliposomal nucleotides
modify the rate of ferricytochrome c dissociation from COX,
representing the rate-limiting step of COX activity.

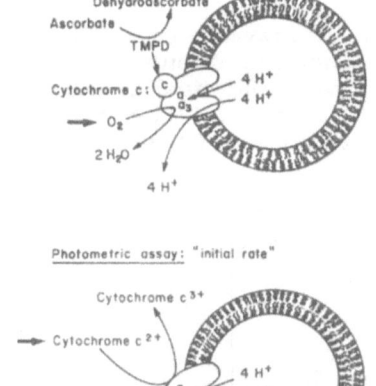

Fig. 5. Schematic presentation of the polarographic and spectrophotometric method of assay for the activity of reconstituted COX. The arrows indicate the measured substrate. TMPD, N,N,N',N'-tetramethyl-p-phenylenediamine; c, cytochrome c; a and a_3, heme a and a_3.

The specific interaction of nucleotides with nucleus-encoded subunits of the reconstituted bovine heart COX was concluded from results obtained with reconstituted COX from <u>Paracoccus denitrificans</u> (fig. 6). The presence of up to 15 mM ATP or ADP within the liposomes had no influence on the kinetics of ferrocytochrome c oxidation (18). Extraliposomal ATP strongly increased the K_m for cytochrome c and the V_{max}. The data are taken to indicate a regulatory influence of intramitochondrial ATP and ADP on the activity of bovine heart COX. A physiological role is suggested, because ADP increases and ATP decreases the activity. So far the binding of the nucleotides to a specific subunit of COX is not known. Previous claims of an interaction of ATP with subunits IV and VIII, as shown by photoaffinity-labelling with radioactive 8-azido-ATP (19) could not be corroborated (20). Nevertheless 8-azido-ATP could be shown to bind specifically to bovine heart COX and to change its kinetic properties (21).

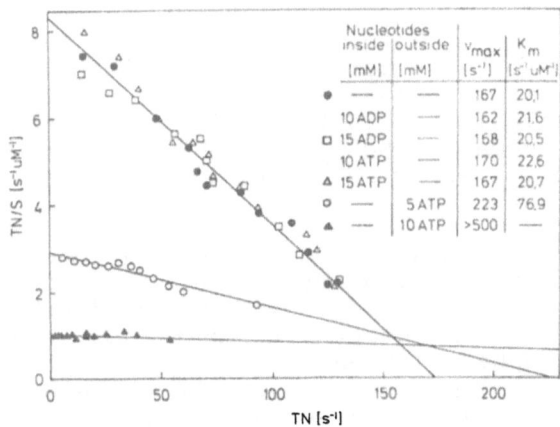

Fig. 6. Influence of intraliposomal ATP or ADP and extraliposomal ATP on the kinetics of ferrocytochrome c oxidation by reconstituted COX from Paracoccus denitrificans. The isolated bacterial enzyme is composed of two subunits only (I and II). The spectrophotometric assay of activity was measured in the presence of uncoupler at 1-80 μM ferrocytochrome concentrations (18).

Different effects of cardiolipin and phosphate on COX isozymes
from liver and heart

Tissue- and developmental-specific isozymes of COX are
based on different isoforms of nucleus-encoded subunits (see 5
and 6 for review). In COX from bovine liver and heart subunits
VIa, VIIa and VIII are different (22-24), whereas the other 10
subunits appear to be identical in both isozymes (24).

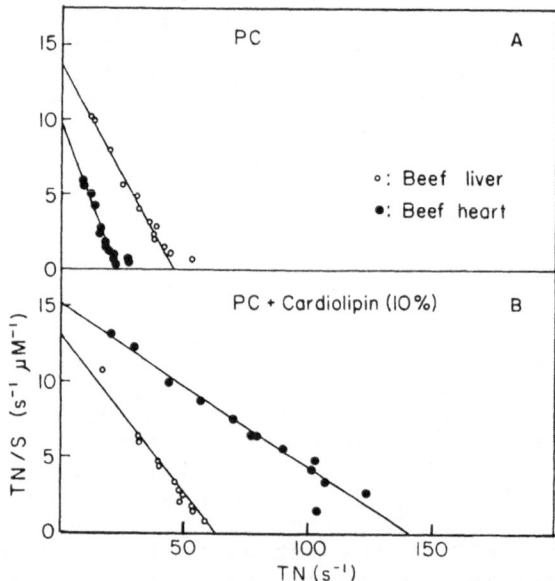

Fig. 7. Effect of liposomal lipids on the kinetics of ferrocytochrome c oxidation by the reconstituted enzyme
from bovine liver and heart. The spectrophotometric
assay of activity was performed with 1-80 uM ferrocytochrome c in the presence of uncoupler. The isolated enzymes were reconstituted either in phosphatidylcholine (PC) or in PC containing 10 % cardiolipin
(25).

Reconstitution of bovine COX with pure phosphatidylcholine
results in a higher activity of the liver as compared to the
heart enzyme (fig. 7A). Inclusion of 10 % cardiolipin in the
liposomal lipids strongly increases the K_m and V_{max} of the
heart enzyme but only slightly that of the liver enzyme (fig.
7B) (25).

An opposite response of the isolated and reconstituted COX
of bovine liver and heart was found with intraliposomal phosphate (fig. 8). Exchange of the intraliposomal Hepes-buffer by
phosphate increased the V_{max} of the heart but decreased the
V_{max} of the liver enzyme. Phosphate increased the K_m for cytochrome c of both isozymes (25). These results can be explained
by the different number of positively charged amino groups of
subunits VIa from liver and heart on both sides of the membrane, as deduced from the two cDNAs of the rat (see below).

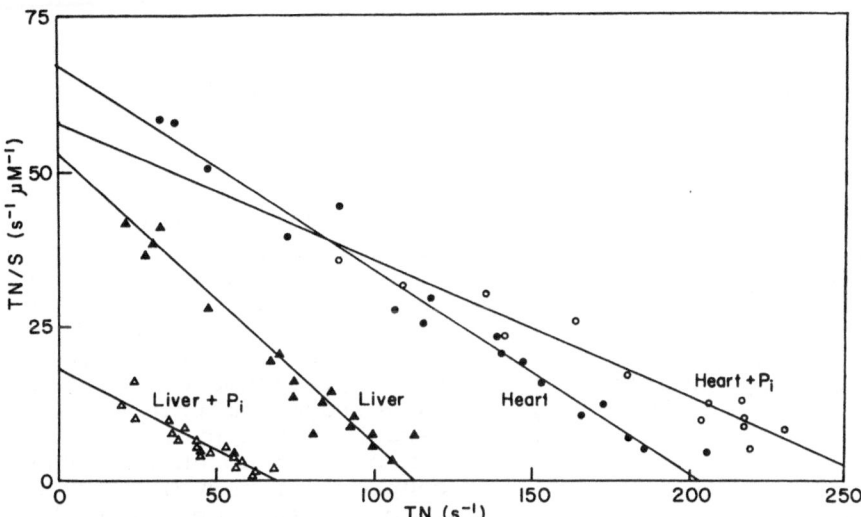

Fig. 8. Different effect of intraliposomal phosphate on the kinetics of ferrocytochrome c oxidation by reconstituted COX from bovine liver and heart. The enzymes were reconstituted in asolectin either in the presence of 100 mM Hepes (closed symbols) or 20 mM phosphate, 200 mM sucrose (open symbols), and the activity was measured by the spectrophotometric method in the presence of uncoupler (25).

Free fatty acids decrease respiratory control of COX from heart but not from liver

The respiratory control index of reconstituted bovine heart COX (ratio of oxygen uptake in the presence and absence of valinomycin + CCCP) is decreased to about 50 % by about 1 uM palmitic or stearic acid, if measured by the photometric method of assay (26). No further decrease is found at free fatty acid concentrations up to 20 uM (fig. 9). Under the same conditions the respiratory control index of the reconstituted bovine liver enzyme is not decreased by free fatty acids (26). It was suggested that in heart (but not in liver) partial uncoupling of oxidative phosphorylation at the third coupling site (complex IV = COX) via free fatty acids could represent a mechanism for nonshivering thermogenesis in particular in those mammals, which lack brown adipose tissue (26). In brown adipose tissue nonshivering thermogenesis is induced by noradrenalin via cAMP and free fatty acids, which open the proton channel of the uncoupling protein in mitochondria (27,28).

Tissue-specific expression of nucleus-encoded genes of COX

By screening cDNA libraries in gt 11 from rat liver and rat heart with antibodies to subunit VIa we could isolate two cDNAs coding for subunits VIa of COX from rat liver and heart (29, corrected heart sequence in 30). The deduced amino acid sequences show only 60 % homology. In fig. 10 are compared the

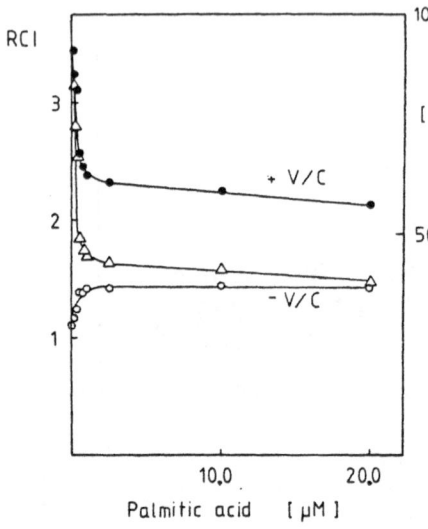

Fig. 9. Influence of free palmitic acid on the respiratory control index (RCI) of reconstituted COX from bovine heart. The RCI (triangles) was calculated from the rate of ferrocytochrome c oxidation (spectrophotometric method of assay) in the presence (closed symbols) or absence of the uncoupler valinomycin + CCCP (VIc) (open symbols) (26).

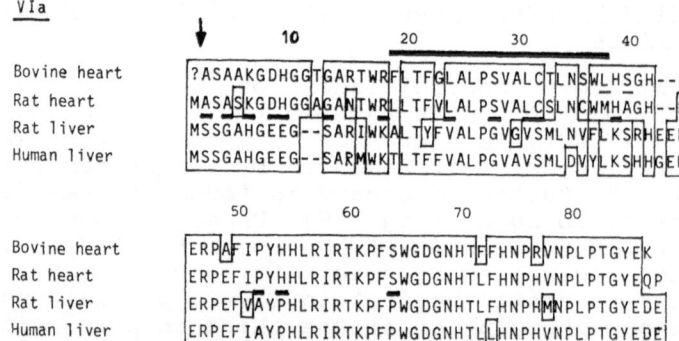

Fig. 10. Amino acid sequences of COX subunits VIa from different tissues and species. The sequences were either deduced from cDNAs (corrected rat heart sequence (30), rat liver (29), human liver (45)) or from protein sequencing (bovine heart (46)). A hydrophobic segment of 20 amino acids is indicated by a balk. The arrow indicates the begin of the mature protein.

amino acid sequences of subunits VIa from bovine heart, rat heart and liver and human liver. The homology between bovine and rat heart is 84 % and that between rat and human liver 87% indicating a larger evolutionary divergence between the rat liver and rat heart subunit VIa than between the rat and bovine heart or rat and human liver subunit VIa.

Although the two amino acid sequences of the isoforms of subunit VIa are only 60 % homologous, they must be bound within the enzyme complex to 12 other subunits, from which 10 are identical in bovine liver and heart (24). In fact a very similar general structure of the two proteins can be deduced from the hydropathy plot shown in fig. 11. Both subunits contain a hydrophobic sequence of 20 amino acids which is suggested to penetrate the membrane.

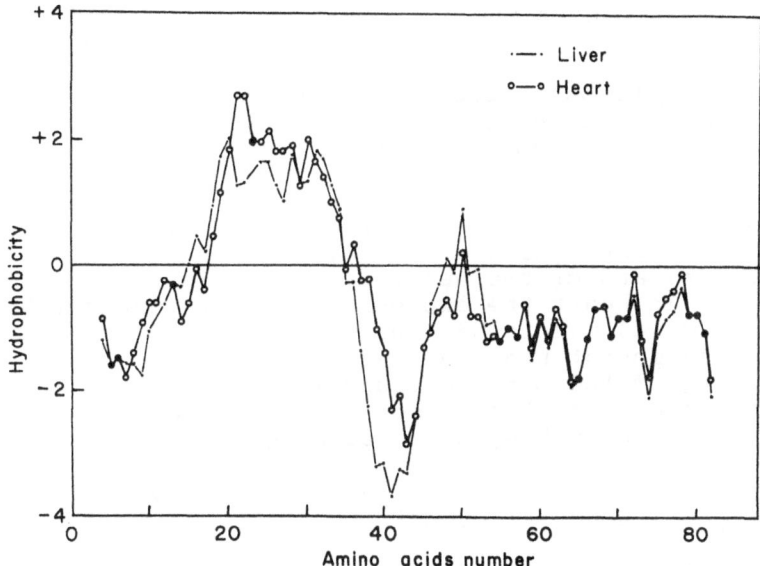

Fig. 11. Hydropathy plot of rat liver and rat heart subunit VIa amino acid sequences. The hydrophathy plot in (29) was redrawn using the corrected amino acid sequence of the rat heart subunit VIa (fig. 10).

The basic cytochrome c (pI = 10.5) interacts with COX via charged groups by electrostatic forces (31). Subunit VIa was found to participate in the binding of cytochrome c to COX from pig liver and kidney but not from pig heart and diaphragm, as shown by cytochrome c protection of labelling of its carboxylic groups with radioactive glycineethylester after activation with a carbodiimide (13,14). This result is supported by 5 additional acidic amino acids found in the rat liver sequence, from which only two are compensated by additional basic amino acids (fig. 12). Unfortunately we do not know the orientation of subunit VIa in the membrane (16), and if the N-terminal or C-terminal arm of the protein interacts with cytochrome c. But on both sides one or two additional negatively charged amino acids occur, respectively, which could interact with cytochrome c. The above described different kinetics of ferrocytochrome c

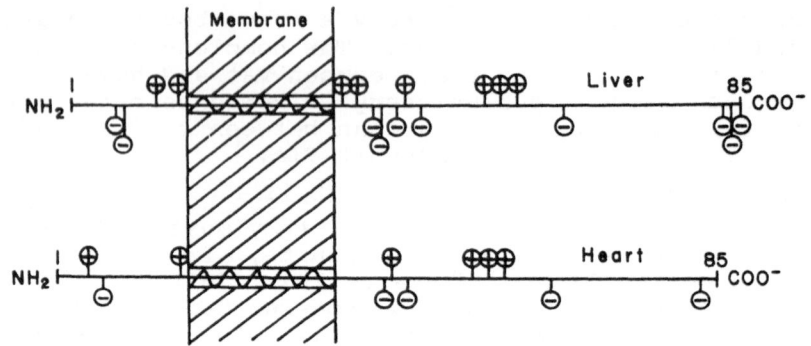

Fig. 12. Charge distribution in subunits VIa from rat liver and
rat heart. The positively (Arg and Lys) and negatively
charged amino acids (Asp and Glu) in the sequences of
the two proteins are indicated.

oxidation by the reconstituted liver and heart enzyme (fig. 7
and 8) may in part be due to the different amount of charged
amino acids in the two subunits VIa.

By hybridization of the cDNA probes for rat liver and
heart subunit VIa with RNA from different rat tissues (Northern
Blot) a specific expression of the liver-type cDNA in liver,
kidney and brain, and of the heart-type cDNA in skeletal muscle
was found (29). In heart the expression of both genes (which do
not cross-hybridize) was found. The expression of the two genes
for subunit VIa in rat heart was verified by N-terminal amino
acid sequencing of the purified subunit VIa from rat heart COX
(32). Also in human heart both corresponding genes were found
to be expressed (E.A. Schon, personal communication).

Expression of COX genes in respiratory defective tissues of patients with mitochondrial myopathies

Mitochondrial myopathies represent a heterogeneous group of
human diseases with mitochondrial dysfunction (33). In many
cases the activity of COX is impaired (34). In chronic progressive external opthalmoplegia (CPEO) a mosaic pattern of
COX-defective fibers is found by histochemical activity stain
with diaminobenzidine of skeletal muscle sections from patients
(35,36) (fig. 13). Fibers lacking COX activity, however, show a
normal succinate dehydrogenase activity as well as immuno-
fluoreszenz with an antiserum to the holoenzyme of rat liver
COX (see fig. 13). The COX defect is bordered by the plasma
membrane of the fibers. Neighbouring fibers show normal COX
activity. By immunohistochemistry of tissue sections from 17
CPEO patients with various monospecific antisera to human COX
subunits a variable pattern of lacking immunoreactivity was
found in each COX-defective fiber (38). A statistic and in each
fiber different mutation of mtDNA was suggested to cause the
defect of COX activity. Recently some cases of CPEO could be
related to deletions of mitochondrial DNA (mtDNA) (39). A
mutation of a nuclear gene coding for a protein involved in
replication of mtDNA, as suggested recently (40), could represent the primary defect for CPEO.

Succinate Dehydrogen   Cytochrome c Oxidase   Immunofluorescence

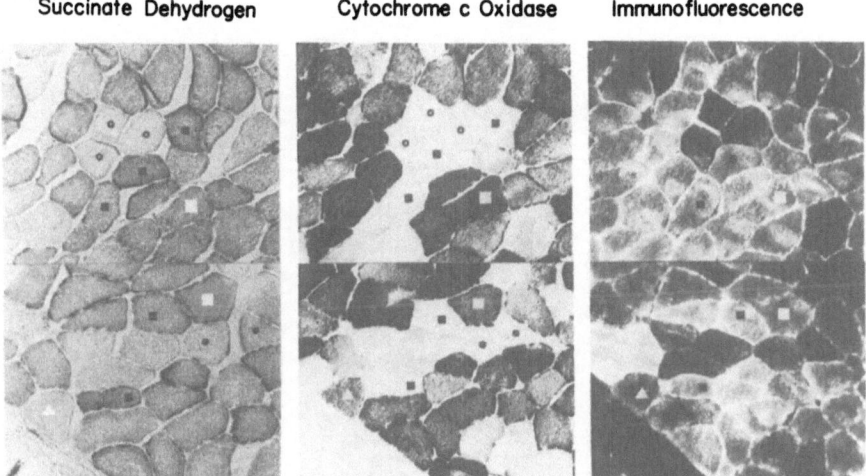

Fig. 13. Histochemical analysis of serial skeletal muscle sec-
        tions of a patient with CPEO described previously in
        (37). Activity stain for succinate dehydrogenase and
        COX as well as immunofluoreszenz with an antiserum to
        rat liver holo-COX and FITC-protein A were done as
        previously described (38). Black squares and circles
        indicate COX-defective fibers.

A different gene defect is apparently involved in fatal
infantile mitochondrial myopathies, where the children die
within the first year of age mainly due to respiratory arrest
(33,41,42). In a systematic immunohistochemical study of a
patient with fatal infantile mitochondrial myopathy with monospecific antisera to human COX subunits a specific lack of
immunoreactivity was found only with an antiserum to subunits
II/III and VIIbc in skeletal muscle sections (43) (fig. 14). In
contrast to CPEO the defective reactivity was seen in all
muscle fibers but not in fibers of the muscle spindle or in
heart, liver and kidney. All other applied antisera to COX
subunits (IV, Vab, VIa, VIbc, VIIa, VIIbc) showed normal
immunoreactivity (43).

Similar results were recently obtained with a monoclonal
antibody which reacts with subunits VIIa and VIIb. In four
patients with fatal infantile mitochondrial myopathy no
immunoreactivity was obtained with this antibody while all
other applied antisera showed normal immunoreactivity (H.-J.
itschler and E. Schon, unpublished results). Surprisingly
skeletal muscle sections of patients with Kearns-Sayre-
ndrome, which could also be related to deletions of mtDNA
(44), did also reveal no immunoreactivity with the monoclonal
antibody to subunit VIIab. Thus it cannot be excluded that in
fatal infantile mitochondrial myopathy a defective assembly of
the COX enzyme complex may cause the missing respiratory
activity. In contrast to subunit IV which was frequently found
to be present in COX-defective cells of CPEO patients (38),
subunits VIIab may not be stable and be proteolytically
degraded if assembly of the complex is impaired.

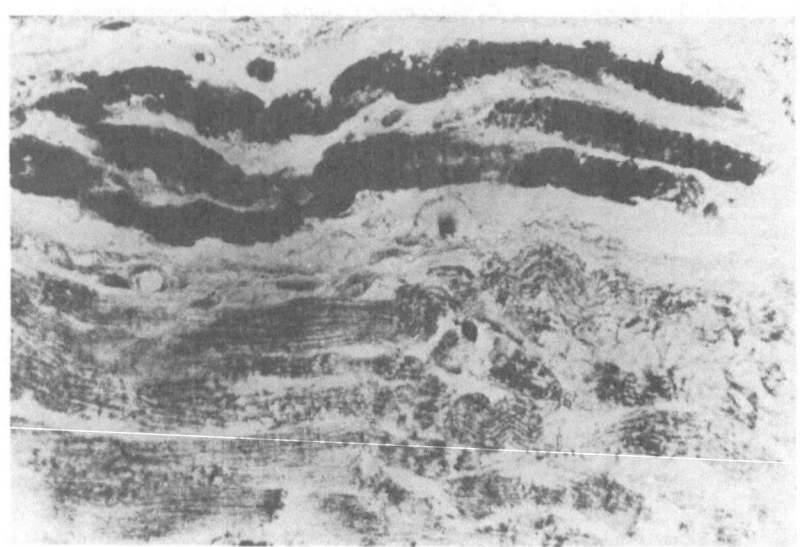

Fig. 14. Immunocytochemical reaction (with diaminobenzidine) of
a skeletal muscle section of a patient with fatal
infantile myopathy with an antiserum to human heart
COX subunits VIIbc (43). Whereas all fibers of the
muscle show no enzyme reaction, the muscle spindle
fibers gave a strong COX reaction.

References

1) P. Mitchell, Nature 191, 144-148 (1961).
2) D. G. Nicholls, Bioenergetics, An Introduction to the
    Chemiosmotic Theory, Academic Press, New York (1982).
3) M. Erecinska and D.F. Wilson, J. Membr. Biol. 70, 1-14
    (1982).
4) L. Kuhn-Nentwig and B. Kadenbach, Eur. J. Biochem. 149, 147-
    158 (1985).
5) B. Kadenbach, L. Kuhn-Nentwig and U. Büge, Curr. Top.
    Bioenerg. 15, 113-161 (1987).
6) B. Kadenbach, J. Bioenerg. Biomembr. 18, 39-54 (1986).
7) E. Bombelka, F.-W. Richter, A. Stroh and B. Kadenbach,
    Biochem. Biophys. Res. Commun. 140, 1007-1014 (1986).
8) G.C.M. Steffens, R. Biewald and G. Buse, Eur. J. Biochem.
    164, 295-300 (1987).
9) M. Raitio, T. Jalli and M. Saraste, EMBO J. 6, 2825-2833
    (1987).
10) J. Jarausch and B. Kadenbach, Eur. J. Biochem. 146, 211-217
    (1985).
11) J. Jarausch and B. Kadenbach, Eur. J. Biochem. 146, 219-225
    (1985).
12) L. Kuhn-Nentwig and B. Kadenbach, Eur. J. Biochem. 153, 101-
    104 (1985).
13) B. Kadenbach and A. Stroh, FEBS Lett. 173, 374-380
    (1984).

14) B. Kadenbach, A. Stroh, M. Ungibauer, L. Kuhn-Nentwig, U. Büge and J. Jarausch, Methods Enzymol. 126, 22-45 (1986).

15) M. Wikström, M. Saraste and T. Penttilä, in: The Enzymes of Biological Membranes, A. Martonosi, ed., Plenum Pubbl. Corp., New York, pp. 111-148 (1985).

16) Y.-Z. Zhang, M.A. Lindorfer and R.A. Capaldi, Biochemistry 27, 1389-1394 (1988).

17) F.-J. Hüther and B. Kadenbach, Biochem. Biophys. Res. Commun. 147, 1268-1275 (1987).

18) F.-J. Hüther and B. Kadenbach, Biochem. Biophys. Res. Commun. 153, 525-534 (1988).

19) C. Montecucco, G. Schiavo, and R. Bisson, Biochem. J. 234, 241-243 (1986).

20) F.-J. Hüther, J. Berden and B. Kadenbach, J. Bioenerg. Biomembr. 20, 503-515 (1988).

21) F.-J. Hüther and B. Kadenbach, FEBS Lett. 207, 89-94 (1986).

22) B. Kadenbach, R. Hartmann, R. Glanville and G. Buse, FEBS Lett. 138, 236-238 (1982).

23) B. Kadenbach, Angew. Chem. Int. Ed. Engl. 22, 275-282 (1983).

24) W. Yanamura, W. Zhang, Y.-Z. Takamiya and R.A. Capaldi, Biochemistry 27, 4909-4914 (1988).

25) U. Büge and B. Kadenbach, Eur. J. Biochem. 161, 383-390 (1986).

26) C. Thiel and B. Kadenbach, FEBS Lett. 251, 270-274 (1989).

27) D.G. Nicholls and R.M. Locke, Physiol. Rev. 64, 1-64 (1984).

28) J. Nedergaard and B. Cannon, in: Bioenergetics, L. Ernster, ed., pp. 291-314, Elsevier, Amsterdam (1984).

29) A. Schlerf, M. Droste, M. Winter and B. Kadenbach, EMBO J. 7, 2387-2391 (1988).

30) B. Kadenbach, F.-J. Hüther, U. Büge, A. Schlerf and M.A. Johnson, in: Molecular Basis of Membrane Associated Disease, A. Azzi, Z. Drahota, J. Jaz and S. Papa, eds., pp. 216-227 Springer-Verlag, Berlin (1989).

31) W.H. Koppenol and E. Margoliash, J. Biol. Chem. 257, 4426-4437 (1982).

32) B. Kadenbach, A. Stroh, A. Becker, C. Eckerskorn and F. Lottspeich, Biochim. Biophys. Acta, in press (1989).

33) S. DiMauro, E. Bonilla, M. Zeviani, M. Nakagawa and D.C. DeVivo, Ann. Neurol. 17, 521-538 (1985).

34) S. DiMauro, M. Zeviani, S. Servidei, E. Bonilla, A.F. Miranda, A. Prelle and E.A. Schon, Ann. N. Y. Acad. Sci. 488, 19-32 (1986).

35) M.A. Johnson, D.M. Turnbull, D.J. Dick and H.S.A. Sherratt, J. Neurol. Sci. 60, 31-53 (1983).

36) J. Müller-Höcker, D. Pongratz and G. Hübner, Virch. Arch. Pathol. Anat. 402, 61-71 (1983).

37) G. Spalke, R. Heene and D. Herold, J. Neurol. 209, 9-29 (1975).

38) M.A. Johnson, B. Kadenbach, M. Droste, S.L. Old and D.M. Turnbull, J. Neurol. Sci. 87, 75-90 (1988).

39) C.T. Moraes, S. DiMauro, and M. Zeviani, New Engl. J. Med. 320, 1293-1299 (1989).

40) M. Zeviani, S. Servidei, C. Gellera, E. Bertini, S. DiMauro and S. DiDonato, Nature 339, 309-311 (1989).

41) P.E. Minchom, R.L. Dormer, J.A. Hughes, D. Stansbie, A.R. Cross, G.A.F. Hendry, O.T.G. Jones, M.A. Johnson, H.S.A. Sherratt and D.M. Turnbull, J. Neurol. Sci. 60, 453-463.

42) J. Müller-Höcker, D. Pongratz, J.M.F. Trijbels, W. Endres and G. Hübner, Virch. Arch. Pathol. Anat. 399, 11-23 (1983).

43) J. Müller-Höcker, M. Droste, D. Pongratz and G. Hübner, Human Pathol. 20, 666-672 (1989).

44) M. Zeviani, C.T. Moraes, S. DiMauro, H. Nakase, E. Bonilla, E.A. Schon and L.P. Rowland, Neurology 38, 1339-1346 (1988).

45) R. Rizzuto, H. Nakase, B. Darras, U. Francke, G.M. Fabrizi, T. Mengel, F. Walsh, B. Kadenbach, S. DiMauro and E.A. Schon, J. Biol. Chem. 264, 10595-10600 (1989).

46) L. Meinecke and G. Buse, Biol. Chem. Hoppe-Seyler 366, 687-694 (1985).

MITOCHONDRIAL DNA MUTATIONS AS AN ETIOLOGY OF HUMAN DEGENERATIVE DISEASES

Takayuki Ozawa, Masashi Tanaka, Wataru Sato, Kinji Ohno, Satoru Sugiyama
Makoto Yoneda, Tomoko Yamamoto, Kazuki Hattori, Shin-ichiro Ikebe
Michinari Tashiro, and Ko Sahashi

Department of Biomedical Chemistry, Faculty of Medicine
University of Nagoya, Nagoya 466, Japan

SUMMARY

Because mitochondrial DNA (mtDNA) is exclusively maternally transmitted, mutations of mtDNA are implicated to be the cause of maternally inherited diseases. Recent extensive studies have clarified three types of mtDNA mutations in several human diseases.

Type A, homoplasmy: A base transition from guanine at nucleotide number 11,778 to adenine converts the 340th arginine to histidine in the ND4 subunit of Complex I. This maternally inherited mutation was found both in American pedigrees and in Japanese pedigrees of Leber's hereditary optic neuropathy.

Type B, heteroplasmy: Heteroplasmic existence of mutant mtDNA with a multi-gene deletion with the normal-sized mtDNA was found in 1/3 of patients with Kearns-Sayre syndrome (KSS) and chronic progressive external ophthalmoplegia (CPEO). The population of the deleted mtDNA was large enough to be detected by the conventional Southern blot analysis. This group of patients showed early onset of clinical symptoms and no family history. The location of the deletion in each individual was precisely determined by the combination of polymerase chain reaction and S_1 nuclease digestion (PCR plus S_1 method). Direct sequencing of the boundaries of the deletions revealed that the deletions occur between 3~13-bp directly repeated sequences.

Type C, pleioplasmy: Pleioplasmic existence of mutated mtDNAs with various deletions with the normal-size mtDNA was found in 2/3 of patients with mitochondrial myopathy. The population of each mutated mtDNA was so small that the existence could not be detected by the Southern blot analysis but by the PCR amplification. The deletions were confirmed by comparing the shift in the positions of the primers used for amplification with the shift in the sizes of the amplified DNA fragments (primer-shift PCR method). PCR direct sequencing of the one of the deletions in patients with pleioplasmic deletions showed a 13-bp directly repeated sequence which was identical with the sequence found in patients with homoplasmic deletion. The patients with pleioplasmic deletions showed late onset of clinical symptoms. About a half of the patients had a positive family history.

Pleioplasmic mtDNA deletions were also found in patients with exertional myoglobinuria without external ophthalmoplegia and in patients with cardiomyopathy. It is proposed that accumulation of mtDNA mutations and subsequent cytoplasmic segregation of these mutations during life is an important contributor both to the ageing process and to several human degenerative diseases. Since pleioplasmic mtDNA deletions can be analyzed semiquantitatively by using PCR, this type of mutation would be a useful indicator of the levels of accumulation of mtDNA mutations in age-related degenerative disorders.

Bioenergetics, Edited by C. H. Kim and
T. Ozawa, Plenum Press, New York, 1990

413

INTRODUCTION

The biogenesis of a functional mitochondrion in all types of cells is a complex process. The nuclear genome provides the genetic information for most of the several hundreds of proteins required for construction of the two membranes of the organelle. Small though it is, human mitochondrial DNA (mtDNA) makes an essential contribution to this process. mtDNA codes for thirteen hydrophobic polypeptides, which play a central role in energy-transducing function of four multi-subunit enzyme complexes (Complexes I, III, IV, and V) of the oxidative phosphorylation system (Fig. 1), and for two mitochondrial rRNAs and 22 organelle-specific tRNAs, which are essential for the assembly of a functional mitochondrial protein-synthesizing system.[1,2] Because mtDNA is exclusively maternally transmitted,[3] mutations of mtDNA are implicated to be the cause of maternally inherited diseases.[4] It is also proposed that the accumulation of mtDNA mutations and the subsequent cytoplasmic segregation of these mutations during life is an important contributor both to the aging process and to several human degenerative diseases.[5] Therefore, it is essential to identify the putative mutations in these pathophysiological processes.

One of the several unusual features of mtDNA is its capacity for rapid mutation. We proposed that the accumulation of mitochondrial genome mutation during the whole life of an individual is an important contributor to ageing and degenerative diseases.[5] This hypothesis is based on following: the high frequency of gene mutation in mtDNA; the small and economically packed size of the mitochondrial genome and its known information content; the lack of a repair mechanism for mtDNA, unlike nuclear DNA; the accumulation of oxidative mtDNA damage by reactive oxygen species produced in the mitochondria; the established features of the somatic segregation of individual mtDNA genomes during eukaryotic cell division; and findings on molecular genetic lesions underlying several human mitochondrial disorders. Recent reports

Cytosolic side

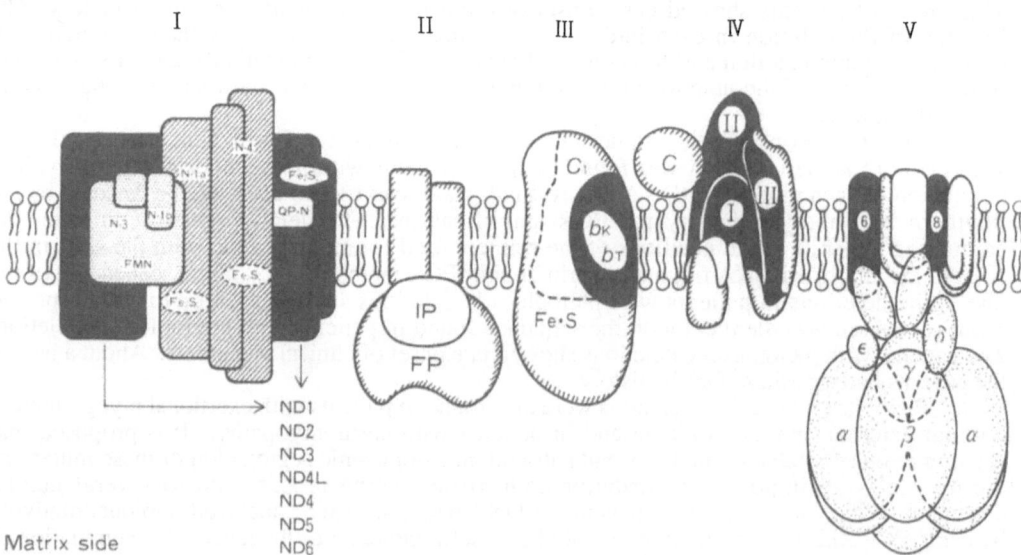

Fig. 1. Model of five complexes in the mitochondrial oxidative phosphorylation system. Dark areas represent subunits encoded for by the mitochondrial DNA: ND1, ND2, ND3, ND4L, ND4, ND5, and ND6 subunits in Complex I; cytochrome *b* subunit in Complex III; CO1, CO2, and CO3 subunits in Complex IV; and subunits 6 and 8 in Complex V.

demonstrate that a number of human neuromuscular diseases have structural abnormalities of mitochondria as their common feature[6,7] and these appear to be caused by mutations in mtDNA. In some cases, the diseases, like mtDNA itself, are maternally inherited. Maternal inheritance of mutated mtDNA was reported by our group in a family of chronic progressive external ophthalmoplegia[8] and in a family of Leber's hereditary optic neuropathy.[9] In others, mutations appear to rise spontaneously during the whole life of an individual.

Recent extensive studies have clarified at least three types of mtDNA mutations in human diseases (Fig. 2). Type A is homoplasmy of mutant mtDNA with point mutation within a single gene. A single base transition, G to A, converting a highly conserved arginine residue to a histidine in the ND5 gene for subunit 4 of Complex I in the electron transfer chain was found in patients with Leber's hereditary optic neuropathy.[9,10] Type B is heteroplasmy of mutant and normal mtDNAs. Southern blot analysis reveals coexistence of the normal-sized mtDNA and mutant mtDNA with a large deletion in patients with Kearns-Sayre syndrome.[8] Type C is pleioplasmy. In this type, multiple mutant mtDNAs with various deletions coexist with normal-sized mtDNA. Multiple deletions of mtDNA can be detected by the gene amplification method using polymerase chain reaction.[11]

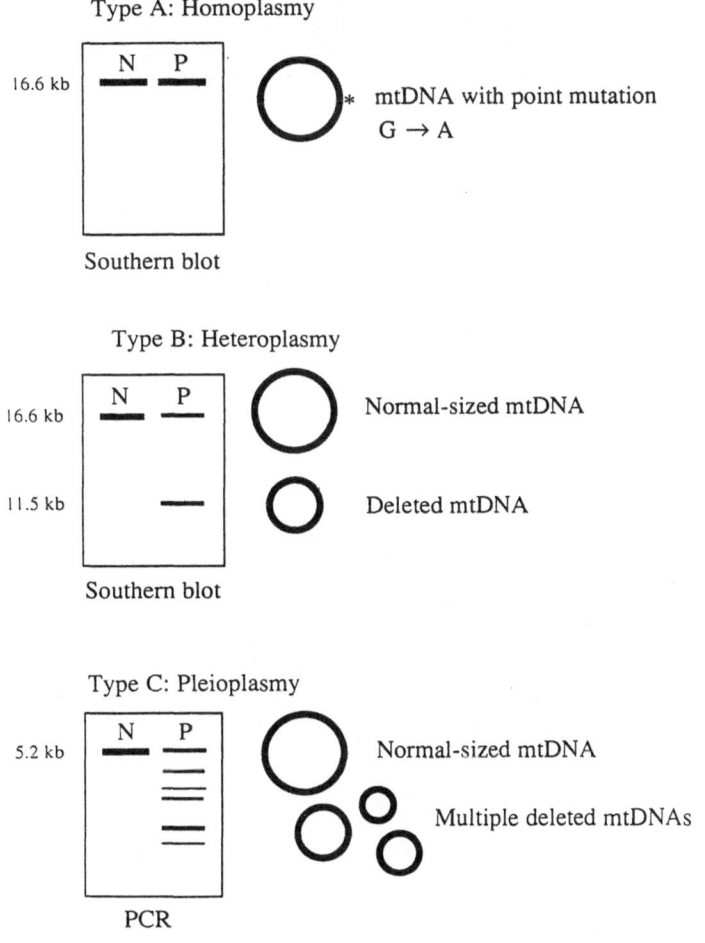

Fig. 2. Three types of mutations of mitochondrial DNA.

RESULTS AND DISCUSSION

Homoplasmy: Leber's Hereditary Optic Neuropathy

Fig. 3 shows the Southern blot analysis of mtDNA from a Japanese family with Leber's hereditary optic neuropathy.[9] Wallace an co-workers[10] have pointed out a base transition from G at nucleotide number 11,778 to A converts the highly conserved 340th arginine to histidine in the ND4 subunit of Complex I in patients with Lever's disease. The base transition can be detected as a site loss of a restriction enzyme, *Sfa*NI. In the proband (indicated with an arrow), a 1.6-bp DNA fragment resulting from the site loss was detected. But the mutation was not transmitted to his children, in whom two fragments of 0.92 and 0.68 kb were transmitted from their normal mother. The analysis shows the two nephews of the proband have the mutation specific for Leber's disease. Careful follow up of these two boys is recommended. Because this maternally inherited mutation was found both in the American pedigrees and in Japanese pedigrees, we could confirm that this mutation is responsible for the disorder.

Heteroplasmy: Kearns-Sayre Syndrome and CPEO

The second type of mtDNA mutation is heteroplasmic deletion.[8] Heteroplasmic existence of mutant mtDNA with a multi-gene deletion with the normal-sized mtDNA was found in 1/3 of patients with Kearns-Sayre syndrome (KSS) and chronic progressive external ophthalmoplegia (CPEO).[12-14] To investigate the mechanism of mtDNA deletion, we analyzed the crossover

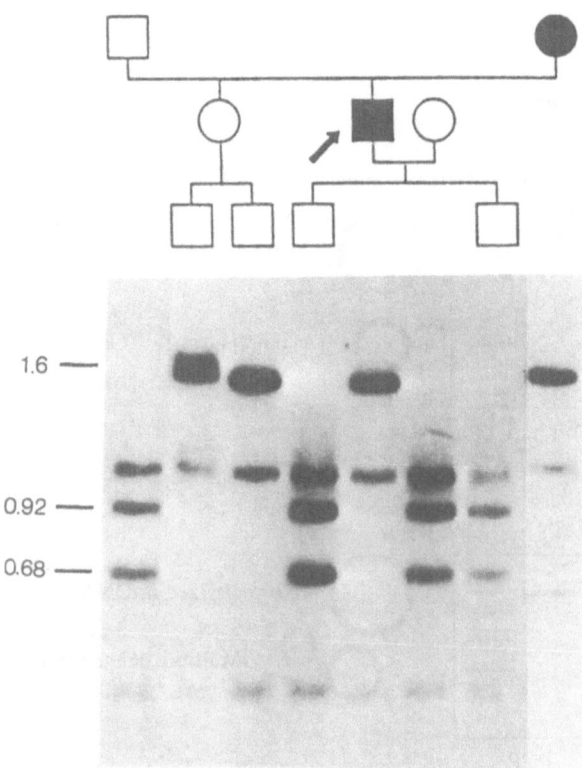

Fig. 3. Southern blot analysis of mitochondrial DNA in a Japanese family with Leber's optic neuropathy.

sequences of mtDNA in the skeletal muscles of five patients with mitochondrial myopathy.[15] We localized the deleted region using the combination of polymerase chain reaction and S_1 nuclease digestion.[16] Then, we directly sequenced the crossover regions of the deleted mitochondrial DNA without cloning[15] (Fig. 4).

In Patient 1, a 7-bp directly repeated sequence of 5'-ATCCCCA-3' was found at the boundaries of deleted segment spanning 7,039 bp between the ATPase6 and the cytochrome *b* genes. In Patients 2, a 13-bp sequence of 5'-ACCTCCCTCACCA-3' was found in the boundaries of deleted segment spanning 4,977 bp between the ATPase8 and the ND5 genes. In Patient 3, a 3-bp sequence of 5'-CCT-3' was found in the boundaries of deleted segment spanning 3,717 bp between the ATPase 6 and the ND5 genes. In Patients 4 and 5 with multiple deletions, one of the mutant mitochondrial DNA was amplified. Direct sequencing of the mutant mtDNA revealed the same directly repeated sequence as in Patient 2. These directly repeated sequences may contribute to mitochondrial DNA deletions in human degenerative diseases.

Therefore, we can distinguish two types of mtDNA deletions. Type 1 deletion results in frame shift at the junction of the two genes. In Type 2 deletion, there is no frame shift at the junction of the two genes. The sequences predict hybrid proteins. If they are translated, the effect of hybrid proteins on the molecular assembly of the oxidative phosphorylation complexes would be greater than that of the abnormally short proteins, because the hybrid proteins would affect two complexes at the same time. It should be examined whether the message for these abnormally short protein and hybrid proteins could be translated into proteins and whether they could disturb the molecular assembly of these energy-transducing complexes or not. We hypothesize that the hybrid protein of two genes perhaps capable of providing an abnormal mitogenic stimulus to mitochondria resulting in deregulated proliferation observed in CPEO.

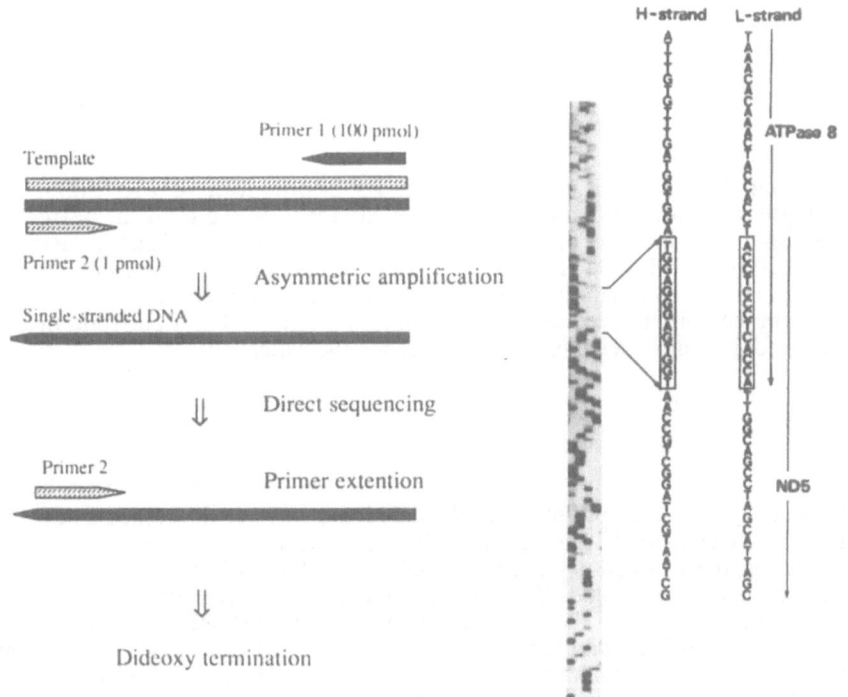

Fig. 4. Principle of PCR direct sequencing and determination of the directly repeated sequence involved in the deletion of mitochondrial DNA in a patient with chronic progressive external ophthalmoplegia.

Pleioplasmy: Multiple mtDNA Deletions in CPEO

The last type of mutation is observed in pleioplasmy.[11] In this type, multiple mutant mtDNAs with various deletions are present in the patient tissues. The population of each mutated mtDNA was so small that the deletions were often not detected by the Southern blot analysis but by the PCR amplification.

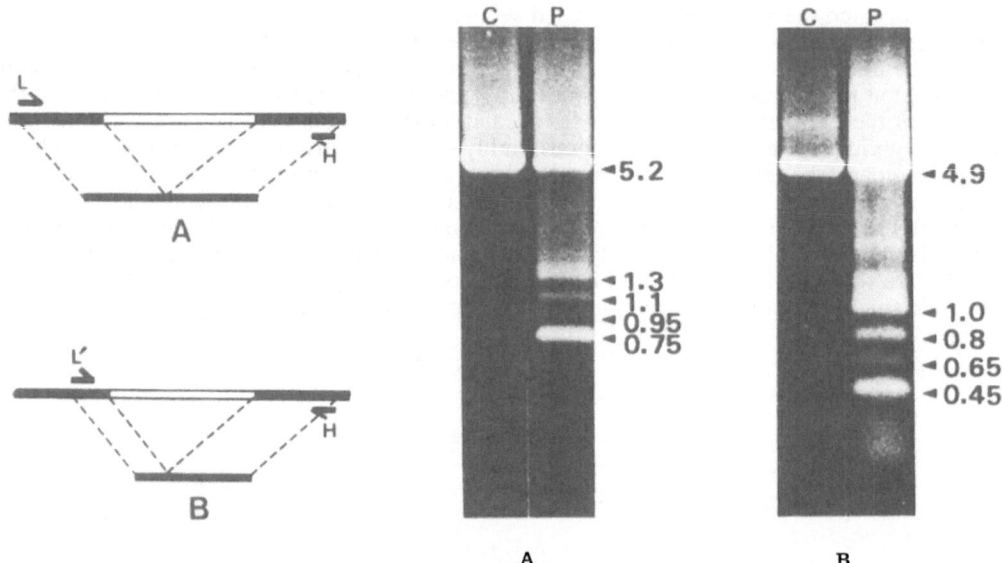

Fig. 5. Principle primer shift PCR and its application to the analysis of multiple deletions of mtDNA in the skeletal muscle of a patient with chronic progressive ophthalmoplegia. A: PCR amplification using primers L820 (L) and H1338 (H). B: PCR amplification using primers L853 (L') and H1338 (H). Sizes of amplified fragments are indicated in kb. A single band derived from the normal-sized mtDNA is seen in the control (C). Four additional bands derived from the deleted mtDNAs are seen in the patient (P).

For the analysis of pleioplasmic mtDNA deletions, we have developed a novel method, the primer shift PCR method.[11] The principle of the method is presented on the left side of Fig. 5. This method can detect small populations of deleted mtDNA, which are undetectable by the conventional Southern blot method. The presence of deleted mtDNAs can be confirmed, if we compare the shift in the sizes of the amplified fragments with the shift in the positions of the primers used for the amplification. We applied the primer shift PCR method for the analysis of skeletal muscle from a patient with familial ocular myopathy. When we used a pair of primers, the distance between which was 5.2-kb on the mitochondrial genome, multiple bands, representing at least four populations of differently deleted mtDNA, were detected in the patient muscle. When we used another pair of primers, the distance between which was 4.9 kb, the size of each band shifted by 0.3 kb as expected. In this patient, age 55, had external ophthalmoplegia from age 49. Multiple mtDNA deletions might be somatically accumulated as an accelerated ageing process.

Table 1. Characteristics of patients.

Patient		Age(year)/ Sex	Age of Onset	Family History	Clinical Diagnosis	Deletion*
1	SK	29 / M	11	–	KSS	Single
2	HA	40 / M	12	–	KSS	Single
3	SO	42 / F	27	–	CPEO	Single
4	YO	15 / F	12	–	CPEO	Single
5	TH	25 / M	21	–	CPEO	Single
6	KK	17 / M	14	–	CPEO	Single
7	KO	32 / F	14	+	CPEO	Multiple
8	KW	74 / M	45	+	KSS	Multiple
9	MS	55 / F	49	+	CPEO	Multiple
10	JW	44 / M	34	+	CPEO	Multiple
11	YO	46 / M	39	–	CPEO	Multiple
12	NT	63 / M	62	–	Myopathy	Multiple
13	MY	67 / F	66	–	CPEO	Multiple
14	TS	54 / F	44	+	CPEO	Multiple
15	MY	39 / F	32	–	CPEO	Multiple

KSS: Kearns-Sayre syndrome
CPEO: chronic progressive external ophthalmoplegia

Multiple mtDNA Deletions As a Cause of Late-Onset Mitochondrial Myopathy

In order to examine whether multiple deletions are involved in other cases of myopathy, we analyzed mitochondrial DNA from biopsied skeletal muscles of 15 patients with chronic progressive external ophthalmoplegia, Kearns-Sayre syndrome, and myopathy by using both the conventional Southern blot method and the gene amplification method (Table 1). In all the patients, either a single deletion or multiple deletions of mitochondrial DNA were detected.

The Southern blot analysis detected two populations of mtDNA, namely the normal-sized mtDNA and deleted mtDNA in Patients 1-7. In Patient 8, two populations of deleted mtDNA were detected. On the other hand, in Patients 9 to 12, no extra mtDNA fragment were observed. Patient 13, 14, and 15 were not examined by the Southern blot method.

In PCR experiments, we used two primer pairs; primers L820 (position 8,201-8,220) and H1338 (position 13,381-13,400), primers L596 (position 5,961-5980) and H12 (position 121-140). All mtDNAs from the patients were examined. Using primers L820 and H1338, we could amplify 5.2-kb fragments from the normal mtDNA. In Patients 7-15, we detected multiple extra fragments along with normal 5.2-kb fragment. In Patient 3, we found one 1.5 kb fragment surrounding the large deletion. The other patients and control detected only 5.2-kb fragment. Using primers L596 and H12, we amplified one fragment surrounding a large deletion or nothing in Patients 1-6. Normally 10.1-kb fragment from the normal mtDNA was too long to be amplified. In Patients 7-15, PCR detected small extra fragments indicating that they had multiple deletions.

The results have demonstrated the presence of two distinct modes of mtDNA deletions in patients with mitochondrial myopathy by the PCR amplification. In the first type, the muscle tissue contains a single population of deleted mtDNA (heteroplasmy). In the second type, the tissue contains multiple populations of deleted mtDNA (pleioplasmy). All the patients with CPEO or KSS analyzed in this study were found to be classified into either types of mtDNA deletions. We confirmed that all the patients with CPEO and KSS had mtDNA deletion.

The patients with the heteroplasmic-type mtDNA deletion were mean 16.2 years old (range 11-27) at the onset of clinical symptoms. The patients with pleioplasmic-type mtDNA deletions were mean 42.3 years (range 14-66) except Patient 7 who had family history, almost over 30 years, at the onset of clinical symptoms. This indicates that patients with multiple deletions affected late-onset type mitochondrial myopathy. Patient 7 had multiple mtDNA deletions and one of the deletions were increased to be detectable by the conventional Southern blot analysis. It suggested that Patient 7 was a transitory type between early-onset and late-onset types of mtDNA deletions, so she showed clinical symptoms before age 30 years with multiple mtDNA deletions.

In nine patients with multiple mtDNA deletions, five patients had family history in contrast to no family histories in patients with single mtDNA deletion. Autosomal dominant inheritance of multiple mtDNA deletions was reported in a family[17]. Patient 7 was the daughter from mother with a large deletion detected by conventional southern blot analysis.[8] Patients with multiple mtDNA deletions had only mild symptoms, it suggested that an unknown factor inducing mtDNA deletions were inherited to the next generation.

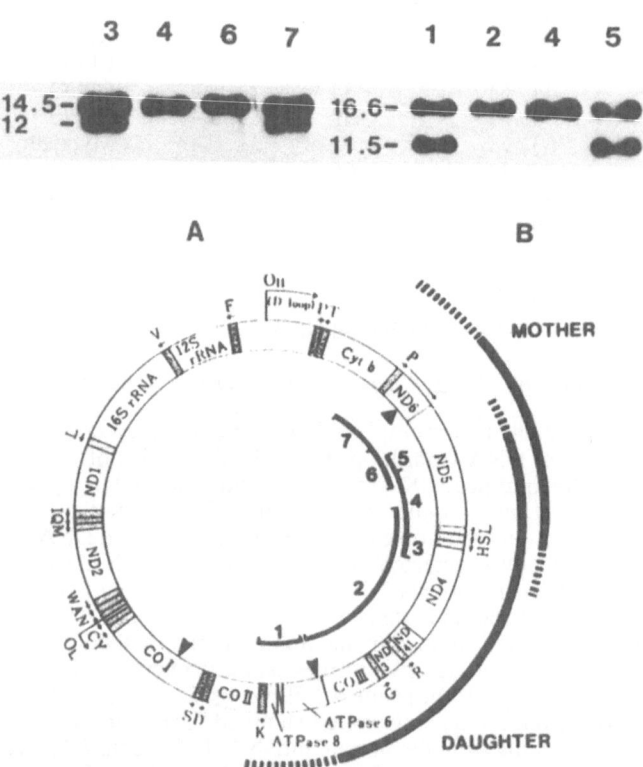

Fig. 6. Differently deleted mitochondrial DNA in a mother and her daughter with chronic progressive external ophthalmoplegia. The number in each lane in the upper panels shows the number of probes the position of which is indicated inside the circle in the lower panel. Upper panel A shows the Southern blot analysis of mitochondrial DNA from the skeletal muscle of the mother digested with *Pst*I. In addition to the normal-sized DNA (14.5 kb) mutant DNA of 12 kb hybridized to probes 3 and 7, but to neither probes 4 nor 6. Upper panel B shows the Southern blot analysis of mitochondrial DNA from the daughter digested with *Bam*HI. In addition to the normal-sized DNA (16.5 kb) mutant DNA of 11.5 kb hybridized to probes 1 and 5, but to neither probes 2 nor 4. Lower panel shows the physical map of mitochondrial DNA and the deleted regions of mitochondrial DNA in the mother and the daughter. Solid line, defined deleted region of mitochondrial DNA; and dotted line, undefined deleted region.

Maternal Transmission of Deleted mtDNA

In the previous study,[8] we observed mtDNA deletions in both a mother and daughter with CPEO; although the deletions overlapped substantially, they were not identical, as shown in Fig. 6. This observation is consistent with the possibility of somatic extensions of a smaller lesion, which is perhaps genetically transmitted from the mother to the daughter (Fig. 7).

Alternatively, we can speculate that pleioplasmic mtDNA deletions might be responsible for the difference in the location of the deletions between the mother and daughter. Therefore, we examined the mtDNAs from the mother and the daughter. Fig. 8A shows the PCR amplification, in which the primers were selected so that the deleted mtDNA from the daughter can be amplified, as indicated by the arrowhead. In the mother (M), in addition to the band of 0.5 kb that is identical with the band in the daughter, several bands were detected, suggesting that multiple mtDNA deletions was present in the mother. In Fig. 8B, primers were selected so that the mother's deleted mtDNA can be detected. In the mother (M), the deleted mtDNA was amplified as a 1.5-kb band. In the daughter (D), no corresponding band was detected, except a common 0.3-kb band. These results suggest that the mother had pleioplasmic mtDNA deletions and that one of the multiple deletions in the mother was maternally transmitted and increased its population in the daughter.

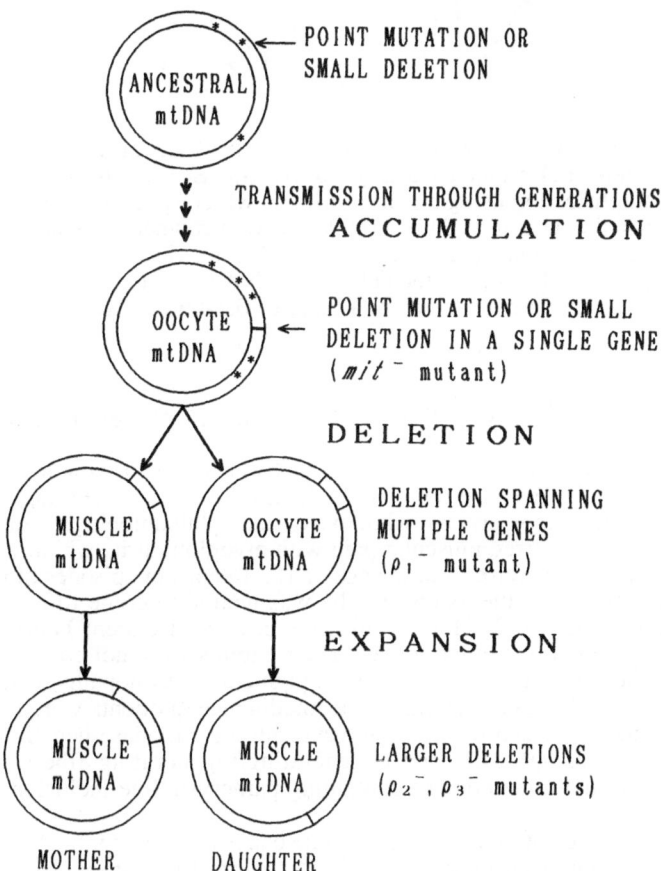

Fig. 7. A hypothesized mechanism of differently deleted mitochondrial DNA in the mother and the daughter with chronic progressive external ophthalmoplegia.

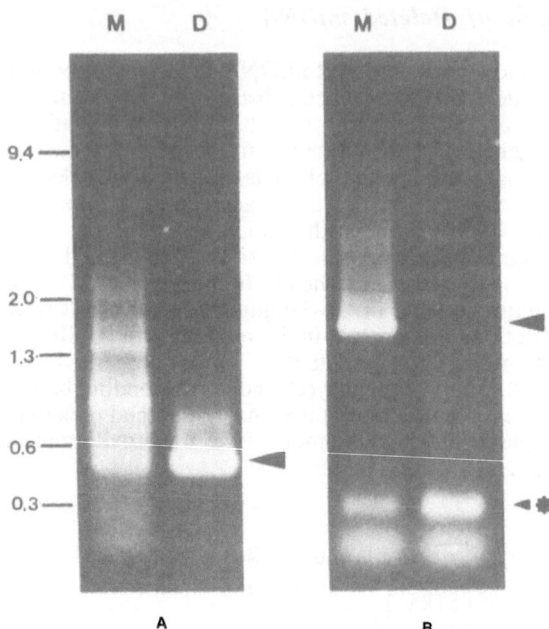

Fig. 8.   PCR analysis of the deletions in the mother and the daughter who had different deletions as analyzed by the Southern blot method. Panel A: PCR amplification using a pair of primers which specifically amplify the deleted mtDNA in the daughter (D) as a 0.5-kb band (arrowhead). Multiple bands were observed in the mother (M). Panel B: PCR amplification using a pair of primers which specifically amplify the deleted mtDNA in the mother (M) as a 1.6-kb band (arrowhead). No bands, except a band of 0.3-kb (*), were amplified in the daughter (D).

Multiple mtDNA Deletions As a Cause of Exertional Myoglobinuria

Moraes et al. described deletions of mitochondrial DNA in patients with progressive external ophthalmoplegia.[14] We have found multiple deletions of muscle mitochondrial DNA in patients with recurrent myoglobinuria but without external ophthalmoplegia. Patient 1 (26-year-old male) experienced generalized muscular pain with myoglobinuria, which was provoked by alcohol intake or strenuous exercise from the age of 18. Recurrent episodes of rhabdomyolysis wasted his skeletal muscles to the extent that he was unable to come to a standing position without using his arms. Patient 2 (21-year-old male, brother of Patient 1) had similar clinical features. Another brother was asymptomatic. Their parents were not consanguineous. Both patients showed neither external ophthalmoplegia, pigmentary retinopathy, cardiac conduction block, perceptive hearing loss, nor ataxia. Their intelligence quotients were full. The muscle biopsies revealed abundant ragged-red fibers and cytochrome-*c*-oxidase-negative fibers. Central migration of nuclei, fiber splitting, and degenerating/regenerating fibers were observed. Accumulation of abnormal mitochondria containing paracrystalline inclusions were found in electron microscopy.

Southern blot hybridization disclosed multiple bands of deleted mitochondrial DNAs and a 16.6-kb band of normal-sized mitochondrial DNA (Fig. 9). Although the blot indicated several common bands between the patients, a 7-kb band was dominant in Patient 1 and a 13-kb band was dominant in Patient 2. Gene amplification by the polymerase chain reaction method detected 2.1-kb and 1.4-kb fragments, indicating 3.8-kb and 3.5-kb deletions, in Patient 1, and 1.0-kb and 0.8-kb fragments, indicating 3.9-kb and 4.1-kb deletions, in Patient 2. Both experiments identified multiple populations of deleted mitochondrial DNA. In contrast to the multiple deletions starting at the D-loop region in autosomal dominant cases described by Zeviani et al.[17], the deletions detected in our cases were present on the outside of the D-loop region.

Recurrent myoglobinuria is a fairly common disorder, and frequently induces acute renal failure. Defects in the enzymes involved in glycolysis, glycogenolysis, and lipid metabolism have been reported to cause exertional myoglobinuria.[18] The histochemical findings in our cases were completely different from those of the established forms of exertional myoglobinuria. Our study demonstrates that multiple mitochondrial DNA deletions, which lead to mitochondrial respiratory enzyme defects, provide a novel genetic entity in recurrent myoglobinuria.

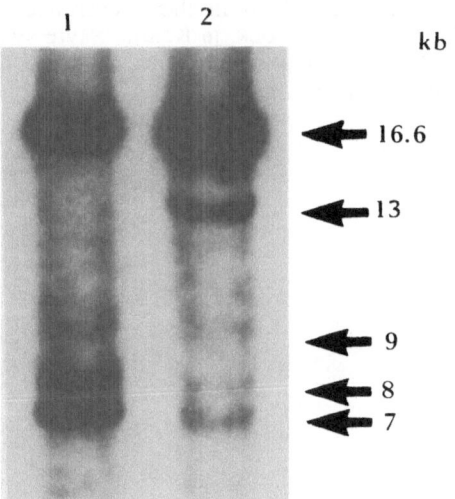

Fig. 9. Southern blot analysis of the skeletal muscle DNA from brothers with exertional myoglobinuria due to multiple deletions of mtDNA. Southern blotting detected several bands of deleted mtDNA in both patients along with the normal sized mtDNA (16.6 kb). It is demonstrated that the two brothers had different populations of deleted mtDNAs.

Large mitochondrial DNA deletions have been described in many cases of Kearns-Sayre syndrome and chronic progressive external ophthalmoplegia,[8,12-14] Recently, multiple populations of largely deleted mitochondrial DNAs have been described in familial cases of chronic progressive external ophthalmoplegia.[11,17] An important feature of the present study is that multiple large deletions of mitochondrial DNA were found in recurrent myoglobinuria, a myopathy without ophthalmoplegia. Exertional myoglobinuria is an skeletal muscle disorder that sometimes causing acute renal failure.

The presence of multiple mitochondrial DNA deletions in the present cases of recurrent myoglobinuria supports our recent hypothesis that the accumulation of mitochondrial DNA mutations is an important factor in several degenerative diseases.[5]

Cardiac Conduction Block

PCR technique can be used for clinical diagnosis of cardiac conduction block using only two milligrams of biopsied cardiac muscle.[19] Fig. 10 shows the PCR analysis of the mtDNA from the skeletal muscle (S) and the heart muscle (H) of a patient with heart block which necessitated an external pacemaker. The analysis demonstrated that the two tissues possessed different populations of deleted mtDNA.

The upper panel of Fig. 11 shows the cytochrome *c* oxidase activity staining of the longitudinal section of the skeletal muscle of a patient with chronic progressive external ophthalmoplegia. Cytochrome-*c*-oxidase positive portions and cytochrome-*c*-oxidase negative portions were segmentally distributed. The patient had a large deletion of mtDNA involving the gene for CO3 subunit of cytochrome *c* oxidase (Complex IV). This observation suggests that abnormal and normal mitochondria forms a mosaic pattern within a single fiber. As shown in the lower panel of Fig. 11, it is hypothesized that the presence of tissues with different electrophysiological properties is involved in the occurrence of conduction block and arrhythmias. Therefore, the conduction block, in Kearns-Sayre syndrome as well as in other cases with isolated conduction block, may be explained by segmental distribution of deleted mtDNA in the cardiac conduction system.

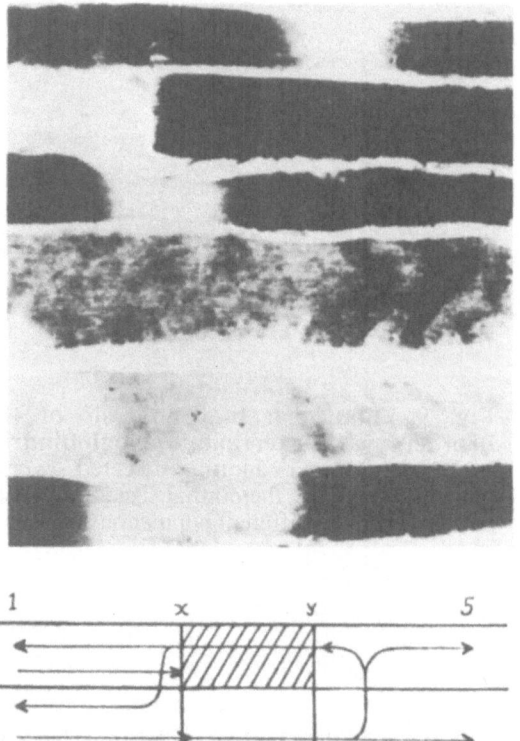

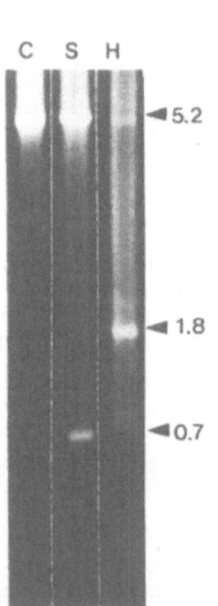

Fig. 10. PCR analysis of mtDNA deletions in the tissues of a patient with conduction block. C, control. S, skeletal muscle. H, heart muscle biopsied from a patient with atrioventricular conduction block.

Fig. 11. Segmental distribution of abnormal mitochondria as a cause of conduction block and arrhythmia. Upper panel, cytochrome *c* oxidase staining of longitudinal section of the skeletal muscle of a patient with chronic progressive external ophthalmoplegia due to mtDNA deletion. Lower panel, a diagram of reentry published by Schmitt and Erlanger.

Multiple mtDNA Deletions As a Cause of Cardiomyopathy

The cardiomyopathies are often of unknown etiology, and are recognized as a significant cause of morbidity and mortality. We previously described defects in subunits of mitochondrial respiratory chain Complex I (NADH-ubiquinone oxidoreductase) in two patients with hypertrophic cardiomyopathy associated with encephalomyopathy[7,20]. Deleted mitochondrial DNA (mtDNA) has been described in patients with chronic progressive external ophthalmoplegia[8] and Kearns-Sayre syndrome[13]. Recently, we demonstrated multiple deletions of mtDNA in a patient with external ophthalmoplegia[11] and in patients with recurrent myoglobinuria[7] by using the polymerase chain reaction (PCR). We proposed a hypothesis that accumulation of mtDNA mutations is an important contributor in several degenerative diseases[8]. To examine the possibility that some cases of cardiomyopathies are caused by multiple mtDNA deletions, we have studied the autopsied cardiac tissues from five patients with cardiomyopathies.

Patient 1, a 47-year-old female, had hypertrophic cardiomyopathy and died from cardiac failure. Patient 2, a 47-year-old male, had familial conduction block and developed dilated cardiomyopathy. His mother and sister had atrioventricular block. Patient 3, a 53-year-old male, had hypertrophic cardiomyopathy and died from cerebral infarction. Patient 4, a 40-year-old female, had dilated cardiomyopathy that was found after delivery of her third child. Patient 5, a 59-year-old male, had diabetes mellitus and died from cardiac infarction.

Total DNA was extracted from the cardiac tissues. In the Southern blot analysis, no deletions of mtDNA could be detected in any of the patients. By the PCR method, multiple deletions of mtDNA were detected in Patients 1-3, but were not detected in Patients 4, 5, nor in

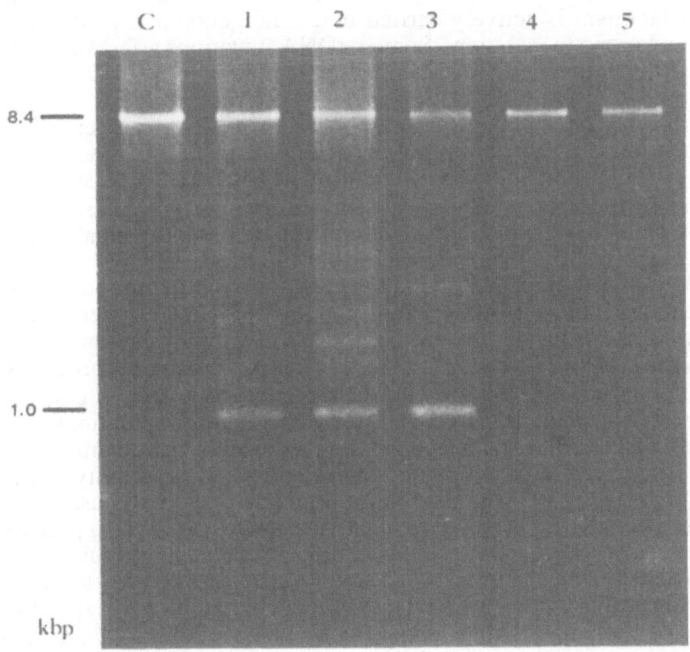

Fig. 12. PCR amplification of cardiac muscle mtDNA in cardiomyopathy. Lane C, controls; lanes 1-5, Patients 1-5. Primers used were L853 (position 8,531 to 8,550, 5'-ACGAAAATCTGTTCGCTTCA-3') and H38 (position 400 to 381, 5'-AAATTTGAAATCTGGTTAGG-3'). Amplified fragments were detected by the ethidium bromide fluorescent staining of the gel.

the controls (Fig. 12). The dominant population of mtDNA with the 7.4-kb deletion in Patients 1-3 was amplified as a 1.0-kb fragment by using two primers, the distance between them being 8.4 kb. We further confirmed the multiple mtDNA deletions by using the primer shift PCR method[11], in which the shift in sizes of amplified fragments is compared with the shift in positions of primers used for amplification.

In this experiment, we have found multiple deletions of mtDNA, which could not be detected by the conventional Southern blot analysis. By using the PCR, we could detect multiple deletions of mtDNA in three of the five patients with cardiomyopathies. This suggests that multiple mtDNA deletions are an important genetic defect, not only in mitochondrial myopathies but also in cardiomyopathies. We propose "mitochondrial cardiomyopathy due to multiple mtDNA deletions" as a novel disease entity among cardiomyopathies.

Müller-Höcker[21] demonstrated that the number of cytochrome-*c*-oxidase deficient cardiomyocytes increases with age. Because cytochrome-c-oxidase negative fibers are a common finding in the skeletal muscle of patients with mtDNA deletions, the increase in the number of cytochrome-*c*-oxidase deficient cardiomyocytes with age may be related to the increase in the population of deleted mtDNA.

PERSPECTIVE

We have proposed a hypothesis that accumulation of mutations of mtDNA in somatic cells, which results in decreased bioenergetic capacity, is an important contributor to several degenerative diseases and ageing process.[5] Four fundamental facts support the hypothesis. Firstly, the small, beautiful and essential genome, mtDNA, is situated inside the mitochondria where oxygen metabolism is actively carried out. The genome is continuously exposed to oxygen radicals, and prone to mutation. Some mtDNA mutations affect bioenergetic functions. Secondly, because the human mtDNA has no introns, mutations of mtDNA directly affect the function of the oxidative phosphorylation system. Thirdly, mtDNA is situated inside the mitochondria where oxygen metabolism is actively carried out. The genome is continuously exposed to oxygen radicals, and prone to mutation. Fourthly, there are multiple copies of mtDNA in each cell. Accumulation of mtDNA mutations results in a state of coexistence of normal mtDNA and various mutated mtDNAs, which is called pleioplasmy.[11] Segregation of mutated mtDNA among cells results in mosaicism of bioenergetically normal cells and bioenergetically defective cells.

Among the various mutations that are accumulated in the mitochondrial genome, we propose that deletions of mtDNA play the most important role in degenerative disorders and in decreased function of the brain and the skeletal and cardiac muscles in senescence. Exposure of mtDNA to oxygen radicals results in formation of oxidized products, such as 8-hydroxydeoxyguanosine.[22] Deletion of mtDNA may occur as the consequence of such oxidative damage of the mitochondrial genome. The small populations of multiple point mutations resulted from oxidative damage are hard to be analyzed quantitatively. In contrast, small populations of deleted mtDNAs can be detected at a high sensitivity and can be analyzed semiquantitatively using the PCR method. Further study must be focused on the analysis of small populations of deleted mtDNA in degenerative disorders and ageing process.

Degeneration of neurons is the most remarkable changes in senile dementia of Alzheimer-type (SDAT) and Parkinson's disease. Accumulation of a toxic substance in the mitochondria has been postulated as the cause of degeneration of the specific neurons. For example, 1-methyl-4-phenylpyridinium ion (MPP$^+$), a potent inhibitor of mitochondrial NADH-ubiquinone oxidoreductase (Complex I), induces degeneration of dopaminergic neurons in the substantia nigra and causes experimental parkinsonism. However, the mechanisms responsible for the degeneration of specific neurons in each type of degenerative diseases of the central nervous system have not been fully elucidated. We have recently reported that the amount of Complex I subunits was markedly decreased in the striatum of the patients with Parkinson's disease.[23] The changes in the amount of the enzyme protein could result from mutations of mtDNA. It is hypothesized that the accumulation of mtDNA mutations may be accelerated by an unknown toxic substance.

ACKNOWLEDGMENTS

We thank Dr. Ikuya Nonaka for histochemistry of the muscle tissue. This work was supported in part by the Grants-in-Aid for General Scientific Research (62570128) to M.T. and for Scientific Research on Priority Areas (Bioenergetics, 63617002) to T.O. from the Ministry of Education, Science and Culture, Japan and by Grant 88-02-39 from National Center for Nervous, Mental and Muscular Disorders of the Ministry of Health and Welfare, Japan to T.O.

REFERENCES

1. Anderson, S., Bankier, A.T., Barrell, B.G., de Bruijn, M.H.L., Coulson, A.R., Drouin, J., Eperon, I.C., Nierlich, D.P., Roe, B.A., Sanger, F., Schreier, P.H., Smith, A.J.H., Staden, R., and Young, I.G. (1981) *Nature* **290**, 457-465.

2. Chomyn, A., Mariottini, P., Cleeter, M.W.J., Ragan, C.I., Matsuno-Yagi, A., Hatefi, Y., Doolittle, R.F., and Attardi, G. (1985) *Nature* **314**, 592-597.

3. Giles, R.E., Blanc, H., Cann, H.M., and Wallace, D.C. (1980) *Proc. Natl. Acad. Sci. USA* **77**, 6715-6719.

4. Egger, J., and Wilson, J. (1983) *N. Engl. J. Med.* **309**, 142-146.

5. Linnane, A.W., Marzuki, S., Ozawa, T., and Tanaka, M. (1989) *Lancet* **i**, 642-645.

6. Tanaka, M., Miyabayashi, S., Nishikimi, M., Suzuki, H., Shimomura, Y., Ito, K., Narisawa, K., Tada, M., and Ozawa, T. (1988) *Pediatr. Res.* **24**, 447-457.

7. Yoneda, M., Tanaka, M., Nishikimi, M., Suzuki, H., Tanaka, K., Nishizawa, M. Atsumi, T., Ohama, E., Horai, S., Ikuta, F., Miyatake, T.,and Ozawa, T. (1989) *J. Neurol. Sci.* **92**, 143-158, 1989

8. Ozawa, T., Yoneda, M., Tanaka, M., Ohno, K., Sato, W., Suzuki, H., Nishikimi, M., Yamamoto, M., Nonaka, I., and Horai, S. (1988) *Biochem. Biophys. Res. Commun.* **154**, 1240-1247.

9. Yoneda, M., Tsuji, S., Yamauchi, T., Inuzuka, T., Miyatake, T. and Ozawa, T. (1989) *Lancet* **i**, 1076-1077.

10. Wallace, D.C., Singh, G., Lott, M.T., Hodge, J.A., Shurr, T.G., Lezza, A.M.S., Elsas II, L.J., and Nikoskelainen, E.K. (1989) *Science* **242**, 1427-1430.

11. Sato, W., Tanaka, M., Ohno, K., Yamamoto, T., Takada, G. and Ozawa, T. (1989) *Biochem. Biophys. Res. Commun.* **162**, 664-72.

12. Holt, I.J., Harding, A.E., Morgan-Hughes, J.A. (1988) *Nature* **331**, 717-19.

13. Zeviani, M., Moraes, C.T., DiMauro, S., Nakase, H., Bonilla, E., Schon, E.A., and Rowland, L.P. (1988) *Neurology* **38**, 1339-1346.

14. Moraes, C.T., DiMauro, S., Zeviani, M., et al. (1989) *N. Engl. J . Med .* **320**, 1293-1299.

15. Tanaka, M., Sato, W., Ohno, K., Yamamoto, T., and Ozawa, T. (1989) *Biochem. Biophys. Res. Commun.* **164**, 156-163.

16. Tanaka-Yamamoto, T., Tanaka, M., Ohno, K., Sato, W., Horai, S., and Ozawa, T. (1989) *Biochim. Biophys. Acta* **1009**, 151-155.

17. Zeviani, M., Servidei, S., Gellera, C., Bertini, E., DiMauro, S., and DiDonato, S. (1989) *Nature* **339**, 309-311.

18. Rowland, L.P. (1984) In: Rowland LP ed. Merritt's textbook of neurology. 7th ed. Philadelphia, Lea and Febiger, 585-590.

19. Sato, W., Tanaka, M., Sugiyama, S., Hattori, K., Ito, T., Kawaguchi, H., Onozuka, H., Yasuda, H., Ito, K., Tashiro, K., Ohno, K., Yamamoto, T., Takada, G., and Ozawa, T. (submitted)

20. Tanaka, M., Nishikimi, M., Suzuki, H., Ozawa, T., Nishizawa, M., Tanaka, K., and Miyatake, T. (1986) *Biochem. Biophys. Res. Commun.* **140**, 88-93.

21. Müller-Höcker, J. (1989) *Am J Pathol* **134**, 1167-1173.

22. Richter, C., Park, J.-W., and Ames, B.N. (1988) *Proc. Natl. Acad. Sci. USA* **85**, 6465-6467.

23. Mizuno, Y., Ohta, S., Tanaka, M., Takamiya, S., Suzuki, K., Sato, T., Oya, H., Ozawa, T., and Kagawa, Y. (1989) *Biochem. Biophys. Res. Commun.* **163**, 1450-1455.

MITOCHONDRIAL MYOPATHIES:

MORPHOLOGICAL APPROACH TO MOLECULAR ABNORMALITIES

Takeshi Sato, Shinji Nakamura*, Hiroko Hirawake,
Etsuko Uchida, Yasunori Ishigaki, Koichi Seki,
Ryo Kobayashi**, Satoshi Horai*** and Takayuki
Ozawa****

Departments of Neurology, *Pathology & **Forensic
Medicine, Juntendo University School of Medicine
Tokyo, ***National Institute of Genetics, Mishima
and ****Department of Biomedical Chemistry, Nagoya
University, Nagoya, Japan

SUMMARY

Cytochrome c oxidase (CCO) activity was shown by biochemical and histochemical examination to be decreased in the skeletal muscles of twelve
patients with mitochondrial myopathies, especially in 8 chronic progress-
sive external ophthalmoplegia (CPEO) cases, which included 2 Kearns-Sayre
syndrome and 6 ocular myopathy patients. In 4 MELAS patients, NADH
cytochrome c reductase activity was decreased. Immunocytochemical examination, using anti-CCO, anti-complex I and III rabbit sera revealed
that CCO was stained more weakly in the muscle fibers of one of the CPEO
patients than in those of the control.

Immuno-electron microscopic examination of CCO, complex I and III,
using a method of gold labeling, was also performed. Extensive labeling
by gold particles, representing the localization of respiratory enzymes,
could be seen in close vicinity to the cristae and inner mitochondrial
membrane of normal shaped mitochondria. The concentration of gold particles was markedly decreased in one of the CPEO patients.

To detect the localization of mitochondrial DNA or mRNA, in situ
hybridization was performed on human biopsied muscles using a ^{35}S labeled
mitochondrial DNA probe. The wide distribution of autoradiographic grains
for mRNA over the sarcoplasm of all muscle fibers was correlated with the
distribution of immuno-stained mitochondria.

Southern blotting revealed large deletions of mitochondrial DNA in
six of the patients with CPEO. In one of these patients, in situ
hybridization showed a marked decrease in density of autoradiographic
grains in muscle fibers.

INTRODUCTION

    Mitochondrial myopathies[1] are defined either by the presence of "ragged-red fibers" in biopsied muscle specimens stained with modified Gomori trichrome stain or by biochemical abnormalities.

    The clinical features of patients with mitochondrial myopathies are varied and include generalized muscle weakness, external ophthalmoplegia, peripheral neuropathy and symptoms of the central nervous system, such as deafness, mental retardation, seizures, ataxia, stroke-like episodes and retinal pigmentation [2,3]. Some patients also have cardiac conduction defects. The age of onset varies from childhood to adulthood. Some mitochondrial myopathies show only external ophthalmoplegia with or without generalized muscle weakness.

    Mitochondrial myopathies associated with various central nervous system symptoms have been reported as mitochondrial encephalomyopathies. In adult cases, there are three main types of mitochondrial encephalomyopathies.

    (1) Kearns-Sayre syndrome is characterized by onset before the age of 15 and progressive external ophthalmoplegia, pigment retinopathy, and one or more of the following: heart block; increased protein concentration in the cerebrospinal fluid; ataxia; deafness; optic nerve atrophy; seizure; and peripheral neuropathy. Many incomplete forms of this syndrome are present and some of these are manifest only in chronic progressive external ophthalmoplegia (CPEO) without generalized muscular weakness[4].

    (2) Mitochondrial encephalomyopahty with lactic acidosis and stroke-like episode (MELAS) [5] is characterized by episodic vomiting; seizures; proximal limb weakness; and recurrent stroke-like symptoms, such as hemiparesis, hemianopsia, or cortical blindness. The onset of MELAS in most patients occurs between 10 and 20 years of age.

    (3) Myoclonus epilepsy with ragged-red fibers (MERRF) [6] is dominated by myoclonus epilepsy, ataxia and limb weakness. The neuropathological findings of MERRF indicate multisystemic involvement, including the pallido-lysian system; spinal cord lesions, resembling Friedreich ataxia; and degeneration of the substantia nigra, cerebellar cortex, inferior olivary nucleus, locus ceruleus, gracile and cuneate nucleus and pontine tegmentum[7].

    In a recent study, Holt et al.[8] reported that deletions of mitochondrial DNA in biopsied muscle specimens were found in 9 of 25 patients with mitochondrial myopathies. Zeviani et al. [9] subsequently found deletions of mitochondrial DNA in seven patients with Kearns-Sayre syndrome. A more recent study by Moraes et al.[10], in which Southern blotting was performed on muscle DNA from 123 patients with different mitochondrial myopathies or encephalopathies, found deletions of mitochondrial DNA in 32 patients with CPEO. They concluded that deletions of muscle mitochondrial DNA are implicated in the symptoms of ophthalmoplegia.

    To determine whether any correlation could be determined between molecular abnormality and morphological abnormalities of respiratory enzymes, we performed an immuno-electron microscopic study and in situ hybridization of mitochondrial mRNA in biopsied muscle specimens in patients with mitochondrial encephalomyopathies.

MATERIALS AND METHODS

    Patients: This study consisted of 10 patients with CPEO; 7 patients with MELAS; 3 patients with MERRF; and 3 patients with other types of myopathies.

<u>Biochemistry</u>: Mitochondria were isolated form frozen muscle biopsied specimens by the method of Makinen and Lee[11]. Spectrophotometric assays were used to measure cytochrome c oxidase, succinate cytochrome c reductase, rotenon sensitive NADH cytochrome c reductase, complex II, and citrate synthase[12][14].

<u>Histochemistry</u>: After freezing the biopsied muscles immediately in liquid nitrogen serial cryostat sections 10 um thick were cut in order to demonstrate CCO activity using the method of Seligman et al.[15].

<u>Immuno-electron microscopy</u>: To observe the ultrastructural localization of complexes I and III and CCO in the mitochondrion, the muscle biopsied specimens were fixed with 4.0% paraformaldehyde and 0.1% glutaraldehyde in 0.1M phosphate buffer, pH 7.4. After being rinsed, the fixed specimens were dehydrated in graded N, N-dimethylformamide (DMF), then embedded in methyl methacrylate at -20° C and polymerized by ultraviolet light.

The ultrathin sections were mounted on nickel grids and labeled as follows: first, the grids were blocked with 2.0% normal goat serum and albumin in PBS for 5 min; second, the sections were incubated with antisera for 1.5 h at 37°C; third, after being washed in PBS,the incubated sections wer labeled with immunogold complex using colloidal gold labeled protein A (8nm) or anti-rabbit IgG goat serum (5nm, Janssen, Belgium) for 40 min at room temperature[16][18]. The sections were then washed again, and rinsed in distilled water. The sections were then fixed by OsO4 vapor, stained with uranyl acetate and observed by an Hitachi H-500 electron microscope. The density of gold particles in the mitochondria was measured by a two-dimensional imaging analyzer (Nikon).

<u>In situ hybridization</u>: Cryostat sections of frozen biopsied muscle specimen were fixed for 5 min in 4% paraformaldehyde. After being rinsed, sections were stored at -70°C until use. A cloned mitochondrial cDNA probe (bp No. 5274-6203) was subcloned into the pUC19 vector. The DNA was excised from pUC19 by using EcoRI and HindIII restriction endonuclease, purified by agarose gel electrophoresis, and labeled by a random priming method using 35S-dCTP and 35S-dATP. The sections were pretreated with 0.2N HCl and pepsin (1mg/ml). After being treated with glysine (2mg/ml), the sections were dehydrated.

Slides were hybridized in a humidified box at 45°C for 18 h with the ^{35}S labeled mitochondrial DNA probe. The hybridization buffer contained 0.6 M sodium chloride, 10 mM Tris HCl (pH 7.4), 0.02% Ficoll, 0.02% polyvinylpyrolidine, 0.02% bovine serum albumin, 1 mM EDTA, 250 µg/ml tRNA, 125 µg/ml salmon sperm DNA, 10% (wt/vol) dextran sulfate, and 40% (vol/vol) deionized formamide[20][21].

After hybridization, the sections were rinsed 4 times, 30 min each, in 50% formamide + 2xSSC (0.3M NaCl + 30mM sodium citrate), and 3 times at 37°C, 20 min in 2xSSC, followed by treatment with the single strand specific nuclease S-1 (Behringer, Indianapolis, IN) at a concentration of 25-500 units/ml for 30 min at 37°C. Sections were dehydrated in an alcohol series in which 0.3 M ammonium acetate replaced water. The sections were dipped in Konica NR-M2 emulsion, and the autoradiograms were exposed at 4°C for 10 days. The autoradiograms were developed in a Konica-Konidor developer and fixed in Fuji Fix; the sections were stained with hematoxylin.

Mitochondrial DNA: Mitochondrial DNA (mtDNA) of morph 1, typically found in Japanese [22], was prepared from placenta as decribed by Drouin et al.[23] with modifications.   Sixteen different small fragments of mtDNA obtained by digestion with appropriate restriction enzymes were cloned into pUC19 vectors (Horai, unpublished).

Southern blot analysis: Total DNAs (50 ng) extracted from the frozen biopsied muscles were digested with restriction enzymes (2-3 units), and the digests were electrophoretically separated in 0.75% agarose gel. After being denatured, the DNA was transferred onto Gene Screen Plus membranes (Du Pont-New England Nuclear). Hybridization was carried out under conditions recommended by the manufacturers using 32P-labeled probe DNA (2-3 x 10^5 cpm/ml). The membranes were subjected to autoradiography for 6-24 h at room temperature.   The human mitochondrial DNA and the mtDNA fragments excised from the recombinant pUC19 plasmids were labeled with [^{32}-P] cCTP (3000 Ci/mmol), using multiprime labeling systems (Amersham, England), then employed as probes.

A segment of the patient's mtDNA encompassing the deletions was amplified by polymerase chain reaction (PCR) [24], with pair of synthesied single stranded oligonucleotide primers corresponding to mtDNA sequences located upstream and downstream of the breakpoint. Thirty cycles of amplification were carried out with Taq polymerase (Perkin-Elmer Cetus); denaturation at 94°C for 15 sec; annealing at 45°C for 15 sec, and extension at 72°C for 120 sec in a computer-controlled alternating temperature water bath. A part of the preoducts was electrophoresed on 1% agarose gel stained with ethidium bromide.

RESULTS AND DISCUSSION

Biochemistry. Biochemical analysis of respiratory enzymes of extracted mitochondria from biopsied muscles in patients with CPEO, involving Kearns-Sayre syndrome and ocular myopathy, and MELAS are shown in Figure 1. In the CPEO patients, mainly CCO and succinate cytochrome c

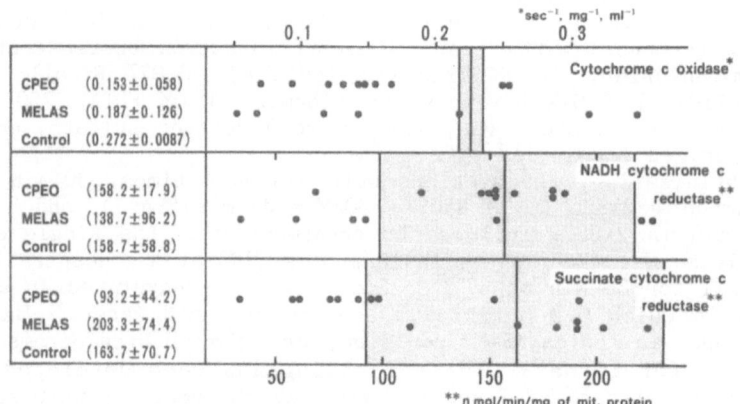

Fig. 1. Mitochondrial Enzyme Activity in Biopsied Muscles.

reductase activities are decreased. 4 out of the 7 patients with MELAS showed decreased NADH cytochrome c reductase activity. Numerous cases with CCO deficiency have been reported; their clinical features are varied . Partial deficiency of CCO revealed by histochemical examination has been reported in patients with CPEO[25] . There are, however, few studies which have reported deficiencies of CCO by biochemical analysis in patients with CPEO. But our biochemical results clearly demonstrated that deficiencies of CCO and succinate cytochrome c reductase are common in patients with CPEO.

Immuno-electron microscopy. Ultrastructural localization of mitochondrial enzymes was investigated by immuno-electron microscopy using colloidal-gold labeled antibodies. As shown in Fig. 2, extensive labeling by gold particles can been seen in close vicinity to the cristae and inner mitochondrial membrane of normal shaped-mitochondria, representing the localization of respiratory enzymes.

In patients with mitochondrial myopathies (CPEO and MELAS), accumulation of abnormal mitochondria containing paracrystalline inclusions was noted in the subsarcolemmal space of muscle fiber. Positively reacted gold particles were seen also in the cristae in these abnormal mitochondria. But the presence of only a few particles in the center of the paracrystalline structures suggests a lack of respiratory enzymes in these inclusions.

In one of the patients with CPEO, many abnormal giant mitochondria can be observed in the muscle fiber (Fig. 3). Gold particles are present in the inner mitochondrial membrane. However, only a few particles are noted in the matrix and paracrystalline inclusions.

The concentration of colloidal gold in the mitochondria was measured by a two-dimensional imaging analyzer. The rate of depletion of the density of gold particles by anti-CCO antibody was about 58% in the CPEO patient with a defect of this enzyme (Table 1). The results of our study indicate that the decrease in the amount of enzyme protein is directly related to the reduction in enzyme activity[26] . In our previous report, a

TABLE 1

Density of gold particles in mitochondria of a patient with CPEO

	Complex I	Complex III	Complex IV
Patient	46.8	219.2	58.2[*]
Control	78.8	372.0	122.1

(Number of gold particles/μm)

Activities of mit.respiratory enzymes

	ComplexI-III	ComplexII-III	ComplexIV
Patient	177	90	0.084[*]
Control	198.4±89.4	243.0±117.8	0.271±0.07

(n mol/min/mg of mit.protein)   (sec⁻¹,mg⁻¹,ml⁻¹)

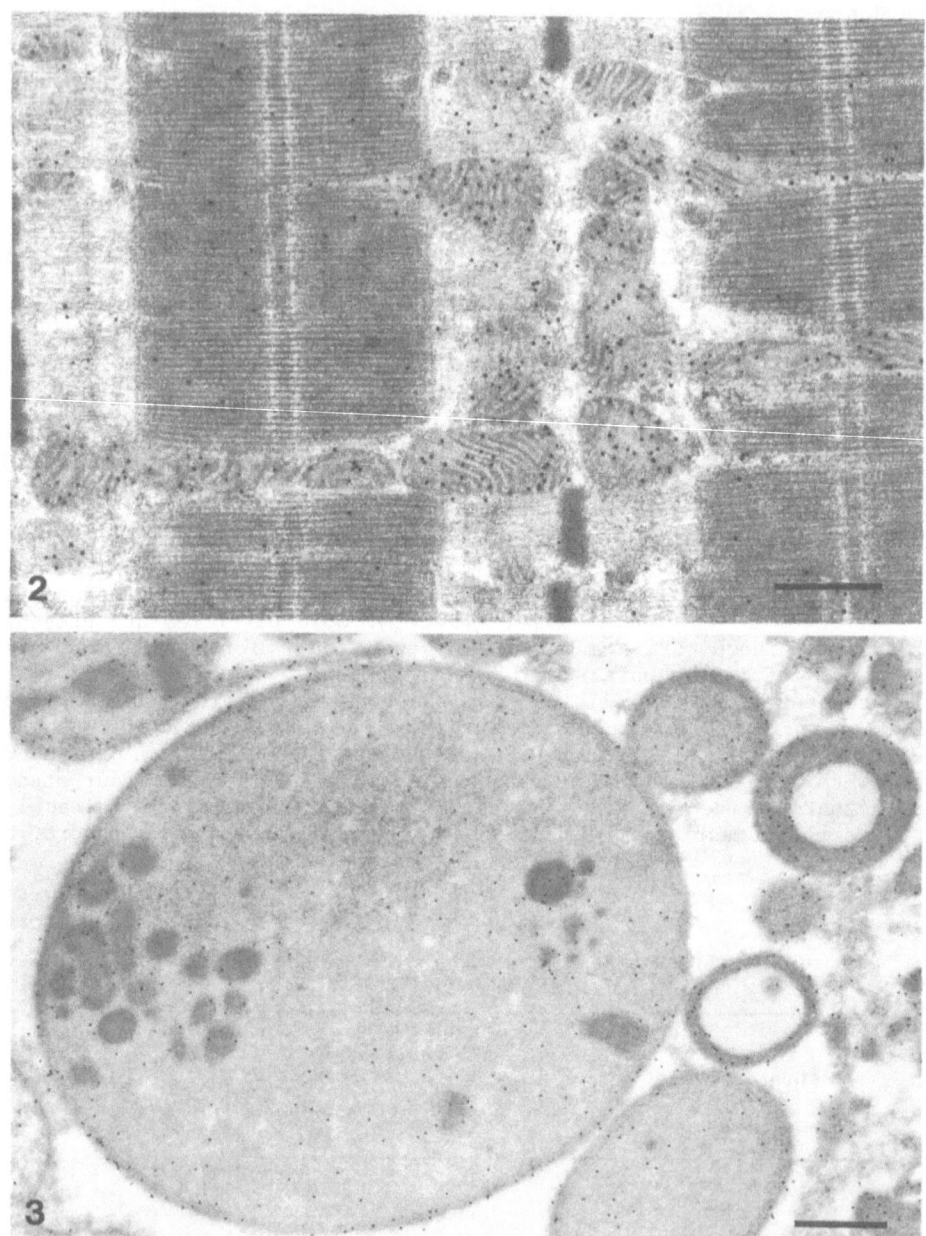

Fig. 2.   Electron micrograph of mitochondria in normal skeletal muscle, showing ultrastructural distribution of colloidal gold particles (8 nm) for complex III immunoreactivity. Bar = 0.5 μm.

Fig. 3. High magnification of giant mitochondria in a patient with CPEO, showing dense colloidal gold particles (5 nm) in the inner mitochondrial membrane. Matrix and paracrystalline inclusions are devoid of gold particles. Bar = 0.5 μm.

similar correlation was found between the density of gold particles and enzyme activity in Mottled mice, a murine model of the kinky hair syndrome, which is characterized by congenital defects in copper metabolism and a defect in CCO [27].

In situ hybridization. To detect the localization of mitochondrial DNA or mRNA, in situ hybridization was carried out in human biopsied muscles. The sections were hybridized by a 35S-labeled mitochondrial DNA probe.

To confirm the specificity of in situ hybridization, the localization of mitochondrial DNA was compared with the cellular localization of immuno-stained mitochondrial enzymes in human kidney tissue. Hybridization reactions showed a higher labeling of the tubules than of the glomerulus of the kidney. Immuno-staining of the mitochondrial enzymes, are clearly seen in the proximal and distal tubules indicate that the reaction in the glomerulus is weaker than that in the tubules. The

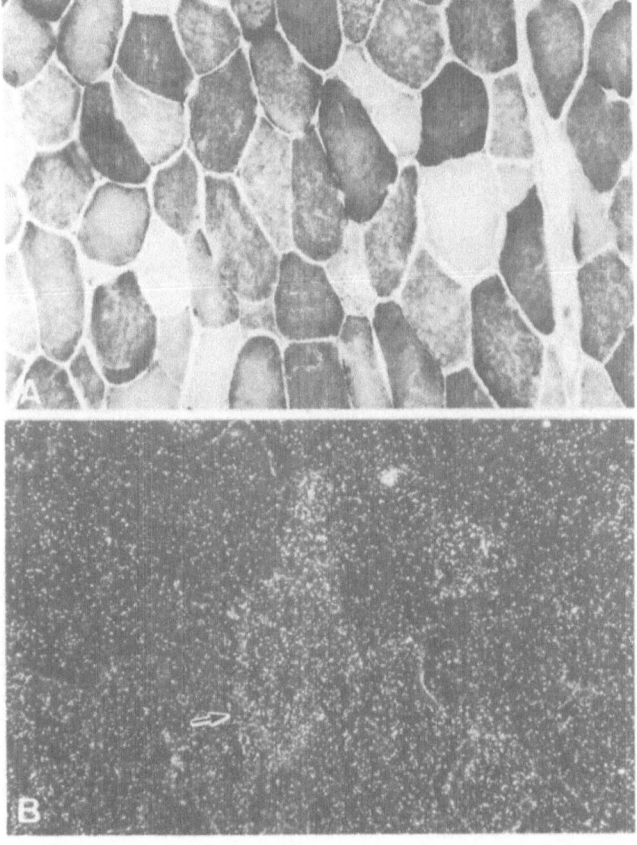

Fig. 4.   (A) Histochemical staining of cytochrome c oxidase activity shows scattered negative fibers in a patient with MELAS. (X 180)
(B)   Dark-field photomicrograph of an autoradiogram of serial section of 4A, in which mitochondrial mRNA are hybridized with a tritiated mitochondrial cDNA probe. Autoradiographic grains are widely distributed over the sarcoplasm and accumulation of grains can been seen in "ragged-red" fiber (arrow). (X 300)

TABLE 2

mit.DNA deletions in three mitochondrial myopathies

Groups	Moraes, 1989		Sato, 1989	
	No.	mtDNA deletions	No.	mtDNA deletions
1: CPEO				
KSS	18	15	3	2
Ocular myopathy	44	17	5	4
2: Defined mit.encephalomyopathies				
MERRF	2	0	1	0
MELAS	13	0	4	0
Leigh	9	0		
3: Undefined mit.encephalomyopathies				
Cong.lactic acidosis	6	0		
Others	8	0	3	0

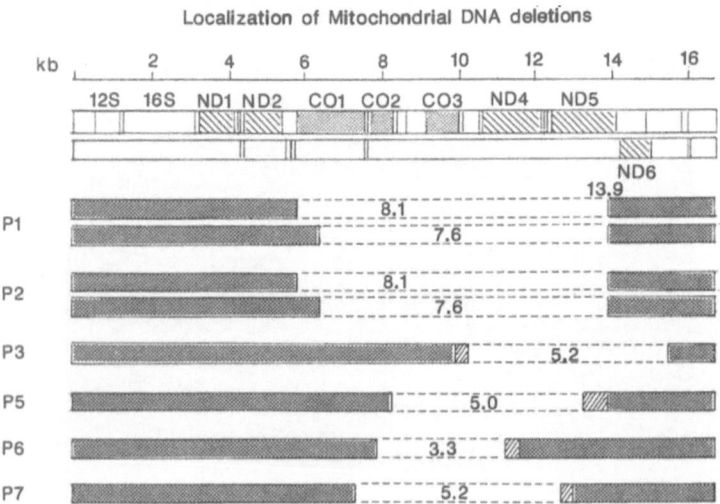

Localization of Mitochondrial DNA deletions

Fig. 5. The maps at the top represent the heavy (upper line) and light (lower line) strands of linealized mitochondrial DNA, showing the genes for the subunits of respiratory chain enzymes [10]. Muscle mitochondrial DNAs from 6 CPEO patients are depicted by the hatched regions.

similarity in cellular localization of reaction products between both methods comfirms that in situ hybridization clearly determines the localization of mitochondrial DNA or mRNA[21].

Hybridization reactions for mRNA of mitochondria were widely distributed over the sarcoplasm of all muscle fibers. These muscle fibers were stained normal by immunohistochemistry using anti-CCO antibody. However, in the patient with CPEO, hybridization reaction showed a marked decrease in density of autoradiografic grains in about 90% of all muscle fibers. This decreased reactivity of CCO, revealed by histochemical and immunohistochemical examination, corresponds to the decreased density of autoradiographic grains in muscle fibers. Our results suggest that mRNA processing from polycistronic RNA or the stability of mRNA may be affected. Further in situ hybridization is required using another probe encompassing the deleted region of mitochondrial DNA.

Deletions of mitochondrial DNA. Southern blot analysis was carried out on DNA extracted from the skeletal muscles from 8 patients with CPEO, 4 with MELAS, one with MERRF and 3 with other types of mitochondrial myopathies. Large deletions of mitochondrial DNA were found only in 2 patients with incomplete forms of Kearns-Sayre syndrome and 4 with ocular myopathy (Table 2). No large deletions were found in either MELAS, MERRF or any other myopathies. The results of Moraes's report[10] and our study indicate that similar large deletions are present only in patients with CPEO. These results suggest that each of the three kinds of mitochondrial myopathies may have a different pathogenesis.

Mitochondrial DNA deletions were found in patients with ocular myopathy who had no generalized muscular weakness, suggesting that the coexistence of mutant and normal mitochondrial DNA may not disturb the function of skeletal muscles in these patients.

The size of the deletions in the 6 CPEO patients ranged from 4.1 to 7.6 kb. Shon et al.[28] have reported that there are some common sites of deletions, "hot spots," in the mitochondrial genome. Our results also revealed that there are similar hot spots in the mitochondrial DNA of the 6 CPEO patients. Fig. 5 more clearly reveals the site of the DNA deletions by PCR in 5 of the 6 patients. Direct DNA sequencing of the affected genomes in our patients is necessary to clarify why there are large deletions of mitochondrial DNA.

Acknowledgments. We are indebted to Drs. Yutaka Oya, Shinzaburo Takamiya and Kiyoshi Kita (Department of Parasitology, Juntendo University) and Douglas Barton for their assistance and advice. This work was supported in part by a Grant-in-Aid for General Scientific Research (61480203) and for Scientific Research on Priority Areas from the Ministry of Education and by a Grant (62-2-05) from the National Center of Neurology and Psychiatry (NCNP) of the Ministry of Health and Welfare, Japan.

References

1. Y. Shapira, S. Harel and A. Russell, Mitochondrial encephalopathy: a group of neuromuscular disorders with defects in oxidative metabolism, Isr J Med Sci 13: 161 (1977).

2.   J. A. Morgan-Hughes, D. J. Hayes, J. B. Clark et al., Mitochondrial encephalomyopathies biochemical studies in two cases revealing defects in the respiratory chain, Brain 105:553 (1982).

3.   S. DiMauro, E. Bonilla, M. Zeviani et al., Mitochondrial myopathies, Ann Neurol 17: 521 (1985).

4.   D. A. Drachman, Ophthalmoplegia plus, the neurodegenerative disorders associated with progressive external ophthalmoplegia, Arch Neurol 18: 654 (1968).

5.   S. G. Pavlakis, P. C. Phillips, S. DiMauro et al., Mitochondrial myopathy, encephalopathy, lactic acidosis and stroke-like episodes (MELAS): a distinctive clinical syndrome, Ann Neurol 16: 481 (1984).

6.   N. Fukuhara, Myoclonus epilepsy and mitochondrial myopathy, in: "Mitochondrial Pathology in Muscle Diseases," G. Scarlato & C. Cerri eds., Piccin Medical Books, Padova (1983), pp88.

7.   S. Takeda, K. Wakabayashi, E. Ohama et al., Neuropathology of myoclonus epilepsy associated ragged-red fiber, Acta Neuropathol 75: 433 (1988).

8.   I. J. Holt, A. E. Harding and J. A. Morgan-Hughes, Deletions of mitochondrial DNA in patients with mitochondrial myopathies, Nature 331: 717 (1988).

9.   M. Zeviani, C. T. Moraes, S. DiMauro et al., Deletions of mitochondrial DNA in Kearns-Sayre syndrome, Neurology 38: 1339 (1988).

10. C. T. Moraes, S. DiMauro, M. Zeviani et al., Mitochondrial DNA deletions in progressive external ophthalmoplegia and Kearns-Sayre syndrome, N Eng J Med 320: 1293 (1989).

11. M. W. Makinen and C. Lee, Biochemical studies of skeletal muscle mitochondria. 1. Microanalysis of cytochrome content, oxidative and phoshorylative activities of mammalian skeletal muscle mitochondria, Arch Biochem Biophys 126: 75 (1968).

12. Y. Hatefi and J. S. Rieske, The preparation and properties of DPNH-cytochrome c reductase (complex I - III of the respiratory chain), Methods Enzymol 10: 225 (1967).

13. T. E. King, Preparation of succinate-cytochrome c reductase, Methods Enzymol 10: 216 (1967).

14. I. Sekuzu, Y. Orii and K. Ohnishi, Purification and assay of cytochromes, Protein, nucleic acid, enzyme (Tokyo) 10: 1610 (1965).

15. A. M. Seligman, M. J. Karnovsky, H. L. Wasserkrug et al., Non-droplet ultrastructural demonstartion of cytochrome oxidase activity with a polymerising osmiophilic reagent, diaminobenzidine, J Cell Biol 38:1 (1968).

16. J. Roth, M. Bendayan and L. Orci, Ultrastructural localization of intracellular antigens by the use of protein A-gold complex, J Histochem Cytochem 26: 1074 (1978).

17. S. Kobayashi, T. Uchida, T. Ohaishi et al., Met-Enkephalin-Arg-Gly--Leu-like immunoreactivity in adrenal chromaffin cells and carotid body chief cells of the dog and monkey, Biomedical Recearch 4: 201 (1983).

18. T. Sato, M. Anno, S. Nakamura et al., Colloidal-gold labeling of electron transport enzymes in mitochondrial myopathy for immuno-electron microscopy, Shinkei Kenkyu No Shinpo 31: 646 (1987).

20. J. V. Priestley, M. A. Hynes, V. K. M. Han et al., In situ hybridization using 32P labelled oligodeoxyribonucleotides for the cellular localisation of mRNA in neuronal and endocrine tissue, Histochem 89: 467 (1988).

21. T.Sato, S.Nakmura, Y. Ishigaki et al., Defect in cytochrome c oxidase

mRNA in chronic progressive external ophthalmoplegia: Detection by in situ hybridization, Ann Neurol 26: 145 (abstruct), (1989).

22. S. Horai and E. Matsunaga, Mitochondrial DNA polymorphism in Japanese, Hum Genet 72: 105 (1986).

23. J. Drouin, Cloning of human mitochondrial DNA in Escherichia coli, J Mol Biol 140: 15 (1980).

24. R. K. Saiki, S. Scharf, F. Faloona et al., Enzymatic amplification of beta - globin genomic sequences and restriction site analysis for diagnosis of sickle cell anemia, Science 230: 1350 (1985).

25. M. A. Johnson, D. M. Turnbull, D. J. Dick et al., A partial deficiency of cytochrome c oxidase in chronic progressive external ophthalmoplegia, J Neurol Sciences 60: 31 (1983).

26. D. L. Song, T. Sato, H. Ujike et al., Ultrastructural immunocytochemical localization of electron-transport enzymes in mitochondrial myopathy, Clin Neurol (Japan) 29: 405 (1989).

27. K. Seki, T. Sato, Y. Ishigaki et al., Decreased activity of cytochrome c oxidase in the macular mottled mouse: an immuno-electron microscopic study, Acta Neuropathol 77: 465 (1989).

28. E. A. Schon, R. Rizzuto, C. T. Moraes et al., A direct repeat is a hotspot for large-scale deletion of human mitochondrial DNA, Science 244: 346 (1989).

S₁ NUCLEASE ANALYSIS AND DIRECT SEQUENCING OF DELETED

MITOCHONDRIAL DNA IN MYOPATHIC PATIENTS:

ROLE OF DIRECTLY REPEATED SEQUENCES IN DELETION

Masashi Tanaka, Wataru Sato, Kinji Ohno, Tomoko Yamamoto, and Takayuki Ozawa

Department of Biomedical Chemistry, Faculty of Medicine
University of Nagoya, Nagoya 466, Japan

SUMMARY

Heteroplasmy of the normal-sized and the deleted mitochondrial DNA (mtDNA) has been observed in mitochondrial myopathy. To investigate the mechanism of mtDNA deletion, we analyzed the crossover sequences of mtDNA in the skeletal muscles of five patients with mitochondrial myopathy. We localized the deleted region using the combination of polymerase chain reaction and S₁ nuclease digestion. Then, we directly sequenced the crossover regions of the deleted mitochondrial DNA without cloning. In Patient 1, a 7-bp directly repeated sequence of 5'-ATCCCCA-3' was found at the boundaries of deleted segment spanning 7,039 bp between the ATPase 6 and the cytochrome *b* genes. In Patients 2, a 13-bp sequence of 5'-ACCTCCCTC ACCA-3' was found in the boundaries of deleted segment spanning 4,977 bp between the ATPase 8 and the ND5 genes. In Patient 3, a 3-bp sequence of 5'-CCT-3' was found in the boundaries of deleted segment spanning 3,717 bp between the ATPase 6 and the ND5 genes. In Patients 4 and 5 with multiple deletions, one of the mutant mitochondrial DNA was amplified. Direct sequencing of the mutant mtDNA revealed the same directly repeated sequence as in Patient 2. These directly repeated sequences may contribute to mitochondrial DNA deletions in human degenerative diseases.

INTRODUCTION

Human mitochondrial genome is a closed circular DNA of 16,569 bp,[1] which encodes for thirteen subunits of the oxidative phosphorylation complexes and for two rRNAs and 22 tRNAs of the mitochondrial protein synthesizing system.[2] Mitochondrial DNA (mtDNA) is exclusively maternally inherited,[3] and is reported to be highly prone to deleterious mutations.[4] It is proposed that mtDNA mutation is an important contributor to aging and degenerative diseases.[5]

Deletions of mtDNA have been reported in various human diseases, *i.e.*, mitochondrial myopathy,[6] Kearns-Sayre syndrome,[7,8] familial chronic progressive external ophthalmoplegia,[9] encephalomyopathy,[10] and Pearson's marrow/pancreas syndrome.[11] Mapping of the deletions in a number of patients[6,8,9] revealed apparently identical deletions in several patients, suggesting nonrandom occurrence of deletions. There may be "hot spots" on the mtDNA that are more prone to deletion than other regions of mtDNA.

In order to elucidate the molecular genetic basis of the hot spots of human mtDNA deletion, it is necessary to localize and to determine the nucleotide sequence the putative hot spots in the mtDNA among many patients with mitochondrial myopathy. For this purpose, we developed a method, which utilizes the combination of the polymerase chain reaction (PCR) and the S_1 nuclease analysis.[13] This PCR plus S_1 method enabled us to choose primers for sequencing of the boundaries of the deletions. Then we have applied the asymmetric PCR amplification method[14] to direct sequencing of the deleted mtDNA from patients with myopathy.[15]

EXPERIMENTAL PROCEDURES

Patients. Five patients, Patient 1: a 21-year-old male, Patient 2: a 17-year-old female, Patient 3: a 42-year-old female, Patient 4: a 32-year-old female, and Patient 5: a 39-year-old female, had external ophthalmoplegia but had no pigmentary retinopathy. The Southern blot analysis of mtDNA from Patient 4 was reported previously.[9]

Preparation of DNA. Total DNA was extracted from 5 mg of the frozen muscles as described previously.[15]

Synthesis of Primers. Primers for PCR were synthesized using a Shimadzu model NS-1 DNA synthesizer and an Applied Biosystems model 380B DNA synthesizer and purified on NENSORB Cartridges from Du Pont-NEN. The base sequences of the oligonucleotides are shown in Table 1.

Primary PCR Amplification. PCR amplification was carried out on 1 μl of the DNA solution (ca. 10 ng of total DNA) in a final volume of 100 μl which included 200 μM of each dNTP, 2.5 units of *Taq* DNA polymerase (Ampli*Taq*, Cetus) and PCR buffer (50 mM Tris-HCl, pH 8.3, containing 50 mM KCl, 1.5 mM MgCl2, and 0.01% gelatin) with 1 μM each of primers.[16] The reactions were carried out for a total of 35 cycles, with the use of a Perkin-Elmer/Cetus Thermal Cycler. The cycle time were as follows: denaturation 15 sec at 94°C; annealing, 15 sec at 45°C; and primer extension, 80 sec at 72°C.

Heteroduplex formation. Two mtDNA fragments (500 ng each) amplified with PCR were mixed and precipitated with ethanol. The pellet was dissolved in 100 μl of 10 mM Tris-HCl (pH 7.5), 7 mM MgCl2 , 60 mM NaCl and denatured at 95°C for 10 min. For heteroduplex formation the samples were cooled down to 25°C at a speed of 1°C/sec in the Thermal Cycler.[13]

TABLE I

Synthesized primers used for PCR

Primer*	Sequence 5'→ 3'	Complementary site**
L 568	CAAACACTTAGTTAACAGCT	5,681 to 5,700
L 790	TGAACCTACGAGTACACCGA	7,901 to 7,920
L 820	TTCATGCCCATCGTCCTAGA	8,201 to 8 220
L 853	ACGAAAATCTGTTCGCTTCA	8,531 to 8,550
L 909	ACACTTATCATCTTCACAAT	9,091 to 9,110
L 962	GCATCAGGAGTATCAATCAC	9,621 to 9,640
L1108	ATAACATTCACAGCCACAGA	11,081 to 1,100
L1451	CTATTAAACCCATATAACCT	14,511 to 4,530
H 60	AAACATTTTCAGTGTATTGC	620 to 601
H1136	TATCTTTACTATAAAAGCTA	11,380 to 11,361
H1363	CAGGTCAACCTCGCTTCCCC	13,650 to 13,631
H1338	TCTTGTTCATTGTTAAGGTT	13,400 to 13,381
H1479	GGAGGTCGATGAATGAGTGG	14,810 to 14,791
H1619	ACTTGCTTGTAAGCATGGG	16,209 to 16,191

*Primers L568 - L1619 are used for amplification of the light strand of mtDNA and H60 - H1619 are used for amplification of the heavy strand of mtDNA.

**Numbering of mtDNA is according to Anderson et al. [1]

S_1 Nuclease Digestion. The mixture after heteroduplex formation was precipitated with ethanol. The pellet was dissolved in 20 μl of the S_1 nuclease reaction mixture[17] containing 50 mM sodium acetate, 1 mM zinc acetate, 250 mM NaCl, 50 μg/ml bovine serum albumin and 5 units of S_1 nuclease from Takara Shuzo was added. The mixture was incubated at 37°C for 20 min. The reaction was stopped by addition of EDTA to a final concentration of 5 mM. The digested fragments were separated by electrophoresis on 2% agarose gels and detected fluorographically after staining with ethidium bromide.[13]

Asymmetric PCR Amplification. PCR reamplification was carried out on 2.5 μl of the primary PCR product in a final volume of 100 μl which included the reagents described above with 0.01 μM of one primer and 1 μM of another primer essentially according to the method of Gyllensten and Erlich.[14] PCR was performed for a total of 35 cycles as above. The PCR product containing single-stranded DNA was precipitated with polyethylene glycol to remove dNTP and primers.[18] This step is essential for the reduction of the background in the subsequent sequencing.[15]

Direct DNA Sequencing. DNA was sequenced by the Sanger's dideoxynucleotide chain termination method[19] using the incorporation of α-[32P]dCTP as the radiolabeling extension method as described previously.[15]

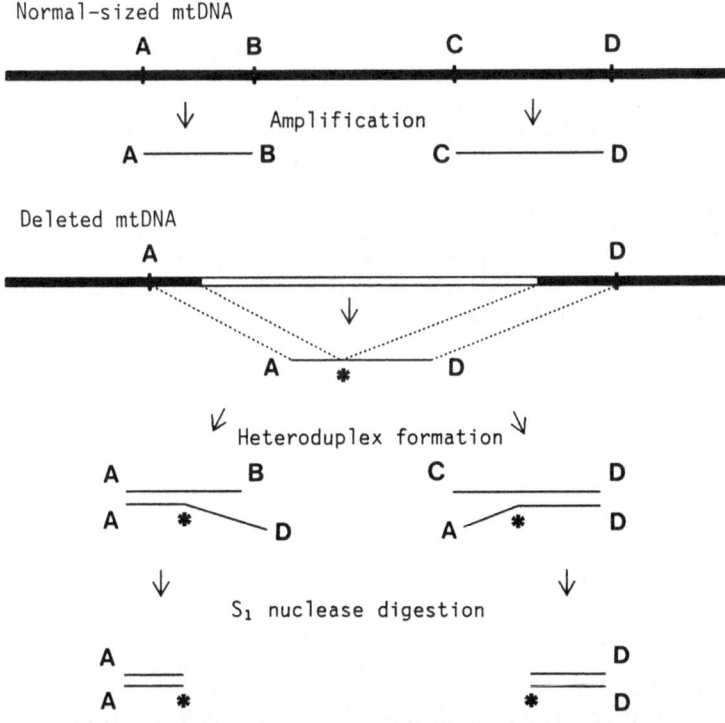

Fig. 1. **Schematic presentation of the PCR plus S_1 method. Open box indicates the region of the deletion in mutant DNA.** Three fragments A-B, C-D, and A-D are amplified with each pair of the appropriate primers. The asterisk indicates the crossover point of the deletion. The fragments A-B and A-D are mixed and a heteroduplex is formed. Only the part from point A to the asterisk is complementary and is protected against S_1 nuclease digestion. The fragments C-D and A-D are subjected to the same procedure for defining the other end of the deletion.

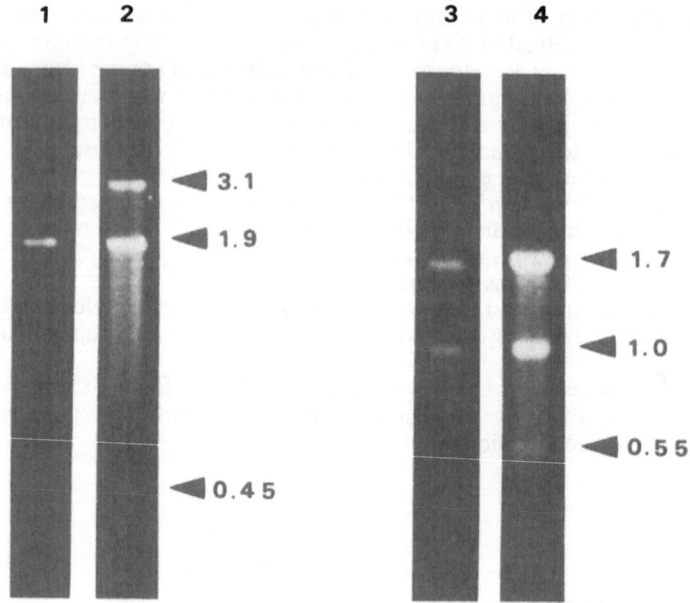

Fig. 2. **Electrophoretic patterns of PCR-amplified mtDNA fragments after heteroduplex formation and S_1 nuclease digestion.** Two bands in lane 1 indicate fragments A-B (3.1 kb) and A-D (1.9 kb), amplified using the pairs of primers L820 and H1136 and of primers L820 and H60, respectively. A band with the size of 0.45 kb appeared after S_1 nuclease digestion. The size of the band indicates the distance from the position of primer L820 to the beginning point of the deletion. Fragments A-B (1.7 kb) and C-D (1.0 kb) in lane 3 are amplified using the pairs of primers L1451 and H1619 and of primers L820 and H1619, respectively. The band with the size of 0.55 kb is shown in lane 4. The size indicates the distance form the end point of the deletion to the position of primer H1619.

RESULTS

PCR Plus S_1 Analysis of Deleted mtDNA

PCR Amplification Using Various Primers. In order to examine the region of deletion by PCR, we tried to amplify fragments from the deleted mtDNA using various pairs of primers surrounding the deleted region. The analysis of the mtDNA deletion in Patient 1 is dipicted here. Using the primers L820 and H60, the distance between which is 9.0 kb, we amplified a fragment of 2.0 kb, consistent with a 7.0 kb deletion. Similarly, using the primers L568 and H1619, the distance between which is 10.5 kb, we amplified a fragment of 3.5 kb, consistent with a 7.0 kb deletion. However, no fragments were amplified using other primers, namely L962, L1108, H1479, or H1338, suggesting that the deleted mtDNA in this patient lost the complementary sites of these primers. These results indicate that the size of deletion was approximately 7 kb and that it extended from at least position 9,621 to position 14,810 with lower and upper limits of 8,201 to 16,209.

Principle of PCR Plus S_1 Analysis. We further analyzed the deleted region using the PCR plus S_1 method, the principle of which is schematically shown in Fig. 1. Fragment A-B which includes the starting point of the deletion is amplified from the normal-sized mtDNA and fragment A-D which includes both ends of the deletion is amplified from the mutant mtDNA. These fragments are mixed and subjected to heteroduplex formation. The complementary region of the heteroduplex formed between fragments A-D and A-B as well as the homoduplexes formed from the self-reannealing of each fragment are protected against S_1 nuclease. The size of the complementary region of the heteroduplex indicates the distance between point A and the starting point of the deletion (asterisks in the figure) in the mutant mtDNA. Similarly, the end point of the deletion can be determined by S_1 nuclease analysis of the heteroduplex formed from fragments C-D and A-D.

PCR Plus S_1 Analysis of Deleted mtDNA. Fig. 2 shows the electrophoretic patterns of fragments before and after the S_1 nuclease digestion. For determination of the starting point of the deletion, a fragment of 3.2 kb (position 8,201-11,380) which was amplified from the normal-sized mtDNA using the primers L820 and H1136 was mixed with a fragment of 1.9 kb which was amplified from the deleted mtDNA using the primers L820 and H60 (Fig. 2, lane 1). Then these fragments were subjected to heteroduplex formation. It is essential to cool down the mixture quickly in order to obtain a sufficient amount of heteroduplexes. After the S_1 nuclease digestion, a fragment of 0.45 ± 0.05 kb appeared (lane 2), assuming the error in estimation of the fragment size to be less than 10%. Since the primer L820 starts at position 8,201, the start point of the deletion should be located at position $8,650 \pm 50$ bp within the ATPase subunit 6 gene. To determine the end point of the deletion, a 1.7 kb fragment (position 14,511-16,209) which was amplified from the normal-sized mtDNA using the primers of L1451 and H1619 was mixed with a 1.0 kb fragment which was amplified from the deleted mtDNA using the primers of L820 and H1619 (Fig. 2, lane 3). A fragment of 0.55 ± 0.06 kb appeared after the S_1 nuclease digestion (lane 4). Since the primer H1619 starts at position 16,209, the end point of the deletion should be located at position $15,660 \pm 60$ within the cytochrome *b* gene.

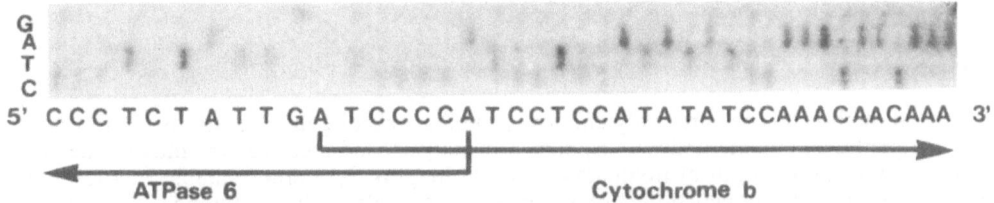

Fig. 3. **PCR amplification and direct sequencing of the deleted mtDNA from the skeletal muscle of Patient 1.** Shown is a portion of an autoradiograph of sequencing gel of the amplified DNA from the deleted mtDNA.

Direct Sequencing of Deleted mtDNA

Crossover sequence in Patient 1. According to the result of the PCR plus S_1 analysis, sequencing primer L853 was chosen for the direct sequencing of the deleted mtDNA. The crossover sequence was found to be a directly repeated sequence containing four cytosines (5'-ATCCCCA-3'), flanked by AT-rich regions (Fig. 3). The directly repeated sequence was located in the boundaries of the deletion between the ATPase 6 gene and the cytochrome *b* gene.

When the wild-type sequences of the ATPase 6 and the cytochrome *b* genes were compared each other without adding a gap, almost no sequence homologies were obtained except in the crossover sequence. But when they were compared with addition of a gap, additional homologies were found in 11 of 12 bases adjacent to the 3'-side of the directly repeated sequence. Therefore, the deletion spanned 7,039 bp, probably starting from position 8,624 at the 3'-side of the directly repeated sequence within the ATPase 6 gene, and ends at position 15,662 at the 3'-end of the directly repeated sequence within the cytochrome *b* gene (Fig. 4).

Crossover sequence in Patient 2. The deleted mtDNA of Patient 2 was amplified by the PCR method using primers L790 and H1479, the distance between which was 6.9 kb. A single band of 1.9 kb was amplified consistently with a 5.0-kb deletion. Direct sequencing using primer H1363 revealed that the crossover sequence was found to be a 13-bp directly repeated sequence of 5'-ACCTCCCTCACCA-3' (Fig. 4), which was located in the boundaries of the deletion between the ATPase 8 gene and the ND5 gene. The deletion spanned 4,977 bp from position 8,483 to position 13,459. The directly repeated sequence was flanked by AT-rich regions. When the wild-type sequences of the ATPase 8 and the ND5 genes were compared each other, no significant homologies were found in the region surrounding the directly repeated sequence.

Crossover sequence in Patient 3. The deleted mtDNA of Patient 3 was amplified by the PCR method using primers L820 and H1338, the distance between which was 5.2 kb. A single band of 1.5 kb was amplified consistently with a 3.7-kb deletion. Direct sequencing using primer L909 revealed that the crossover sequence was a 3-bp directly repeated sequence of 5'-CCT-3' (Fig. 4), which was located in the boundaries of the deletion between the ATPase 6 gene and the ND5 gene. The deletion spanned 3,717 bp from position 9,192 to position 12,908. When the wild-type sequences of the ATPase 6 and the ND5 genes were compared each other, homologies were found in the regions adjacent to the directly repeated sequence.

Crossover sequence in patients with multiple mtDNA deletions. In Patient 4, the Southern blot analysis showed apparently a single mtDNA deletion (6), but the PCR method detected multiple mtDNA deletions. In Patient 5, multiple deletions, namely multi-types of mutated mtDNAs, were also detected by the PCR method. We selectively amplified a mutated mtDNA of relatively large population in Patients 4 and 5 with the same primers used for Patient 2. Direct sequencing of the PCR product demonstrated that the deletion was identical with the 4,977-bp deletion found in Patient 2.

DISCUSSION

PCR method has been recently employed in diagnosis of genetic diseases. In the present study, we combined this method with S_1 nuclease analysis, and localized the mtDNA deletions in patients with mitochondrial myopathy. The PCR plus S_1 method enabled us to choose appropriate primers to sequence the crossover region directly without cloning. The crossover sequences were determined in five patients in this study.

The PCR plus S_1 method has the following advantages over the Southern blot method. 1) This method requires no radioisotopes, and can be performed in ordinary clinical laboratories. 2) The new method is fast. Using the Southern blot method, it takes at least a couple of days to determine the deleted region. Using the new method, we can complete the analysis of the deleted region within 5 h: 2 h for amplification, 1 h for heteroduplex formation, 20 min for S_1 nuclease digestion, and 1 h for electrophoresis and detection. 3) This method is so sensitive that total DNA extracted from only 5 mg of muscle tissue is sufficient for the determination of the deleted region of mtDNA. Therefore, this method is of value especially when only a small amount of clinically biopsied sample is available. 4) This method is accurate. The deleted region can be determined within ±60 bp. Because the experimental error in estimation of electrophoresed fragments depends on the sizes of fragments obtained after S_1 nuclease digestion, the accuracy in determination of the deleted regions can be improved by choosing appropriate pairs of primers so that smaller fragments are obtained after S_1 nuclease digestion. 5) The present method is simple, because we can selectively and directly amplify fragments from the deleted mtDNA or from the normal-sized mtDNA without separating the two populations.

We directly sequenced the deleted mtDNAs from patients by using the asymmetric PCR method. The results have demonstrated that three different types of mtDNA deletions occur

Type 1 deletion   Patient 1: 7,039-bp deletion

8,624
↓

Wild-type sequence
ATPase 6 gene   5'-CCCCCTCTATTG**ATCCCCA**-<u>CCTCCAAATATCTCA</u>-3'
                              * *   * ******* ****** *** *   *
Cytochrome *b* gene   5'-<u>ATCCTAGCAATA</u>**ATCCCCA**TCCTCCATATACCAAA-3'
                                           ↑
15,662

Deleted mtDNA sequence

hybrid gene 5'-CCC CCT CTA TTG **ATC CCC A**TC CTC CAT ATA TCC AAA-3'
hybrid protein -Pro Pro Leu Leu Ile Pro Ile Leu His Met Ser Lys-

Type 2 deletion   Patients 2, 4, and 5: 4,977-bp deletion

8,483
↓

Wild-type sequence
ATPase 8 gene   5'-AACTACCACCT**ACCTCCCTCACCA**<u>AAAGCCCATAAA</u>-3'
                           * *   *   * *************   * *   *
ND5 gene   5'-<u>CTCTCACTTCA</u>**ACCTCCCTCACCA**TTGGCAGCCTAG-3'
                                        ↑
13,459

Deleted mtDNA sequence

hybrid gene 5'-AAC TAC CAC CT**A CCT CCC TCA CCA** TTG GCA GCC TAG-3'
hybrid protein -Asn Tyr His Leu Pro Pro Ser Pro Leu Ala Ala **stop**

Type 3 deletion   Patient 3: 3,717-bp deletion

9,192
↓

Wild-type sequence
ATPase 6 gene   5'-CTTCTAGTAAGCCTCTA**CCT**<u>GCACGACAACACATAA</u>-3'
                             *** * *   *   *** ***   **** ***   *
ND5 gene   5'-<u>GCCTTAGCATGATTTAT</u>**CCT**ACACTCCAACTCATGA-3'
                                     ↑
12,908

Deleted mtDNA sequence

hybrid gene 5'-CTT CTA GTA AGC CTC TA**C CTA** CAC TCC AAC TCA TGA-3'
hybrid protein -Leu Leu Val Ser Leu Tyr Leu His Ser Asn Ser Trp-

Fig. 4. Comparison of the nucleotide sequences of the wild-type genes with those of the hybrid genes in the deleted mtDNAs from Patients 1-5. The sequence of the wild-type genes are shown above the sequence of the hybrid gene resulted from deletion of mtDNA in each patient. Bold letters indicate the directly repeated sequences. Underlined is the excised sequence in mtDNA deletion in each patient. Numerals indicate the nucleotide numbers (1).

between the directly repeated sequences present in the wild-type mtDNA (Figs. 3 and 4). The first type of deletion, a 7,038-bp deletion between the ATPase 6 gene and the cytochrome *b* gene, was found only in Patient 1 out of 8 patients with large deletions studied in our laboratory. We found the 7-bp directly repeated sequence, 5'-ATCCCCA-3', in the boundaries of this deletion (Fig. 3). In contrast, the second type of deletion, a 4,977-bp deletion between the ATPase 8 gene and the ND5 gene, was found in Patients 2, 3, and 4. The 13-bp directly repeated sequence, 5'-ACCTCCCTCACCA-3' was found in the boundaries of this deletion (Fig. 4). The same deletion was reported in a patient with Pearson's marrow/pancreas syndrome by Rotig et al.[12] and in 19 out of 29 patients with Kearns-Sayre syndrome and progressive external ophthalmoplegia by Schon et al.[20] The third type of deletion, a 3,717-bp deletion between the ATPase 6 gene and the ND5 gene, was found only in Patient 5 in our laboratory. The 3-bp directly repeated sequence, 5'-CCT-3' was found in the boundaries of this deletion (Fig. 4). Because additional homologies were found adjacent to the directly repeated sequences in Patients 1 and 5, the deletions might be promoted by not only the short directly repeated sequences but also by the homologous sequences surrounding them.

As shown in Fig. 4, the sequence of the mutant mtDNA with the first type of deletion forms an open reading frame which predicts a 12-kDa hybrid protein composed of 32 amino acid residues from the N-terminal side of the ATPase subunit 6 and of 75 amino acid residues from the C-terminal side of the cytochrome *b* protein. The second type of deletion introduces a stop codon 12 nucleotides aside from the boundary, resulting in a premature termination of the ND5 protein. Therefore, the sequence of the mutant mtDNA predicts a 5-kDa abnormal protein composed of 42 amino acid residues from the N-terminal side of the ATPase subunit 8 and of 3 amino acid residues resulted from a frame shift in the ND5 gene. The sequence of the mutant mtDNA with the third type of deletion forms an open reading frame which predicts a 69-kDa hybrid protein composed of 222 amino acid residues from the N-terminal side of the ATPase subunit 6 and of 412 amino acid residues from the C-terminal side of the ND5 protein. Because the ATPase subunits 6 and 8 in Complex V, the ND5 protein in Complex I, and the cytochrome *b* protein in Complex III play central roles in energy transducing function of each complex, it should be examined whether the message for these hybrid proteins could be translated into proteins and whether they could disturb the molecular assembly of these energy-transducing complexes or not.

Human mtDNA is very small and economically packed; the rRNA, tRNA, and the protein-coding genes are organized in the mtDNA continuously with no non-coding sequences, except for the D-loop region involved in the regulation of the replication and translation of the mtDNA.[21] Therefore, any mutation in the mtDNA will involve a functionally important part of the genome. In the present study, three types of directly repeated sequences responsible for the deletions were found to be located within the protein-coding genes.[22] In contrast with the recombination of nuclear genes which occurs mainly between non-coding regions, recombination of mtDNA will occur inevitably between the protein-coding genes. The recombinational events in mtDNA, which we could designate as pseudo-recombination, might result in formation of hybrid genes and thereby affect the mitochondrial bioenergetic function.

We have recently described multiple deletions of mtDNA in a patient with familial ocular myopathy.[16] In the present paper, we found multiple mtDNA deletions in Patients 3 and 4. The boundary sequence of one of the deleted mtDNA in these patients was identical with that of the second type of deletion. Therefore, it is likely that directly repeated sequences are involved not only in the single mtDNA deletion but also in the multiple mtDNA deletions observed in patients with myopathy. Zeviani et al. described multiple mtDNA deletions starting exclusively at the D-loop region in a family with autosomal dominant mitochondrial myopathy.[23] In contrast, we found multiple mtDNA deletions outside the D-loop region both in the present patients and in the previous patient with familial ocular myopathy.[16] Therefore, the mechanism responsible for the multiple mtDNA deletions described here apparently differs from the mechanism proposed by Zeviani et al. for the multiple mtDNA deletions in their cases.[23]

We have proposed a hypothesis that accumulation of somatic mitochondrial gene mutations will eventually lead to accumulation of partially or grossly bioenergetic defective cells with time as a key facet of the processes of ageing and degenerative diseases.[5] The directly repeated sequences at the boundaries of deletions reported here might contribute to the hypothesized accumulation of mtDNA deletions associated with these processes. Since directly repeated sequences of relatively short base pairs, similar to the sequences reported here, are abundant in mtDNA, these sequences could provide numerous chances for mtDNA mutations.

ACKNOWLEDGMENTS

This work was supported in part by the Grants-in-Aid for General Scientific Research (62570128) to M.T. and for Scientific Research on Priority Areas (Bioenergetics, 63617002) to T.O. from the Ministry of Education, Science and Culture, Japan and by Grant 88-02-39 from National Center for Nervous, Mental and Muscular Disorders of the Ministry of Health and Welfare, Japan to T.O.

REFERENCES

1. Anderson, S., Bankier, A.T., Barrell, B.G., de Bruijn, M.H.L., Coulson, A.R., Drouin, J., Eperon, I.C., Nierlich, D.P., Roe, B.A., Sanger, F., Schreier, P.H., Smith, A.J.H., Staden, R., and Young, I.G. (1981) *Nature* **290**, 457-465
2. Chomyn, A., Mariottini, P., Cleeter, M.W.J., Ragan, C.I., Matsuno-Yagi, A., Hatefi, Y., Doolittle, R.F., and Attardi, G. (1985) *Nature* **314**, 592-597
3. Gyles, R.E., Blanc, H., Cann, H.M., and Wallace, D.C. (1980) *Proc. Natl. Acad. Sci. USA* **77**, 6715-6719.
4. Neckelmann, N., Li, K., Wade, R.P., Shuster, R., and Wallace, D.C. (1980) *Proc. Natl. Acad. Sci. USA* **87**, 7580-7584.
5. Linnane, A.W., Marzuki, S., Ozawa, T., and Tanaka, M. (1989) *Lancet i*, 642-645.
6. Holt, I.J., Harding, A.E., and Morgan-Hughes, J.A. (1988) *Nature* **331**, 717-19.
7. Lestienne, P., and Ponsot, G. (1988) *Lancet i*, 885.
8. Zeviani, M., Moraes, C.T., DiMauro, S., Nakase, H., Bonilla, E., Schon, E.A., and Rowland, L.P. (1988) *Neurology* **38**, 1339-1346.
9. Ozawa, T., Yoneda, M., Tanaka, M., Ohno, K., Sato, W., Suzuki, H., Nishikimi, M., Yamamoto, M., Nonaka, I., and Horai, S. (1988) *Biochem. Biophys. Res. Commun.* **154**, 1240-1247.
10. Noer, A.S., Marzuki, S., Trounce, I., and Byrne, E. (1988) *Lancet ii*, 1253-1254.
11. Rotig, A., Colonna, M., Blanche, S., S., Fischer, A., Le Deist, F., Frezal, J., Saudubray, J.-M., and Munnich, A. (1988) *Lancet ii*, 567-568.
12. Rotig, A., Colonna, M., Bonnefont, J.P., Blanche, S., Fischer, A., Saudubray, J.M., and Munnich, A. (1989) *Lancet i*, 902-903.
13. Tanaka-Yamamoto, T., Tanaka, M., Ohno, K., Sato, W., Horai, S., and Ozawa, T. (1989) *Biochim. Biophys. Acta* **1009**, 151-155.
14. Gyllensten, U.B., and Erlich, H.A. (1988) *Proc. Natl. Acad. Sci. USA* **85**, 7652-7656.
15. Tanaka, M., Sato, W., Ohno, K., Yamamoto, T., and Ozawa, T. (1989) *Biochem. Biophys. Res. Commun.* **164**, 156-163.
16. Sato, W., Tanaka, M., Ohno, K., Yamamoto, T., Takada, G., and Ozawa, T. (1989) *Biochem. Biophys. Res. Commun.* **162**, 664-672.
17. Taber, S. (1987) *Current Protocols in Molecular Biology* (Ausubel, F.M., Brent, R., Kingston, R.E., Moore, D.D., Seidman, J. G., Smith, J.A., and Struhl, K., eds) Vol. 1, 3.12.2, John Wiley & Sons, New York.
18. Perbel, B. (1988) *Practical Guide to Molecular Cloning*, pp. 636-640. John Wiley & Sons, New York.
19. Sanger, F., Nicklen, S., and Coulson, A.R. (1977) *Proc. Natl. Acad. Sci. USA* **74**, 5463-5467.
20. Schon, E.A., Rizzuto, R., Moraes, C.T., Nakase, H., Zeviani, M., and DiMauro, S. (1989) *Science* **244**, 346-349.
21. Attardi, G. (1981) *Trends Biochem. Sci.* **89**, 100-103.
22. Vnencak-Jones, C.L., Phillips, J.A.III, Chen, E.Y., and Seeburg, P.H. (1988) *Proc. Natl. Acad. Sci. USA* **85**, 5615-5619.
23. Zeviani, M., Servidei, S., Gellera, C., Bertini, E., DiMauro, S., and DiDonato, S. (1989) *Nature* **339**, 309-311.

THIRTY YEARS OF MITOCHONDRIAL PATHOPHYSIOLOGY:

FROM LUFT'S DISEASE TO OXYGEN TOXICITY§

Lars Ernster* and Chuan-pu Lee**

*Department of Biochemistry, Arrhenius Laboratories, Stockholm
University, S-106 91 Stockholm, Sweden, and **Department of
Biochemistry, Wayne State University, School of Medicine
Detroit, MI 48201, USA

INTRODUCTION

Over the last three decades there has been a growing interest in
mitochondrial pathophysiology. Beginning with the first report on a clinical syndrome related to mitochondrial dysfunction, published in 1959
by Ernster, Ikkos and Luft (1) and later commonly referred to as Luft's
disease, the number of reports on mitochondrial diseases has been increasing
at a progressive rate and is now up in the several hundreds (Fig. 1).

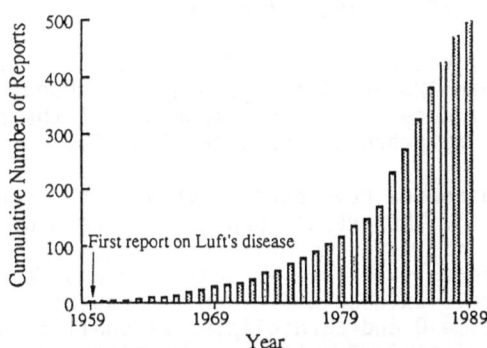

Fig. 1. Reports on mitochondrial diseases 1959-1989.
The numbers are based mainly on papers quoted in refs. 2-7,
and are probably incomplete, in particular those for 1987-1989
(open bars).

---

§This paper is dedicated to Professor Kunio Yagi on the occasion of his
70th birthday.

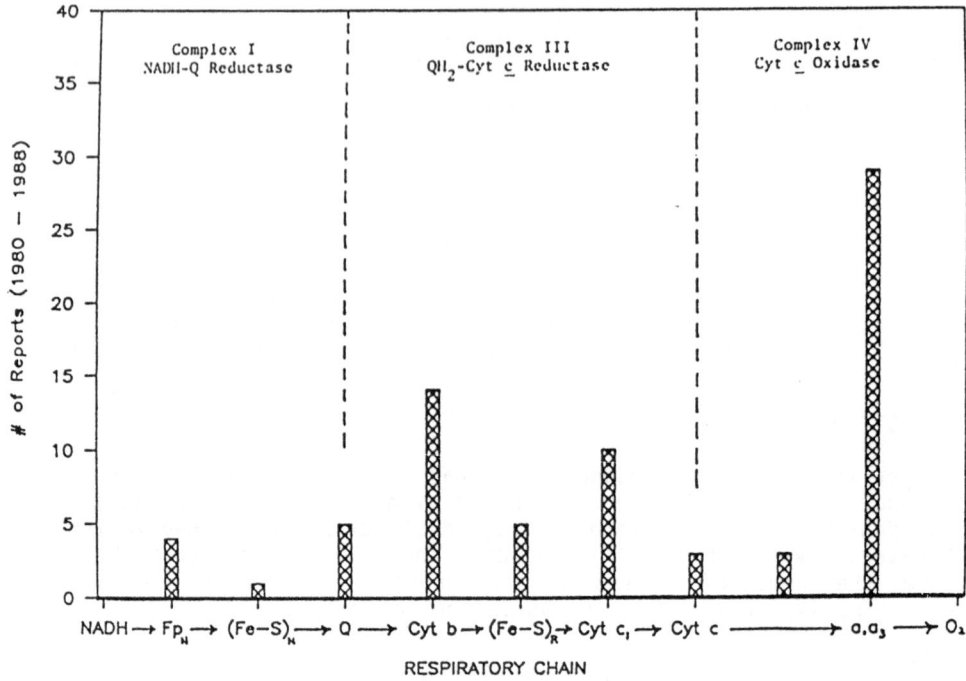

Fig. 2. Mitochondrial anomalies related to the respiratory chain descri-
     bed between 1980 and 1988. (From ref. 8).

     A recent classification of mitochondrial defects of clinical inte-
rest contain more than 120 entries (6). Many mitochondrial anomalies
described in the last decade concern components of the respiratory
chain, in particular cytochrome $\underline{c}$ oxidase (Fig. 2).

     Mitochondrial diseases have been covered in numerous reviews (cf.
2-8 and refs. therein). The present paper will therefore be focused on
certain specific topics, with emphasis on the role of oxygen toxicity in
mitochondrial pathogenesis. First, we shall discuss recent information
concerning cytochrome $\underline{c}$ oxidase deficiencies, especially the possible
involvement of coenzyme Q and cardiolipin. We shall then review some
other aspects of cardiolipin function, suggesting a role of this phospholipid in mitochondrial signalling. Finally, we shall revisit Luft's
disease and discuss its etiology from the perspective of current knowledge concerning the regulation of mitochondrial function and biogenesis.

CYTOCHROME C OXIDASE DEFICIENCIES: POSSIBLE INVOLVEMENT OF COENZYME Q
AND CARDIOLIPIN

     There is a vast literature on mitochondrial diseases classified as
cytochrome $\underline{c}$ oxidase deficiencies, and the subject has been covered in
numerous review articles (cf., e.g., refs. 8-13). The Kearns-Sayre
syndrome (KSS), chronic progressive external opthalmoplegia (CPEO),
myoclonus epilepsy with ragged red fibers (MERRF), and mitochondrial encephalomyopathy with lactic acidosis and stroke-like episodes (MELAS),
are examples of pathologic conditions that are attributed to cytochrome
$\underline{c}$ oxidase deficiencies. Patients suffering from these diseases have been
shown to possess reduced levels and/or activities of cytochrome c oxida-

se in the mitochondria of various tissues. In some instances, deficiencies of nuclear-encoded subunits of cytochrome c oxidase could be demonstrated (10-12), with variations in extent and in the subunit(s) concerned among different patients and tissues. In other cases, deletions in mitochondrial DNA were found (14-18), varying in size between 1.3 and 7.6 kilobases, and including the genes of one or several of the mitochondrially encoded subunits of cytochrome c oxidase, along with other mitochondrial genes. In one instance (19), maternal inheritance of CPEO was established, with partly overlapping but different extents of mitochondrial DNA deletion. The extent of DNA deletion could also vary in the same patient from one tissue to another (20; see also Table I).

Table I

Extents of deletion of mitochondrial DNA
in different tissues from patients with
Kearns-Sayre Syndrome
(From ref. 20)

Tissue	Percent DNA deleted
Brain	67
Liver	53
Heart	64
Skeletal Muscle	83

Studies with isolated skeletal muscle mitochondria from patients suffering from KSS (8,21,22) revealed a strong decrease in the content of cytochrome c oxidase (cytochrome a + a_3) and a marked increase in the content of cytochrome c_1 + c (in particular the latter), with no change in cytochrome b and flavin (Table II). The respiratory rates with both

Table II

Respiratory pigment content of isolated skeletal
muscle mitochondria from healthy subjects and
patients with Kearns-Sayre Syndrome
(From ref. 8)

Pigment*	Content (nmoles/mg protein)**	
	Healthy subjects	Patients
Cyt. a_3	0.52 ± 0.06	0.20 ± 0.05
Cyt. a	0.34 ± 0.04	0.07 ± 0.04
Cyt. c + c_1	0.53 ± 0.03	0.68 ± 0.08
Cyt. b	0.29 ± 0.04	0.25 ± 0.05
Fp	1.03 ± 0.16	0.83 ± 0.27

*Dithionite-reducible
**Values are means ± SE for 3 healthy subjects and
  4 patients.

succinate and $\overline{\text{NAD}}$-linked substrates were diminished by about 75% as compared to normal controls. However, the low cytochrome c oxidase levels
of the patients' mitochondria were not rate-limiting for the observed
overall rates of respiration, as could be deduced from measurements of
the cytochrome c oxidase activities using ascorbate + PMS as substrate.
The rate-limiting step was instead found to be located between cytochrome
c and cytochrome c oxidase, as revealed by the high steady-state levels
of reduced cytochrome c during State-3 respiration with succinate and
NAD-linked substrates. It thus appeared that the defect responsible for
the low rates of respiration with the latter substrates was in the interaction of cytochrome c and cytochrome c oxidase.

Another circumstance of interest in this connection is that there
have been several reports in recent years (23-27) describing decreased
levels of coenzyme Q in the serum and in skeletal muscle mitochondria of
KSS patients, and a favorable response of these patients to administration of coenzyme Q. As follows from the biochemical observations summarized above, coenzyme Q is unlikely to be rate-limiting for the respiratory activity of these mitochondria, at least not in its capacity as
component of the electron-transport system. An alternative possibility
is that the respiratory defect of the mitochondria may be due to a lack
of antioxidant capacity normally provided by the reduced form of coenzyme Q (28).

Support for this notion comes from recent studies of the cytochrome
c oxidase activity of rat skeletal muscle mitochondria after exposure of
rats to tourniquet ischemia of one of their hind limbs, followed by
reperfusion (29-31). It has been known from earlier work (32) that
ischemia-reperfusion may lead to oxidative damage arising from
superoxide-radical generation through xanthine oxidase, and that this
damage may involve, among other consequences, a peroxidation of membrane
phospholipids, in particular cardiolipin (cf. ref. 31). The latter is a

Table III

Effects of ischemia and reperfusion on the major mitochondrial
phospholipids.
The values are mean ± SD (n = 10 in each group) and are
expressed as percentage of control levels.
(From ref. 29)

	Phosphatidyl choline	Phosphatidyl ethanolamine	Cardiolipin
Control	100 ± 17	100 ± 18	100 ± 25
2 h ischemia	98 ± 17	97 ± 17	104 ± 26
2 h ischemia + 1 h reperfusion	102 ± 17	95 ± 17	96 ± 24
4 h ischemia	95 ± 16	97 ± 17	75 ± 19
4 h ischemia + 1 h reperfusion	99 ± 14	86 ± 15	55 ± 14

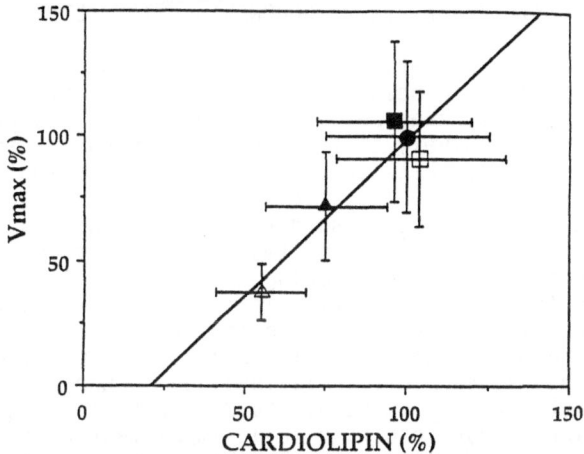

Fig. 3. Correlation between the cardiolipin content and the V_{max} of cytochrome c oxidase in rat skeletal muscle in normal controls (●) and after exposure to 2 h ischemia (■), 2 h ischemia followed by 1 h reperfusion (□), 4 h ischemia (▲), and 4 h ischemia followed by 1 h reperfusion (Δ). Values are expressed as percentages of control levels. r = 0.94, p < 0.02, n = 10 in each group. (From ref. 29, with permission).

component of the mitochondrial inner membrane and is essential for cytochrome c oxidase activity (33). Based on this information (cf. also ref. 34), and on a detailed investigation of the interaction of cardiolipin with cytochrome c and cytochrome c oxidase (29-31), these studies have shown that (a) cardiolipin is destroyed selectively among mitochondrial phospholipids in the course of ischemia-reperfusion (Table III); and (b) this is accompanied by an inactivation of cytochrome c oxidase (Fig. 3), due to a diminished ability of the enzyme to interact with cytochrome c, i.e., an effect similar to that found in KSS. Interestingly, the opposite effect, i.e., an increased activity of cytochrome c oxidase, was found in skeletal muscle mitochondria from rats after endurance training (35), which had been shown to result in an elevation of the coenzyme Q content of skeletal muscle (36).

In the light of these observations it seems conceivable that the cytochrome c oxidase deficiency found in KSS - and perhaps in related syndromes as well - may be the consequence of an oxidative damage resulting from a diminished antioxidant capacity due to insufficient coenzyme Q contents. This would lead, on one hand, to a deletion of certain especially exposed segments - "hot spots" (cf. ref. 9) - of the mitochondrial DNA, including some gene(s) of cytochrome c oxidase, and resulting, secondarily, in a diminished content of cytochrome c oxidase as a whole; and, on the other hand, to a peroxidative destruction of cardiolipin, and thereby a diminished capacity of the residual cytochrome c oxidase to interact with cytochrome c. The latter point could be tested directly by determining the level of cardiolipin in skeletal muscle mitochondria from KSS patients and, if this is found to be diminished, by attempting to raise it through the administration of coenzyme Q.

CARDIOLIPIN: ITS POSSIBLE ROLE IN MITOCHONDRIAL SIGNALLING

Cardiolipin, or diphosphatidyl glycerol, is unique among the phospholipids of animal cells in being synthesized in the mitochondria - all other phospholipids are synthesized in the endoplasmic reticulum - and in occurring almost exclusively in the inner mitochondrial membrane (cf. ref. 37). It is distributed between the outer and inner leaflets of the phospholipid bilayer, with its bulk facing the matrix surface of the inner membrane (38). Cardiolipin also differs from other phospholipids in being strongly acidic - it contains two phosphate groups and no nitrogenous substituents - and in containing a relatively large proportion of polyunsaturated fatty acids, mostly linoleic acid. The latter feature probably is responsible for the selective sensitivity of cardiolipin to agents causing lipid peroxidation as discussed in the previous section. As also discussed above, cardiolipin is involved in the interaction of cytochrome c and cytochrome c oxidase, its destruction - e.g. by ischemia-reperfusion - thus resulting in an impairment of cell respiration. Peroxidation of cardiolipin has also been reported to result in an inactivation of mitochondrial ATPase (39), and there are indications that cardiolipin may influence the interaction of the F_1 component of ATPase with both the F_0 moiety of the enzyme (40,41) and the ATPase-inhibitor protein (42-44).

A recent report (45) indicates another, rather unexpected, potential function of cardiolipin, namely its possible involvement in the targeting of precursor proteins across the mitochondrial membranes. It was shown (45) that cardiolipin incorporated into liposomes specifically binds the precursor forms of several mitochondrial proteins, including malate dehydrogenase, adrenodoxin and cytochrome P-450 (SCC). The binding was specific with respect to both the signal peptide of the precursor proteins and to cardiolipin; minor alterations in either molecule obliterated the binding. The polar head of the cardiolipin molecule, including the two phosphate groups and the free hydroxy group in the glycerol, were essential. These requirements were similar to those earlier found for the interaction of cardiolipin with anticardiolipin antibodies (46,47). It was suggested (45) that a fraction of cardiolipin in mitochondria may be located at the contact sites between the inner and outer membranes - where the import of protein into mitochondria appears to occur (cf. ref. 48) - and serve as a recognition site for proteins destined to mitochondria. The outer-membrane protein porin, which has no cleavable signal peptide but which is believed to be taken up and located at the contact sites between the two membranes, was also found to bind to cardiolipin-containing liposomes (45).

Still another distinctive feature of cardiolipin among mitochondrial phosphoipids is its antigenicity. It has been known for some time that cardiolipin can give rise to antibodies that give false-positive tests for syphilis and that are found in patients suffering from systemic lupus erythomatosus (49). These antibodies were also shown to bind to DNA (50), especially when the latter was denatured by unfolding (dDNA), a finding that was attributed to a structural similarity between the phosphodiester regions of cardiolipin and dDNA (Fig. 4). More recently it was demonstrated (51) that, indeed, monoclonal antibodies raised against cardiolipin react with DNA. At present there is a rapidly growing literature concerning cardiolipin antibodies and various autoimmune diseases (52-67). This development, together with current information regarding the occurrence of a number of antimitochondrial antibodies directed against various mitochondrial proteins (cf. refs. 68, 69 for reviews), raises the question as to the mechanisms that are responsible

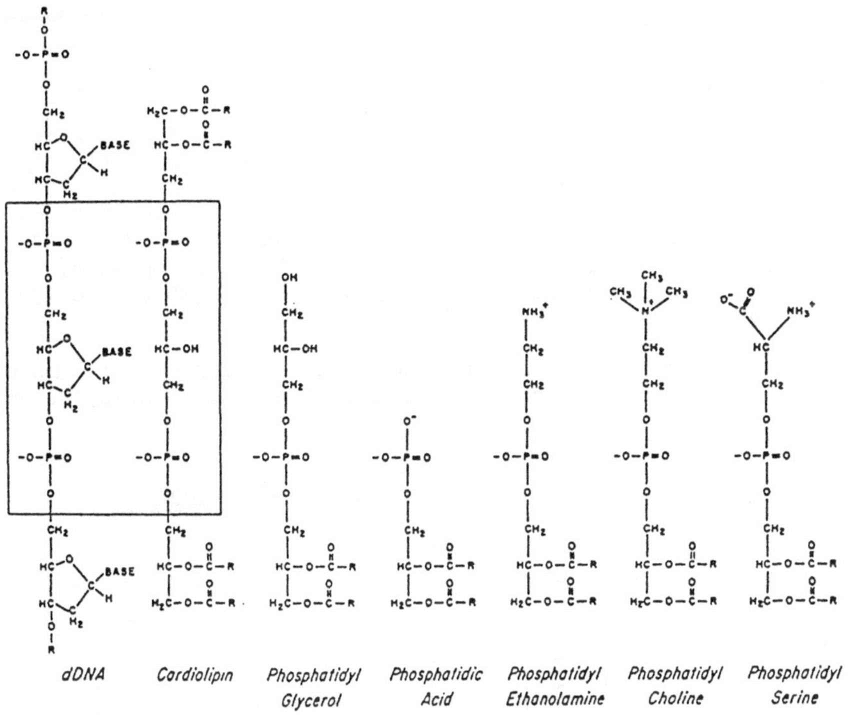

Fig. 4. Structural relationship between denatured DNA (dDNA) and cardiolipin. (From ref. 50).

for the antigenicity of cardiolipin and these proteins. How are these
components released from mitochondria and how do they elicit an immune
response that apparently fulfills a physiological purpose but may lead,
in certain cases, to pathologic conditions manifested as autoimmune diseases?

   At the present state of knowledge one can only speculate about these
questions. A simple answer would be that these antigens actually origi-
nate from bacteria but are identical or similar to mitochondrial components. An alternative, and biologically more appealing, explanation is
that an oxidative damage to a mitochondrial constituent, e.g. cardiolipin or DNA, serves as a signal to the lysosomes to break down the functionally handicapped organelle, with a release of its contents and an
activation of the immune system. The antigenicity of mitochondrial con-
stituents is not surprising if one considers that these are either synthesized in the mitochondria (such as cardiolipin or mitochondrial DNA
and its gene products) or are covalently modified nuclear gene products
(i.e. proteins altered either by the removal of signal peptides or, in
the case of cytochromes c_1 and c, by the addition of heme). It is of
great interest in this connection that there is now evidence for the occurrence of a maternally inherited antigen in mouse which can specifically activate a cell-surface histocompatibility antigen complex
triggering T-cell activation and cell killing (70). This maternally inherited antigen has recently been shown to be attributable to allelic
variation at a single residue of the ND1 protein, i.e., the product of
the mitochondrial gene coding for subunit 1 of NADH-ubiquinol oxidoreductase (71). The mechanisms by which this antigen is released from
the mitochondrion and is targeted to the cell surface are of obvious
relevance to the concept of mitochondrial signalling discussed above.

LUFT'S DISEASE REVISITED

      Before concluding this paper, it seems appropriate to go back to
Luft's disease and discuss its possible etiology from the perspective of
current knowledge concerning the regulation of mitochondrial function
and biogenesis. Luft's disease is a rare syndrome; over 30 years only
two cases have been described, a 35-year old Swedish woman (1,72,73) and
a 19-year old girl from Jordan (74-77). In both cases, the patient suf-
fered from an extremely severe hypermetabolism (BMR 80-270% above normal) of non-thyroid origin, with intense perspiration, polyphagia,
polydipsia without polyuria, decreased body weight and general fatigu-
able weakness as their main symptoms.

      Some of the main biochemical findings with isolated skeletal muscle
mitochondria are illustrated in Figs. 5 and 6. The mitochondria showed a

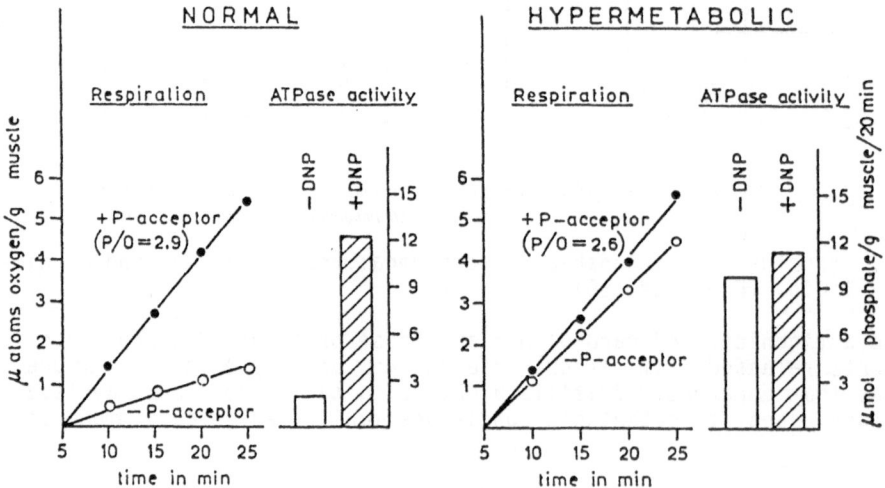

Fig. 5. "Loose-coupling" of skeletal muscle mitochondria from a patient
        with severe hypermetabolism of nonthyroid origin (Luft's
        disease). (From ref. 1).

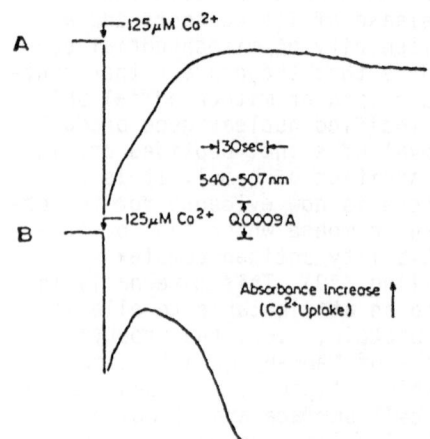

Fig. 6. Calcium uptake by skele-
        tal muscle mitochondria
        from a normal control
        (A) and a patient with
        Luft's disease (B).
        (From ref. 75).

nearly maximal rate of respiration in the absence of phosphate acceptor but had an almost normal phosphorylating efficiency (expressed as P/O ratio) in its presence (Fig. 5). They also exhibited a high ATPase activity, which was only slightly stimulated by the uncoupler 2,4-dinitrophenol (cf. Fig. 5). These features of "loosely-coupled respiration", i.e. a deficient respiratory control with a partially maintained ability to synthesize ATP, adequately account for the clinical symptoms of high heat production and caloric demand at rest, with a limited capacity for work performance. In addition it was found (73) that the respiration of the mitochondria was insensitive to oligomycin although the latter did inhibit ATP synthesis. This finding indicated an "energy leak" above the level of the phosphorylating system. A similar conclusion was suggested by the observation, made in the second case of Luft's disease, that the mitochondria lacked the ability to accumulate Ca^{2+} (Fig. 6), suggesting an energy-dissipating futile cycle of Ca^{2+} uptake and release. Another possibility that has been considered (78), and that could be explored if and when a new case of the disease would be encountered, is that the skeletal muscle mitochondria of these patients may contain a functional "uncoupling protein", also known as "thermogenin" (79), which is normally only present in brown adipose tissue and is instrumental in nonshivering thermogenesis in newborn and hibernating animals.

Morphologically, the mitochondria revealed some striking structural anomalies, characterized, i.a., by a large accumulation of mitochondria especially in the perinuclear region of the muscle cells, and by the oc-

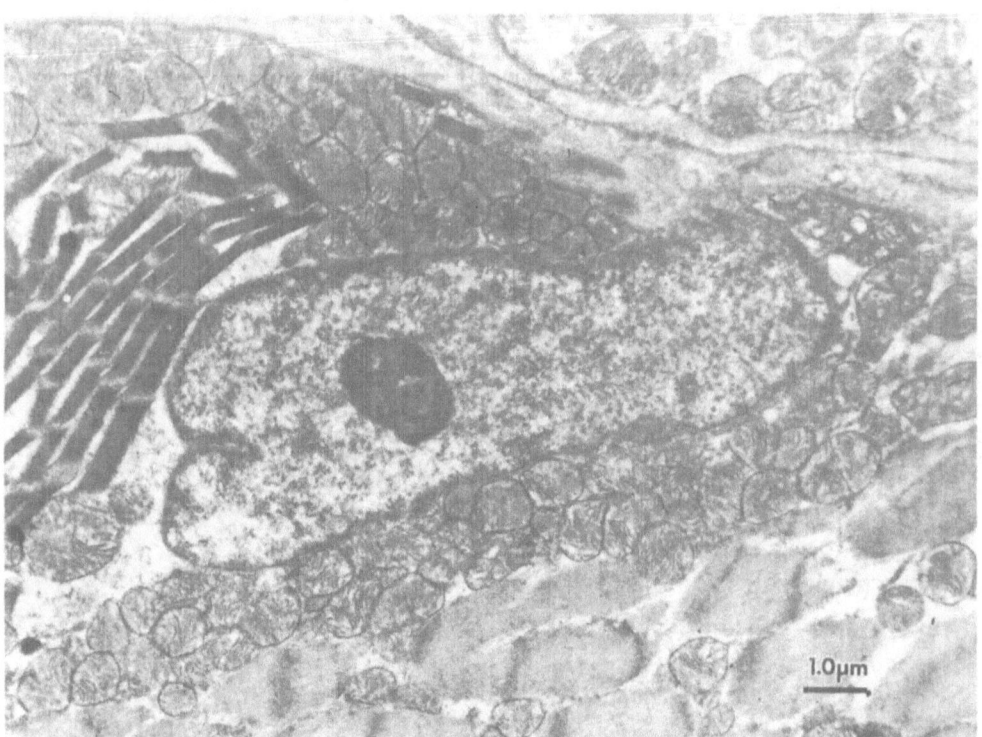

Fig. 7. Electron micrograph of a muscle specimen from a patient with Luft's disease. (From ref. 72).

currence of vast paracrystalline inclusions possibly consisting of lipofuscin granules (Fig. 7).   Paracrystalline inclusions in skeletal muscle
mitochondria have been described in various myopathies (cf. ref. 80) and
have recently been shown to consist presumably of protein (81). In some
instances, these mitochondria exhibited a loose-coupling similar to that
found in Luft's disease, but the patients did not have symptoms of hypermetabolism (cf. ref. 80). An extreme proliferation of mitochondria has
been described in oncocytomas (82-85), and found to be accompanied by a
loosely coupled state of the isolated mitochondria (82), but, again, the
clinical implication of the latter observation remains unclear. In general, it appears that Luft's disease is so far the only instance of a
clear correlation between lack of respiratory control at the biochemical
and clinical levels, and illustrates in a striking manner the critical
role of mitochondrial respiratory control in the energy metabolism of
the living organism.

     As already discussed above, the etiological background of Luft's
disease remains unknown. An observation that may be of interest in this
connection is that the skeletal muscle mitochondria of the first patient
(the second was not tested in this respect) contained normal levels of
coenzyme Q (72,73). This is in contrast to other instances of increased
mitochondrial population and respiratory activity of skeletal muscle
such as those found after thyroid hormone treatment (86-88) or endurance
training (36). In the former case it was in fact shown (87) that the increase in coenzyme Q content occurred <u>after</u> the enhancement of mitochondrial respiratory activity and thus was an effect rather than a
cause of the latter. Both in this case and in that of endurance training
- and, in general, whenever there is an activation of mitochondrial biogenesis and thereby an increase in tissue respiration - this seems to
require an enhancement of endogenous antioxidant capacity. This is
brought about in the first place by an elevation of the cellular level
of coenzyme Q, which in its reduced form has been shown to serve as a
powerful antioxidant (cf. ref. 28), inhibiting lipid peroxidation, and
thereby preventing membrane, protein and DNA damage and the formation of
lipofuscin aggregates.

     Such a scenario might provide an etiological basis for Luft's
disease if one assumes, as has been proposed (89), that the lack of
respiratory control originates from a pathologically enhanced proliferation of mitochondria, where the formation of a component necessary for
maintaining a tight coupling between respiration and phosphorylation
does not keep pace with the overall rate of mitochondrial biogenesis.
Coenzyme Q may be such a component. To investigate this relationship
further it will be necessary to learn more about factors that regulate
mitochondrial biogenesis (48,90) and coenzyme Q biosynthesis (91-93).
Luft's disease, and mitochondrial disorders in general, may open new
insights into these fundamental problems.

ACKNOWLEDGEMENTS

     We thank our colleagues Björn Afzelius, Karl-E. Arfors, Ann-
Christin Bylund-Fellenius and Bassam Soussi for stimulating discussions,
Ann Nielsen and Kerstin Nordenbrand for their valuable help in preparing
this manuscript, and the Swedish Medical and Natural-Science Research
Councils and the National Institutes of Health and Muscular Distrophy
Association, USA, for financial support of the work quoted from our
laboratories.

REFERENCES

1. Ernster, L., Ikkos, D., and Luft, R., Nature 184:1851-1854 (1959).

2. DiMauro, S., in: Handbook of Clinical Neurology, Vol. 17 (Vinken, P.J. and Bruyn, G.W., eds.) North-Holland, Amsterdam, 1979, pp. 175-234.

3. Carafoli, E., and Roman, I., Mol. Aspects Med. 3:295-429 (1980).

4. Morgan-Hughes, J.A., in: Recent Advances in Clinical Neurology, 3rd edn. (Matthews, W.B., and Glaser, J.H., eds.), Churchill, Livingstone, Edinburgh, 1982, pp. 1-46.

5. Wallace, D.C., in: Medical and Experimental Mammalian Genetics: A Perspective (McKusick, V.A. et al., eds.) Vol. 23, Alan Liss, New York, 1985, pp. 137-190.

6. Scholte, H.R., J. Bioenerg. Biomembr. 20:161-191 (1988).

7. Azzi, A., Drahota, Z., and Papa, S. (eds.), Molecular Basis of Membrane-Associated Diseases. Springer, Berlin, 1989.

8. Lee, C.P., Martens, M.E., Peterson, P.L., and Tsang, S.H., in: Molecular Basis of Membrane-Associated Diseases (Azzi, A., et al., eds.). Springer, Berlin, 1989, pp. 166-182.

9. DiMauro, S., Lombes, A., Nakase, H., Zeviani, M., Rizzutto, R., Moraes, C.T., and Schon, E.A., in: Molecular Basis of Membrane-Associated Diseases (Azzi, A. et al., eds.) Springer, Berlin, 1989, pp. 157-166.

10. Johnson, M.A., in: Molecular Basis of Membrane-Associated Diseases (Azzi, A., et al., eds.). Springer, Berlin, 1989, pp. 200-215.

11. Kadenbach, B., Hüther, F.-J., Büge, U., Scherf, A., and Johnson, M.A., in: Molecular Basis of Membrane-Associated Diseases (Azzi, A., et al., eds.). Springer, Berlin, 1989, pp. 216-227.

12. Glerum, D.M., Robinson, B.H., and Capaldi, R., in: Molecular Basis of Membrane-Associated Diseases (Azzi, A., et al., eds.). Springer, Berlin, 1989, pp. 228-238.

13. Sinjorgo, K.M.C., Muijsers, A.O., Scholte, H.R., van Kuilenburg, A.B.P., van Beeumen, J.J., Brouwer-Kelder, E.M., Berden, J.A., Bohuis, P.A., and Tager, J.M., in: Molecular Basis of Membrane-Associated Diseases (Azzi, A., et al., eds.). Springer, Berlin, 1989, pp. 239-253.

14. Holt, I.J., Harding, A.E., and Morgan-Hughes, J.A., Nature, 331: 717-719 (1988).

15. Holt, I.J., Cooper, J.M., Morgan-Hughes, J.A., and Harding, A.E., Lancet, 1:1462 (1988).

16. Zeviani, M., Moraes, C.T., DiMauro, S., Nakase, H., Bonilla, E., Schon, E.A., and Rowland, L.P., Neurology, 38:1339-1346 (1988).

17. Lestienne, P., and Ponsot, G., Lancet, 1:885 (1988).

18. Moraes, C.T., DiMauro, S., Zeviani, M., Lombes, A., Shanske, S.,
    Miranda, A.F., Nakase, H., Bonilla, E., Werneck, L.C., Servidei, S.,
    Nonaka, I., Koga, Y., Spiro, A.J., Brownell, K.W., Schmidt, B.,
    Schotland, D.L., Zupang, M., DeVivo, D.C., Schon, E.A., and Rowland,
    L.P., N. Engl. J. Med., 320:1293-1299 (1989).

19. Ozawa, T., Yoneda, M., Tanaka, M., Ohno, K., Sato, W., Suzuki, H.,
    Nishikimi, M., Yamamoto, M., Nonaka, I., and Horai, S., Biochem.
    Biophys. Res. Comm., 154:1240-1247 (1988).

20. Hurko, O., Johns, D.R., Rutledge, S.L., Stine, O.C., Peterson, P.L.,
    Martens, M.E., Drachman, D.B., Brown, R.H., and Lee, C.P., Pediatric
    Research, in press.

21. Lee, C.P., Martens, M.E., Peterson, P.L., and Hatfield, J.S., in:
    Cytochrome Systems: Molecular Biology and Bioenergetics. (Papa, S.,
    Chance, B., and Ernster, L., eds.), Plenum, New York, 1987, pp.
    407-414.

22. Martens, M.E., Peterson, P.L., Lee, C.P., Nigro, M.A., Hart, Z.,
    Glasberg, M., Hatfield, J.S., and Chang, C.H., Ann. Neurol., 24:
    630-637 (1988).

23. Ogasahara, S., Yorifuji, S., Nishikawa, Y., Takahashi, M., Wada, K.,
    Hazama, T., Nakamura, S., Kono, N., and Tarui, S., Neurology, 35:
    372-377 (1985).

24. Ogasahara, S., Nishikawa, Y., Yorifuji, S., Soga, F., Nakamura, Y.,
    Takahashi, M., Hashimoto, S., Kono, N., and Tauri, S., Neurology,
    36:45-53 (1986).

25. Takahashi, M., Ogasahara, S., Nishikawa, Y., Yorifuji, S., Nakamura,
    y., Soga, F., Ueno, S., and Tarui, S., in: Biomedical and Clinical
    Aspects of Coenzyme Q, Vol. 5 (Folkers, K., and Yamamura, Y., eds.),
    Elsevier, Amsterdam, 1986, pp. 345-352.

26. Bresolin, N., Bet, L., Binda, A., Moggio, M., Comi, G., Nador, F.,
    Ferrante, C., Carenzi, A., and Scarlato, G., Neurology, 38:892-899
    (1988).

27. Zierz, S., Jahns, G., and Jerusalem, F., J. Neurol. 236:97-101
    (1989).

28. Beyer, R.E., and Ernster, L., in: Biochemistry, Bioenergetics and
    Clinical Aspects of Ubiquinone (Lenaz, G., ed.), Taylor & Francis,
    London, in press.

29. Soussi, B., Idström, J.-P., Scherstén, T., and Bylund-Fellenius,
    A.-C., Acta Physiol. Scand., 138:107-114 (1990).

30. Soussi, B., Bylund-Fellenius, A.-C., Scherstén, T., and Ångström,
    J., Biochem. J., 265:227-232 (1990).

31. Soussi, B., Ph.D. Thesis, University of Gothenburg, Sweden, 1989.

32. Granger, D.N., Rutili, G., and McCord, J.M., Gastroenterology, 81:
    22-29 (1981).

33. Vik, S.B., Georgevich, G., and Capaldi, R., Proc. Natl. Acad. Sci. USA, 78:1458-1460 (1981).

34. Idström, J.-P., Soussi, B., Elander, A., and Bylund-Fellenius, A.-C., Amer. J. Physiol, in press.

35. Soussi, B., Idström, J.-P., Scherstén, T., and Bylund-Fellenius, A.-C., Acta Physiol. Scand., 135:373-379 (1989).

36. Beyer, R.E., Morales-Corral, P.G., Ramp, B.J., Kreitman, K.R., Falzon, M.J., Rhee, S.Y.S., Kuhn, T.W., Stein, M., Rosenwasser, M.J., and Cartwright, K.J., Arch. Biochem. Biophys., 234:323-329 (1984).

37. Tzagoloff, A., Mitochondria, Plenum, New York, 1982.

38. Nilsson, O., and Dallner, G., Biochim. Biophys. Acta, 464:453-458 (1977).

39. Santiago, E., López-Moratalla, N., and Segovia, J.L., Biochem. Biophys. Res. Comm., 53:439-445 (1973).

40. Toson, G., Contessa, A.R., and Bruni, A., Biochem. Biophys. Res. Commun., 48:341-347 (1972).

41. Lee, C.P., Huang, C.H., and Cierkosz, B.I.T., in: Membrane Proteins in Transport and Phosphorylation (Azzone, G.F., Klingenberg, M., Quagliariello, E., and Siliprandi, N., eds.), North Holland, Amsterdam, 1974, pp. 161-170.

42. Bruni, A., and Bigon, E., Biochim. Biophys. Acta, 357:333-343 (1974).

43. Dabbeni-Sala, F., Furlan, R., Pitotti, A., and Bruni, A., Biochim. Biophys. Acta, 374:77-86 (1974).

44. Ernster, L., Sandri, G., Hundal, T., Carlsson, C., and Nordenbrand, K., BBA Library, 14:209-222 (1977).

45. Ou, W.-J., Ito, A., Inoue, K., and Omura, T., J. Biochem., 103:589-595 (1988).

46. Inoue, K., Suhara, Y., and Nojima, S., Chem. Pharm. Bull. 11:1150-1156 (1963).

47. Inoue, K., and Nojima, S., Chem. Pharm. Bull., 16:76-81 (1968).

48. Attardi, G., and Schatz, G., Ann. Rev. Cell Biol., 4:289-333 (1988).

49. Pangborn, M.C., Proc. Soc. Exp. Biol. Med., 48:484 (1941).

50. Lafer, E.M., Rauch, J., Andrzejewski, C., Mudd, D., Furie, B., Furie, B., Schwartz, R.S., and Stollar, B.D., J. Exp. Med., 153:897-909 (1981).

51. Rauch, J., Tannenbaum, H., Stollar, B.D., and Schwartz, R.S., Eur. J. Immunol., 14:529-534 (1984).

52. Khamashta, M.A., Harris, E.N., Gharavi, A.E., Derue, G., Gil, A., Vásquez, J.J., and Hughes, G.R.V., Ann. Rheum. Dis., 47:849-854 (1988).

53. Asherson, R.A., Gibson, Evans, D.W., Baguley, E., and Hughes, G.R.V., Ann. Rheum. Dis., 47:947-953 (1988).

54. Asherson, R.A., and Lubbe, W.F., J. Rheumatol., 15:539-543 (1988).

55. Deleze, M., Oria, C.V., and Alarconsegovia, D., J. Rheumatol., 15:611-615 (1988).

56. Alarconsegovia, D., J. Rheumatol., 15:890-893 (1988).

57. Asherson, R.A., Hughes, G.R.V., and Derksen, R.H.W.M., J. Urol., 140:1028-1029 (1988).

58. Cheng, H.M., and Yap, S.F., Arthritis Rheum., 31:1211 (1988).

59. Agopian, M.S., Boctor, F.N., and Peter, J.B., Arthritis Rheum., 31:1212 (1988).

60. Exner, T., and Koutts, T., Aust. NZ J. Med., 18:669-673 (1988).

61. Nagasawa, T., Takeuchi, K., and Yamada, H., Biochem. Biophys. Res. Comm., 155:1001-1007 (1988).

62. Cowchook, S., Fort, J., Munoz, S., Norberg, R., and Maddrey, W., Clin. Exp. Immunol., 73:289-294 (1988).

63. Cooper, P., Strauss, R.C., Jordan, E.R., Przybojewski, J.Z., and Nel, N., Clin. Exp. Immunol., 74:254-288 (1988).

64. Korn, S., Huppert, A., Spitzer, S., and DeHoratius, R.J., J. Rheumatol., 15:1281-1283 (1988).

65. Shergy, W.J., Kredich, D.W., and Pisetsky, D.S., J. Rheumatol., 15:1389-1394 (1988).

66. Cheng, M.H., and Ngeow, F.Y., Med. Sci. Res., 16:1011-1012 (1988).

67. Loizou, S., Byron, M.A., Englert, H.J., David, J., and Hughes, G.R.V., Q.J. Med., 68:525-532 (1988).

68. Carafoli, E., Ann. NY Acad. Sci., 488:1-18 (1988).

69. Baum, H., and Fusconi, M., in: Molecular Basis of Membrane-Associated Diseases (Azzi, A., et al., eds.). Springer, Berlin, 1989, pp. 323-341.

70. Fischer-Lindahl, K., Trends in Genetics, 1:135-139 (1985).

71. Loveland, B., Wang, C.-R., Yonekawa, H., Hermel, E., and Fischer-Lindahl, K., Cell, 60:971-980 (1990).

72. Luft, R., Ikkos, D., Palmieri, G., Ernster, L., and Afzelius, B., J. Clin. Invest. 41:1776-1804 (1962).

73. Ernster, L., and Luft, R., Exp. Cell Res., 32:26-35 (1963).

74. DiMauro, S., Schotland, D.L., Lee, C.P., Bonilla, E., and Conn, H., Trans. Amer. Acad. Neurol. 66:265-267 (1972).

75. DiMauro, S., Schotland, D.L., Bonilla, E., Lee, C.P., DiMauro, S. and Scarpa, A., in: Mitochondrial Myopathies: Which and How Many? (Milhorat, E., ed.) Excerpta Medica, Amsterdam, 1975, pp. 506-515.

76. DiMauro, S., Bonilla, E., Lee, C.P., Schotland, D.L., Scarpa, A., Conn, H., and Chance, B., J. Neurol. Sci., 27:217-232 (1976).

77. Bonilla, E., Schotland, D.L., DiMauro, S., and Lee, C.P., J. Ultrastruct. Res., 58:1-9 (1977).

78. Ernster, L., EBEC Short Reports, Prague, 1986, p. 76.

79. Nedergaard, J., and Cannon, B., in: Bioenergetics, New Comprehensive Biochemistry, Vol. 9 (Ernster, L., ed.), Elsevier, Amsterdam, pp. 291-314.

80. Morgan-Hughes, J.A., in: Skeletal Muscle Pathology (Mastaglia, F.L., and Walton, J., eds.) Churchill-Livingstone, Edinburgh, 1982, pp. 309-339

81. Farrants, G.W., Hovmöller, S., and Stadhouders, A.D.M., Muscle and Nerve, 11:45-55 (1988).

82. Schiefer, H.-G., Hübner, G., and Kleinsasser, O., Virchov. Arch. (Zellpath.), 1:230-239 (1968).

83. Tandler, B., and Shipkey, F.H., J. Ultrastruct. Res., 11: 292-305 (1964).

84. Tandler, B., Hutter, R.V.P., and Erlandson, R.A., Lab. Invest., 23:567-580 (1970).

85. Sun, C.N., White, H.J., and Thompson, W.W., Arch. Pathol. 99:208-214 (1975).

86. Tata, J.R., Ernster, L., Lindberg, O., Arrhenius, E., Pedersen, S., and Hedman, R., Biochem. J., 86:408-428 (1963).

87. Pedersen, S., Tata, J.R., and Ernster, L., Biochim. Biophys. Acta, 69:407-409 (1963).

88. Gustafsson, R., Tata, J.R., Lindberg, O., and Ernster, L., J. Cell Biol., 26:555-578 (1965).

89. Ernster, L., and Luft, R., Adv. Metabolic Disorders, 1:95-123 (1964).

90. Nelson, B.D., Mutvei, A., Joste, V., Wielburski, A., and Kuzela, S., Chemica Scripta, 27B:239-244 (1987).

91. Kalén, A., Norling, B., Appelkvist, E.-L., and Dallner, G., Biochim. Biophys. Acta, 926:70-78 (1987).

92. Kalén, A., Appelkvist, E.-L., Chojnacki, T., and Dallner, G., J. Biol. Chem., 265:1158-1164 (1990).

93. Kalén, A., Appelkvist, E.-L., and Dallner, G., Lipids, 2:579-584 (1989).

Appendix

PARTICIPANTS

Yasuhiro Anraku, Department of Biology, Faculty of Science, University of Tokyo, 113 Tokyo, Japan

Angelo Azzi, Institut für Biochemie, und Molekularbiologie, Universität Bern, Bühlstrasse 28, CH-3012 Bern, Switzerland

Moo Bae, Department of Biology, College of Arts and Sciences, Ewha Womans University, Seoul. Korea

Claude Balny, Institut National de la Sante et de la Recherche Medicale, U 128, C.N.R.S., B.P. 5051, 34033 Montpellier, Cedex, France

Jan A. Berden, E.C. Slater Institute for Biochemical Research, University of Amsterdam, 1018 TV Amsterdam, The Netherlands

Ernesto Carafoli, Department of Biochemistry, Swiss Federal Institute of Technology, Universitätstr. 16, ETH-Zentrum, 8092 Zurich, Switzerland

Young Bog Chae, Korea Research Institute of Chemical Technology, P.O. Box 9 Daedeog-Danji, Daejeon, Korea

Young-Nam Cha, Department of Pharmacology and Toxicology, Medical College, Inha University, Inchon 160, Korea

Britton Chance, Department of Biochemistry and Biophysics, School of Medicine, University of Pennsylvania, Philadelphia, PA 19104, U.S.A.

Boji Cheng, Department of Biochemistry, Beijing Medical University, Beijing, China

Kum Hee Cho, College of Pharmacy, Ewha Womans University, Seoul, Korea

Eun Yong Chung, Department of Chemistry, College of Arts and Sciences, Ewha Womans University, Seoul, Korea

Kun Hoe Chung, Korea Research Institute of Chemical Technology, P.O. Box 9, Daedeog-Danji, Daejeon, Korea

Kimitoshi Denda, Department of Life Sciences, Tokyo Institute of Technology, Yokohama 227, Japan

Lars Ernster, Department of Biochemistry, University of Stockholm, Stockholm, Sweden S10691

Taketomo Fujiwara, Department of Life Sciences, Faculty of Science, Tokyo Institute of Technology, O-Okayama, Meguro-Ku, Tokyo 151, Japan

Masamitsu Futai, Department of Organic Chemistry and Biochemistry, Institute of Scienctific, and Industrial Research, Osaka University, Osaka 567, Japan

Jong Sik Hah, Department of Physiology, College of Medicine, Ewha Womans University, Seoul 120-750, Korea

K. S. Han, R & D Center, MIWON Co., Ltd., C.P.O. Box 2156, Seoul Korea

Moon Hi Han, Genetic Engineering Center, Korea Advanced Institute of Science and Technology, P.O. Box 150, Seoul, Korea

Tadao Hashimoto, Department of Physiological Chemistry, Osaka University Medical School, Osaka 530, Japan

Youssef Hatefi, Department of Basic and Clinical Research, Scripps Clinics and Research Foundation, 10666 N. Torrey Pines Road, La Jolla, CA 92037, U.S.A.

Hajime Hirata, Department of Biochemistry, Jichi Medical School, 329-04 Tochigi-Ken, Japan

Satoshi Horai, Department of Human Genetics, National Instutite of Genetics, Mishima, Shizuoka, 411, Japan

Bernard Kadenbach, Fachbereich Chemie Biochemie, Philips-Universität, D-3550 Marburg, Federal Republic of Germany

Yasuo Kagawa, Department of Biochemistry, Jichi Medical School, 329-04 Tochigi-Ken, Japan

Kiyoshi Kawai, Department of Food and Nutritional Science, Chukyo Women's University, Ohbu 474, Aichi, Japan

Chong H. Kim, Department of Biology, Rensselaer Polytechnic Institute, Troy. NY 12180, U.S.A.

Choon Mi Kim, Laboratory of Biochemistry, College of Pharmacy, Ewha Womans University, Seoul, Korea

Hae-Jin Kim, Department of Toxicology, College of Pharmacy, Ewha Womans University, Seoul, Korea

Jung-Sun Kim, Chemical Analysis Laboratory, Korea Institute of Science and Technology, Seoul, Korea

Tsoo E. King, Institute for Structural and Functional Studies, University City Science Center,, Suite 300, 3401 Market Street, Philadelphia, PA 19104, U.S.A.

Horst Kleinkauf, Institute of Biochemestry and Molecular Biology, Technical University of Berlin, D-1000 Berlin 10 (West), Federal Republic of Germany

Martin Klingenberg, Institute for Physical Biochemistry, University of Munich, Goethestrasse 33, 8000 Munich 2, Federal Republic of Germany

Arnost Kotyk, Institute of Physiology, Czechoslovak Academy of Sciences, 1420 20 PRAHA 4-KRC, Videnska 1083, Czechoslovakia

Hye-Gyu Lee, Department of Toxicology, College of Pharmacy, Ewha Womans University, Seoul, Korea

Ihn Rhan Lee, Department of Medicinal Chemistry, College of Pharmacy, Ewha Womans University, Seoul, Korea

Keun-Bae Lee, Department of Biochemistry, Chosun University Medical School, Kwangju. Korea

Sang Sup Lee, Department of Biochemistry, College of Pharmacy, Seoul National University, Seoul, Korea

Bun-Sam Lim, R & D Center, MIWON Co., Ltd., C.P.O. Box 2156, Seoul, Korea

Qi-Shui Lin, Shanghai Institite of Biochemistry, 20 Yue-Yang Road, Shanghai 200031, China

Anthony W. Linnane, Department of Biochemistry, Monash University, 31 Clayton, Melbourne, Australia

Emanuel Margoliash, Department of Biochemistry, Northwestern University, Evanston, IL 60201, U.S.A.

Sangkot Marzuki, Department of Biochemistry, Monash University, 31 Clayton, Melbourne, Australia

Vincent Massey, Department of Biological Chemistry, Medical Science - I Box 034, The University of Michigan, Ann Arbor, MI 48109-0606, U.S.A.

Hiroshi Matsubara, Department of Biology, Faculty of Science, Osaka University, Osaka 560, Japan

Yasuo Mukohata, Department of Biology, Faculty of Science, University of Nagoya, Nagoya 466, Japan

Hirozo Oh-oka, Department of Biology, Faculty of Science, Osaka University, Osaka 560, Japan

Yutaka Orii, Department of Medicine, Faculty of Public Health, University of Kyoto, Kyoto 606, Japan

Takayuki Ozawa, Department of Biomedical Chemistry, Faculty of Medicine, University of Nagoya, Nagoya 466, Japan

Sergio Papa, Institute of Medical Biochemistry and Chemistry, University of Bari, Piazza G. Cesare, 70124, Bari, Italy

Yong-Nam Roh, Department of Toxicology, Ewha Womans University, Seoul, Korea

Yhun Y. Sheen, Department of Toxicology, College of Pharmacy, Ewha Womans University, Seoul, Korea

Yoshiharu Shimomura, Laboratory of Biochemistry, Institute of Health and Sports Sciences University of Tsukuba, Ibaraki 305, Japan

Sang-Hee Shin, Department of Medicinal Chemistry, College of Pharmacy, Ewha Womans University, Seoul, Korea

Sung-Hee Shin, College of Pharmacy, Ewha Womans University, Seoul, Korea

E. C. Slater, Department of Biochemistry, The University of Southhampton, Medical and Biological Sciences Bldg., Southhampton S09 3TU, England

Pill-Soon Song, Department of Chemistry, University of Nebraska, Lincoln, NE 68588, U.S.A.

Hiroshi Suzuki, Department of Biomedical Chemistry, Faculty of Medicine, University of Nagoya, Nagoya 466, Japan

Masashi Tanaka, Department of Biomedical Chemistry, Faculty of Medicine, University of Nagoya, Nagoya 466, Japan

Chen-Lu Tsou, Institute of Biophysics, Academia Sinica, Beijing, China

Sadao Wakabayashi, Department of Biology, Faculty of Science, Osaka University, Osaka 560, Japan

Kunio Yagi, Institute of Applied Biochemistry, Yagi Memorial Park, Mitake, Gifu 505-01 Japan

Takashi Yonetani, Department of Biophysics and Biochemistry, School of Medicine, University of Pennsylvania, Philadelphia, PA 19104, U.S.A.

Kwan-sa You, Biomedical Chemistry Laboratory, R & D Center, Cheil Sugar Co., Ltd., Seoul, Korea

Byung Sul Yu, Department of Biochemistry, College of Pharmacy, Seoul National University, Seoul, Korea

Hea Chung Yun, Department of Toxicology, College of Pharmacy, Ewha Womans University, Seoul, Korea

Jai Soon Yun, Department of Pharmacology, College of Pharmacy, Ewha Womans University, Seoul, Korea

Myung-Eun Su, Laboratory of Organic Synthesis, College of Pharmacy, Ewha Womans University, Seoul, Korea

AUTHOR INDEX

SUBJECT INDEX